NATIONAL RENOVATION & INSURANCE REPAIR ESTIMATOR

edited by Jonathan Russell

Includes inside the back cover:

Inside the back cover of this book you'll find a software download certificate. To access the download, follow the instructions printed there. The download includes the *National Estimator*, an easy-to-use estimating program with all the cost estimates in this book. The software will run on PCs using Windows XP, Vista, 7, 8, or 10 operating systems.

Quarterly price updates on the Web are free and automatic all during 2020. You'll be prompted when it's time to collect the next update. A connection to the Web is required.

Download all of Craftsman's most popular costbooks for one low price with the Craftsman Site License. http://www.craftsmansitelicense.com

Craftsman Book Company
6058 Corte del Cedro, Carlsbad, CA 92011

Cover design by: Jennifer Johnson

 ISBN 978-1-57218-359-9 Published October 2019 for the year 2020.

contents

about this book

WHAT'S NEW IN 2020

A robust new construction and remodeling market heading into 2020, bolstered by a multi-year stream of above-normal catastrophes, is making it increasingly difficult to find and keep enough skilled construction labor. Many contractors are building at full capacity, with room to grow, but cost pressures are eroding margins and making it difficult to expand. Now more than ever it's crucial to monitor your costs and embrace new technologies that make estimating faster and easier, with comprehensive and up-to-date pricing information that informs those systems. To access updated pricing for all construction types, go to www.craftsman-book.com.

A TOOL

This book is a tool, and like all tools it can be misused. It is an excellent tool for the renovation and repair pro- fessional. It is not a substitute for experience, skill, and knowledge.

Prices in this book are based on research of actual jobs and successful estimates. They represent an average of the typical conditions. Estimators should compare the conditions described in this book with actual conditions on site and adjust the price accordingly.

UNIQUE TO RENOVATION AND REPAIR WORK

This book is compiled specifically for the unique problems and conditions found in renovation and repair work. It is not a new construction cost book.

Renovation and repair work involve completely different circumstances than those found in new construction.

For example, the renovation or repair professional must work around existing conditions including home contents, access problems, out-of-plumb or out-of-square buildings, outdated materials, existing conditions that violate current building codes.

New-construction professionals have the luxury of placing items in a logical order, but renovation, remodel, or repair professionals must deal with conditions as they find them.

This means that joists have to be replaced in an existing floor system, paint has to be applied in a room where stain-grade base and carpeting are already installed, structures may have to be braced, contents have to be moved or worked around and materials and installation techniques must be matched.

DETERMINING COSTS

All costs in this book are based on typical conditions and typical problems found when remodeling or repairing a structure.

This means a door takes 10 to 15 minutes longer to install than it would in the ideal circumstances found in new construction.

Stairs are more difficult to install around pre-existing walls, wall framing takes longer when walls are typically splicing into existing work, and so on.

Some prices in this book will very closely match prices used in new construction. Other prices will reflect the complex conditions found in renovation and repair and will be dramatically different.

For example, using this book's stair building prices to estimate stair work in a series of 150 tract homes will result in an estimate that is far too high.

THE ART OF ESTIMATING

Estimating is part art, part science. Estimators must consider many factors, including access, crew productivity, special techniques, special abilities, temperament of the owner, and how busy the company is.

A contractor who is desperate for work will estimate much lower than a contractor who is swamped with work.

All of these factors — and many other similar ones — cannot be included in this or any other price book. They are part of the art of estimating.

The science of estimating, which includes prices, typical techniques, and materials, is included in this book.

This book is designed to make the science of estimating easier, which allows you to spend much more time focusing on the art of estimating, where your skill is crucial to the success of your company.

GENERAL VS. SPECIFIC

It is important to note that the more specific the estimator is, the more accurate the final estimate will be.

For example, when an estimator calculates all electrical costs for a typical home using a square foot cost, it may not be as accurate as if the estimator priced each fixture, outlet, and appliance hook-up.

Since the square foot price is based on a typical installation, it will not be exact for a home that is atypical in any way — for example, one with special outdoor lighting or with an expensive crystal fixture in the entry.

The more specific the item, the more exact the prices. The more general an item, the more assumptions must be made.

To help ensure the accuracy of your estimates, we describe any assumptions made when determining general items.

For example, the Rough Carpentry chapter contains a square foot price for estimating 2" by 4" wall framing. To help you make sure that this price will work for you, we describe our assumptions: the stud centers, the number of openings, headers, corners, plates and so forth, that would typically be found in a wall.

rounding

This book rounds hourly wage rates and the material, labor, and equipment components of a unit price.

These prices are rounded to "three significant digits." This means that prices under three digits (including two to the right of the decimal) are not rounded. Prices four digits and larger are rounded to the third digit from the left.

For example:

.23 is not rounded
2.33 is not rounded
23.33 is rounded to **23.30**
233.33 is rounded to **233.00**
2,333.33 is rounded to **2,330.00**
23,333.33 is rounded to **23,300.00**

In most cases, the square foot price will apply, but you will always want to carefully consider items that are more general, and if needed, adjust them to fit the conditions.

In the case above, the estimator may want to use the square foot price for average fixtures, then add an allowance for the crystal fixture and the outdoor lighting.

TIME AND MATERIAL CHARTS

Almost all chapters include time and material charts at the end. These time and material charts are designed to show you the materials used, waste, labor rates, labor burden costs, and labor productivity.

When materials with a range of sizes appear, only the small and large size are usually listed.

When materials with a range of qualities appear, only the low and high prices are usually listed.

These charts are designed to give you accurate detail on the exact prices used. When prices change, this book does not become obsolete. Compare current prices with those used and factor accordingly.

MATERIAL COSTS

National average material costs are compiled from surveys of suppliers throughout the country.

Costs for some materials, such as clay tile, building stone and hardwood, will vary a great deal from region to region.

For example, clay tile plants are located near naturally occurring clay sources. Because clay tiles are heavy, the further the tiles have to be shipped, the more expensive the tiles will be. The user of this book must be aware of local price variations.

Materials commonly found in every city are priced based on local delivery. In most cases this will be delivery no greater than 20 miles away from a local source. However, many rural areas have lumber yards that will deliver to a wider area at no additional charge.

Materials that are not commonly available locally, like hand-carved moldings or historical wallpaper, include shipping costs anywhere in North America. Estimators in Hawaii, Alaska and remote areas of Canada should add for additional shipping costs when applicable.

Material waste is often indicated with the items, but it's always a good idea to check the time and material charts for the exact waste calculated for all the components of an item.

Waste indicates material that is discarded during installation. It does not include waste that occurs when materials are taken to storage and ruined, run over at the job site, spilled, improperly cut, or damaged due to mishandling.

These types of occurrences are kept to a minimum by every good contractor, but will still occur on any job site.

Another common waste issue in renovation and repair is when a contractor must buy a minimum quantity for a small repair. For example, to replace a six-inch section of base, it is necessary to buy a piece of base that is eight-feet long or longer. In these cases, use the minimum price.

Material prices may not be listed with the time and material charts in some chapters. In these chapters, little new information would be provided by the materials chart so the space is saved for other information.

For example, a materials chart in the Appliance chapter will not provide an estimator with any new information.

The materials component of the unit price for an oven as listed in the main body of the chapter will tell the estimator how much the oven and connections cost.

Relisting these appliances in a materials chart wastes valuable space.

LABOR COSTS

Labor costs are national average rates that usually are consistent with union labor wages.

See the time and material charts for specific benefit costs and labor rates.

Crew labor rates are an average hourly rate for each member of the crew. For example, a masonry crew might consist of a mason, a mason's helper, and a hod carrier.

The hourly rate for this crew is the average cost of all three. In other words, the hourly rate is for 20 minutes work by the mason, 20 minutes work by the mason's helper, and 20 minutes work by the hod carrier.

Separation of labor in renovation and insurance repair work is much more difficult than is separation of labor on large commercial construction projects.

On a typical repair or renovation job a carpenter may participate in demolition, frame walls, set doors, set toilets, install electrical outlets and fixtures, and do a little painting.

In the jobs analyzed for this cost book, well over 40 percent of the demolition work was done by skilled workers. This is because demolition is often selective, requiring a skilled worker to ensure that additional damage does not occur.

Many renovation and repair companies are relatively small, so skilled workers participate in all phases of construction.

These realities are reflected in the labor costs used in this book. This means that a demolition laborer's hourly rate may seem higher than is normal for an unskilled worker.

The time and material charts show all items that are built into each labor rate.

Some contractors may not provide health insurance or retirement plans to some or all of their workers. Estimators can "back-out" these expenses from the labor costs.

It is critical that estimators examine the Workers' Compensation costs calculated in the wage rates. Some states have Workers' Compensation rates that are double, triple, or even quadruple the national average rates used in this book.

Workers' Compensation rates should be adjusted to match local conditions.

Labor productivity is based on observation of work performed in renovation and repair conditions.

These conditions differ from new construction in many ways, but a few of the most common are:

❶ difficulty matching existing work, ❷ access problems, ❸ materials that must be more carefully shaped and attached than is typical in new construction, ❹ out-of-plumb or out-of-square structures, ❺ reinforcing, ❻ more trips and effort are required to find materials, ❼ much more travel time is required because most jobs will have a relatively small amount of work in some trades, ❽ more vehicles are required by the renovation or repair contractor because many tradespeople are often traveling between jobs, compared to new construction where crews may spend weeks or months on one job, and because crews tend to be smaller and each crew may need a vehicle, ❾ more unexpected problems, ❿ more restrictions in established neighborhoods.

Labor productivity is based on a clean job site where tools are put away and secured at the end of each day.

Depending on the trade, 20 to 30 minutes per eight hours is allowed for clean-up and putting away tools. Normally, skilled workers spend half as much time cleaning up as do unskilled workers.

As is typical in new construction, labor includes unpacking materials, in some cases unloading materials from a truck on site, some travel to pick up minor materials (e.g. a forgotten tube of caulk, or a forgotten tool), typical breaks, lay-out, planning, discussion, coordination, mobilization (many companies meet at a central location each morning to receive instructions), recording hours (including specific information needed for job costing), occasional correction of mistakes in installation, and so forth.

Supervision is not included in these costs but should not generally be required. This is because each crew includes a skilled tradesperson who normally would not require supervision beyond the normal dispatch and mobilization discussed previously.

EQUIPMENT COSTS

Equipment costs are included only when equipment will be used that is not typically a part of the tools used by the majority of renovation or repair contractors.

For example, each carpenter should have a worm-drive saw, miter box, compressor, nail guns, and so forth. These types of tools are not included in the equipment costs.

However, equipment like cranes, backhoes, concrete saws, and jack hammers are not assumed to be part of the equipment and tools owned by a typical renovation or repair contractor. When these are needed, equipment rates are included in the unit price.

Equipment costs include the typical cost to rent the equipment from a local equipment rental shop. When applicable, they also include fuel or blade costs.

Check each item to determine if it includes delivery to the job site or operator labor. Also be careful to note minimum costs for work where rented equipment is needed.

MARKUP

Prices in this book do not include markup. Insurance repair markup is almost always 20 percent: 10 percent for overhead and 10 percent for profit.

In renovation work on historical structures, markup may be as high as 30 percent, although markup over 20 percent may be "hidden" inside the unit cost. Typical remodeling work markup varies from 15 percent to 25 percent. The most common markup for all types of work is 20 percent.

THE COST LINES

The cost tables in each section of this manual consist of individual tasks or items followed by a description. Beneath the description is a list of the items to be replaced or removed. For instance, under Appliances on page 17, you'll see Electric range, followed by a list of different ranges. Looking across, you'll see five columns of numbers and symbols. Let me explain how to read the numbers and symbols in those columns.

First let's look at the column headed Craft@Hrs. The Craft@Hrs column shows the recommended crew and manhours per unit for installation. For example, 2A in the Craft@Hrs column means that we recommend a crew of one appliance installer. The crew composition, with the cost per hour, is listed on page 12.

The manhours (following the @ symbol) is our estimate of the crew time required for installation (or demolition) of each unit. Manhours are listed in hundredths of an hour rather than minutes because it's easier to add, subtract, multiply and divide hundredths of a unit. For example, if the Craft@Hrs column shows 2A@.250, the Labor Cost column will show $15.15. That's the labor cost per unit for a crew of one appliance installer at $60.60 per hour multiplied by .250 manhours, rounded to the nearest penny. The unit is listed right after the Craft@Hrs, and may be the cost per square foot (sf), linear foot (lf), each (ea), or another unit of measurement.

The crew costs include the basic wage, taxable fringe benefits (vacation pay), Workers' Compensation insurance, liability insurance, taxes (state and federal unemployment, Social Security and Medicare), and typical nontaxable fringe benefits such as medical insurance and retirement. A breakdown of these expenses is included as a percentage in the footnote beneath the Labor table at the end of each section.

If your hourly crew cost is much lower or much higher, you can adjust your totals. For example, if your hourly labor cost is 25 percent less, reduce the labor figures in the cost tables by 25 percent to find your local cost.

The Material column shows your material cost for the item described under the heading.

The Total column is the sum of the Material and Labor cost columns.

CHARTS

Material charts show the material description, the material cost priced per typical unit purchased, the gross coverage, the typical waste, the net coverage after waste has been subtracted, and the resulting materials price including waste and often converted to a different unit of measure.

Equipment charts show the cost to rent equipment, the amount of work that can be done with the equipment per a period of time, and the resulting unit price.

Labor charts show the base wage, then add all additional costs that are based on wage. More information is always listed below this chart. It's important to note the "True" wage before adding labor related expenses. The true wage is the wage rate plus an allowance for vacation time. Since all the other costs must be paid even when an employee is on vacation, it is important that they become a component in the cost calculations when estimating work.

Labor productivity charts show the description of the work, the laborer or crew who will do the work, the average cost per man hour, the productivity and the resulting cost per unit.

REGIONAL DIFFERENCES

Construction techniques vary from region to region. Different climates and different local customs provide a variety of unique regional methods.

For example, in southern Florida it is common to build the first floor of a home from concrete block capped with a grade beam. This method won't be found in Colorado.

Similarly, coral stone walls won't be commonly found in Denver, although they are widely used in Miami.

Slate roofs are common on historical homes and newer custom homes in Philadelphia but are virtually nonexistent in Rapid City.

Homes in the south often include screened porches which aren't nearly so common in the west.

A Georgia home is much more likely to include a series of architecturally-correct columns with Corinthian capitals than is a home in Minnesota.

A Hawaii home may be built entirely from treated wood, when an Arizona home only uses treated lumber when it contacts dirt or concrete.

Many regional materials and techniques are priced in this book. Keep in mind that you should not use these prices if the item is not common to your area.

NATIONAL ESTIMATOR '19

The software download in the back of this book has all the information that appears in the printed book, but with one advantage. The National Estimator program makes it easy to copy and paste these costs into an estimate, or bid, and then add whatever markup you select. Quarterly price updates on the Web are free and automatic all during 2020. You'll be prompted when it's time to collect the next update. A connection to the Web is required.

To access the software download, follow the instructions printed on the certificate in the back of the book. The software will run on PCs using Windows XP, Vista, 7, 8, or 10 operating systems.

When you've installed the National Estimator program, click Help on the menu bar to see a list of topics that will get you up and running. Or, go online to www.craftsman-book.com, click on "Support," then "Tutorials" to view an interactive video for National Estimator.

If you have any problems using National Estimator, we'll be glad to help. Free telephone assistance is available from 8 a.m. until 5 p.m. Pacific time, Monday through Friday (except holidays). Call 760-438-7828.

WE WANT TO HEAR FROM YOU

If you have questions or concerns about this cost book, let us hear from you. Send letters to:

Craftsman Book Company
NRI Editor
6058 Corte del Cedro
Carlsbad, CA 92011

abbreviations

ABS....Acrylonitrile butadiene styrene
ac...................... alternating current
bf..................................board foot
Btu....................British thermal units
CEC..... California Earthquake Code
..........................(also see page 56)
cf...................................cubic foot
cfm...............cubic foot per minute
ci.................................. cubic inch
cy..................................cubic yard
ea.. each
FUTA.......... Federal Unemployment
.... Compensation Act tax
gal.......................................gallon
GFCI..ground fault circuit interrupter
gph.....................gallon(s) per hour
gpm................gallon(s) per minute
hp.............................. horsepower
hr(s)...................................hour(s)
IMC....... intermediate metal conduit
kd................................... kiln dried
kv..................................kilovolt(s)
kva......................1,000 volt amps
kw.................................kilowatt(s)
lb(s).................................pound(s)
lf.................................. lineal foot
li.................................. lineal inch
m............................. one thousand
mbf.....................1,000 board feet
mBtu.......1,000 British thermal units
mh................................man hour
mi..mile
mlf...................... 1,000 linear feet
mm............................millimeter(s)
mo......................................month
mph........................miles per hour
msf....................1,000 square feet
no. ....number
oc.................................. on center
oz.. ounce
pr.. pair
psi.............pounds per square inch
PVC....................polyvinyl chloride
qt.. quart
R/L...................... random length(s)
R/W/L....random widths and lengths
RSC.................. rigid steel conduit
S1S2E....... surfaced 1 side, 2 edges
S2S....................... surfaced 2 sides
S4S....................... surfaced 4 sides
SBS................. styrene butyl styrene
sf................................ square foot
sh...sheet
si................................ square inch
sq........................100 square feet
st... step
sy...............................square yard
t&g.............tongue-&-groove edge
TV...................................television
UBC........... Uniform Building Code
UL............. Underwriters' Laboratory
vlf......................vertical linear foot
wk..week
w/... with
x..................................by or times

SYMBOLS

/.. per
-................................through or to
@... at
%..................................... percent
$................................U.S. dollars
'..feet
"..inches
#........................ pound or number

acknowledgements

The editor wishes to gratefully acknowledge the contribution of the following: 18th Century Hardware Co., Inc. - A-Ball Plumbing Supply - A&Y Lumber - American Building Restoration Chemicals, Inc. - American Custom Millwork, Inc. - American Society for Testing and Materials (ASTM) - Anderson Windows - Anthony Lombardo Architectural Paneling Inc. - Anthony Wood Products Incorporated - The Antique Hardware Store - Architectural Components, Inc. - Architectural Woodwork Institute - The Balmer Architectural Art Studios - Bathroom Hardware, Inc. - Bathroom Machineries - Bendix Mouldings, Inc. - Brass Reproductions - Brick Institute of America - C & H Roofing, Inc. - C. G. Girolami & Sons - Caradco - Cataumet Sawmills - CertainTeed - Chadsworth Incorporated - Chelsea Decorative Metal Company - Classic Accents, Inc. - Classic Ceilings - CMW, Inc. - Conant Custom Brass - Conklin Metal Industries - Craftsman Lumber Company - Crown City Hardware Co. - Cumberland General Stores - Cumberland Woodcraft Co. Inc. - Designs in Tile - Dimension Hardwood - Donnell's Clapboard Mill - Driwood Moulding - Eisenhart Wallcoverings - Empire Wood Works - Raymond Enkeboll Designs - James Facenelli, CGR #28 - Focal Point Architectural Products - Four Seasons Sun Rooms - Futurbilt - Garland - Gates Moore - General Electric Company - George Taylor Specialties Co. - Goodwin Heart Pine Company - Gougeon Brothers, Inc. - Grand Era, Inc. - Granville Manufacturing Company - Groff & Hearne Lumber - Hampton Decor - Harris-Tarkett, Inc. - Heatway - Heritage Vinyl Products - Michael Higuera - Italian Tile Center - J. G. Braun Company - Jeffries Wood Works, Inc. - Jennifer's Glass Works - JGR Enterprises - Johnson Paint Company, Inc. - Joseph Biunno Ltd - Kenmore Industries - King's Chandelier Company - Kraftmaid Cabinetry, Inc. - David Lawrence, illustrator - Lasting Impression Doors - Lehman's - Linoleum City, Inc. - Ludowici-Celadon, Inc. - MCA, Inc. - Millwork Specialties - The Millworks, Inc. - Mountain Lumber Co. - National Oak Flooring Manufacturers Association (NOFMA) - National Wood Flooring Association - Northwest Energy, Inc. - Oak Flooring Institute - The Old Fashioned Milk Paint Company, Inc. - Ole Fashion Things - Omega Too - Pagliacco Turning & Milling - Pella Corporation - Permaglaze - Piedmont Home Products, Inc. - Piedmont Mantel & Millwork Ltd - Pinecrest - Radiantec - The Readybuilt Products Company - Rejuvenation Lamp & Fixture, Co. - Runtal Radiators - The Saltbox - Salvage One Architectural Artifacts - Silverton Victorian Millworks - Stairways Inc. - Steptoe and Wife Antiques Ltd - Stromberg's Architectural Stone - The Structural Slate Co. - Sunbilt - Supradur Manufacturing Corporation - Taylor Door - Brian Thaut - Touchstone Woodworks - Turncraft - USG Company - Vande Hey's Roofing Tile Co., Inc. - Velux-America Inc. - Vixen Hill Manufacturing Co. - W. F. Norman Corporation - Watercolors, Inc. - WELco - Western Wood Products Association - Williamsburg Blacksmiths - Windy Hill Forge - Wolverine Technologies - The Wood Factory - Worthington Group Ltd

area modification factors

Construction costs are higher in some cities than in other cities. Use the factors on this and the following page to adapt the costs listed in this book to your job site. Increase or decrease your estimated total project cost by the percentage listed for the appropriate city in this table to find your estimated building cost.

These factors were compiled by comparing the actual construction cost of residential, institutional and commercial buildings in communities throughout the United States. Because these factors are based on completed project costs, they consider all construction cost variables, including labor, equipment and material cost, labor productivity, climate, job conditions and markup.

Use the factor for the nearest or most comparable city. If the city you need is not listed in the table, use the factor for the appropriate state. Note that these location factors are composites of many costs and will not necessarily be accurate when estimating the cost of any particular part of a building. But when used to modify all estimated costs on a job, they should improve the accuracy of your estimates.

ALABAMA	**-4%**
Anniston	-8%
Auburn	-4%
Bellamy	5%
Birmingham	2%
Dothan	-7%
Evergreen	-10%
Gadsden	-9%
Huntsville	-1%
Jasper	-8%
Mobile	-2%
Montgomery	-2%
Scottsboro	-4%
Selma	-5%
Sheffield	0%
Tuscaloosa	-4%

ALASKA	**23%**
Anchorage	26%
Fairbanks	27%
Juneau	19%
Ketchikan	18%
King Salmon	23%

ARIZONA	**-4%**
Chambers	-8%
Douglas	-8%
Flagstaff	-7%
Kingman	-5%
Mesa	3%
Phoenix	3%
Prescott	-6%
Show Low	-7%
Tucson	-5%
Yuma	2%

ARKANSAS	**-7%**
Batesville	-9%
Camden	-2%
Fayetteville	-4%
Fort Smith	-7%
Harrison	-12%
Hope	-8%
Hot Springs	-13%
Jonesboro	-9%
Little Rock	-3%
Pine Bluff	-11%
Russellville	-4%
West Memphis	-2%

CALIFORNIA	**9%**
Alhambra	8%
Bakersfield	2%
El Centro	0%
Eureka	7%
Fresno	-2%
Herlong	9%
Inglewood	9%
Irvine	13%
Lompoc	3%
Long Beach	9%
Los Angeles	8%
Marysville	9%
Modesto	1%
Mojave	5%
Novato	18%
Oakland	24%
Orange	12%
Oxnard	2%
Pasadena	9%
Rancho Cordova	4%
Redding	-3%
Richmond	17%
Riverside	4%
Sacramento	3%
Salinas	1%
San Bernardino	2%
San Diego	8%
San Francisco	27%
San Jose	17%
San Mateo	21%
Santa Barbara	7%
Santa Rosa	16%
Stockton	4%
Sunnyvale	20%
Van Nuys	8%
Whittier	8%

COLORADO	**1%**
Aurora	7%
Boulder	4%
Colorado Springs	0%
Denver	8%
Durango	-1%
Fort Morgan	-2%
Glenwood Springs	4%
Grand Junction	0%
Greeley	5%
Longmont	2%
Pagosa Springs	-4%
Pueblo	0%
Salida	-6%

CONNECTICUT	**8%**
Bridgeport	6%
Bristol	12%
Fairfield	9%
Hartford	11%
New Haven	7%
Norwich	3%
Stamford	12%
Waterbury	6%
West Hartford	5%

DELAWARE	**2%**
Dover	-4%
Newark	6%
Wilmington	4%

DISTRICT OF COLUMBIA	
Washington	12%

FLORIDA	**-5%**
Altamonte Springs	-3%
Bradenton	-6%
Brooksville	-7%
Daytona Beach	-9%
Fort Lauderdale	2%
Fort Myers	-6%
Fort Pierce	-10%
Gainesville	-9%
Jacksonville	-2%
Lakeland	-8%
Melbourne	-8%
Miami	1%
Naples	-2%
Ocala	-12%
Orlando	1%
Panama City	-11%
Pensacola	-8%
Saint Augustine	-2%
Saint Cloud	-2%
St Petersburg	-6%
Tallahassee	-6%
Tampa	-1%
West Palm Beach	-2%

GEORGIA	**-4%**
Albany	-6%
Athens	-5%
Atlanta	12%
Augusta	-2%
Buford	-2%
Calhoun	-9%
Columbus	-3%
Dublin/Fort Valley	-8%
Hinesville	-6%
Kings Bay	-10%
Macon	-4%
Marietta	4%
Savannah	-4%
Statesboro	-11%
Valdosta	-1%

HAWAII	**20%**
Aliamanu	22%
Ewa	20%
Halawa Heights	20%
Hilo	20%
Honolulu	22%
Kailua	22%
Lualualei	20%
Mililani Town	20%
Pearl City	20%
Wahiawa	20%
Waianae	20%
Wailuku (Maui)	20%

IDAHO	**-9%**
Boise	-5%
Coeur d'Alene	-10%
Idaho Falls	-9%
Lewiston	-11%
Meridian	-9%
Pocatello	-10%
Sun Valley	-8%

ILLINOIS	**4%**
Arlington Heights	14%
Aurora	14%
Belleville	0%
Bloomington	-1%
Carbondale	-4%
Carol Stream	14%
Centralia	-3%
Champaign	-2%
Chicago	15%
Decatur	-7%
Galesburg	-4%
Granite City	3%
Green River	5%
Joliet	13%
Kankakee	-3%
Lawrenceville	-6%
Oak Park	18%
Peoria	6%
Peru	2%
Quincy	16%
Rockford	3%
Springfield	0%
Urbana	-4%

INDIANA	**-3%**
Aurora	-5%
Bloomington	-2%
Columbus	-4%
Elkhart	-4%
Evansville	4%
Fort Wayne	-1%
Gary	8%
Indianapolis	4%
Jasper	-8%
Jeffersonville	-5%
Kokomo	-8%
Lafayette	-5%
Muncie	-8%
South Bend	-2%
Terre Haute	-3%

IOWA	**-3%**
Burlington	1%
Carroll	-11%
Cedar Falls	-4%

Cedar Rapids 2%
Cherokee 1%
Council Bluffs -1%
Creston 1%
Davenport 1%
Decorah -8%
Des Moines 5%
Dubuque -4%
Fort Dodge -3%
Mason City -3%
Ottumwa -6%
Sheldon -7%
Shenandoah -14%
Sioux City 5%
Spencer -7%
Waterloo -3%

KANSAS 0%
Colby -8%
Concordia -12%
Dodge City -4%
Emporia 8%
Fort Scott -6%
Hays -13%
Hutchinson -6%
Independence 29%
Kansas City 5%
Liberal 14%
Salina -7%
Topeka -1%
Wichita -4%

KENTUCKY -4%
Ashland -4%
Bowling Green -5%
Campton -11%
Covington 2%
Elizabethtown -10%
Frankfort 7%
Hazard -10%
Hopkinsville -5%
Lexington 1%
London -7%
Louisville 2%
Owensboro -4%
Paducah 0%
Pikeville -8%
Somerset -11%
White Plains -4%

LOUISIANA 2%
Alexandria 4%
Baton Rouge 10%
Houma 4%
Lafayette 8%
Lake Charles 13%
Mandeville -3%
Minden -5%
Monroe -8%
New Orleans 2%
Shreveport -4%

MAINE -5%
Auburn -4%
Augusta -5%
Bangor -6%
Bath -6%
Brunswick -1%
Camden -10%
Cutler -7%
Dexter -4%
Northern Area -8%
Portland 2%

MARYLAND 2%
Annapolis 8%
Baltimore 7%
Bethesda 13%
Church Hill -4%
Cumberland -8%
Elkton -5%
Frederick 7%
Laurel 8%
Salisbury -6%

MASSACHUSETTS 12%
Ayer 6%
Bedford 15%
Boston 37%
Brockton 20%
Cape Cod 4%
Chicopee 7%
Dedham 18%
Fitchburg 11%
Hingham 19%
Lawrence 14%
Nantucket 9%
New Bedford 6%
Northfield 2%
Pittsfield 1%
Springfield 8%

MICHIGAN 1%
Battle Creek -1%
Detroit 7%
Flint -4%
Grand Rapids 1%
Grayling -7%
Jackson -1%
Lansing 0%
Marquette 3%
Pontiac 12%
Royal Oak 7%
Saginaw -5%
Traverse City -2%

MINNESOTA -1%
Bemidji -6%
Brainerd -3%
Duluth 2%
Fergus Falls -10%
Magnolia -8%
Mankato -4%
Minneapolis 13%
Rochester -1%
St Cloud 2%
St Paul 12%
Thief River Falls -2%
Willmar -6%

MISSISSIPPI -6%
Clarksdale -9%
Columbus 0%
Greenville -14%
Greenwood -10%
Gulfport -6%
Jackson -3%
Laurel -7%
McComb -11%
Meridian 3%
Tupelo -7%

MISSOURI -3%
Cape Girardeau -5%
Caruthersville -7%
Chillicothe -4%
Columbia -4%
East Lynne 4%
Farmington -8%
Hannibal -2%
Independence 5%
Jefferson City -5%
Joplin -6%
Kansas City 6%
Kirksville -15%
Knob Noster 3%
Lebanon -12%
Poplar Bluff -10%
Saint Charles 1%
Saint Joseph -1%
Springfield -8%
St Louis 8%

MONTANA -3%
Billings -2%
Butte -3%
Fairview 12%
Great Falls -6%
Havre -9%
Helena -2%
Kalispell -6%
Miles City -7%
Missoula -6%

NEBRASKA -8%
Alliance -10%
Columbus -7%
Grand Island -8%
Hastings -9%
Lincoln -4%
McCook -9%
Norfolk -10%
North Platte -6%
Omaha 0%
Valentine -15%

NEVADA 1%
Carson City -4%
Elko 9%
Ely -3%
Fallon 0%
Las Vegas 3%
Reno -1%

NEW HAMPSHIRE -1%
Charlestown -5%
Concord -3%
Dover 1%
Lebanon -3%
Littleton -6%
Manchester 2%
New Boston 3%

NEW JERSEY 9%
Atlantic City 4%
Brick 2%
Dover 9%
Edison 13%
Hackensack 10%
Monmouth 12%
Newark 11%
Passaic 12%
Paterson 7%
Princeton 10%
Summit 16%
Trenton 7%

NEW MEXICO -8%
Alamogordo -11%
Albuquerque -3%
Clovis -11%
Farmington -1%
Fort Sumner -2%
Gallup -7%
Holman -10%
Las Cruces -8%
Santa Fe -8%
Socorro -14%
Truth or Consequences -8%
Tucumcari -8%

NEW YORK 6%
Albany 7%
Amityville 9%
Batavia 1%
Binghamton -2%
Bronx 10%
Brooklyn 7%
Buffalo 1%
Elmira -3%
Flushing 15%
Garden City 15%
Hicksville 14%
Ithaca -5%
Jamaica 14%
Jamestown -7%
Kingston -4%
Long Island 30%
Montauk 7%
New York (Manhattan) 31%
New York City 31%
Newcomb 0%
Niagara Falls -6%
Plattsburgh -1%
Poughkeepsie 1%
Queens 17%
Rochester 2%
Rockaway 10%
Rome -4%
Staten Island 8%
Stewart -5%
Syracuse 2%
Tonawanda -1%
Utica -6%
Watertown -1%
West Point 6%
White Plains 14%

NORTH CAROLINA -4%
Asheville -7%
Charlotte 7%
Durham 0%
Elizabeth City -8%
Fayetteville -6%
Goldsboro 0%
Greensboro -3%
Hickory -8%
Kinston -9%
Raleigh 3%
Rocky Mount -7%
Wilmington -6%
Winston-Salem -5%

NORTH DAKOTA 4%
Bismarck 3%
Dickinson 15%
Fargo 0%
Grand Forks -1%
Jamestown -4%
Minot 9%
Nekoma -10%
Williston 21%

OHIO 0%
Akron 1%
Canton -2%
Chillicothe -2%
Cincinnati 3%
Cleveland 3%
Columbus 5%
Dayton 1%
Lima -5%
Marietta -5%
Marion -6%

- Newark 3%
- Sandusky -3%
- Steubenville 1%
- Toledo 7%
- Warren -5%
- Youngstown -3%
- Zanesville -1%

OKLAHOMA -5%

- Adams -10%
- Ardmore -1%
- Clinton -3%
- Durant -11%
- Enid -4%
- Lawton -8%
- McAlester -7%
- Muskogee -8%
- Norman -4%
- Oklahoma City -3%
- Ponca City -1%
- Poteau -7%
- Pryor -6%
- Shawnee -8%
- Tulsa 0%
- Woodward 5%

OREGON -3%

- Adrian -12%
- Bend -5%
- Eugene -3%
- Grants Pass -5%
- Klamath Falls -8%
- Pendleton -3%
- Portland 10%
- Salem -2%

PENNSYLVANIA -1%

- Allentown 3%
- Altoona -8%
- Beaver Springs -5%
- Bethlehem 4%
- Bradford -8%
- Butler -2%
- Chambersburg -7%
- Clearfield -3%
- DuBois -10%
- East Stroudsburg -5%
- Erie -6%
- Genesee -4%
- Greensburg -4%
- Harrisburg 3%
- Hazleton -3%
- Johnstown -9%
- Kittanning -6%
- Lancaster -1%
- Meadville -9%
- Montrose -4%
- New Castle -3%
- Philadelphia 11%
- Pittsburgh 6%
- Pottsville -8%
- Punxsutawney -3%
- Reading 2%
- Scranton 1%
- Somerset -9%
- Southeastern 8%
- Uniontown -6%
- Valley Forge 11%
- Warminster 11%
- Warrendale 5%
- Washington 8%
- Wilkes Barre -1%
- Williamsport -2%
- York -1%

RHODE ISLAND 5%

- Bristol 5%
- Coventry 5%
- Cranston 6%
- Davisville 5%
- Narragansett 5%
- Newport 5%
- Providence 6%
- Warwick 5%

SOUTH CAROLINA -1%

- Aiken 4%
- Beaufort -2%
- Charleston -1%
- Columbia -2%
- Greenville 8%
- Myrtle Beach -8%
- Rock Hill -6%
- Spartanburg -4%

SOUTH DAKOTA -6%

- Aberdeen -7%
- Mitchell -6%
- Mobridge -9%
- Pierre -10%
- Rapid City -8%
- Sioux Falls -1%
- Watertown -4%

TENNESSEE -2%

- Chattanooga 2%
- Clarksville 1%
- Cleveland -1%
- Columbia -7%
- Cookeville -8%
- Jackson -2%
- Kingsport -5%
- Knoxville -2%
- McKenzie -8%
- Memphis 1%
- Nashville 2%

TEXAS 5%

- Abilene -2%
- Amarillo -2%
- Arlington 1%
- Austin 12%
- Bay City 39%
- Beaumont 18%
- Brownwood -8%
- Bryan 8%
- Childress -14%
- Corpus Christi 18%
- Dallas 6%
- Del Rio 0%
- El Paso -7%
- Fort Worth 2%
- Galveston 24%
- Giddings 6%
- Greenville 3%
- Houston 26%
- Huntsville 26%
- Longview 1%
- Lubbock -7%
- Lufkin 8%
- McAllen -6%
- Midland 10%
- Palestine 2%
- Plano 7%
- San Angelo -6%
- San Antonio 8%
- Texarkana -8%
- Tyler -7%
- Victoria 12%
- Waco -3%
- Wichita Falls -9%
- Woodson -3%

UTAH -3%

- Clearfield 0%
- Green River -3%
- Ogden -9%
- Provo -6%
- Salt Lake City 1%

VERMONT -5%

- Albany -7%
- Battleboro -4%
- Beecher Falls -8%
- Bennington -6%
- Burlington 4%
- Montpelier -4%
- Rutland -7%
- Springfield -6%
- White River Junction -5%

VIRGINIA -4%

- Abingdon -9%
- Alexandria 10%
- Charlottesville -6%
- Chesapeake -4%
- Culpeper -5%
- Farmville -12%
- Fredericksburg -5%
- Galax -10%
- Harrisonburg -6%
- Lynchburg -9%
- Norfolk -2%
- Petersburg -3%
- Radford -9%
- Reston 7%
- Richmond 2%
- Roanoke -9%
- Staunton -7%
- Tazewell -6%
- Virginia Beach -3%
- Williamsburg -3%
- Winchester 4%

WASHINGTON 0%

- Clarkston -8%
- Everett 2%
- Olympia -2%
- Pasco 1%
- Seattle 11%
- Spokane -3%
- Tacoma 2%
- Vancouver 3%
- Wenatchee -6%
- Yakima -5%

WEST VIRGINIA -5%

- Beckley -5%
- Bluefield 0%
- Charleston 4%
- Clarksburg -7%
- Fairmont -11%
- Huntington -4%
- Lewisburg -14%
- Martinsburg -5%
- Morgantown -4%
- New Martinsville -9%
- Parkersburg 1%
- Romney -7%
- Sugar Grove -8%
- Wheeling 5%

WISCONSIN 0%

- Amery -1%
- Beloit 5%
- Clam Lake -8%
- Eau Claire -2%
- Green Bay 3%
- La Crosse 0%
- Ladysmith -2%
- Madison 8%
- Milwaukee 6%
- Oshkosh 4%
- Portage 0%
- Prairie du Chien -7%
- Wausau -3%

WYOMING -1%

- Casper 1%
- Cheyenne/Laramie -2%
- Gillette 3%
- Powell -3%
- Rawlins 8%
- Riverton -6%
- Rock Springs 1%
- Sheridan -3%
- Wheatland -3%

CANADIAN AREA MODIFIERS

These figures assume an exchange rate of $1.00 Canadian to $0.76 U.S.

ALBERTA AVERAGE... 13%

- Calgary 14%
- Edmonton 14%
- Fort McMurray 12%

BRITISH COLUMBIA AVERAGE 7%

- Fraser Valley 6%
- Okanagan 6%
- Vancouver 9%

MANITOBA AVERAGE... 0%

- North Manitoba 0%
- Selkirk 0%
- South Manitoba 0%
- Winnipeg 0%

NEW BRUNSWICK AVERAGE -13%

- Moncton -13%

NEWFOUNDLAND/ LABRADOR AVERAGE... -3%

NOVA SCOTIA AVERAGE -8%

- Amherst -8%
- Nova Scotia -7%
- Sydney -8%

ONTARIO AVERAGE..... 7%

- London 7%
- Thunder Bay 6%
- Toronto 7%

QUEBEC AVERAGE..... -1%

- Montreal -1%
- Quebec City -1%

SASKATCHEWAN AVERAGE 4%

- La Ronge 3%
- Prince Albert 2%
- Saskatoon 5%

crews

CRAFT CODE	AVG. COST /HR	CREW COMPOSITION
1A	$53.20	acoustic ceiling installer
2A	$60.60	appliance installer
3A	$72.30	appliance refinisher
4A	$62.30	awning installer
5A	$47.10	awning installer's helper
6A	$54.70	awning installer awning installer's helper
1B	$38.20	cleaning laborer
2B	$85.20	post & beam carpenter
3B	$60.70	post & beam carpenter's helper
4B	$73.00	post & beam carpenter post & beam carpenter's helper
1C	$67.60	carpenter
5C	$50.40	carpenter's helper
6C	$59.00	carpenter carpenter's helper
1D	$47.20	demolition laborer
2D	$54.90	drywall hanger drywall hanger's helper
3D	$62.60	drywall hanger
4D	$47.20	drywall hanger's helper
5D	$63.10	drywall taper
6D	$57.60	drywall hanger drywall hanger's helper drywall taper
5E	$44.00	excavation laborer
6E	$66.50	excavation laborer equipment operator
7E	$72.40	electrician
8E	$53.90	electrician's helper
9E	$63.20	electrician electrician's helper
1F	$63.60	concrete form installer
2F	$46.30	concrete laborer
3F	$55.00	concrete form installer concrete laborer
4F	$59.00	carpenter (fence installer) carpenter's helper
5F	$65.50	painter
6F	$63.40	concrete finisher
7F	$52.40	concrete finisher's helper
8F	$57.90	concrete finisher concrete finisher's helper
9F	$56.50	concrete form installer concrete laborer concrete finisher concrete finisher's helper
1G	$88.80	compaction grouting specialist
2G	$67.60	compaction grouting specialist concrete laborer
1H	$57.40	hazardous materials laborer
2H	$70.50	HVAC installer
3H	$52.40	wallpaper hanger
1I	$49.00	insulation installer
2I	$56.30	cabinet installer laborer
3I	$65.00	flooring installer
4I	$48.30	flooring installer's helper
5I	$56.70	flooring installer flooring installer's helper
6I	$67.40	paneling installer
7I	$50.40	paneling installer's helper
2L	$44.90	cabinet installer's helper
3L	$41.40	masking & moving laborer
1M	$71.10	mason
2M	$65.20	mason's helper
3M	$50.40	hod carrier
4M	$62.20	mason mason's helper hod carrier / laborer
5M	$68.20	mason mason's helper
6M	$115.00	stone carver
7M	$48.60	mobile home repair specialist
1O	$88.90	equipment operator
2O	$62.70	concrete saw operator
4P	$64.70	plasterer
5P	$58.30	plasterer's helper
6P	$61.50	plasterer plasterer's helper
7P	$77.40	plumber
8P	$58.90	paneling installer paneling installer's helper
9P	$59.10	plumber's helper
3R	$66.70	retaining wall installer
4R	$78.90	roofer
5R	$66.90	roofer's helper
6R	$72.90	roofer roofer's helper
1S	$54.10	susp. ceiling installer susp. ceiling installer's helper
2S	$61.70	susp. ceiling installer
3S	$46.40	susp. ceiling installer's helper
4S	$66.50	siding installer
5S	$49.70	siding installer's helper
6S	$58.10	siding installer siding installer's helper
7S	$59.70	security system installer
8S	$76.10	swimming pool installer
9S	$57.50	water extractor
1T	$62.80	tile layer
7Z	$45.90	mildew remediation specialist
8Z	$28.50	mildew remediation assistant
9Z	$37.20	mildew remediation specialist mildew remediation assistant

Acoustic Ceilings

	Craft@Hrs	Unit	Material	Labor	Total
Minimum charge.					
for acoustic ceiling tile work	1A@1.25	ea	24.10	66.50	90.60

1/2" on strips. 12" x 12" tiles stapled in place. Includes 12" x 12" ceiling tiles, staples, and installation labor. Does not include furring strips. Includes 3% waste.

	Craft@Hrs	Unit	Material	Labor	Total
1/2" ceiling tiles on furring strips					
replace, smooth face	1A@.025	sf	1.47	1.33	2.80
replace, fissured face	1A@.025	sf	1.88	1.33	3.21
replace, textured face	1A@.025	sf	2.19	1.33	3.52
replace, patterned face	1A@.025	sf	2.30	1.33	3.63
remove only	1D@.012	sf	—	.57	.57

1/2" on flat ceiling. 12" x 12" tiles glued in place. Includes 12" x 12" ceiling tiles, glue, and installation labor. Includes 3% waste.

	Craft@Hrs	Unit	Material	Labor	Total
1/2" ceiling tiles on flat ceiling					
replace, smooth face	1A@.029	sf	1.47	1.54	3.01
replace, fissured face	1A@.029	sf	1.88	1.54	3.42
replace, textured face	1A@.029	sf	2.19	1.54	3.73
replace, patterned face	1A@.029	sf	2.30	1.54	3.84
remove only	1D@.016	sf	—	.76	.76

5/8" on strips. 12" x 12" tiles stapled in place. Includes 12" x 12" ceiling tiles, staples, and installation labor. Does not include furring strips. Includes 3% waste.

	Craft@Hrs	Unit	Material	Labor	Total
5/8" ceiling tiles on furring strips					
replace, smooth face	1A@.025	sf	1.88	1.33	3.21
replace, fissured face	1A@.025	sf	2.19	1.33	3.52
replace, textured face	1A@.025	sf	2.43	1.33	3.76
replace, patterned face	1A@.025	sf	2.59	1.33	3.92
remove only	1D@.012	sf	—	.57	.57

5/8" on flat ceiling. 12" x 12" tiles glued in place. Includes 12" x 12" ceiling tiles, glue, and installation labor. Includes 3% waste.

	Craft@Hrs	Unit	Material	Labor	Total
5/8" ceiling tiles on flat ceiling					
replace, smooth face	1A@.029	sf	1.88	1.54	3.42
replace, fissured face	1A@.029	sf	2.19	1.54	3.73
replace, textured face	1A@.029	sf	2.43	1.54	3.97
replace, patterned face	1A@.029	sf	2.59	1.54	4.13
remove only	1D@.016	sf	—	.76	.76

3/4" on strips. 12" x 12" tiles stapled in place. Includes 12" x 12" ceiling tiles, staples, and installation labor. Does not include furring strips. Includes 3% waste.

	Craft@Hrs	Unit	Material	Labor	Total
3/4" ceiling tiles on furring strips					
replace, smooth face	1A@.025	sf	2.19	1.33	3.52
replace, fissured face	1A@.025	sf	2.43	1.33	3.76
replace, textured face	1A@.025	sf	2.53	1.33	3.86
replace, patterned face	1A@.025	sf	2.80	1.33	4.13
remove	1D@.012	sf	—	.57	.57

	Craft@Hrs	Unit	Material	Labor	Total

3/4" on flat ceiling. 12" x 12" tiles glued in place. Includes 12" x 12" ceiling tiles, glue, and installation labor. Includes 3% waste.

	Craft@Hrs	Unit	Material	Labor	Total
3/4" ceiling tiles on flat ceiling					
replace, smooth face	1A@.029	sf	2.19	1.54	3.73
replace, fissured face	1A@.029	sf	2.43	1.54	3.97
replace, textured face	1A@.029	sf	2.53	1.54	4.07
replace, patterned face	1A@.029	sf	2.80	1.54	4.34
remove	1D@.016	sf	—	.76	.76

Additional costs for acoustical ceiling tile.

	Craft@Hrs	Unit	Material	Labor	Total
add for 3/4" tiles with fire rating	—	sf	.41	—	.41
add for aluminum-coated tiles	—	sf	.60	—	.60

Furring strips. Includes 1" x 2" furring strips, nails, construction adhesive as needed, and installation labor. Furring strips placed as fireblocking is included. Includes 4% waste.

	Craft@Hrs	Unit	Material	Labor	Total
replace 1" x 2" furring strips 12" on center	1A@.007	sf	.36	.37	.73
remove	1D@.009	sf	—	.42	.42

Repair loose ceiling tile.

	Craft@Hrs	Unit	Material	Labor	Total
repair loose acoustic tile	1A@.253	ea	.16	13.50	13.66

Angled install for ceiling tile.

add 24% for diagonal install
add 31% for chevron install
add 40% for herringbone install

Time & Material Charts (selected items)
Acoustic Ceiling Materials

	Craft@Hrs	Unit	Material	Labor	Total
1/2" thick acoustic tile (per 12" x 12" tile)					
smooth face, ($1.23 each, 1 sf, 3% waste)	—	sf	1.26	—	1.26
fissured face, ($1.66 each, 1 sf, 3% waste)	—	sf	1.71	—	1.71
textured face, ($1.90 each, 1 sf, 3% waste)	—	sf	1.95	—	1.95
patterned face, ($2.05 each, 1 sf, 3% waste)	—	sf	2.12	—	2.12
5/8" thick acoustic tile (per 12" x 12" tile)					
smooth face, ($1.66 each, 1 sf, 3% waste)	—	sf	1.71	—	1.71
fissured face, ($1.90 each, 1 sf, 3% waste)	—	sf	1.95	—	1.95
textured face, ($2.20 each, 1 sf, 3% waste)	—	sf	2.26	—	2.26
patterned face, ($2.30 each, 1 sf, 3% waste)	—	sf	2.36	—	2.36
3/4" thick acoustic tile (per 12" x 12" tile)					
smooth face, ($1.90 each, 1 sf, 3% waste)	—	sf	1.95	—	1.95
fissured face, ($2.20 each, 1 sf, 3% waste)	—	sf	2.26	—	2.26
textured face, ($2.27 each, 1 sf, 3% waste)	—	sf	2.33	—	2.33
patterned face, ($2.50 each, 1 sf, 3% waste)	—	sf	2.57	—	2.57

Acoustic Ceiling Labor

Laborer	base wage	paid leave	true wage	taxes & ins.	total
Acoustic ceiling installer	$30.10	2.35	$32.45	20.75	$53.20
Demolition laborer	$25.90	2.02	$27.92	19.28	$47.20

Paid Leave is calculated based on two weeks paid vacation, one week sick leave, and seven paid holidays. Employer's matching portion of **FICA** is 7.65 percent. **FUTA** (Federal Unemployment) is .8 percent. **Worker's compensation** for the acoustic ceiling trade was calculated using a national average of 12.50 percent. **Unemployment insurance** was calculated using a national average of 8 percent. **Health insurance** was calculated based on a projected national average for 2020 of $1,280 per employee (and family when applicable) per month. Employer pays 80 percent for a per month cost of $1,024 per employee. **Retirement** is based on a 401(k) retirement program with employer matching of 50 percent. Employee contributions to the 401(k) plan are an average of 6 percent of the true wage. **Liability insurance** is based on a national average of 12.0 percent.

	Craft@Hrs	Unit	Material	Labor	Total
Acoustic Ceiling Labor Productivity					
Repair of ceiling tile					
repair loose tile	1A@.253	ea	—	13.50	13.50
Installation of ceiling tile					
tiles on furring strips	1A@.025	sf	—	1.33	1.33
tiles on flat ceiling	1A@.029	sf	—	1.54	1.54
furring strips	1A@.007	sf	—	.37	.37

Appliances

	Craft@Hrs	Unit	Material	Labor	Total
Minimum charge for appliances.					
for appliance work	2A@1.05	ea	18.80	63.60	82.40
Gas cook top.					
replace, standard grade	7P@1.99	ea	686.00	154.00	840.00
replace, high grade	7P@1.99	ea	903.00	154.00	1,057.00
replace, with grill / griddle	7P@1.99	ea	1,210.00	154.00	1,364.00
remove	1D@.615	ea	—	29.00	29.00
remove for work, then reinstall	7P@2.20	ea	5.43	170.00	175.43
Gas range, free standing.					
replace, economy grade	7P@2.05	ea	719.00	159.00	878.00
replace, standard grade	7P@2.05	ea	965.00	159.00	1,124.00
replace, high grade	7P@2.05	ea	1,320.00	159.00	1,479.00
replace, with grill / griddle	7P@2.10	ea	1,770.00	163.00	1,933.00
remove	1D@.618	ea	—	29.20	29.20
remove for work, then reinstall	7P@2.80	ea	5.76	217.00	222.76
Space-saver gas range, free-standing.					
replace, standard grade	7P@1.99	ea	937.00	154.00	1,091.00
replace, high grade	7P@1.99	ea	1,210.00	154.00	1,364.00
remove	1D@.618	ea	—	29.20	29.20
remove for work, then reinstall	7P@2.80	ea	5.76	217.00	222.76

	Craft@Hrs	Unit	Material	Labor	Total
Gas range, free-standing with double oven.					
replace, standard grade	7P@2.05	ea	1,530.00	159.00	1,689.00
replace, high grade	7P@2.05	ea	1,820.00	159.00	1,979.00
replace, deluxe grade	7P@2.05	ea	2,170.00	159.00	2,329.00
remove	1D@.618	ea	—	29.20	29.20
remove for work, then reinstall	7P@2.80	ea	5.76	217.00	222.76
Restaurant-style gas range. Some common manufacturers are Viking, Thermador and Garland. Also, be aware of "mimicked" restaurant styles from prominent manufacturers of home kitchen ranges that sell from $1,000 to $2,500.					
replace, standard grade	7P@2.88	ea	3,910.00	223.00	4,133.00
replace, high grade	7P@2.88	ea	8,570.00	223.00	8,793.00
replace, deluxe grade	7P@2.88	ea	13,600.00	223.00	13,823.00
remove	1D@1.58	ea	—	74.60	74.60
remove for work, then reinstall	7P@3.55	ea	8.16	275.00	283.16
Gas wall oven.					
replace, economy grade	7P@1.79	ea	705.00	139.00	844.00
replace, standard grade	7P@1.79	ea	919.00	139.00	1,058.00
replace, high grade	7P@1.79	ea	1,160.00	139.00	1,299.00
replace, deluxe grade	7P@1.79	ea	1,440.00	139.00	1,579.00
remove	1D@.630	ea	—	29.70	29.70
remove for work, then reinstall	7P@2.48	ea	6.81	192.00	198.81
Gas double wall oven.					
replace, economy grade	7P@2.22	ea	1,540.00	172.00	1,712.00
replace, standard grade	7P@2.22	ea	1,810.00	172.00	1,982.00
replace, high grade	7P@2.22	ea	2,280.00	172.00	2,452.00
replace, deluxe grade	7P@2.22	ea	2,830.00	172.00	3,002.00
remove	1D@.782	ea	—	36.90	36.90
remove for work, then reinstall	7P@2.70	ea	8.16	209.00	217.16
Electric cook top.					
replace, economy grade	7E@1.32	ea	370.00	95.60	465.60
replace, standard grade	7E@1.32	ea	543.00	95.60	638.60
replace, high grade	7E@1.32	ea	681.00	95.60	776.60
replace, cook top with grill / griddle	7E@1.32	ea	816.00	95.60	911.60
remove	1D@.612	ea	—	28.90	28.90
remove for work, then reinstall	7E@2.21	ea	2.71	160.00	162.71
Solid-disk electric cook top.					
replace, standard grade	7E@1.32	ea	855.00	95.60	950.60
replace, high grade	7E@1.32	ea	965.00	95.60	1,060.60
replace, with grill / griddle	7E@1.32	ea	1,050.00	95.60	1,145.60
remove	1D@.612	ea	—	28.90	28.90
remove for work, then reinstall	7E@2.21	ea	2.71	160.00	162.71

	Craft@Hrs	Unit	Material	Labor	Total
Flat-surface radiant electric cook top.					
replace, standard grade	7E@1.32	ea	1,030.00	95.60	1,125.60
replace, high grade	7E@1.32	ea	1,160.00	95.60	1,255.60
replace, cook top with grill / griddle	7E@1.32	ea	1,250.00	95.60	1,345.60
remove	1D@.612	ea	—	28.90	28.90
remove for work, then reinstall	7E@2.21	ea	2.71	160.00	162.71
Modular electric cook top.					
replace, two coil (Calrod) burners	7E@.287	ea	421.00	20.80	441.80
replace, two solid disk burners	7E@.287	ea	503.00	20.80	523.80
replace, two flat-surface radiant burners	7E@.287	ea	570.00	20.80	590.80
replace, griddle	7E@.287	ea	600.00	20.80	620.80
replace, grill	7E@.287	ea	639.00	20.80	659.80
replace, downdraft unit	7E@.287	ea	313.00	20.80	333.80
add for telescopic downdraft unit	—	ea	195.00	—	195.00
remove	1D@.215	ea	—	10.10	10.10
Remove for work and reinstall.					
Modular cooking unit	7E@.344	ea	—	24.90	24.90
Modular downdraft unit	7E@.394	ea	2.76	28.50	31.26
Electric range, free-standing.					
replace, economy grade	2A@1.05	ea	517.00	63.60	580.60
replace, standard grade	2A@1.05	ea	744.00	63.60	807.60
replace, high grade	2A@1.05	ea	1,050.00	63.60	1,113.60
replace, deluxe grade	2A@1.05	ea	1,370.00	63.60	1,433.60
replace, cook top with grill / griddle	2A@1.05	ea	1,550.00	63.60	1,613.60
remove	1D@.505	ea	—	23.80	23.80
remove for work, then reinstall	2A@1.72	ea	5.43	104.00	109.43
Solid-disk electric range, free-standing.					
replace, standard grade	2A@1.05	ea	990.00	63.60	1,053.60
replace, high grade	2A@1.05	ea	1,100.00	63.60	1,163.60
replace, deluxe grade	2A@1.05	ea	1,310.00	63.60	1,373.60
replace, cook top with grill / griddle	2A@1.05	ea	1,610.00	63.60	1,673.60
remove	1D@.505	ea	—	23.80	23.80
remove for work, then reinstall	2A@1.72	ea	5.43	104.00	109.43
Flat-surface radiant electric range, free-standing.					
replace, standard grade	2A@1.05	ea	1,180.00	63.60	1,243.60
replace, high grade	2A@1.05	ea	1,310.00	63.60	1,373.60
replace, deluxe grade	2A@1.05	ea	1,610.00	63.60	1,673.60
replace, cook top with grill / griddle	2A@1.05	ea	2,450.00	63.60	2,513.60
remove	1D@.505	ea	—	23.80	23.80
remove for work, then reinstall	2A@1.72	ea	5.43	104.00	109.43

	Craft@Hrs	Unit	Material	Labor	Total
Drop-in or slide-in electric range.					
replace, standard grade	2A@1.05	ea	1,030.00	63.60	1,093.60
replace, high grade	2A@1.05	ea	1,370.00	63.60	1,433.60
replace, deluxe grade	2A@1.05	ea	1,580.00	63.60	1,643.60
replace, cook top with grill / griddle	2A@1.05	ea	1,970.00	63.60	2,033.60
remove	1D@.505	ea	—	23.80	23.80
remove for work, then reinstall	2A@1.72	ea	6.81	104.00	110.81
Space-saver electric range, free-standing.					
replace, standard grade	2A@.980	ea	1,380.00	59.40	1,439.40
replace, high grade	2A@.980	ea	1,440.00	59.40	1,499.40
remove	1D@.505	ea	—	23.80	23.80
remove for work, then reinstall	2A@1.72	ea	5.43	104.00	109.43
High-low electric range with microwave high.					
replace, standard grade	2A@1.90	ea	440.00	115.00	555.00
replace, high grade	2A@1.90	ea	590.00	115.00	705.00
replace, deluxe grade	2A@1.90	ea	913.00	115.00	1,028.00
remove	1D@.883	ea	—	41.70	41.70
remove for work, then reinstall	2A@2.11	ea	6.81	128.00	134.81
Electric range, free-standing with double oven.					
replace, standard grade	2A@.980	ea	1,300.00	59.40	1,359.40
replace, high grade	2A@.980	ea	1,580.00	59.40	1,639.40
replace, deluxe grade	2A@.980	ea	2,120.00	59.40	2,179.40
remove	1D@.505	ea	—	23.80	23.80
remove for work, then reinstall	2A@1.72	ea	2.71	104.00	106.71
Electric wall oven.					
replace, economy grade	7E@1.55	ea	760.00	112.00	872.00
replace, standard grade	7E@1.55	ea	879.00	112.00	991.00
replace, high grade	7E@1.55	ea	1,110.00	112.00	1,222.00
replace, deluxe grade	7E@1.55	ea	1,390.00	112.00	1,502.00
remove	1D@.652	ea	—	30.80	30.80
remove for work, then reinstall	7E@2.50	ea	2.71	181.00	183.71
Electric double wall oven.					
replace, economy grade	7E@2.72	ea	1,110.00	197.00	1,307.00
replace, standard grade	7E@2.72	ea	1,400.00	197.00	1,597.00
replace, high grade	7E@2.72	ea	2,010.00	197.00	2,207.00
replace, deluxe grade	7E@2.72	ea	2,680.00	197.00	2,877.00
remove	1D@.774	ea	—	36.50	36.50
remove for work, then reinstall	7E@2.78	ea	4.10	201.00	205.10

	Craft@Hrs	Unit	Material	Labor	Total
Warming drawer.					
replace, standard grade	7E@.977	ea	907.00	70.70	977.70
replace, high grade	7E@.977	ea	1,090.00	70.70	1,160.70
replace, deluxe grade	7E@.977	ea	1,540.00	70.70	1,610.70
remove	1D@.444	ea	—	21.00	21.00
remove for work, then reinstall	7E@1.66	ea	10.90	120.00	130.90
Range hood. Under-cabinet range hoods. Economy and standard grades may be non-vented. Stainless steel hoods usually range from high to custom grade. Designer hoods are usually deluxe or custom grade.					
replace, economy grade	2A@1.33	ea	88.50	80.60	169.10
replace, standard grade	2A@1.33	ea	146.00	80.60	226.60
replace, high grade	2A@1.33	ea	180.00	80.60	260.60
replace, deluxe grade	2A@1.33	ea	218.00	80.60	298.60
replace, custom grade	2A@1.33	ea	286.00	80.60	366.60
remove	1D@.422	ea	—	19.90	19.90
remove for work, then reinstall	2A@1.58	ea	2.71	95.70	98.41
Range hood, oversized. Under-cabinet range hoods. Economy and standard grades may be non-vented. Stainless steel hoods usually range from high to custom grade. Designer hoods are usually deluxe or custom grade.					
replace, standard grade	2A@1.46	ea	475.00	88.50	563.50
replace, high grade	2A@1.46	ea	571.00	88.50	659.50
replace, deluxe grade	2A@1.46	ea	698.00	88.50	786.50
replace, custom grade	2A@1.46	ea	918.00	88.50	1,006.50
remove	1D@.422	ea	—	19.90	19.90
remove for work, then reinstall	2A@2.04	ea	2.71	124.00	126.71
Range hood, slide-out. Under-cabinet slide-out range hoods.					
replace, standard grade	2A@1.58	ea	646.00	95.70	741.70
replace, high grade	2A@1.58	ea	780.00	95.70	875.70
replace, deluxe grade	2A@1.58	ea	896.00	95.70	991.70
remove	1D@.492	ea	—	23.20	23.20
remove for work, then reinstall	2A@2.84	ea	2.71	172.00	174.71
Chimney range hood. Island-mount or wall-mount vented chimney hood. The most common finish (and the most popular) is stainless steel, but a wide range of finishes are available. Copper chimney hoods usually fall in the deluxe to custom deluxe grades.					
replace, economy grade	2A@2.62	ea	1,130.00	159.00	1,289.00
replace, standard grade	2A@2.62	ea	1,540.00	159.00	1,699.00
replace, high grade	2A@2.62	ea	1,990.00	159.00	2,149.00
replace, deluxe grade	2A@2.88	ea	2,520.00	175.00	2,695.00
replace, custom grade	2A@2.88	ea	3,460.00	175.00	3,635.00
replace, custom deluxe grade	2A@3.04	ea	4,850.00	184.00	5,034.00
remove	1D@.787	ea	—	37.10	37.10
remove for work, then reinstall	2A@4.00	ea	2.71	242.00	244.71
Downdraft ventilation system. Stand-alone downdraft ventilation systems used with cook tops and ranges. These units are often telescopic and retract when not in use. Do not use these prices for cook tops and ranges that have integrated downdraft systems.					
replace, standard grade	2A@1.87	ea	671.00	113.00	784.00
replace, high grade	2A@1.87	ea	806.00	113.00	919.00
replace, deluxe grade	2A@1.87	ea	1,130.00	113.00	1,243.00
remove	1D@.576	ea	—	27.20	27.20
remove for work, then reinstall	2A@3.04	ea	2.71	184.00	186.71

	Craft@Hrs	Unit	Material	Labor	Total
Microwave oven. Countertop microwave.					
replace, economy grade	2A@.290	ea	135.00	17.60	152.60
replace, standard grade	2A@.290	ea	364.00	17.60	381.60
replace, high grade	2A@.290	ea	526.00	17.60	543.60
replace, deluxe grade	2A@.290	ea	748.00	17.60	765.60
replace, custom grade	2A@.290	ea	902.00	17.60	919.60
remove	1D@.182	ea	—	8.59	8.59
remove for work, then reinstall	2A@.345	ea	—	20.90	20.90
Microwave, under cabinet. Microwave oven mounted to wall and/or cabinets. Also called an "over-the-range" microwave.					
replace, standard grade	2A@.950	ea	485.00	57.60	542.60
replace, high grade	2A@.950	ea	753.00	57.60	810.60
replace, deluxe grade	2A@.950	ea	894.00	57.60	951.60
remove	1D@.346	ea	—	16.30	16.30
remove for work, then reinstall	2A@.677	ea	—	41.00	41.00
Dishwasher.					
replace, economy grade	2A@1.85	ea	305.00	112.00	417.00
replace, standard grade	2A@1.85	ea	467.00	112.00	579.00
replace, high grade	2A@1.85	ea	753.00	112.00	865.00
replace, deluxe grade	2A@1.85	ea	1,100.00	112.00	1,212.00
replace, custom deluxe grade	2A@1.85	ea	1,780.00	112.00	1,892.00
remove	1D@.649	ea	—	30.60	30.60
remove for work, then reinstall	2A@2.32	ea	—	141.00	141.00
Dishwasher, convertible (portable).					
replace, standard grade	2A@1.18	ea	549.00	71.50	620.50
replace, high grade	2A@1.18	ea	694.00	71.50	765.50
replace, deluxe grade	2A@1.18	ea	882.00	71.50	953.50
remove	1D@.388	ea	—	18.30	18.30
remove for work, then reinstall	2A@1.56	ea	—	94.50	94.50
Dishwasher, space-saver under-sink .					
replace, standard grade	2A@2.63	ea	811.00	159.00	970.00
replace, high grade	2A@2.63	ea	954.00	159.00	1,113.00
remove	1D@.732	ea	—	34.60	34.60
remove for work, then reinstall	2A@3.22	ea	6.81	195.00	201.81
Garbage disposal.					
replace, 1/3 hp	7P@1.51	ea	140.00	117.00	257.00
replace, 1/2 hp, standard grade	7P@1.51	ea	173.00	117.00	290.00
replace, 1/2 hp, high grade	7P@1.51	ea	269.00	117.00	386.00
replace, 3/4 hp, standard grade	7P@1.51	ea	275.00	117.00	392.00
replace, 3/4 hp, high grade	7P@1.51	ea	409.00	117.00	526.00
replace, 1 hp, standard grade	7P@1.51	ea	375.00	117.00	492.00
replace, 1 hp, high grade	7P@1.51	ea	413.00	117.00	530.00
remove	1D@.628	ea	—	29.60	29.60
remove for work, then reinstall	7P@2.51	ea	6.81	194.00	200.81

	Craft@Hrs	Unit	Material	Labor	Total
Refrigerator / freezer, side-by-side.					
24 to 29 cf capacity refrigerator / freezer					
replace, economy grade	2A@.813	ea	1,280.00	49.30	1,329.30
replace, standard grade	2A@.813	ea	1,610.00	49.30	1,659.30
replace, high grade	2A@.813	ea	2,380.00	49.30	2,429.30
replace, deluxe grade	2A@.813	ea	2,750.00	49.30	2,799.30
replace, custom deluxe grade	2A@.813	ea	4,250.00	49.30	4,299.30
remove	1D@.794	ea	—	37.50	37.50
remove for work, then reinstall	2A@1.77	ea	6.81	107.00	113.81
19 to 23 cf capacity refrigerator / freezer					
replace, economy grade	2A@.813	ea	1,480.00	49.30	1,529.30
replace, standard grade	2A@.813	ea	1,660.00	49.30	1,709.30
replace, high grade	2A@.813	ea	2,410.00	49.30	2,459.30
replace, deluxe grade	2A@.813	ea	2,830.00	49.30	2,879.30
replace, custom deluxe grade	2A@.813	ea	4,950.00	49.30	4,999.30
remove	1D@.976	ea	—	46.10	46.10
remove for work, then reinstall	2A@1.77	ea	6.81	107.00	113.81
Refrigerator with top freezer.					
21 to 25 cf capacity					
replace, economy grade	2A@.700	ea	885.00	42.40	927.40
replace, standard grade	2A@.700	ea	1,290.00	42.40	1,332.40
replace, high grade	2A@.700	ea	1,550.00	42.40	1,592.40
replace, deluxe grade	2A@.700	ea	3,020.00	42.40	3,062.40
replace, custom deluxe grade	2A@.700	ea	5,070.00	42.40	5,112.40
remove	1D@.365	ea	—	17.20	17.20
remove for work, then reinstall	2A@1.42	ea	6.81	86.10	92.91
17 to 20 cf capacity					
replace, economy grade	2A@.700	ea	638.00	42.40	680.40
replace, standard grade	2A@.700	ea	1,130.00	42.40	1,172.40
replace, high grade	2A@.700	ea	1,390.00	42.40	1,432.40
replace, deluxe grade	2A@.700	ea	1,900.00	42.40	1,942.40
replace, custom deluxe grade	2A@.700	ea	4,760.00	42.40	4,802.40
remove	1D@.365	ea	—	17.20	17.20
remove for work, then reinstall	2A@1.42	ea	6.92	86.10	93.02

Refrigerator with bottom freezer. Higher grades sometimes have French-style refrigerator doors. Deluxe and custom deluxe grades may have French-style refrigerator doors and/or armoire-style freezer drawers.

	Craft@Hrs	Unit	Material	Labor	Total
21 to 25 cf capacity					
replace, economy grade	2A@.700	ea	932.00	42.40	974.40
replace, standard grade	2A@.700	ea	1,360.00	42.40	1,402.40
replace, high grade	2A@.700	ea	1,660.00	42.40	1,702.40
replace, deluxe grade	2A@.700	ea	3,230.00	42.40	3,272.40
replace, custom deluxe grade	2A@.700	ea	5,360.00	42.40	5,402.40
remove	1D@.365	ea	—	17.20	17.20
remove for work, then reinstall	2A@1.42	ea	6.81	86.10	92.91
17 to 20 cf capacity					
replace, economy grade	2A@.700	ea	673.00	42.40	715.40
replace, standard grade	2A@.700	ea	1,200.00	42.40	1,242.40
replace, high grade	2A@.700	ea	1,450.00	42.40	1,492.40
replace, deluxe grade	2A@.700	ea	2,000.00	42.40	2,042.40
replace, custom deluxe grade	2A@.700	ea	5,030.00	42.40	5,072.40
remove	1D@.365	ea	—	17.20	17.20
remove for work, then reinstall	2A@1.42	ea	6.81	86.10	92.91

	Craft@Hrs	Unit	Material	Labor	Total
Refrigerator / freezer, counter-depth. Also called cabinet-depth refrigerators.					
23 to 26 cf capacity					
replace, standard grade	2A@.733	ea	2,960.00	44.40	3,004.40
replace, high grade	2A@.733	ea	3,750.00	44.40	3,794.40
replace, deluxe grade	2A@.733	ea	4,570.00	44.40	4,614.40
replace, custom deluxe grade	2A@.733	ea	5,270.00	44.40	5,314.40
remove	1D@.548	ea	—	25.90	25.90
remove for work, then reinstall	2A@1.67	ea	—	101.00	101.00
19 to 22 cf capacity					
replace, standard grade	2A@.733	ea	2,680.00	44.40	2,724.40
replace, high grade	2A@.733	ea	3,650.00	44.40	3,694.40
replace, deluxe grade	2A@.733	ea	4,160.00	44.40	4,204.40
replace, custom deluxe grade	2A@.733	ea	4,950.00	44.40	4,994.40
remove	1D@.548	ea	—	25.90	25.90
remove for work, then reinstall	2A@1.67	ea	—	101.00	101.00
Refrigerator / freezer, cabinet-match. Refrigerator and freezer doors made to match kitchen cabinetry.					
Add for refrigerator / freezer doors made to match cabinets					
replace	—	ea	600.00	—	600.00
Refrigerator drawer. Cabinet-match or stainless steel drawer face. Per unit installed in lower cabinets, with two to three drawers per unit.					
replace, standard grade	2A@1.49	ea	2,960.00	90.30	3,050.30
replace, high grade	2A@1.49	ea	3,670.00	90.30	3,760.30
remove	1D@.599	ea	—	28.30	28.30
remove for work, then reinstall	2A@2.11	ea	—	128.00	128.00
Refrigerator / freezer refreshment center.					
Add for refreshment center in refrigerator / freezer					
replace	2A@1.88	ea	305.00	114.00	419.00
Refreshment center plumbing. With supply lines and hook-up to appliances. For installation of supply lines only, see Plumbing on page 320.					
Disconnect water supply line for work, then reinstall					
replace	7P@.283	ea	.95	21.90	22.85
Refreshment center plumbing hook-up					
replace, in existing building	7P@2.22	ea	37.80	172.00	209.80
replace, in new construction	7P@1.37	ea	35.40	106.00	141.40
Refinish refrigerator.					
replace, typical	3A@2.21	ea	43.30	160.00	203.30
replace, refinish and recondition (replace gaskets)	3A@2.69	ea	100.00	194.00	294.00
Washing machine, top loading.					
replace, economy grade	2A@1.49	ea	471.00	90.30	561.30
replace, standard grade	2A@1.49	ea	624.00	90.30	714.30
replace, high grade	2A@1.49	ea	879.00	90.30	969.30
replace, deluxe grade	2A@1.49	ea	1,270.00	90.30	1,360.30
replace, custom deluxe grade	2A@1.49	ea	1,730.00	90.30	1,820.30
remove	1D@.421	ea	—	19.90	19.90
remove for work, then reinstall	2A@2.04	ea	—	124.00	124.00

	Craft@Hrs	Unit	Material	Labor	Total
Washing machine, front loading.					
replace, economy grade	2A@1.49	ea	782.00	90.30	872.30
replace, standard grade	2A@1.49	ea	1,150.00	90.30	1,240.30
replace, high grade	2A@1.49	ea	1,450.00	90.30	1,540.30
replace, deluxe grade	2A@1.49	ea	2,110.00	90.30	2,200.30
replace, custom deluxe grade	2A@1.49	ea	2,750.00	90.30	2,840.30
remove	1D@.421	ea	—	19.90	19.90
remove for work, then reinstall	2A@2.04	ea	—	124.00	124.00
Washing machine, steam.					
replace, standard grade	2A@1.49	ea	1,730.00	90.30	1,820.30
replace, high grade	2A@1.49	ea	1,900.00	90.30	1,990.30
replace, deluxe grade	2A@1.49	ea	2,360.00	90.30	2,450.30
remove	1D@.421	ea	—	19.90	19.90
remove for work, then reinstall	2A@2.04	ea	—	124.00	124.00
Washing machine and dryer, stacked. Also called unitized, Spacemaker by GE, and sometimes referred to as combo units, but most manufacturers refer to these space-saver styles as stacked or stackable units.					
with electric dryer					
replace, standard grade	2A@2.04	ea	1,130.00	124.00	1,254.00
replace, high grade	2A@2.04	ea	1,580.00	124.00	1,704.00
replace, deluxe grade	2A@2.04	ea	1,780.00	124.00	1,904.00
remove	1D@.542	ea	—	25.60	25.60
remove for work, then reinstall	2A@2.88	ea	—	175.00	175.00
with gas dryer					
replace, standard grade	2A@2.54	ea	1,210.00	154.00	1,364.00
replace, high grade	2A@2.54	ea	1,730.00	154.00	1,884.00
replace, deluxe grade	2A@2.54	ea	1,880.00	154.00	2,034.00
remove	1D@.533	ea	—	25.20	25.20
remove for work, then reinstall	2A@3.22	ea	—	195.00	195.00
Clothes dryer, electric. Electric dryer with up to 20' of vent pipe. Deluxe and custom deluxe grades include steam dryers.					
replace, economy grade	2A@.932	ea	493.00	56.50	549.50
replace, standard grade	2A@.932	ea	701.00	56.50	757.50
replace, high grade	2A@.932	ea	1,090.00	56.50	1,146.50
replace, deluxe grade	2A@.932	ea	1,320.00	56.50	1,376.50
replace, custom deluxe grade	2A@.932	ea	1,730.00	56.50	1,786.50
remove	1D@.400	ea	—	18.90	18.90
remove for work, then reinstall	2A@1.46	ea	—	88.50	88.50
Clothes dryer, gas. Gas dryer with up to 20' of vent pipe. Deluxe and custom deluxe grades include steam dryers.					
replace, economy grade	2A@1.94	ea	569.00	118.00	687.00
replace, standard grade	2A@1.94	ea	779.00	118.00	897.00
replace, high grade	2A@1.94	ea	1,160.00	118.00	1,278.00
replace, deluxe grade	2A@1.94	ea	1,400.00	118.00	1,518.00
replace, custom deluxe grade	2A@1.94	ea	1,830.00	118.00	1,948.00
remove	1D@.488	ea	—	23.00	23.00
remove for work, then reinstall	2A@2.59	ea	—	157.00	157.00

	Craft@Hrs	Unit	Material	Labor	Total
Trash compactor. Use high grade for compactor with cabinet-match door.					
replace, economy grade	2A@1.09	ea	631.00	66.10	697.10
replace, standard grade	2A@1.09	ea	749.00	66.10	815.10
replace, high grade	2A@1.09	ea	1,090.00	66.10	1,156.10
remove	1D@.288	ea	—	13.60	13.60
remove for work, then reinstall	2A@1.39	ea	.95	84.20	85.15

Time & Material Charts (selected items)

Appliance Materials

See Appliance material prices above.

Appliance Labor

Laborer	base wage	paid leave	true wage	taxes & ins.	total
Appliance installer	$36.20	2.82	$39.02	21.58	$60.60
Appliance refinisher	$41.70	3.25	$44.95	27.35	$72.30
Electrician	$43.80	3.42	$47.22	25.18	$72.40
Plumber	$46.10	3.60	$49.70	27.70	$77.40
Demolition laborer	$25.90	2.02	$27.92	19.28	$47.20

Paid leave is calculated based on two weeks paid vacation, one week sick leave, and seven paid holidays. Employer's matching portion of **FICA** is 7.65 percent. **FUTA** (Federal Unemployment) is .8 percent. **Worker's compensation** was calculated using a national average of 7.36 percent for the appliance installer; 15.00 percent for the appliance refinisher; 8.08 percent for the electrician; 11.24 percent for the plumber; and 14.26 percent for the demolition laborer. **Unemployment insurance** was calculated using a national average of 8 percent. **Health insurance** was calculated based on a projected national average for 2020 of $1,280 per employee (and family when applicable) per month. Employer pays 80 percent for a per month cost of $1,024 per employee. **Retirement** is based on a 401(k) retirement program with employer matching of 50 percent. Employee contributions to the 401(k) plan are an average of 6 percent of the true wage. **Liability insurance** is based on a national average of 12.0 percent.

Appliances Labor Productivity

	Craft@Hrs	Unit	Material	Labor	Total
Install appliances					
gas cook top	7P@1.99	ea	—	154.00	154.00
gas range	7P@2.05	ea	—	159.00	159.00
space-saver gas range	7P@1.99	ea	—	154.00	154.00
gas range with double oven	7P@2.05	ea	—	159.00	159.00
restaurant-style gas range	7P@2.88	ea	—	223.00	223.00
gas wall oven	7P@1.79	ea	—	139.00	139.00
gas double wall oven	7P@2.22	ea	—	172.00	172.00
electric cook top	7E@1.32	ea	—	95.60	95.60
modular cooking unit	7E@.287	ea	—	20.80	20.80
modular electric cook top downdraft unit	7E@.287	ea	—	20.80	20.80
electric range	2A@1.05	ea	—	63.60	63.60
drop-in or slide-in electric range	2A@1.05	ea	—	63.60	63.60
space-saver electric range	2A@.980	ea	—	59.40	59.40
electric wall oven	7E@1.55	ea	—	112.00	112.00
electric double wall-oven	7E@2.72	ea	—	197.00	197.00
range hood	2A@1.33	ea	—	80.60	80.60
oversized range hood	2A@1.46	ea	—	88.50	88.50
dishwasher	2A@1.85	ea	—	112.00	112.00
convertible (portable) dishwasher	2A@1.18	ea	—	71.50	71.50
space-saver under-sink dishwasher	2A@2.63	ea	—	159.00	159.00

	Craft@Hrs	Unit	Material	Labor	Total
garbage disposal	7P@1.51	ea	—	117.00	117.00
microwave oven	2A@.290	ea	—	17.60	17.60
microwave oven under cabinet	2A@.950	ea	—	57.60	57.60
side-by-side refrigerator / freezer	2A@.813	ea	—	49.30	49.30
refrigerator with freezer top or bottom	2A@.700	ea	—	42.40	42.40
disconnect refrig. water lines for work, reinstall	2A@.283	ea	—	17.10	17.10
refreshment center plumbing hook-up in existing building	7P@2.22	ea	—	172.00	172.00
refreshment center plumbing hook-up in new construction	7P@1.37	ea	—	106.00	106.00
trash compactor	2A@1.09	ea	—	66.10	66.10
clothes washing machine	2A@1.49	ea	—	90.30	90.30
clothes dryer, electric	2A@.932	ea	—	56.50	56.50

Awnings

Free-standing carport. 40 psf. Aluminum with baked enamel finish. Includes 4 posts for carports under 200 sf, 6 to 8 posts for carports over 200 sf.

	Craft@Hrs	Unit	Material	Labor	Total
Free-standing aluminum carport					
replace, 150 sf or less	6A@.077	sf	20.60	4.21	24.81
replace, 151 to 250 sf	6A@.077	sf	19.30	4.21	23.51
replace, 251 to 480 sf	6A@.077	sf	17.50	4.21	21.71
add for heavy gauge aluminum	—	sf	2.39	—	2.39
remove	1D@.050	sf	—	2.36	2.36
remove for work, then reinstall	6A@.161	sf	.69	8.81	9.50

Attached carport. 40 psf. Aluminum with baked enamel finish. Includes 3 posts for carports and patio covers under 200 sf, 4 to 5 posts for carports and patio covers over 200 sf.

	Craft@Hrs	Unit	Material	Labor	Total
Attached aluminum carport or patio cover					
replace, 150 sf or less	6A@.071	sf	19.90	3.88	23.78
replace, 151 to 250 sf	6A@.071	sf	19.00	3.88	22.88
replace, 251 to 480 sf	6A@.071	sf	17.00	3.88	20.88
add for heavy gauge aluminum	—	sf	2.50	—	2.50
remove	1D@.054	sf	—	2.55	2.55
remove for work, then reinstall	6A@.150	sf	.69	8.21	8.90

Minimum charge.

	Craft@Hrs	Unit	Material	Labor	Total
for carport or patio cover work	6A@1.85	ea	67.70	101.00	168.70

Carport or patio cover parts.

	Craft@Hrs	Unit	Material	Labor	Total
Post					
replace	6A@.687	ea	32.20	37.60	69.80
remove	1D@.280	ea	—	13.20	13.20
remove, for work, then reinstall	6A@.925	ea	2.91	50.60	53.51
Scroll post					
replace, cover post scroll	6A@.722	ea	59.10	39.50	98.60
remove	1D@.280	ea	—	13.20	13.20
remove, for work, then reinstall	6A@.994	ea	1.32	54.40	55.72

	Craft@Hrs	Unit	Material	Labor	Total
Fascia, aluminum					
replace	6A@.080	lf	9.60	4.38	13.98
remove	1D@.028	lf	—	1.32	1.32
remove, for work, then reinstall	6A@.110	lf	—	6.02	6.02
Roof panel					
replace	6A@.622	ea	66.20	34.00	100.20
remove	1D@.488	ea	—	23.00	23.00
remove, for work, then reinstall	6A@1.14	ea	5.30	62.40	67.70
Downspout					
replace	6A@.082	ea	27.80	4.49	32.29
remove	1D@.075	ea	—	3.54	3.54
remove for work, then reinstall	6A@.145	ea	—	7.93	7.93

Door awning. 25 to 40 psf. Aluminum with baked enamel finish. Includes all hardware and adjustable supports.

	Craft@Hrs	Unit	Material	Labor	Total
42" projection door awning					
replace, 2 to 4 lf	6A@.464	lf	83.50	25.40	108.90
replace, 5 to 8 lf	6A@.464	lf	89.40	25.40	114.80
replace, 9 to 12 lf	6A@.464	lf	97.10	25.40	122.50
remove	1D@.248	lf	—	11.70	11.70
remove for work, then reinstall	6A@.873	lf	8.62	47.80	56.42
54" projection door awning					
replace, 2 to 4 lf	6A@.484	lf	94.10	26.50	120.60
replace, 5 to 8 lf	6A@.484	lf	101.00	26.50	127.50
replace, 9 to 12 lf	6A@.484	lf	111.00	26.50	137.50
remove	1D@.250	lf	—	11.80	11.80
remove for work, then reinstall	6A@.878	lf	8.62	48.00	56.62
Minimum charge					
for door awning work	6A@1.85	ea	67.70	101.00	168.70

Window awning. 25 to 40 psf. Aluminum with baked enamel finish. Includes all hardware and adjustable supports.

	Craft@Hrs	Unit	Material	Labor	Total
3' high window awning					
replace, 2 to 4 lf	6A@.424	lf	76.70	23.20	99.90
replace, 5 to 8 lf	6A@.424	lf	59.90	23.20	83.10
replace, 9 to 12 lf	6A@.424	lf	56.40	23.20	79.60
remove	1D@.024	lf	—	1.13	1.13
remove for work, then reinstall	6A@.550	lf	8.62	30.10	38.72
4' high window awning					
replace, 2 to 4 lf	6A@.454	lf	89.80	24.80	114.60
replace, 5 to 8 lf	6A@.454	lf	74.10	24.80	98.90
replace, 9 to 12 lf	6A@.454	lf	69.40	24.80	94.20
remove	1D@.028	lf	—	1.32	1.32
remove for work, then reinstall	6A@.589	lf	8.62	32.20	40.82
5' high window awning					
replace, 2 to 4 lf	6A@.470	lf	119.00	25.70	144.70
replace, 5 to 8 lf	6A@.470	lf	109.00	25.70	134.70
replace, 9 to 12 lf	6A@.470	lf	85.80	25.70	111.50
remove	1D@.030	lf	—	1.42	1.42
remove for work, then reinstall	6A@.595	lf	8.62	32.50	41.12
6' high window awning					
replace, 2 to 4 lf	6A@.479	lf	136.00	26.20	162.20
replace, 5 to 8 lf	6A@.479	lf	131.00	26.20	157.20

	Craft@Hrs	Unit	Material	Labor	Total
replace, 9 to 12 lf	6A@.479	lf	101.00	26.20	127.20
remove	1D@.038	lf	—	1.79	1.79
remove for work, then reinstall	6A@.631	lf	8.62	34.50	43.12
Minimum charge					
for window awning work	6A@1.85	ea	67.70	101.00	168.70

Door or window awning slat. Includes prefinished standard length single slat up to 8' long in slat-style aluminum awning. Can be horizontal- or vertical-slat style. Includes replacement slat and hardware. Does not include painting or finishing slat to match.

	Craft@Hrs	Unit	Material	Labor	Total
replace	6A@.381	ea	23.40	20.80	44.20
remove	1D@.232	ea	—	11.00	11.00

Roll-up awning. Security awning that covers window. Aluminum with baked enamel finish. Includes all hardware and adjustable supports.

	Craft@Hrs	Unit	Material	Labor	Total
replace, 2 to 5 lf	6A@.274	lf	86.10	15.00	101.10
replace, 6 to 12 lf	6A@.274	lf	65.80	15.00	80.80
remove	1D@.409	lf	—	19.30	19.30
remove for work, then reinstall	6A@.599	lf	6.63	32.80	39.43

Canvas awning. Waterproof acrylic duck colorfast fabric with retractable metal frame and hardware.

	Craft@Hrs	Unit	Material	Labor	Total
24" drop canvas awning					
replace, 2 to 3 lf	6A@.484	lf	117.00	26.50	143.50
replace, 4 to 5 lf	6A@.493	lf	111.00	27.00	138.00
30" drop canvas awning					
replace, 2 to 3 lf	6A@.526	lf	136.00	28.80	164.80
replace, 4 to 5 lf	6A@.526	lf	121.00	28.80	149.80
replace, 5 to 6 lf	6A@.526	lf	110.00	28.80	138.80
replace, 7 to 8 lf	6A@.526	lf	97.20	28.80	126.00
replace, 9 to 10 lf	6A@.526	lf	79.90	28.80	108.70
replace, 11 to 12 lf	6A@.526	lf	73.90	28.80	102.70
remove	1D@.324	lf	—	15.30	15.30
remove canvas awning for work, then reinstall	6A@1.15	lf	6.77	62.90	69.67
Minimum charge					
for canvas awning work	6A@1.65	ea	57.10	90.30	147.40

Vinyl awning. Vinyl awning with acrylic coating. Includes metal frame and all hardware.

	Craft@Hrs	Unit	Material	Labor	Total
24" drop vinyl awning					
replace, 2 to 3 lf	6A@.491	lf	70.10	26.90	97.00
replace, 4 to 5 lf	6A@.491	lf	63.90	26.90	90.80
30" drop vinyl awning					
replace, 2 to 3 lf	6A@.520	lf	82.10	28.40	110.50
replace, 4 to 5 lf	6A@.520	lf	71.70	28.40	100.10
replace, 5 to 6 lf	6A@.520	lf	65.40	28.40	93.80
replace, 7 to 8 lf	6A@.520	lf	52.30	28.40	80.70
replace, 9 to 10 lf	6A@.520	lf	47.90	28.40	76.30
replace, 11 to 12 lf	6A@.520	lf	46.50	28.40	74.90
remove	1D@.324	lf	—	15.30	15.30
remove awning for work then reinstall	6A@1.13	lf	6.77	61.80	68.57
Minimum charge					
for vinyl awning work	6A@1.47	ea	38.50	80.40	118.90

	Craft@Hrs	Unit	Material	Labor	Total
Time & Material Charts (selected items)					
Awning Materials					
Aluminum carport					
free-standing, 151 to 250 sf	—	sf	19.50	—	19.50
attached, 151 to 250 sf	—	sf	19.20	—	19.20
Aluminum door awning					
with 42" projection, 5 to 8 lf	—	lf	90.60	—	90.60
with 54" projection, 5 to 8 lf	—	lf	102.00	—	102.00
Aluminum window awning					
3' high, 5 to 8 lf	—	lf	60.80	—	60.80
6' high, 5 to 8 lf	—	lf	132.00	—	132.00
Roll-up aluminum awning					
2 to 5 lf	—	lf	87.40	—	87.40
6 to 12 lf	—	lf	66.60	—	66.60
Canvas awning					
24" drop, 2 to 3 lf	—	lf	118.00	—	118.00
24" drop, 4 to 5 lf	—	lf	112.00	—	112.00
30" drop, 2 to 3 lf	—	lf	137.00	—	137.00
30" drop, 7 to 8 lf	—	lf	98.80	—	98.80
30" drop, 11 to 12 lf	—	lf	74.80	—	74.80
Vinyl awning					
24" drop, 2 to 3 lf	—	lf	71.00	—	71.00
24" drop, 4 to 5 lf	—	lf	65.20	—	65.20
30" drop, 2 to 3 lf	—	lf	83.00	—	83.00
30" drop, 7 to 8 lf	—	lf	52.90	—	52.90
30" drop, 11 to 12 lf	—	lf	47.30	—	47.30

Awning Labor

Laborer	base wage	paid leave	true wage	taxes & ins.	total
Awning installer	$34.90	2.72	$37.62	24.68	$62.30
Awning installer's helper	$25.40	1.98	$27.38	19.72	$47.10
Demolition laborer	$25.90	2.02	$27.92	19.28	$47.20

Paid leave is calculated based on two weeks paid vacation, one week sick leave, and seven paid holidays. Employer's matching portion of **FICA** is 7.65 percent. **FUTA** (Federal Unemployment) is .8 percent. **Worker's compensation** for the awning trade was calculated using a national average of 17.00 percent. **Unemployment insurance** was calculated using a national average of 8 percent. **Health insurance** was calculated based on a projected national average for 2020 of $1,280 per employee (and family when applicable) per month. Employer pays 80 percent for a per month cost of $1,024 per employee. **Retirement** is based on a 401(k) retirement program with employer matching of 50 percent. Employee contributions to the 401(k) plan are an average of 6 percent of the true wage. **Liability insurance** is based on a national average of 12.0 percent.

Awning Installation Crew

install, awnings and patio covers	awning installer	$62.30
install, awnings and patio covers	awning installer's helper	$47.10
awning installation crew	awning installation crew	$54.70

	Craft@Hrs	Unit	Material	Labor	Total
Awning Labor Productivity					
Install aluminum carport or patio cover					
free-standing carport	6A@.077	sf	—	4.21	4.21
attached carport or patio	6A@.071	sf	—	3.88	3.88
Install aluminum door awning					
42" projection	6A@.464	lf	—	25.40	25.40
54" projection	6A@.484	lf	—	26.50	26.50
Install aluminum window awning					
3' high	6A@.424	lf	—	23.20	23.20
4' high	6A@.454	lf	—	24.80	24.80
5' high	6A@.470	lf	—	25.70	25.70
6' high	6A@.479	lf	—	26.20	26.20
Install roll-up aluminum awning					
per lf	6A@.274	lf	—	15.00	15.00
Install canvas awning					
24" drop	6A@.493	lf	—	27.00	27.00
30" drop	6A@.526	lf	—	28.80	28.80
Install vinyl awning					
24" drop	6A@.491	lf	—	26.90	26.90
30" drop	6A@.520	lf	—	28.40	28.40

Bathroom Hardware

Hardware Quality. Here are some "rules of thumb" for determining bathroom hardware quality. Economy: Light-gauge metal, chrome plated. May have some plastic components. Little or no pattern. Standard: Heavier gauge metal, chrome or brass plated with little or no pattern, or wood hardware made of ash or oak. High: Brass, chrome over brass, or nickel over brass with minimal detail, or plated hardware with ornate detail, or European-style curved plastic, or hardware with porcelain components, or wood hardware made of walnut, cherry or similar wood. Deluxe: Brass, chrome over brass, or nickel over brass with ornate detail. Antique Reproduction / Custom: Brass, chrome over brass, or nickel over brass with ornate antique-style detail.

	Craft@Hrs	Unit	Material	Labor	Total
Minimum charge.					
for bathroom hardware work	1C@.563	ea	31.90	38.10	70.00

Complete bath hardware. Bathroom with toilet, sink, and bathtub / shower. Includes towel bar, wash cloth bar, door-mount robe hook, toilet paper dispenser, and cup / toothbrush holder.

	Craft@Hrs	Unit	Material	Labor	Total
replace, economy grade	1C@4.02	ea	161.00	272.00	433.00
replace, standard grade	1C@4.02	ea	300.00	272.00	572.00
replace, high grade	1C@4.02	ea	439.00	272.00	711.00
replace, deluxe grade	1C@4.02	ea	593.00	272.00	865.00
replace, custom grade	1C@4.02	ea	674.00	272.00	946.00
replace, antique reproduction	1C@4.02	ea	1,100.00	272.00	1,372.00
remove	1D@2.22	ea	—	105.00	105.00

	Craft@Hrs	Unit	Material	Labor	Total
1/2 bath hardware. Bathroom with toilet and sink. Includes wash cloth bar, towel ring, and toilet paper dispenser.					
replace, economy grade	1C@2.17	ea	52.50	147.00	199.50
replace, standard grade	1C@2.17	ea	102.00	147.00	249.00
replace, high grade	1C@2.17	ea	147.00	147.00	294.00
replace, deluxe grade	1C@2.17	ea	202.00	147.00	349.00
replace, custom grade	1C@2.17	ea	230.00	147.00	377.00
replace, antique reproduction	1C@2.17	ea	389.00	147.00	536.00
remove	1D@1.60	ea	—	75.50	75.50
Glass bathroom shelf. 21" x 5-1/2".					
replace, glass	1C@.283	ea	55.80	19.10	74.90
replace, glass & brass	1C@.283	ea	78.80	19.10	97.90
remove	1D@.205	ea	—	9.68	9.68
remove for work, then reinstall	1C@.300	ea	—	20.30	20.30
Cup & toothbrush holder.					
replace, standard grade	1C@.169	ea	15.40	11.40	26.80
replace, high grade	1C@.169	ea	25.20	11.40	36.60
remove	1D@.163	ea	—	7.69	7.69
remove for work, then reinstall	1C@.217	ea	—	14.70	14.70
Door-mounted clothes hanger.					
replace, standard grade	1C@.141	ea	13.60	9.53	23.13
replace, high grade	1C@.141	ea	22.80	9.53	32.33
remove	1D@.139	ea	—	6.56	6.56
remove for work, then reinstall	1C@.175	ea	—	11.80	11.80
Robe hook.					
replace, standard grade	1C@.141	ea	15.40	9.53	24.93
replace, high grade	1C@.141	ea	26.10	9.53	35.63
remove	1D@.139	ea	—	6.56	6.56
remove for work, then reinstall	1C@.175	ea	—	11.80	11.80
Soap holder.					
replace, standard grade	1C@.169	ea	29.10	11.40	40.50
replace, high grade	1C@.169	ea	49.40	11.40	60.80
remove	1D@.163	ea	—	7.69	7.69
remove for work, then reinstall	1C@.217	ea	—	14.70	14.70
Recessed soap holder.					
replace, standard grade	1C@.201	ea	30.30	13.60	43.90
replace, high grade	1C@.201	ea	62.80	13.60	76.40
remove	1D@.163	ea	—	7.69	7.69
remove for work, then reinstall	1C@.267	ea	—	18.00	18.00

	Craft@Hrs	Unit	Material	Labor	Total
Soap dispenser.					
replace, standard grade	1C@.182	ea	26.10	12.30	38.40
replace, high grade	1C@.182	ea	58.10	12.30	70.40
remove	1D@.163	ea	—	7.69	7.69
remove for work, then reinstall	1C@.224	ea	—	15.10	15.10
Recessed tissue holder.					
replace, standard grade	1C@.171	ea	39.20	11.60	50.80
replace, high grade	1C@.171	ea	62.60	11.60	74.20
remove	1D@.139	ea	—	6.56	6.56
remove for work, then reinstall	1C@.248	ea	—	16.80	16.80
Toilet paper dispenser.					
replace, economy grade	1C@.188	ea	18.00	12.70	30.70
replace, standard grade	1C@.188	ea	42.50	12.70	55.20
replace, high grade	1C@.188	ea	63.10	12.70	75.80
replace, deluxe grade	1C@.188	ea	76.70	12.70	89.40
remove	1D@.325	ea	—	15.30	15.30
remove for work, then reinstall	1C@.244	ea	—	16.50	16.50
Towel bar.					
replace, economy grade	1C@.393	ea	23.60	26.60	50.20
replace, standard grade	1C@.393	ea	34.50	26.60	61.10
replace, high grade	1C@.393	ea	55.40	26.60	82.00
replace, deluxe grade	1C@.393	ea	82.90	26.60	109.50
remove	1D@.325	ea	—	15.30	15.30
remove for work, then reinstall	1C@.563	ea	—	38.10	38.10
Towel ring.					
replace, economy grade	1C@.238	ea	18.20	16.10	34.30
replace, standard grade	1C@.238	ea	30.30	16.10	46.40
replace, high grade	1C@.238	ea	34.50	16.10	50.60
replace, deluxe grade	1C@.238	ea	38.60	16.10	54.70
remove	1D@.325	ea	—	15.30	15.30
remove for work, then reinstall	1C@.299	ea	—	20.20	20.20
Wash cloth bar.					
replace, economy grade	1C@.393	ea	16.70	26.60	43.30
replace, standard grade	1C@.393	ea	26.10	26.60	52.70
replace, high grade	1C@.393	ea	34.50	26.60	61.10
replace, deluxe grade	1C@.393	ea	46.90	26.60	73.50
remove	1D@.325	ea	—	15.30	15.30
remove for work, then reinstall	1C@.563	ea	—	38.10	38.10
Shower curtain rod.					
replace, stainless steel rod	1C@.113	ea	37.40	7.64	45.04
remove	1D@.089	ea	—	4.20	4.20
remove for work, then reinstall	1C@.130	ea	—	8.79	8.79

	Craft@Hrs	Unit	Material	Labor	Total

Recessed medicine cabinet. Economy and standard grades have a plastic or polystyrene box with two to three shelves, box measures up to 16" wide and 25" tall with mirror doors. Higher grades have a steel box with adjustable shelves, box measures up to 20" wide and 32" tall with chrome-plated, brass or wood trim, bevel-edged mirrors and top-arched or oval doors.

	Craft@Hrs	Unit	Material	Labor	Total
replace, economy grade	1C@1.31	ea	76.00	88.60	164.60
replace, standard grade	1C@1.31	ea	92.50	88.60	181.10
replace, high grade	1C@1.31	ea	132.00	88.60	220.60
replace, deluxe grade	1C@1.31	ea	159.00	88.60	247.60
replace, custom grade	1C@1.31	ea	197.00	88.60	285.60
remove	1D@.609	ea	—	28.70	28.70
remove for work, then reinstall	1C@1.88	ea	—	127.00	127.00

high grade

Surface-mounted medicine cabinet. All grades have two to three adjustable shelves. Box size and material vary with grade. Highest grades have three doors, are larger and have oak trim.

	Craft@Hrs	Unit	Material	Labor	Total
replace, economy grade	1C@1.51	ea	94.00	102.00	196.00
replace, standard grade	1C@1.51	ea	133.00	102.00	235.00
replace, high grade	1C@1.51	ea	227.00	102.00	329.00
replace, deluxe grade	1C@1.51	ea	386.00	102.00	488.00
replace, custom grade	1C@1.51	ea	553.00	102.00	655.00
remove	1D@.650	ea	—	30.70	30.70
remove for work, then reinstall	1C@1.97	ea	—	133.00	133.00

Bathroom mirror.

	Craft@Hrs	Unit	Material	Labor	Total
replace, with stainless steel trim	1C@.106	sf	26.10	7.17	33.27
replace, with brass-finished trim	1C@.106	sf	27.50	7.17	34.67
replace, with wood frame	1C@.106	sf	27.50	7.17	34.67
replace, beveled glass	1C@.106	sf	29.00	7.17	36.17
remove bathroom mirror attached with glue	1D@.041	ea	—	1.94	1.94
remove for work, then reinstall	1C@.169	sf	—	11.40	11.40

Time & Material Charts (selected items)
Bathroom Hardware Materials

See Bathroom Hardware material prices above.

Bathroom Hardware Labor

Laborer	base wage	paid leave	true wage	taxes & ins.	total
Carpenter (installer)	$38.10	2.97	$41.07	26.53	$67.60
Demolition laborer	$25.90	2.02	$27.92	19.28	$47.20

Paid leave is calculated based on two weeks paid vacation, one week sick leave, and seven paid holidays. Employer's matching portion of **FICA** is 7.65 percent. **FUTA** (Federal Unemployment) is .8 percent. **Worker's compensation** for the bathroom hardware trade was calculated using a national average of 17.21 percent. **Unemployment insurance** was calculated using a national average of 8 percent. **Health insurance** was calculated based on a projected national average for 2020 of $1,280 per employee (and family when applicable) per month. Employer pays 80 percent for a per month cost of $1,024 per employee. **Retirement** is based on a 401(k) retirement program with employer matching of 50 percent. Employee contributions to the 401(k) plan are an average of 6 percent of the true wage. **Liability insurance** is based on a national average of 12.0 percent.

	Craft@Hrs	Unit	Material	Labor	Total
Bathroom Hardware Labor Productivity					
Install complete bathroom hardware					
full bathroom	1C@4.02	ea	—	272.00	272.00
1/2 bathroom	1C@2.17	ea	—	147.00	147.00
Install hardware					
soap holder	1C@.169	ea	—	11.40	11.40
recessed tissue holder	1C@.171	ea	—	11.60	11.60
towel bar	1C@.393	ea	—	26.60	26.60
toilet paper dispenser	1C@.188	ea	—	12.70	12.70
towel ring	1C@.238	ea	—	16.10	16.10
wash cloth bar	1C@.393	ea	—	26.60	26.60
recessed medicine cabinet	1C@1.31	ea	—	88.60	88.60
surface-mounted medicine cabinet	1C@1.51	ea	—	102.00	102.00
bathroom mirror	1C@.106	sf	—	7.17	7.17

Cabinets

Cabinet Construction. Stock-grade cabinets are manufactured in standard sizes and warehoused until sold. Semi-custom grade cabinets are available in a wide variety of styles and shapes. Within limits, the manufacturer builds the cabinets to match the kitchen. Custom-grade cabinets are built specifically for the kitchen and include specialty doors, interior features, woods, and construction.

Cabinet Grades. Economy grade: Stock-grade cabinets with flush-face doors. Doors made from veneered particleboard. Standard grade: Stock-grade cabinets with raised panel or cathedral doors. Interior panel may be plywood. Lower grade plastic-laminate face cabinets. Semi-custom cabinets: Semi-custom grade cabinets are available in a wide variety of styles and shapes. Within limits, the manufacturer builds the cabinets to match the kitchen. High grade: Semi-custom cabinets with raised panel or cathedral doors. Higher grade plastic-laminate face and foil-face cabinets. Deluxe grade: Semi-custom cabinets with raised panel or cathedral doors. Door corners may be miter cut. May include special slide-out drawers, pull-out baskets, glass doors, or foil-face cabinets. Materials include cherry, pecan, and Shaker-style maple or pine. Custom grade: Custom cabinets with raised panel or cathedral doors. May include special slide-out drawers, pull-out baskets, mullion or leaded glass doors. Materials include cherry, pecan, and Shaker-style maple or pine. Custom deluxe grade: Same as Custom Grade, may have some curved wood cabinets and more custom features.

Foil-Face Cabinets. Foil-faced cabinets (also called thermo foil) are coated with rigid polyvinyl chloride (PVC) that has been heated and pressed. The interior core is usually medium density particleboard. Currently, there is no good way to repair scratched or dented foil-face cabinets. Although colors do not fade from foil-faces, it is almost impossible to replace doors or other parts with colors that will match. Foil-face cabinets are high to custom deluxe quality depending on the selection of interior features and the complexity of the design.

	Craft@Hrs	Unit	Material	Labor	Total
Minimum charge.					
for cabinet work	2I@2.67	ea	68.80	150.00	218.80

Lower kitchen cabinet. Includes prefabricated, prefinished lower cabinets assembled in shop or on site. Includes shims, backing, screws, attachment hardware, and installation labor. Installation includes cutting the cabinet to fit site conditions. This price is an overall allowance when replacing all or most cabinets in a kitchen and is based on a typical mixture of drawer units, door units, and corner units.

	Craft@Hrs	Unit	Material	Labor	Total
replace, economy grade	2I@.360	lf	108.00	20.30	128.30
replace, standard grade	2I@.360	lf	135.00	20.30	155.30
replace, high grade	2I@.360	lf	174.00	20.30	194.30
replace, deluxe grade	2I@.360	lf	245.00	20.30	265.30
replace, custom grade	2I@.360	lf	351.00	20.30	371.30
replace, custom deluxe grade	2I@.360	lf	451.00	20.30	471.30
remove	1D@.262	lf	—	12.40	12.40
remove for work, then reinstall	2I@.636	lf	—	35.80	35.80

Upper kitchen cabinet. Includes prefabricated, prefinished upper cabinets assembled in shop or on site. Includes shims, screws, backing, attachment hardware, and installation labor. Installation includes cutting the cabinet to fit site conditions. This price is an overall allowance when replacing all or most cabinets in a kitchen and is based on a typical mixture of door units, corner units, and over refrigerator units.

	Craft@Hrs	Unit	Material	Labor	Total
replace, economy grade	2I@.349	lf	93.90	19.60	113.50
replace, standard grade	2I@.349	lf	125.00	19.60	144.60
replace, high grade	2I@.349	lf	153.00	19.60	172.60
replace, deluxe grade	2I@.349	lf	227.00	19.60	246.60
replace, custom grade	2I@.349	lf	324.00	19.60	343.60
replace, custom deluxe grade	2I@.349	lf	414.00	19.60	433.60
remove	1D@.236	lf	—	11.10	11.10
remove for work, then reinstall	2I@.611	lf	—	34.40	34.40

Lower kitchen island cabinet. Includes prefabricated, prefinished lower island cabinets assembled in shop or on site. Includes shims, backing, screws, attachment hardware, and installation labor. Installation includes cutting the cabinet to fit site conditions. This price is an overall allowance when replacing all or most cabinets in a kitchen.

	Craft@Hrs	Unit	Material	Labor	Total
replace, economy grade	2I@.371	lf	111.00	20.90	131.90
replace, standard grade	2I@.371	lf	140.00	20.90	160.90
replace, high grade	2I@.371	lf	181.00	20.90	201.90
replace, deluxe grade	2I@.371	lf	255.00	20.90	275.90
replace, custom grade	2I@.371	lf	365.00	20.90	385.90
replace, custom deluxe grade	2I@.371	lf	470.00	20.90	490.90
remove	1D@.253	lf	—	11.90	11.90
remove for work, then reinstall	2I@.653	lf	—	36.80	36.80

Upper kitchen island cabinet. Includes prefabricated, prefinished upper island cabinets assembled in shop or on site. Includes shims, backing, screws, attachment hardware, and installation labor. Installation includes cutting the cabinet to fit site conditions. This price is an overall allowance when replacing all or most cabinets in a kitchen.

	Craft@Hrs	Unit	Material	Labor	Total
replace, economy grade	2I@.359	lf	98.00	20.20	118.20
replace, standard grade	2I@.359	lf	131.00	20.20	151.20
replace, high grade	2I@.359	lf	160.00	20.20	180.20
replace, deluxe grade	2I@.359	lf	235.00	20.20	255.20
replace, custom grade	2I@.359	lf	339.00	20.20	359.20
replace, custom deluxe grade	2I@.359	lf	431.00	20.20	451.20
remove	1D@.238	lf	—	11.20	11.20
remove for work, then reinstall	2I@.633	lf	—	35.60	35.60

Full-height utility cabinet. Includes prefabricated, prefinished full-height cabinet assembled in shop or on site, shims, screws, and installation labor. Any cabinet that is 5'6" or taller is considered a full-height cabinet. Includes shims, backing, attachment hardware and cutting the cabinet to fit site conditions.

	Craft@Hrs	Unit	Material	Labor	Total
replace, economy grade	2I@.593	lf	258.00	33.40	291.40
replace, standard grade	2I@.593	lf	323.00	33.40	356.40
replace, high grade	2I@.593	lf	396.00	33.40	429.40
replace, deluxe grade	2I@.593	lf	505.00	33.40	538.40
replace, custom grade	2I@.593	lf	553.00	33.40	586.40
replace, custom deluxe grade	2I@.593	lf	621.00	33.40	654.40
remove	1D@.331	lf	—	15.60	15.60
remove for work, then reinstall	2I@.706	lf	—	39.70	39.70

Full-height built-in oven cabinet. Includes prefabricated, prefinished full-height cabinet assembled in shop or on site, shims, screws, and installation labor. Any cabinet that is 5'6" or taller is considered a full-height cabinet. Includes shims, backing, attachment hardware and cutting the cabinet to fit site conditions.

	Craft@Hrs	Unit	Material	Labor	Total
replace, economy grade	2I@.592	lf	245.00	33.30	278.30
replace, standard grade	2I@.592	lf	307.00	33.30	340.30
replace, high grade	2I@.592	lf	375.00	33.30	408.30
replace, deluxe grade	2I@.592	lf	482.00	33.30	515.30
replace, custom grade	2I@.592	lf	525.00	33.30	558.30
replace, custom deluxe grade	2I@.592	lf	590.00	33.30	623.30
remove	1D@.331	lf	—	15.60	15.60
remove for work, then reinstall	2I@.706	lf	—	39.70	39.70

Full-height built-in double oven cabinet. Includes prefabricated, pre-finished full-height cabinet assembled in shop or on site, shims, screws, and installation labor. Any cabinet that is 5'6" or taller is considered a full-height cabinet. Includes shims, backing, attachment hardware and cutting the cabinet to fit site conditions.

	Craft@Hrs	Unit	Material	Labor	Total
replace, economy grade	2I@.592	lf	237.00	33.30	270.30
replace, standard grade	2I@.592	lf	297.00	33.30	330.30
replace, high grade	2I@.592	lf	362.00	33.30	395.30
replace, deluxe grade	2I@.592	lf	464.00	33.30	497.30
replace, custom grade	2I@.592	lf	509.00	33.30	542.30
replace, custom deluxe grade	2I@.592	lf	569.00	33.30	602.30
remove	1D@.331	lf	—	15.60	15.60
remove for work, then reinstall	2I@.706	lf	—	39.70	39.70

Cabinet drawer fronts. Includes drawer front and attachment screws only. Front is attached to existing drawer frame. Does not include staining, painting, or finishing to match.

	Craft@Hrs	Unit	Material	Labor	Total
replace, drawer front	2I@.118	ea	20.80	6.64	27.44
remove	1D@.144	ea	—	6.80	6.80

flat panel

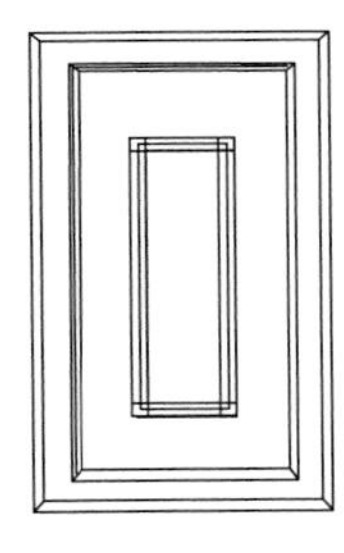

raised panel

raised panel cathedral

with leaded glass

Cabinet doors. Includes cabinet door only, milled to attach to existing hardware. Does not include staining, painting, or finishing to match.

	Craft@Hrs	Unit	Material	Labor	Total
replace, flush-face	2I@.252	ea	25.40	14.20	39.60
replace, flat panel	2I@.252	ea	35.40	14.20	49.60
replace, raised panel	2I@.252	ea	52.50	14.20	66.70
replace, raised panel cathedral	2I@.252	ea	66.70	14.20	80.90
replace, with glass	2I@.252	ea	80.60	14.20	94.80
replace, with leaded glass	2I@.252	ea	161.00	14.20	175.20
remove	1D@.267	ea	—	12.60	12.60

	Craft@Hrs	Unit	Material	Labor	Total

Refinish cabinets. Cabinets are stripped and restained. Edge surfaces that are finished with 1/4" hardwood plywood are replaced.

	Craft@Hrs	Unit	Material	Labor	Total
Replace only					
cabinet drawer front	5F@.219	ea	2.28	14.30	16.58
cabinet door	5F@.513	ea	5.12	33.60	38.72
cabinet door with glass	5F@.516	ea	2.40	33.80	36.20
cabinets & doors, reface plywood ends	5F@1.13	lf	20.20	74.00	94.20
island cabinets and doors, reface plywood ends and faces	5F@1.47	lf	21.20	96.30	117.50

Cabinet repair. Gouge or hole is filled, then "grained" by the painter to look like natural wood grain, matched to the finish of the cabinets.

	Craft@Hrs	Unit	Material	Labor	Total
Replace only					
repair gouge or hole in cabinet, grain patch to match	5F@1.19	ea	9.92	77.90	87.82

Plastic laminate countertop. Includes plastic laminate, medium-density fiberboard core, glue, and installation labor. Flat-laid countertops can be fabricated on site. Post-formed countertops are formed in shop. Flat-laid tops include a plastic-laminate backsplash up to 6" tall glued to the wall with the top and side edges finished with aluminum trim. Post-formed backsplashes are built in. Most commonly known by the trade name "Formica". Add **6%** for installation on bathroom vanity. Post-formed countertops have integrated backsplash with a rounded cove transition. Flat-laid countertops butt into wall with a flat edge.

	Craft@Hrs	Unit	Material	Labor	Total
replace, post-formed	2I@.286	sf	5.67	16.10	21.77
replace, flat-laid	2I@.309	sf	5.53	17.40	22.93
remove	1D@.132	sf	—	6.23	6.23

Cultured marble countertop. Includes prefabricated cultured marble countertop and installation labor. Poured to custom size in shop and transported to site. Sometimes large pieces are joined on site. Includes built-in or separate backsplash up to four inches tall. Add **6%** for installation on bathroom vanity.

	Craft@Hrs	Unit	Material	Labor	Total
replace	2I@.314	sf	25.40	17.70	43.10
remove	1D@.132	sf	—	6.23	6.23
add for integrated sink	—	ea	223.00	—	223.00

Cultured granite countertop. Includes prefabricated cultured granite countertop and installation labor. Poured to custom size in shop and transported to site. Sometimes large pieces are joined on site. Includes built-in or separate backsplash up to four inches tall. Add **6%** for installation on bathroom vanity.

	Craft@Hrs	Unit	Material	Labor	Total
replace	2I@.314	sf	34.20	17.70	51.90
remove	1D@.132	sf	—	6.23	6.23
add for integrated sink	—	ea	283.00	—	283.00

Cultured onyx countertop. Includes prefabricated cultured onyx countertop and installation labor. Poured to custom size in shop and transported to site. Sometimes large pieces are joined on site. Includes built-in or separate backsplash up to four inches tall. Add **6%** for installation on bathroom vanity.

	Craft@Hrs	Unit	Material	Labor	Total
replace	2I@.314	sf	34.20	17.70	51.90
remove	1D@.132	sf	—	6.23	6.23
add for integrated sink	—	ea	289.00	—	289.00

Solid-surface countertop. Includes custom-poured solid-surface countertop and installation labor. Some assembly of the custom-built sections on site is often necessary. Includes solid-surface backsplash. Commonly known by the trade name Corian. Other trade names include Avonite, Surell, Gibraltar, and Fountainhead. Add **9%** for installation on bathroom vanity. The following features vary with the countertop's grade: color options, alternating color layering, use of wood strip / pattern inlays and complexity of edge detail.

	Craft@Hrs	Unit	Material	Labor	Total
replace, economy grade	2I@.331	sf	155.00	18.60	173.60
replace, standard grade	2I@.331	sf	225.00	18.60	243.60
replace, high grade	2I@.331	sf	313.00	18.60	331.60

	Craft@Hrs	Unit	Material	Labor	Total
replace, deluxe grade	2I@.331	sf	379.00	18.60	397.60
replace, custom grade	2I@.331	sf	421.00	18.60	439.60
remove countertop	1D@.132	sf	—	6.23	6.23

Granite countertop. Includes polished, filled and sealed natural granite countertop cut to fit site conditions, and installation labor. Heavy sections are joined on site. Add **6%** for installation on bathroom vanity.

	Craft@Hrs	Unit	Material	Labor	Total
replace	2I@.414	sf	242.00	23.30	265.30
remove	1D@.132	sf	—	6.23	6.23

Stainless steel countertop. Includes custom-formed stainless steel countertop with integrated backsplash and installation labor.

	Craft@Hrs	Unit	Material	Labor	Total
replace	2I@.378	sf	38.50	21.30	59.80
remove	1D@.132	sf	—	6.23	6.23

Butcher-block countertop. Includes prefabricated and prefinished butcher-block countertop with hardware, glue, caulk, screws, and installation labor. Assembly on site or in shop. Some onsite cutting to fit is included. Includes prefinished hardwood or similar backsplash, typically 1/2" to 5/8" thick and up to 6" tall. Add **6%** for installation on bathroom vanity.

	Craft@Hrs	Unit	Material	Labor	Total
replace	2I@.314	sf	34.20	17.70	51.90
remove	1D@.132	sf	—	6.23	6.23

Solid wood maple or oak countertop. Includes prefabricated and prefinished solid-wood countertop with hardware, glue, caulk, screws, and installation labor. Assembly on site or in shop. Some onsite cutting to fit is included. Includes prefinished hardwood or similar backsplash, typically 1/2" to 5/8" thick and up to 6" tall. Add **6%** for installation on bathroom vanity.

	Craft@Hrs	Unit	Material	Labor	Total
replace	2I@.292	sf	11.80	16.40	28.20
remove	1D@.132	sf	—	6.23	6.23
remove for work, then reinstall	2I@.447	sf	—	25.20	25.20

Tile countertop. Includes tile, mortar, grout, sealer, and installation labor. Tile is thinset over cement backerboard except for deluxe grade which is set over a full mortar bed. Also includes edge tile and tile backsplash. Add **6%** for installation on bathroom vanity. For all grades except deluxe the tile is set in a thinset mortar bed over cement backerboard.

	Craft@Hrs	Unit	Material	Labor	Total
replace, economy grade tile	1T@.222	sf	18.30	13.90	32.20
replace, standard grade tile	1T@.222	sf	26.90	13.90	40.80
replace, high grade tile	1T@.222	sf	34.20	13.90	48.10
replace, deluxe grade tile	1T@.222	sf	53.90	13.90	67.80
remove	1D@.229	sf	—	10.80	10.80

Repair tile countertop. Includes selective hand removal of a single tile without damaging surrounding tiles, scraping and removal of old mortar, placement of a new tile, mortar, and grout. Matching tile color is usually not possible unless an original tile is available.

	Craft@Hrs	Unit	Material	Labor	Total
replace tile to match	1T@.813	ea	45.20	51.10	96.30
reaffix loose tile	1T@.553	ea	28.40	34.70	63.10
minimum charge for tile countertop work	1T@1.96	ea	31.30	123.00	154.30

Bathroom vanity cabinet. Includes prefabricated, prefinished bathroom vanity assembled in shop or on site, shims, screws, backing, attachment hardware, and installation labor. Installation includes cutting the cabinet to match plumbing, wall shape, and other site conditions.

	Craft@Hrs	Unit	Material	Labor	Total
replace, economy grade	2I@.371	lf	107.00	20.90	127.90
replace, standard grade	2I@.371	lf	135.00	20.90	155.90
replace, high grade	2I@.371	lf	175.00	20.90	195.90
replace, deluxe grade	2I@.371	lf	244.00	20.90	264.90
replace, custom grade	2I@.371	lf	348.00	20.90	368.90
replace, deluxe grade	2I@.371	lf	450.00	20.90	470.90
remove	1D@.270	lf	—	12.70	12.70
remove for work, then reinstall	2I@.653	lf	—	36.80	36.80

Time & Material Charts (selected items)
Cabinet Materials

See Cabinet material prices above.

Cabinet Labor

Laborer	base wage	paid leave	true wage	taxes & ins.	total
Cabinet installer	$38.10	2.97	$41.07	26.53	$67.60
Painter	$37.50	2.93	$40.43	25.07	$65.50
Tile layer	$34.60	2.70	$37.30	25.50	$62.80
Laborer	$24.00	1.87	$25.87	19.03	$44.90
Demolition laborer	$25.90	2.02	$27.92	19.28	$47.20

Paid leave is calculated based on two weeks paid vacation, one week sick leave, and seven paid holidays. Employer's matching portion of **FICA** is 7.65 percent. **FUTA** (Federal Unemployment) is .8 percent. **Worker's compensation** was calculated using a national average of 17.21 percent for the cabinet installer and helper; 14.50 percent for the painter; and 19.41 percent for the tile layer. **Unemployment insurance** was calculated using a national average of 8 percent. **Health insurance** was calculated based on a projected national average for 2020 of $1,280 per employee (and family when applicable) per month. Employer pays 80 percent for a per month cost of $1,024 per employee. **Retirement** is based on a 401(k) retirement program with employer matching of 50 percent. Employee contributions to the 401(k) plan are an average of 6 percent of the true wage. **Liability insurance** is based on a national average of 12.0 percent.

Cabinet crew

install cabinets & countertop	cabinet installer	$67.60
install cabinets & countertop	laborer	$44.90
cabinet installation	crew	$56.30

Cabinet Labor Productivity

	Craft@Hrs	Unit	Material	Labor	Total
Install kitchen cabinets					
lower kitchen cabinets	2I@.360	lf	—	20.30	20.30
upper kitchen cabinets	2I@.349	lf	—	19.60	19.60
lower kitchen island cabinets	2I@.371	lf	—	20.90	20.90
upper kitchen island cabinets	2I@.359	lf	—	20.20	20.20
full-height cabinet	2I@.593	lf	—	33.40	33.40
Install cabinet drawer fronts & doors					
drawer front	2I@.118	ea	—	6.64	6.64
door	2I@.252	ea	—	14.20	14.20
Refinish cabinets					
cabinet drawer front	5F@.219	ea	—	14.30	14.30
cabinet door	5F@.513	ea	—	33.60	33.60
cabinet door with glass	5F@.516	ea	—	33.80	33.80
cabinets & doors, reface plywood ends	5F@1.13	lf	—	74.00	74.00
island cabinets & doors, reface plywood ends & faces	5F@1.47	lf	—	96.30	96.30
Cabinet repair					
repair gouge or hole in cabinet, grain to match	5F@1.19	ea	—	77.90	77.90

	Craft@Hrs	Unit	Material	Labor	Total
Install countertop					
post-formed plastic laminate	2I@.286	sf	—	16.10	16.10
flat-laid plastic laminate	2I@.309	sf	—	17.40	17.40
cultured material	2I@.314	sf	—	17.70	17.70
solid-surface	2I@.331	sf	—	18.60	18.60
granite	2I@.414	sf	—	23.30	23.30
stainless-steel	2I@.378	sf	—	21.30	21.30
butcher block	2I@.314	sf	—	17.70	17.70
solid wood	2I@.292	sf	—	16.40	16.40
tile (ceramic)	1T@.222	sf	—	13.90	13.90
Install bathroom vanity cabinet					
bathroom vanity cabinet	2I@.371	lf	—	20.90	20.90

Cleaning

Unless otherwise noted, all cleaning prices are for items typically smoke-stained. For lightly stained items deduct **10%**. For heavily stained items add **15%**.

	Craft@Hrs	Unit	Material	Labor	Total
Minimum charge.					
for cleaning work	1B@1.80	ea	2.64	68.80	71.44
Acoustic tile.					
clean acoustic ceiling tiles	1B@.014	sf	.02	.53	.55
Appliances.					
clean cook top	1B@.876	ea	1.19	33.50	34.69
clean range	1B@1.54	ea	2.12	58.80	60.92
clean space-saver range	1B@1.45	ea	1.99	55.40	57.39
clean high-low range	1B@2.02	ea	2.77	77.20	79.97
clean restaurant-style gas range	1B@2.11	ea	2.91	80.60	83.51
clean wall oven	1B@1.27	ea	1.71	48.50	50.21
clean double wall oven	1B@2.11	ea	2.91	80.60	83.51
clean modular electric cook top unit	1B@.440	ea	.68	16.80	17.48
clean drop-in or slide-in range	1B@1.40	ea	1.99	53.50	55.49
clean range hood	1B@.428	ea	.53	16.30	16.83
clean oversize range hood	1B@.455	ea	.68	17.40	18.08
clean dishwasher	1B@1.00	ea	1.29	38.20	39.49
clean microwave	1B@.486	ea	.68	18.60	19.28
clean side-by-side refrigerator	1B@1.80	ea	2.50	68.80	71.30
clean over-under refrigerator	1B@1.62	ea	2.23	61.90	64.13
clean trash compactor	1B@.561	ea	.81	21.40	22.21
clean clothes washing machine	1B@.642	ea	.94	24.50	25.44
clean clothes dryer	1B@.475	ea	.68	18.10	18.78
Awnings.					
clean aluminum or steel carport or patio cover	1B@.019	sf	.02	.73	.75
clean aluminum or steel door or window awning	1B@.193	lf	.25	7.37	7.62

	Craft@Hrs	Unit	Material	Labor	Total
Bathroom hardware.					
clean complete bathroom, fixtures and hardware	1B@4.01	ea	5.52	153.00	158.52
clean complete 1/2 bath, fixtures and hardware	1B@2.57	ea	3.56	98.20	101.76
clean bathroom hardware (per piece)	1B@.246	ea	.33	9.40	9.73
clean medicine cabinet	1B@.570	ea	.79	21.80	22.59
clean bathroom mirror	1B@.026	sf	.03	.99	1.02
Cabinets.					
clean lower cabinets	1B@.446	lf	.64	17.00	17.64
clean upper cabinets	1B@.418	lf	.60	16.00	16.60
clean lower island cabinets	1B@.455	lf	.65	17.40	18.05
clean upper island cabinets	1B@.514	lf	.72	19.60	20.32
clean full-height cabinets	1B@.747	lf	1.02	28.50	29.52
Countertops.					
clean plastic laminate countertop	1B@.035	sf	.04	1.34	1.38
clean cultured stone countertop	1B@.038	sf	.04	1.45	1.49
clean solid-surface countertop	1B@.039	sf	.04	1.49	1.53
clean granite countertop	1B@.043	sf	.04	1.64	1.68
clean stainless steel countertop	1B@.040	sf	.04	1.53	1.57
clean wood countertop (butcher block or solid)	1B@.045	sf	.05	1.72	1.77
clean tile countertop	1B@.062	sf	.06	2.37	2.43
clean bathroom vanity cabinet	1B@.346	lf	.43	13.20	13.63
Columns.					
clean column	1B@.158	lf	.19	6.04	6.23
clean pilaster	1B@.087	lf	.09	3.32	3.41
clean capital	1B@1.38	ea	1.91	52.70	54.61
clean ornate capital (Corinthian)	1B@2.46	ea	3.38	94.00	97.38
clean column base	1B@.856	ea	1.18	32.70	33.88
clean pilaster base	1B@.459	ea	.66	17.50	18.16
Concrete.					
clean wall	1B@.014	sf	.02	.53	.55
clean floor	1B@.012	sf	.02	.46	.48
clean step (per step)	1B@.222	lf	.30	8.48	8.78
Doors.					
clean jamb and casing (per lf)	1B@.020	lf	.02	.76	.78
clean jamb and casing (per door)	1B@.349	ea	.43	13.30	13.73
clean folding door	1B@.260	ea	.34	9.93	10.27
clean louvered folding door	1B@.438	ea	.64	16.70	17.34
clean bypassing door	1B@.462	ea	.66	17.60	18.26
clean louvered bypassing door	1B@.623	ea	.89	23.80	24.69
clean French door	1B@.817	ea	1.11	31.20	32.31
clean standard flush door	1B@.438	ea	.64	16.70	17.34
clean pocket door	1B@.447	ea	.65	17.10	17.75
clean panel door	1B@.459	ea	.66	17.50	18.16
clean transom	1B@.230	ea	.31	8.79	9.10
clean batten door	1B@.479	ea	.68	18.30	18.98
clean storm door	1B@.454	ea	.65	17.30	17.95

	Craft@Hrs	Unit	Material	Labor	Total
clean exterior door side lite	1B@.370	ea	.50	14.10	14.60
clean Dutch door	1B@.462	ea	.66	17.60	18.26
clean fan lite	1B@.339	ea	.42	12.90	13.32
clean cafe doors	1B@.352	ea	.44	13.40	13.84
Sliding patio doors.					
clean 6' wide sliding patio door	1B@1.06	ea	1.41	40.50	41.91
clean 8' wide sliding patio door	1B@1.20	ea	1.59	45.80	47.39
clean 12' wide sliding patio door	1B@1.38	ea	1.85	52.70	54.55
Garage doors.					
clean 8' wide garage door	1B@1.56	ea	2.12	59.60	61.72
clean 9' wide garage door	1B@1.65	ea	2.23	63.00	65.23
clean 10' wide garage door	1B@1.74	ea	2.35	66.50	68.85
clean 12' wide garage door	1B@1.83	ea	2.50	69.90	72.40
clean 14' wide garage door	1B@1.93	ea	2.64	73.70	76.34
clean 16' wide garage door	1B@2.04	ea	2.77	77.90	80.67
clean 18' wide garage door	1B@2.16	ea	2.91	82.50	85.41
clean garage door opener	1B@.965	ea	1.29	36.90	38.19
Drywall or plaster.					
clean wall	1B@.013	sf	.02	.50	.52
clean ceiling	1B@.014	sf	.01	.53	.54
clean ceiling acoustic texture	1B@.017	sf	.01	.65	.66
Electrical.					
clean outlet or switch	1B@.084	ea	.09	3.21	3.30
clean breaker panel	1B@.693	ea	.96	26.50	27.46
clean circuit breaker	1B@.227	ea	.31	8.67	8.98
clean door bell or chime button	1B@.034	ea	.04	1.30	1.34
clean bathroom exhaust fan	1B@.290	ea	.37	11.10	11.47
clean bathroom exhaust fan with heat lamp	1B@.321	ea	.40	12.30	12.70
clean kitchen exhaust fan	1B@.299	ea	.38	11.40	11.78
clean whole-house exhaust fan	1B@.352	ea	.44	13.40	13.84
clean intercom system station	1B@.193	ea	.25	7.37	7.62
clean detector	1B@.185	ea	.24	7.07	7.31
clean thermostat	1B@.211	ea	.27	8.06	8.33
Light fixtures.					
clean light fixture	1B@.447	ea	.65	17.10	17.75
clean bathroom light bar (per light)	1B@.160	ea	.22	6.11	6.33
clean chandelier, typical detail	1B@1.28	ea	1.79	48.90	50.69
clean chandelier, ornate detail	1B@2.00	ea	2.77	76.40	79.17
clean chandelier, very ornate detail	1B@2.99	ea	4.07	114.00	118.07
clean crystal chandelier, typical detail	1B@4.47	ea	6.18	171.00	177.18
clean crystal chandelier, ornate detail	1B@5.80	ea	7.96	222.00	229.96
clean crystal chandelier, very ornate detail	1B@7.82	ea	10.80	299.00	309.80
clean fluorescent light fixture	1B@.518	ea	.74	19.80	20.54
clean ceiling fan	1B@.580	ea	.81	22.20	23.01
clean ceiling fan with light	1B@.686	ea	.94	26.20	27.14
clean recessed spot light fixture	1B@.392	ea	.53	15.00	15.53

	Craft@Hrs	Unit	Material	Labor	Total
clean strip light (per spot)	1B@.195	ea	.25	7.45	7.70
clean exterior flood light fixture (per spot)	1B@.207	ea	.27	7.91	8.18
clean exterior light fixture	1B@.486	ea	.68	18.60	19.28
clean exterior post light fixture	1B@.580	ea	.81	22.20	23.01

Fencing.

	Craft@Hrs	Unit	Material	Labor	Total
clean 4' high wood fence	1B@.064	lf	.07	2.44	2.51
clean 6' high wood fence	1B@.078	lf	.08	2.98	3.06
clean 8' high wood fence	1B@.100	lf	.10	3.82	3.92
clean 4' high chain-link fence	1B@.044	lf	.05	1.68	1.73
clean 6' high chain-link fence	1B@.050	lf	.05	1.91	1.96
clean 8' high chain-link fence	1B@.058	lf	.06	2.22	2.28
clean 4' high vinyl fence	1B@.056	lf	.06	2.14	2.20
clean 6' high vinyl fence	1B@.067	lf	.07	2.56	2.63
clean 8' high vinyl fence	1B@.082	lf	.09	3.13	3.22
clean 60" high ornamental iron fence	1B@.078	lf	.08	2.98	3.06
clean 72" high ornamental iron fence	1B@.069	lf	.07	2.64	2.71

Finish carpentry.

	Craft@Hrs	Unit	Material	Labor	Total
clean molding	1B@.015	lf	.02	.57	.59
clean interior wood architrave	1B@.037	lf	.04	1.41	1.45
clean exterior wood architrave	1B@.042	lf	.04	1.60	1.64
clean exterior door surround	1B@.082	lf	.09	3.13	3.22
clean exterior window surround	1B@.072	lf	.07	2.75	2.82
clean closet shelf and brackets	1B@.038	lf	.04	1.45	1.49
clean closet organizer	1B@.030	sf	.03	1.15	1.18
clean linen closet shelves	1B@.036	sf	.04	1.38	1.42
clean closet rod	1B@.020	lf	.02	.76	.78
clean bookcase	1B@.029	sf	.03	1.11	1.14
clean fireplace mantel	1B@.977	ea	1.31	37.30	38.61
clean ceiling with exposed beams	1B@.028	sf	.03	1.07	1.10
clean coffered ceiling	1B@.037	sf	.04	1.41	1.45
clean wall niche	1B@.227	ea	.31	8.67	8.98
clean gingerbread trim	1B@.025	lf	.03	.96	.99
clean gingerbread bracket	1B@.193	ea	.25	7.37	7.62
clean gingerbread corbel	1B@.214	ea	.27	8.17	8.44
clean gingerbread spandrel	1B@.193	lf	.25	7.37	7.62
clean gingerbread cornice	1B@.171	lf	.23	6.53	6.76
clean gingerbread gable ornament	1B@.440	ea	.64	16.80	17.44
clean gingerbread finial	1B@.359	ea	.44	13.70	14.14
clean porch post	1B@.042	lf	.04	1.60	1.64

Fireplaces.

	Craft@Hrs	Unit	Material	Labor	Total
clean screen	1B@1.12	ea	1.52	42.80	44.32
clean door	1B@1.06	ea	1.41	40.50	41.91
clean marble face	1B@.047	sf	.05	1.80	1.85
clean brick face	1B@.054	sf	.06	2.06	2.12
clean stone face	1B@.058	sf	.06	2.22	2.28
clean tile face	1B@.056	sf	.06	2.14	2.20
clean marble hearth	1B@.095	lf	.10	3.63	3.73
clean brick hearth	1B@.106	lf	.11	4.05	4.16
clean stone hearth	1B@.112	lf	.12	4.28	4.40
clean tile hearth	1B@.106	lf	.11	4.05	4.16

	Craft@Hrs	Unit	Material	Labor	Total
Flooring.					
clean carpet	1B@.008	sf	.01	.31	.32
clean wool carpet	1B@.013	sf	.01	.50	.51
clean carpet cove	1B@.018	lf	.02	.69	.71
add per step for carpet cleaning	1B@.155	ea	.19	5.92	6.11
clean stone floor	1B@.009	sf	.01	.34	.35
clean tile floor	1B@.008	sf	.01	.31	.32
clean tile base	1B@.010	lf	.01	.38	.39
clean and wax vinyl floor	1B@.009	sf	.01	.34	.35
clean vinyl cove	1B@.010	lf	.01	.38	.39
clean and wax wood floor	1B@.011	sf	.01	.42	.43
HVAC.					
clean and deodorize ducts (per lf of duct)	1B@.156	lf	.19	5.96	6.15
clean furnace	1B@2.95	ea	4.01	113.00	117.01
clean heat pump	1B@1.28	ea	1.79	48.90	50.69
clean humidifier	1B@.946	ea	1.29	36.10	37.39
clean through-wall AC unit	1B@1.12	ea	1.52	42.80	44.32
clean evaporative cooler	1B@1.38	ea	1.92	52.70	54.62
clean evaporative cooler grille	1B@.250	ea	.33	9.55	9.88
clean heat register	1B@.240	ea	.32	9.17	9.49
clean cold-air return cover	1B@.243	ea	.32	9.28	9.60
Masonry.					
clean brick wall	1B@.025	sf	.03	.96	.99
clean block wall	1B@.023	sf	.02	.88	.90
clean slump block wall	1B@.025	sf	.03	.96	.99
clean fluted block wall	1B@.029	sf	.03	1.11	1.14
clean glazed block wall	1B@.028	sf	.03	1.07	1.10
clean split-face block wall	1B@.032	sf	.03	1.22	1.25
clean split-rib block wall	1B@.030	sf	.03	1.15	1.18
clean stone wall	1B@.033	sf	.03	1.26	1.29
clean glazed structural tile wall	1B@.024	sf	.02	.92	.94
clean glass block wall	1B@.024	sf	.02	.92	.94
clean pavers	1B@.025	sf	.03	.96	.99
clean stone veneer panels	1B@.025	sf	.03	.96	.99
clean cultured stone veneer panels	1B@.025	sf	.03	.96	.99
clean stone architrave	1B@.045	lf	.05	1.72	1.77
clean stone trim stones	1B@.040	lf	.04	1.53	1.57
Paneling.					
clean wood paneling	1B@.017	sf	.02	.65	.67
clean wood paneling with moldings or onlays	1B@.021	sf	.02	.80	.82
clean panel wall	1B@.035	sf	.04	1.34	1.38
clean pegboard	1B@.018	sf	.02	.69	.71
clean rough-sawn wood paneling	1B@.023	sf	.02	.88	.90
Stucco.					
clean stucco	1B@.024	sf	.02	.92	.94
clean stucco architrave	1B@.046	lf	.05	1.76	1.81
clean tongue-and-groove paneling	1B@.020	sf	.02	.76	.78

	Craft@Hrs	Unit	Material	Labor	Total
Plaster. (See Drywall heading in this section for costs to clean interior plaster walls.)					
clean plaster molding	1B@.035	lf	.04	1.34	1.38
clean plaster architrave	1B@.042	lf	.04	1.60	1.64
clean ceiling medallion	1B@.556	ea	.78	21.20	21.98
clean ceiling rose	1B@.946	ea	1.27	36.10	37.37
Plumbing.					
clean sink faucet	1B@.475	ea	.67	18.10	18.77
clean shower faucet and head	1B@.580	ea	.81	22.20	23.01
clean tub faucet	1B@.499	ea	.71	19.10	19.81
clean tub faucet and shower head	1B@.620	ea	.88	23.70	24.58
clean wet bar sink	1B@.391	ea	.55	14.90	15.45
clean bathroom sink	1B@.418	ea	.60	16.00	16.60
clean kitchen sink, per bowl	1B@.428	ea	.60	16.30	16.90
clean laundry sink, per bowl	1B@.447	ea	.65	17.10	17.75
clean visible sink supply lines	1B@.257	ea	.34	9.82	10.16
clean toilet	1B@.665	ea	.93	25.40	26.33
clean bidet	1B@.693	ea	.96	26.50	27.46
clean toilet seat	1B@.180	ea	.24	6.88	7.12
clean porcelain enamel finish tub	1B@1.00	ea	1.33	38.20	39.53
clean porcelain enamel finish tub with whirlpool jets	1B@1.28	ea	1.79	48.90	50.69
clean fiberglass finish tub	1B@1.20	ea	1.65	45.80	47.45
clean fiberglass finish tub with whirlpool jets	1B@1.63	ea	2.23	62.30	64.53
clean tub and shower combination	1B@1.63	ea	2.23	62.30	64.53
clean fiberglass or metal shower stall	1B@1.17	ea	1.59	44.70	46.29
clean glass shower stall	1B@1.47	ea	2.06	56.20	58.26
clean tub surround	1B@.545	ea	.74	20.80	21.54
clean shower door	1B@.879	ea	1.19	33.60	34.79
clean sliding glass tub door	1B@1.18	ea	1.65	45.10	46.75
clean folding plastic bathtub door	1B@1.20	ea	1.65	45.80	47.45
clean water heater	1B@.946	ea	1.29	36.10	37.39
clean claw-foot tub faucet	1B@.498	ea	.68	19.00	19.68
clean claw-foot tub faucet and shower conversion	1B@.681	ea	.94	26.00	26.94
clean free-standing water feeds	1B@.284	ea	.37	10.80	11.17
clean pedestal sink	1B@.513	ea	.74	19.60	20.34
clean pill-box toilet	1B@.817	ea	1.11	31.20	32.31
clean low-tank toilet	1B@1.06	ea	1.41	40.50	41.91
clean high-tank toilet	1B@1.38	ea	1.92	52.70	54.62
clean antique tub	1B@1.17	ea	1.59	44.70	46.29
Siding.					
clean fiberglass corrugated siding	1B@.014	sf	.02	.53	.55
clean aluminum or vinyl siding	1B@.015	sf	.02	.57	.59
clean wood lap siding	1B@.015	sf	.02	.57	.59
clean vertical wood siding (board-on-board, etc.)	1B@.015	sf	.02	.57	.59
clean cement fiber shingle siding	1B@.017	sf	.02	.65	.67
clean shake or wood shingle siding	1B@.019	sf	.02	.73	.75
clean hardboard or plywood siding	1B@.015	sf	.02	.57	.59

	Craft@Hrs	Unit	Material	Labor	Total
clean tongue-and-groove siding	1B@.016	sf	.02	.61	.63
clean metal or vinyl fascia	1B@.016	sf	.02	.61	.63
clean wood fascia	1B@.016	sf	.02	.61	.63
clean metal or vinyl soffit	1B@.014	sf	.01	.53	.54
clean wood soffit	1B@.015	sf	.02	.57	.59
clean rough-sawn wood soffit	1B@.018	sf	.02	.69	.71
clean shutter	1B@.359	ea	.44	13.70	14.14
Stairs.					
clean wood stair tread	1B@.163	ea	.22	6.23	6.45
clean spiral stair balustrade	1B@.113	lf	.12	4.32	4.44
clean stair balustrade	1B@.114	lf	.12	4.35	4.47
clean disappearing attic stair	1B@2.40	ea	3.29	91.70	94.99
clean stair bracket	1B@.310	ea	.39	11.80	12.19
Suspended ceiling.					
clean suspended ceiling grid	1B@.010	sf	.02	.38	.40
Tile. (See Cabinet heading in this section for costs to clean tile countertops. See Flooring heading for costs to clean tile floors and base.)					
clean tile shower	1B@1.38	ea	1.92	52.70	54.62
clean tile tub surround	1B@1.80	ea	2.42	68.80	71.22
Wallpaper.					
clean vinyl wallpaper	1B@.015	sf	.02	.57	.59
clean paper wallpaper	1B@.017	sf	.02	.65	.67
clean grass or rice cloth wallpaper	1B@.025	sf	.03	.96	.99
Windows.					
clean window to 8 sf	1B@.408	ea	.56	15.60	16.16
clean window 9 to 14 sf	1B@.475	ea	.67	18.10	18.77
clean window 15 to 20 sf	1B@.580	ea	.81	22.20	23.01
clean window 21 to 30 sf	1B@.856	ea	1.18	32.70	33.88
clean window, per sf	1B@.030	sf	.03	1.15	1.18
clean skylight to 8 sf	1B@.428	ea	.60	16.30	16.90
clean skylight 9 to 14 sf	1B@.498	ea	.71	19.00	19.71
clean skylight 15 to 20 sf	1B@.619	ea	.88	23.60	24.48
clean skylight 21 to 30 sf	1B@.899	ea	1.22	34.30	35.52
clean skylight per sf	1B@.032	sf	.03	1.22	1.25
add to clean multiple, small panes	—	%	—	15.0	—
Mirrors. (Also see Bathroom Hardware heading in this section for costs to clean other bathroom mirrors.)					
clean wall mirror	1B@.027	sf	.03	1.03	1.06
Final construction clean-up.					
broom clean	1B@.002	sf	.02	.08	.10

Time & Material Charts (selected items)
Cleaning Materials

See Cleaning material prices above.

Cleaning Labor

Laborer	base wage	paid leave	true wage	taxes & ins.	total
Cleaning laborer	$20.90	1.63	$22.53	15.67	$38.20

Paid leave is calculated based on two weeks paid vacation, one week sick leave, and seven paid holidays. Employer's matching portion of **FICA** is 7.65 percent. **FUTA** (Federal Unemployment) is .8 percent. **Worker's compensation** for the cleaning trade was calculated using a national average of 9.24 percent. **Unemployment insurance** was calculated using a national average of 8 percent. **Health insurance** was calculated based on a projected national average for 2020 of $1,280 per employee (and family when applicable) per month. Employer pays 80 percent for a per month cost of $1,024 per employee. **Retirement** is based on a 401(k) retirement program with employer matching of 50 percent. Employee contributions to the 401(k) plan are an average of 6 percent of the true wage. **Liability insurance** is based on a national average of 12.0 percent.

Cleaning Labor Productivity	Craft@Hrs	Unit	Material	Labor	Total
Clean appliances					
cook top	1B@.876	ea	—	33.50	33.50
range	1B@1.54	ea	—	58.80	58.80
wall oven	1B@1.27	ea	—	48.50	48.50
range hood	1B@.428	ea	—	16.30	16.30
dishwasher	1B@1.00	ea	—	38.20	38.20
microwave	1B@.486	ea	—	18.60	18.60
side-by-side refrigerator	1B@1.80	ea	—	68.80	68.80
over-under refrigerator	1B@1.62	ea	—	61.90	61.90
trash compactor	1B@.561	ea	—	21.40	21.40
Clean awnings					
aluminum or steel carport or patio	1B@.019	sf	—	.73	.73
aluminum or steel door or window awning	1B@.193	lf	—	7.37	7.37
Clean bathroom hardware					
bathroom hardware (per piece)	1B@.246	ea	—	9.40	9.40
medicine cabinet	1B@.570	ea	—	21.80	21.80
Clean cabinets					
lower cabinets	1B@.446	lf	—	17.00	17.00
upper cabinets	1B@.418	lf	—	16.00	16.00
full-height cabinets	1B@.747	lf	—	28.50	28.50
plastic laminate countertop	1B@.035	sf	—	1.34	1.34
tile countertop	1B@.062	sf	—	2.37	2.37
Clean concrete					
wall	1B@.014	sf	—	.53	.53
floor	1B@.012	sf	—	.46	.46

	Craft@Hrs	Unit	Material	Labor	Total
Clean door					
folding door (per section)	1B@.260	ea	—	9.93	9.93
bypassing door	1B@.462	ea	—	17.60	17.60
door	1B@.438	ea	—	16.70	16.70
storm door	1B@.454	ea	—	17.30	17.30
Clean drywall or plaster					
wall	1B@.013	sf	—	.50	.50
ceiling	1B@.014	sf	—	.53	.53
ceiling acoustic texture	1B@.017	sf	—	.65	.65
Clean electrical					
light fixture	1B@.447	ea	—	17.10	17.10
bathroom light bar (per light)	1B@.160	ea	—	6.11	6.11
chandelier, typical detail	1B@1.28	ea	—	48.90	48.90
fluorescent light fixture	1B@.518	ea	—	19.80	19.80
ceiling fan	1B@.580	ea	—	22.20	22.20
ceiling fan with light	1B@.686	ea	—	26.20	26.20
Clean fence					
4' high wood	1B@.064	lf	—	2.44	2.44
4' high chain-link	1B@.044	lf	—	1.68	1.68
4' high vinyl	1B@.056	lf	—	2.14	2.14
60" high ornamental iron	1B@.078	lf	—	2.98	2.98
Clean finish carpentry					
molding	1B@.015	lf	—	.57	.57
Clean fireplace					
screen	1B@1.12	ea	—	42.80	42.80
door	1B@1.06	ea	—	40.50	40.50
Clean flooring					
carpet	1B@.008	sf	—	.31	.31
stone floor	1B@.009	sf	—	.34	.34
tile floor	1B@.008	sf	—	.31	.31
vinyl floor and wax	1B@.009	sf	—	.34	.34
wood floor and wax	1B@.011	sf	—	.42	.42
Clean HVAC					
furnace	1B@2.95	ea	—	113.00	113.00
through-wall AC unit	1B@1.12	ea	—	42.80	42.80
evaporative cooler	1B@1.38	ea	—	52.70	52.70
Clean masonry					
brick wall	1B@.025	sf	—	.96	.96
block wall	1B@.023	sf	—	.88	.88
stone wall	1B@.033	sf	—	1.26	1.26
glass block wall	1B@.024	sf	—	.92	.92
stone veneer panels	1B@.025	sf	—	.96	.96
Clean paneling					
wood paneling	1B@.017	sf	—	.65	.65
wood paneling with moldings or onlays	1B@.021	sf	—	.80	.80
panel wall	1B@.035	sf	—	1.34	1.34
rough-sawn wood paneling	1B@.023	sf	—	.88	.88
tongue-and-groove paneling	1B@.020	sf	—	.76	.76

	Craft@Hrs	Unit	Material	Labor	Total
Clean plaster (also see Drywall heading in this section)					
stucco	1B@.024	sf	—	.92	.92
Clean plumbing					
sink faucet	1B@.475	ea	—	18.10	18.10
tub faucet	1B@.499	ea	—	19.10	19.10
tub faucet and shower head	1B@.620	ea	—	23.70	23.70
bathroom sink	1B@.418	ea	—	16.00	16.00
kitchen sink, per bowl	1B@.428	ea	—	16.30	16.30
laundry sink, per bowl	1B@.447	ea	—	17.10	17.10
toilet	1B@.665	ea	—	25.40	25.40
bidet	1B@.693	ea	—	26.50	26.50
porcelain-enamel finish bathtub	1B@1.00	ea	—	38.20	38.20
fiberglass finish bathtub	1B@1.20	ea	—	45.80	45.80
tub and shower combination	1B@1.63	ea	—	62.30	62.30
fiberglass or metal shower stall	1B@1.17	ea	—	44.70	44.70
glass shower stall	1B@1.47	ea	—	56.20	56.20
tub surround	1B@.545	ea	—	20.80	20.80
shower door	1B@.879	ea	—	33.60	33.60
sliding glass bathtub door	1B@1.18	ea	—	45.10	45.10
pedestal sink	1B@.513	ea	—	19.60	19.60
pill-box toilet	1B@.817	ea	—	31.20	31.20
Clean siding					
fiberglass corrugated	1B@.014	sf	—	.53	.53
metal or vinyl siding	1B@.015	sf	—	.57	.57
wood lap siding	1B@.015	sf	—	.57	.57
vertical wood siding (board-on-board, etc.)	1B@.015	sf	—	.57	.57
cement fiber shingle siding	1B@.017	sf	—	.65	.65
shake or wood shingle siding	1B@.019	sf	—	.73	.73
plywood siding	1B@.015	sf	—	.57	.57
tongue-and-groove siding	1B@.016	sf	—	.61	.61
Clean stairs					
wood stair tread	1B@.163	ea	—	6.23	6.23
stair balustrade	1B@.114	lf	—	4.35	4.35
disappearing attic stair	1B@2.40	ea	—	91.70	91.70
Clean tile					
tile shower	1B@1.38	ea	—	52.70	52.70
tile bathtub surround	1B@1.80	ea	—	68.80	68.80
Clean wallpaper					
vinyl wallpaper	1B@.015	sf	—	.57	.57
paper wallpaper	1B@.017	sf	—	.65	.65
grass or rice cloth wallpaper	1B@.025	sf	—	.96	.96
Clean window					
window up to 8 sf	1B@.408	ea	—	15.60	15.60
window 21 to 30 sf	1B@.856	ea	—	32.70	32.70
window, per sf	1B@.030	ea	—	1.15	1.15
skylight up to 8 sf	1B@.428	ea	—	16.30	16.30
skylight 21 to 30 sf	1B@.899	ea	—	34.30	34.30
skylight per sf	1B@.032	ea	—	1.22	1.22

Columns

Column Materials. Composite materials: A combination of marble (or similar material) polymers, and fiberglass or columns made from high-density polyurethane. Pine: Made from laminated staves of Douglas fir or Ponderosa pine. Redwood: Made from laminated staves of redwood. Oak: Made from laminated staves of red oak. Stone: Crushed and reconstructed limestone reinforced with glass fibers. Synthetic stone: Lightweight simulated stone which are either cement or gypsum-based fiberglass. Aluminum: Cast aluminum. Plaster: Made from fibrous plaster and gypsum products reinforced with fiberglass or jute, or steel or sisal for use on interior of structure. Stucco: Made from fibrous plaster products reinforced with fiberglass or jute or steel or sisal for use on exterior of structure.

	Craft@Hrs	Unit	Material	Labor	Total
Minimum charge for columns.					
for column or pilaster work	1C@1.25	ea	56.20	84.50	140.70
12" round columns. See Additional costs for columns and pilasters below to add for taper, fluting, and architecturally correct taper.					
replace, composite materials	1C@.333	lf	60.30	22.50	82.80
replace, pine	1C@.333	lf	63.50	22.50	86.00
replace, redwood	1C@.333	lf	75.70	22.50	98.20
replace, oak	1C@.333	lf	140.00	22.50	162.50
replace, stone	1C@.333	lf	111.00	22.50	133.50
replace, synthetic stone	1C@.333	lf	105.00	22.50	127.50
replace, aluminum	1C@.333	lf	65.70	22.50	88.20
replace, interior plaster	1C@.333	lf	61.10	22.50	83.60
replace, exterior stucco	1C@.333	lf	59.20	22.50	81.70
remove, all material types	1D@.212	lf	—	10.00	10.00
12" round pilasters. Half-round with no taper. See Additional costs for columns and pilasters below to add for taper, fluting, and architecturally correct taper.					
replace, composite materials	1C@.292	lf	38.40	19.70	58.10
replace, pine	1C@.292	lf	40.40	19.70	60.10
replace, redwood	1C@.292	lf	48.40	19.70	68.10
replace, oak	1C@.292	lf	89.70	19.70	109.40
replace, stone	1C@.292	lf	71.50	19.70	91.20
replace, synthetic stone	1C@.292	lf	67.00	19.70	86.70
replace, aluminum	1C@.292	lf	42.20	19.70	61.90
replace, interior plaster	1C@.292	lf	38.80	19.70	58.50
replace, exterior stucco	1C@.292	lf	37.70	19.70	57.40
remove, all material types	1D@.134	lf	—	6.32	6.32
Additional column and pilaster costs.					
add for round fluted column or pilaster	—	%	18.0	—	—
add for tapered column or pilaster consistent taper from bottom to top	—	%	13.0	—	—
add for architecturally correct tapered column or pilaster	—	%	65.0	—	—

	Craft@Hrs	Unit	Material	Labor	Total
12" square column.					
replace, composite materials	1C@.308	lf	79.50	20.80	100.30
replace, pine	1C@.308	lf	83.30	20.80	104.10
replace, redwood	1C@.308	lf	99.90	20.80	120.70
replace, oak	1C@.308	lf	189.00	20.80	209.80
replace, stone	1C@.308	lf	148.00	20.80	168.80
replace, synthetic stone	1C@.308	lf	139.00	20.80	159.80
replace, aluminum	1C@.308	lf	86.80	20.80	107.60
replace, interior plaster	1C@.308	lf	80.80	20.80	101.60
exterior stucco	1C@.308	lf	78.30	20.80	99.10
remove, all material types	1D@.212	lf	—	10.00	10.00
12" square pilaster.					
replace, composite materials	1C@.269	lf	50.50	18.20	68.70
replace, pine	1C@.269	lf	53.70	18.20	71.90
replace, redwood	1C@.269	lf	64.00	18.20	82.20
replace, oak	1C@.269	lf	119.00	18.20	137.20
replace, stone	1C@.269	lf	94.20	18.20	112.40
replace, synthetic stone	1C@.269	lf	88.80	18.20	107.00
replace, aluminum	1C@.269	lf	55.60	18.20	73.80
replace, interior plaster	1C@.269	lf	51.30	18.20	69.50
replace, exterior stucco	1C@.269	lf	49.80	18.20	68.00
remove, all material types	1D@.134	lf	—	6.32	6.32
Column and pilaster repair.					
repair, wood column with wood patch	1C@.706	ea	5.21	47.70	52.91
repair, fluted wood column with wood patch	1C@.863	ea	5.21	58.30	63.51
repair, wood column by replacing section	1C@1.94	ea	9.30	131.00	140.30
repair, fluted wood column by replacing section	1C@2.59	ea	11.50	175.00	186.50
repair, dry rot in wood column with epoxy	1C@2.74	ea	9.30	185.00	194.30
repair, dry rot in fluted wood column with epoxy	1C@3.24	ea	9.30	219.00	228.30
add for column repair if column is removed from structure, includes temporary brace wall and column removal	1C@9.31	ea	93.00	629.00	722.00
repair, minimum charge	1C@2.33	ea	25.50	158.00	183.50
Contemporary capital. Capital up to 4" high. Includes load-bearing plug.					
replace, composite materials	1C@.566	ea	231.00	38.30	269.30
replace, pine	1C@.566	ea	264.00	38.30	302.30
replace, redwood	1C@.566	ea	305.00	38.30	343.30
replace, oak	1C@.566	ea	587.00	38.30	625.30
replace, stone	1C@.566	ea	465.00	38.30	503.30
replace, synthetic stone	1C@.566	ea	432.00	38.30	470.30
replace, aluminum	1C@.566	ea	270.00	38.30	308.30
replace, interior plaster	1C@.566	ea	250.00	38.30	288.30
replace, exterior stucco	1C@.566	ea	245.00	38.30	283.30
remove, all material types	1D@.850	ea	—	40.10	40.10

Corinthian capital. Capital up to 14" high. Includes load-bearing plug.

	Craft@Hrs	Unit	Material	Labor	Total
replace, composite materials	1C@.566	ea	704.00	38.30	742.30
replace, pine	1C@.566	ea	1,240.00	38.30	1,278.30
replace, redwood	1C@.566	ea	1,420.00	38.30	1,458.30
replace, oak	1C@.566	ea	2,770.00	38.30	2,808.30
replace, stone	1C@.566	ea	938.00	38.30	976.30
replace, synthetic stone	1C@.566	ea	871.00	38.30	909.30
replace, aluminum	1C@.566	ea	545.00	38.30	583.30
replace, interior plaster	1C@.566	ea	506.00	38.30	544.30
replace, exterior stucco	1C@.566	ea	493.00	38.30	531.30
remove, all material types	1D@.850	ea	—	40.10	40.10

Doric or Tuscan capital. Capital up to 4" high. Includes load-bearing plug.

	Craft@Hrs	Unit	Material	Labor	Total
replace, composite materials	1C@.566	ea	239.00	38.30	277.30
replace, pine	1C@.566	ea	273.00	38.30	311.30
replace, redwood	1C@.566	ea	316.00	38.30	354.30
replace, oak	1C@.566	ea	607.00	38.30	645.30
replace, stone	1C@.566	ea	484.00	38.30	522.30
replace, synthetic stone	1C@.566	ea	447.00	38.30	485.30
replace, aluminum	1C@.566	ea	278.00	38.30	316.30
replace, interior plaster	1C@.566	ea	259.00	38.30	297.30
replace, exterior stucco	1C@.566	ea	250.00	38.30	288.30
remove, all material types	1D@.850	ea	—	40.10	40.10

Empire capital. Capital up to 7" high. Includes load-bearing plug.

	Craft@Hrs	Unit	Material	Labor	Total
replace, composite materials	1C@.566	ea	447.00	38.30	485.30
replace, pine	1C@.566	ea	796.00	38.30	834.30
replace, redwood	1C@.566	ea	921.00	38.30	959.30
replace, oak	1C@.566	ea	1,740.00	38.30	1,778.30
replace, stone	1C@.566	ea	599.00	38.30	637.30
replace, synthetic stone	1C@.566	ea	556.00	38.30	594.30
replace, aluminum	1C@.566	ea	346.00	38.30	384.30
replace, interior plaster	1C@.566	ea	320.00	38.30	358.30
replace, exterior stucco	1C@.566	ea	313.00	38.30	351.30
remove, all material types	1D@.850	ea	—	40.10	40.10

Erechtheum capital. Capital up to 7" high. Add **50%** for necking. Includes load-bearing plug.

	Craft@Hrs	Unit	Material	Labor	Total
replace, composite materials	1C@.566	ea	436.00	38.30	474.30
replace, pine	1C@.566	ea	775.00	38.30	813.30
replace, redwood	1C@.566	ea	890.00	38.30	928.30
replace, oak	1C@.566	ea	1,670.00	38.30	1,708.30
replace, stone	1C@.566	ea	578.00	38.30	616.30
replace, synthetic stone	1C@.566	ea	540.00	38.30	578.30
replace, aluminum	1C@.566	ea	338.00	38.30	376.30
replace, interior plaster	1C@.566	ea	313.00	38.30	351.30
replace, exterior stucco	1C@.566	ea	304.00	38.30	342.30
remove, all material types	1D@.850	ea	—	40.10	40.10

	Craft@Hrs	Unit	Material	Labor	Total
Roman Ionic capital. Capital up to 5" high. Includes load-bearing plug.					
replace, composite materials	1C@.566	ea	453.00	38.30	491.30
replace, pine	1C@.566	ea	807.00	38.30	845.30
replace, redwood	1C@.566	ea	929.00	38.30	967.30
replace, oak	1C@.566	ea	1,790.00	38.30	1,828.30
replace, stone	1C@.566	ea	605.00	38.30	643.30
replace, synthetic stone	1C@.566	ea	560.00	38.30	598.30
replace, aluminum	1C@.566	ea	351.00	38.30	389.30
replace, interior plaster	1C@.566	ea	324.00	38.30	362.30
replace, exterior stucco	1C@.566	ea	316.00	38.30	354.30
remove, all material types	1D@.850	ea	—	40.10	40.10
Scamozzi capital. Capital up to 5" high. Includes load-bearing plug.					
replace, composite materials	1C@.566	ea	478.00	38.30	516.30
replace, pine	1C@.566	ea	847.00	38.30	885.30
replace, redwood	1C@.566	ea	973.00	38.30	1,011.30
replace, oak	1C@.566	ea	1,890.00	38.30	1,928.30
replace, stone	1C@.566	ea	636.00	38.30	674.30
replace, synthetic stone	1C@.566	ea	589.00	38.30	627.30
replace, aluminum	1C@.566	ea	366.00	38.30	404.30
replace, interior plaster	1C@.566	ea	343.00	38.30	381.30
replace, exterior stucco	1C@.566	ea	335.00	38.30	373.30
remove, all material types	1D@.850	ea	—	40.10	40.10
Temple-of-the-Winds capital. Capital up to 12" high. Includes load-bearing plug.					
replace, composite materials	1C@.566	ea	548.00	38.30	586.30
replace, pine	1C@.566	ea	967.00	38.30	1,005.30
replace, redwood	1C@.566	ea	1,110.00	38.30	1,148.30
replace, oak	1C@.566	ea	2,170.00	38.30	2,208.30
replace, stone	1C@.566	ea	726.00	38.30	764.30
replace, synthetic stone	1C@.566	ea	675.00	38.30	713.30
replace, aluminum	1C@.566	ea	421.00	38.30	459.30
replace, interior plaster	1C@.566	ea	391.00	38.30	429.30
replace, exterior stucco	1C@.566	ea	378.00	38.30	416.30
remove, all material types	1D@.850	ea	—	40.10	40.10
Capital repair.					
repair, wood capital with wood patch	1C@1.06	ea	4.08	71.70	75.78
repair, wood capital by replacing section	1C@3.89	ea	36.60	263.00	299.60
repair, dry rot in wood capital with epoxy	1C@2.08	ea	4.08	141.00	145.08
minimum charge, all material types	1C@2.91	ea	44.90	197.00	241.90
Additional column and pilaster sizes. Add or deduct from the costs for 12" columns, pilasters, capitals and bases. (Bases are built to hold columns of the size indicated.)					
add for spiral column or pilaster	—	%	78.0	—	—
column, pilaster, capital, or base					
deduct for 6" diameter	—	%	- 45.0	—	—
deduct for 8" diameter	—	%	- 35.0	—	—
deduct for 10" diameter	—	%	- 17.0	—	—

	Craft@Hrs	Unit	Material	Labor	Total
add for 14" diameter	—	%	59.0	—	—
add for 16" diameter	—	%	81.0	—	—
add for 18" diameter	—	%	115.0	—	—
add for 20" diameter	—	%	185.0	—	—

Column base.

	Craft@Hrs	Unit	Material	Labor	Total
replace, composite materials	1C@.566	ea	70.30	38.30	108.60
replace, pine	1C@.566	ea	74.50	38.30	112.80
replace, redwood	1C@.566	ea	88.80	38.30	127.10
replace, oak	1C@.566	ea	161.00	38.30	199.30
replace, stone	1C@.566	ea	130.00	38.30	168.30
replace, synthetic stone	1C@.566	ea	122.00	38.30	160.30
replace, aluminum	1C@.566	ea	77.50	38.30	115.80
replace, interior plaster	1C@.566	ea	71.70	38.30	110.00
replace, exterior stucco	1C@.566	ea	68.90	38.30	107.20
remove column or pilaster base	1D@.785	ea	—	37.10	37.10

standard *high* *deluxe* *custom*

Column pedestal. Pedestal framed with 2" x 4" up to 3' high. Outside trimmed with paint-grade pine or poplar. Moldings around top of pedestal and bottom. Grades above standard have panel moldings of increasingly complex design on all four sides.

	Craft@Hrs	Unit	Material	Labor	Total
replace, standard grade	1C@8.48	ea	264.00	573.00	837.00
replace, high grade	1C@9.40	ea	350.00	635.00	985.00
replace, deluxe grade	1C@10.6	ea	518.00	717.00	1,235.00
replace, custom grade	1C@11.7	ea	609.00	791.00	1,400.00
remove column pedestal	1D@.410	ea	—	19.40	19.40

Pilaster pedestal. Pedestal framed with 2" x 4" up to 3' high. Outside trimmed with paint-grade pine or poplar. Moldings around top of pedestal and bottom. Grades above standard have panel moldings of increasingly complex design on all four sides.

	Craft@Hrs	Unit	Material	Labor	Total
replace, standard grade	1C@6.00	ea	161.00	406.00	567.00
replace, high grade	1C@6.67	ea	213.00	451.00	664.00
replace, deluxe grade	1C@7.47	ea	313.00	505.00	818.00
replace, custom grade	1C@8.33	ea	367.00	563.00	930.00
remove pilaster pedestal	1D@.334	ea	—	15.80	15.80

Add for other wood species.

	Craft@Hrs	Unit	Material	Labor	Total
add for stain-grade redwood materials	—	%	12.0	—	—
add for mahogany materials	—	%	16.0	—	—

	Craft@Hrs	Unit	Material	Labor	Total
Time & Material Charts (selected items)					
Columns Materials					
Minimum materials charge for 8' column or pilaster work	—	ea	57.40	—	57.40
12" round column					
composite materials	—	ea	494.00	—	494.00
pine	—	ea	519.00	—	519.00
12" round pilaster					
composite materials	—	ea	313.00	—	313.00
pine	—	ea	328.00	—	328.00
12" square column					
composite materials	—	ea	652.00	—	652.00
pine	—	ea	684.00	—	684.00
12" square pilaster					
composite materials	—	ea	416.00	—	416.00
pine pilaster	—	ea	440.00	—	440.00
Contemporary capital					
composite materials	—	ea	235.00	—	235.00
pine	—	ea	270.00	—	270.00
Corinthian capital					
composite materials	—	ea	719.00	—	719.00
pine	—	ea	1,270.00	—	1,270.00
Doric or Tuscan capital					
composite materials	—	ea	245.00	—	245.00
pine	—	ea	280.00	—	280.00
Empire capital					
composite materials	—	ea	457.00	—	457.00
pine	—	ea	817.00	—	817.00
Erechtheum capital					
composite materials	—	ea	444.00	—	444.00
pine	—	ea	792.00	—	792.00
Roman Ionic capital					
composite materials	—	ea	463.00	—	463.00
pine	—	ea	823.00	—	823.00
Scamozzi capital					
composite materials	—	ea	490.00	—	490.00
pine	—	ea	864.00	—	864.00
Temple-of-the-Winds capital					
composite materials	—	ea	559.00	—	559.00
pine	—	ea	986.00	—	986.00
Column base					
composite materials	—	ea	71.80	—	71.80
pine	—	ea	76.40	—	76.40
Column pedestal					
standard grade	—	ea	270.00	—	270.00
custom grade	—	ea	623.00	—	623.00
Pilaster pedestal					
standard grade	—	ea	165.00	—	165.00
custom grade	—	ea	374.00	—	374.00

Columns Labor

Laborer	base wage	paid leave	true wage	taxes & ins.	total
Carpenter	$38.10	2.97	$41.07	26.53	$67.60
Demolition laborer	$25.90	2.02	$27.92	19.28	$47.20

Paid leave is calculated based on two weeks paid vacation, one week sick leave, and seven paid holidays. Employer's matching portion of **FICA** is 7.65 percent. **FUTA** (Federal Unemployment) is .8 percent. **Worker's compensation** for columns was calculated using a national average of 17.21 percent. **Unemployment insurance** was calculated using a national average of 8 percent. **Health insurance** was calculated based on a projected national average for 2020 of $1,280 per employee (and family when applicable) per month. Employer pays 80 percent for a per month cost of $1,024 per employee. **Retirement** is based on a 401(k) retirement program with employer matching of 50 percent. Employee contributions to the 401(k) plan are an average of 6 percent of the true wage. **Liability insurance** is based on a national average of 12.0 percent.

Columns Labor Productivity

	Craft@Hrs	Unit	Material	Labor	Total
Demolition					
remove column	1D@.212	lf	—	10.00	10.00
remove pilaster	1D@.134	lf	—	6.32	6.32
remove column or pilaster capital	1D@.850	ea	—	40.10	40.10
remove column or pilaster base	1D@.785	ea	—	37.10	37.10
remove column pedestal	1D@.410	ea	—	19.40	19.40
remove pilaster pedestal	1D@.334	ea	—	15.80	15.80
Install column					
round	1C@.333	lf	—	22.50	22.50
square	1C@.308	lf	—	20.80	20.80
Install pilaster					
round	1C@.292	lf	—	19.70	19.70
square	1C@.269	lf	—	18.20	18.20
Install capital					
column or pilaster capital	1C@.566	ea	—	38.30	38.30
Install column or pilaster base					
column or pilaster base	1C@.566	ea	—	38.30	38.30
Repair column or pilaster					
round with wood patch	1C@.706	ea	—	47.70	47.70
fluted with wood patch	1C@.863	ea	—	58.30	58.30
round by replacing section	1C@1.94	ea	—	131.00	131.00
fluted by replacing section	1C@2.59	ea	—	175.00	175.00
round dry rot with epoxy	1C@2.74	ea	—	185.00	185.00
fluted dry rot with epoxy	1C@3.24	ea	—	219.00	219.00
minimum charge	1C@2.33	ea	—	158.00	158.00
Repair wood capital					
with wood patch	1C@1.06	ea	—	71.70	71.70
by replacing section	1C@3.89	ea	—	263.00	263.00
dry rot with epoxy	1C@2.08	ea	—	141.00	141.00
minimum charge	1C@2.91	ea	—	197.00	197.00
Build and install column pedestal					
standard grade	1C@8.48	ea	—	573.00	573.00
custom grade	1C@11.7	ea	—	791.00	791.00
Build and install pilaster pedestal					
standard grade	1C@6.00	ea	—	406.00	406.00
custom grade	1C@8.33	ea	—	563.00	563.00

Concrete

CEC. Items that include CEC in the description are priced according to California earthquake code requirements. Although these Uniform Building Code or similar standards are also required in other areas of the country, they are most commonly associated with efforts initiated in the State of California to improve the construction and engineering of structures in quake zones.

	Craft@Hrs	Unit	Material	Labor	Equip.	Total
Minimum charge.						
for concrete repair work	9F@2.47	ea	99.30	140.00	—	239.30

Epoxy repair. Includes epoxy, equipment, and installation. A pressure pot or similar device is used to inject epoxy into the crack. In structural settings the type of epoxy used must often be determined by an engineer and for long cracks core samples may be necessary to test the effectiveness of the repairs.

	Craft@Hrs	Unit	Material	Labor	Equip.	Total
Concrete crack repair with pressurized epoxy injection, per linear foot						
replace	6F@.111	lf	1.09	7.04	2.31	10.44
Minimum charge for pressurized epoxy injection concrete crack repair						
replace	6F@2.00	ea	103.00	127.00	143.00	373.00
Concrete hole repair with epoxy system, per cubic inch of hole						
replace	6F@.053	ci	.03	3.36	—	3.39
Minimum charge for concrete hole repair with epoxy system						
replace	6F@.501	ea	93.60	31.80	—	125.40
Concrete crack repair with caulking gun epoxy injection, per linear foot						
replace	6F@.025	lf	.66	1.59	—	2.25
Minimum charge for concrete crack repair with epoxy injection						
replace	6F@.501	ea	103.00	31.80	—	134.80
Repair spalled concrete with concrete resurfacer, per square foot						
replace	6F@.009	sf	1.27	.57	—	1.84
Minimum charge to repair spalled concrete						
replace	6F@1.00	ea	62.30	63.40	—	125.70

Footings. Standard 16", 20", and 24" footings include three horizontal lengths of #4 rebar. Standard 32" footing includes four horizontal lengths of #4 rebar. CEC 16" and 20" footings include three horizontal lengths of #5 rebar. CEC 24" and 32" footings include four horizontal lengths of #5 rebar. Tie-in rebar from foundation wall bent continuously into footing is included in the foundation wall prices. Does not include excavation or grading. Includes covering to protect concrete from the weather as needed.

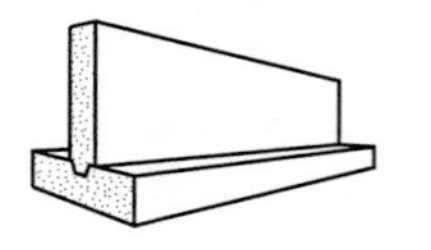

	Craft@Hrs	Unit	Material	Labor	Equip.	Total
Concrete footings including rebar package and forming, per cubic yard						
replace	9F@2.44	cy	189.00	138.00	—	327.00
remove	1D@1.64	cy	—	77.40	89.00	166.40
Concrete footings including CEC rebar package and forming, per cubic yard						
replace	9F@2.94	cy	227.00	166.00	—	393.00
remove	1D@1.96	cy	—	92.50	147.00	239.50

	Craft@Hrs	Unit	Material	Labor	Equip.	Total
16" wide by 10" deep concrete footings with rebar, per linear foot						
replace	9F@.150	lf	8.17	8.48	—	16.65
remove	1D@.121	lf	—	5.71	4.58	10.29
16" wide by 10" deep concrete footings with CEC rebar, per linear foot						
replace	9F@.180	lf	10.40	10.20	—	20.60
remove	1D@.147	lf	—	6.94	5.57	12.51
20" wide by 10" deep concrete footings with rebar, per linear foot						
replace	9F@.159	lf	9.79	8.98	—	18.77
remove	1D@.121	lf	—	5.71	4.58	10.29
20" wide by 10" deep concrete footings with CEC rebar, per linear foot						
replace	9F@.190	lf	12.20	10.70	—	22.90
remove	1D@.147	lf	—	6.94	5.57	12.51
24" wide by 12" deep concrete footings with rebar, per linear foot						
replace	9F@.167	lf	14.00	9.44	—	23.44
remove	1D@.121	lf	—	5.71	4.58	10.29
24" wide by 12" deep concrete footings with CEC rebar, per linear foot						
replace	9F@.197	lf	16.60	11.10	—	27.70
remove	1D@.147	lf	—	6.94	5.57	12.51
32" wide by 14" deep concrete footings with rebar, per linear foot						
replace	9F@.177	lf	20.70	10.00	—	30.70
remove	1D@.121	lf	—	5.71	4.58	10.29
32" wide by 14" deep concrete footings with CEC rebar, per linear foot						
replace	9F@.207	lf	23.20	11.70	—	34.90
remove	1D@.147	lf	—	6.94	5.57	12.51
Minimum charge for concrete footing work						
replace	9F@1.03	ea	73.90	58.20	—	132.10

Foundation. Includes concrete, forms, and installation. Installation is on top of footings with appropriate tie-in rebar, forming, pouring, and finishing concrete with J-bolts and straps. Includes covering to protect concrete from the weather as needed. Standard foundation walls include horizontal and vertical lengths of #4 rebar 24" on center. Vertical rebar is bent continuously into footing. Foundations that follow California earthquake standards include horizontal and vertical lengths of #5 rebar 12" on center.

	Craft@Hrs	Unit	Material	Labor	Equip.	Total
Concrete foundation wall including rebar package and forming, per cubic yard						
replace	9F@2.70	cy	178.00	153.00	46.70	377.70
remove	1D@2.44	cy	—	115.00	111.00	226.00
Concrete foundation wall including CEC rebar package & forming, per cubic yard						
replace	9F@3.13	cy	210.00	177.00	46.70	433.70
remove	1D@2.94	cy	—	139.00	147.00	286.00

	Craft@Hrs	Unit	Material	Labor	Equip.	Total
Concrete foundation wall, per square foot						
replace, 5" wide	9F@.059	sf	2.48	3.33	1.15	6.96
replace, 6" wide	9F@.064	sf	3.01	3.62	1.15	7.78
remove 5" or 6" wide	1D@.073	sf	—	3.45	2.74	6.19
replace, 8" wide	9F@.074	sf	4.00	4.18	1.15	9.33
replace, 10" wide	9F@.084	sf	5.01	4.75	1.15	10.91
remove 8" or 10" wide	1D@.076	sf	—	3.59	3.17	6.76
Rebar package for 5", 6", 8", or 10" wide concrete foundation wall, per square foot						
replace	3F@.037	sf	.34	2.04	—	2.38
CEC rebar package for 5", 6", 8", or 10" wide concrete foundation wall, per square foot						
replace	3F@.067	sf	1.15	3.69	—	4.84
Add for insulating foam stay-in-place foundation wall forms, per square foot						
replace	3F@.045	sf	2.20	2.48	—	4.68
Minimum charge for concrete foundation wall work						
replace	9F@3.00	ea	93.60	170.00	150.00	413.60

Single-pour footing & foundation. Includes concrete, forms, rebar and installation. Installation includes minor finish excavation but not general excavation, forming, pouring, and finishing concrete. Includes covering to protect concrete from the weather as needed.

	Craft@Hrs	Unit	Material	Labor	Equip.	Total
Single-pour footing and foundation wall with rebar package, per cubic yard						
replace	9F@.589	cy	166.00	33.30	57.50	256.80
remove	1D@1.96	cy	—	92.50	100.00	192.50
Single-pour footing and foundation wall with CEC rebar package, per cubic yard						
replace	9F@.713	cy	168.00	40.30	57.50	265.80
remove	1D@2.33	cy	—	110.00	107.00	217.00
Single-pour footing, per linear foot						
with 6" wide by 3' stem wall, replace	9F@.346	lf	17.20	19.50	6.92	43.62
with 8" wide by 3' stem wall, replace	9F@.346	lf	20.40	19.50	6.92	46.82
with 10" wide by 3' stem wall, replace	9F@.346	lf	21.00	19.50	6.92	47.42
Single-pour footing and 3' stem wall, per linear foot, remove	1D@.346	lf	—	16.30	14.20	30.50
Rebar package for single-pour footing 6", 8" or 10" wide and 3' stem wall, per linear foot, replace	3F@.139	lf	2.61	7.65	—	10.26
CEC rebar package for single-pour footing 6", 8" or 10" wide and 3' stem wall, per linear foot, replace	3F@.243	lf	6.23	13.40	—	19.63

	Craft@Hrs	Unit	Material	Labor	Equip.	Total
Single-pour footing, per linear foot						
with 6" wide by 4' stem wall, replace	9F@.437	lf	20.40	24.70	9.21	54.31
with 8" wide by 4' stem wall, replace	9F@.437	lf	24.40	24.70	9.21	58.31
with 10" wide by 4' stem wall, replace	9F@.437	lf	28.60	24.70	9.21	62.51
Single-pour footing and 4' stem wall per linear foot, remove	1D@.422	lf	—	19.90	14.20	34.10
Rebar package for single-pour footing 6", 8" or 10" wide and 4' stem wall, per linear foot, replace	3F@.152	lf	3.00	8.36	—	11.36
CEC rebar package for single-pour footing 6", 8" or 10" wide and 4' stem wall, per linear foot, replace	3F@.255	lf	6.97	14.00	—	20.97
Minimum charge for single-pour footing and stem wall work, replace	9F@3.00	ea	37.90	170.00	158.00	365.90

Drill pier hole. Pier hole from 12" to 24" in diameter. (Most common sizes are 16" and 18".)

	Craft@Hrs	Unit	Material	Labor	Equip.	Total
Tractor-mounted auger, mobilization	10@6.25	ea	—	556.00	687.00	1,243.00
Drill concrete pier hole, hillside residential construction	10@.083	lf	—	7.38	3.50	10.88
minimum charge	10@3.50	ea	—	311.00	388.00	699.00

Pier. Includes concrete, rebar cage, equipment, and installation. Does not include drilling the pier hole or pump truck fees. Rebar cage includes rebar, wire, welding, assembly (usually off-site), hauling to site, and installation in hole. Includes equipment to lift unit from truck and place in hole. The rebar cage is engineered to match California earthquake standards for a pier that is from 12" to 24" in diameter with 16" and 18" being the most common sizes. Rebar includes vertical members the length of the pier and round horizontal members. In practice, rebar cages for piers are often engineered.

	Craft@Hrs	Unit	Material	Labor	Equip.	Total
Concrete pier including rebar cage						
replace	9F@14.9	cy	674.00	842.00	—	1,516.00
remove	1D@.699	cy	—	33.00	82.50	115.50
Concrete pier with up to 36" fiber tube wrapped cap						
replace	9F@1.75	lf	25.40	98.90	—	124.30
remove	1D@.082	lf	—	3.87	9.63	13.50
Above grade concrete pier with fiber tube wrap						
replace	9F@1.81	lf	62.90	102.00	—	164.90
remove	1D@.082	lf	—	3.87	9.63	13.50
Rebar cage for concrete pier (meets CEC requirements)						
replace	3F@.091	lf	59.90	5.01	—	64.91
Concrete piers minimum charge	9F@7.25	ea	492.00	410.00	—	902.00

	Craft@Hrs	Unit	Material	Labor	Equip.	Total

Jacket pier. Includes forming, rebar, epoxy, concrete, equipment and installation. The existing pier top is jackhammered. The new rebar is secured to rebar in the pier exposed by jackhammering and with epoxy in holes drilled in the concrete. The pier top is formed and a concrete jacket is poured.

	Craft@Hrs	Unit	Material	Labor	Equip.	Total
replace, new concrete and rebar ties	9F@4.92	ea	59.10	278.00	54.30	391.40
minimum charge	9F@4.00	ea	149.00	226.00	189.00	564.00

Shallow pier. Includes concrete, stirrup or similar fastener, and installation. Installation includes hand digging a hole up to three feet deep, pouring concrete, finish the top and placing the stirrup. Pre-cast pier includes pre-cast pier, hand excavation, and installation.

	Craft@Hrs	Unit	Material	Labor	Equip.	Total
Shallow concrete pier (less than 3' deep)						
for deck or light structure						
replace	9F@2.94	ea	30.50	166.00	—	196.50
remove	1D@1.14	ea	—	53.80	—	53.80
Pre-cast concrete pier						
for deck or light structure						
replace	2F@.322	ea	35.00	14.90	—	49.90
remove	1D@.233	ea	—	11.00	—	11.00

Grade beam with rebar cage. Prefabricated rebar cage included. All rebar cages are to CEC standards. When removed, grade beams are detached from piers with a jackhammer or concrete saw. The beams are broken into manageable chunks and loaded by crane or backhoe into a truck. (Does not include hauling or dump fees.)

	Craft@Hrs	Unit	Material	Labor	Equip.	Total
Grade beam, per cubic yard						
replace	9F@4.01	cy	962.00	227.00	—	1,189.00
remove	1D@2.33	cy	—	110.00	61.70	171.70
Grade beam, per linear foot						
replace, 12" wide by 18" deep	9F@.264	lf	67.50	14.90	4.62	87.02
replace, 12" wide by 20" deep	9F@.289	lf	68.50	16.30	4.62	89.42
remove, 12" wide by 18" or 20" deep	1D@.174	lf	—	8.21	6.60	14.81
replace, 12" wide by 22" deep	9F@.339	lf	69.70	19.20	4.62	93.52
remove, 12" wide by 22" deep	1D@.195	lf	—	9.20	7.97	17.17
replace, 12" wide by 24" deep	9F@.279	lf	70.60	15.80	4.62	91.02
remove, 12" wide by 24" deep	1D@.195	lf	—	9.20	7.97	17.17
replace, 14" wide by 18" deep	9F@.310	lf	69.00	17.50	4.62	91.12
replace, 14" wide by 20" deep	9F@.370	lf	70.10	20.90	4.62	95.62
remove, 14" wide by 18" or 20" deep	1D@.174	lf	—	8.21	6.60	14.81
replace, 14" wide by 22" deep	9F@.370	lf	71.60	20.90	4.62	97.12
replace, 14" wide by 24" deep	9F@.429	lf	72.40	24.20	4.62	101.22
remove, 14" wide by 22" or 24" deep	1D@.199	lf	—	9.39	8.12	17.51
replace, 16" wide by 18" deep	9F@.550	lf	70.60	31.10	4.62	106.32
replace, 16" wide by 20" deep	9F@.672	lf	71.90	38.00	4.62	114.52
remove, 16" wide by 18" or 20" deep	1D@.216	lf	—	10.20	8.82	19.02
replace, 16" wide by 22" deep	9F@.550	lf	73.20	31.10	4.62	108.92
replace, 16" wide by 24" deep	9F@.672	lf	74.40	38.00	4.62	117.02
remove 16" wide by 22" or 24" deep	1D@.223	lf	—	10.50	9.12	19.62
replace, 18" wide by 18" deep	9F@.310	lf	72.00	17.50	4.62	94.12
replace, 18" wide by 20" deep	9F@.370	lf	73.50	20.90	4.62	99.02
remove 18" wide by 18" or 20" deep	1D@.216	lf	—	10.20	8.79	18.99
replace, 18" wide by 22" deep	9F@.370	lf	75.00	20.90	4.62	100.52
replace, 18" wide by 24" deep	9F@.430	lf	76.40	24.30	4.62	105.32
remove 18" wide by 22" or 24" deep	1D@.223	lf	—	10.50	9.12	19.62
Concrete grade beams, minimum charge	9F@1.03	ea	512.00	58.20	158.00	728.20

	Craft@Hrs	Unit	Material	Labor	Equip.	Total

Jacket grade beam with new concrete and rebar ties. Includes forming, rebar, epoxy, concrete, equipment and installation. The existing grade beam is jackhammered and holes are drilled for rebar. The rebar is secured with epoxy. The grade beam is formed and a concrete jacket is poured.

	Craft@Hrs	Unit	Material	Labor	Equip.	Total
Jacket concrete grade beam						
replace	9F@.537	lf	17.50	30.30	7.19	54.99
minimum charge	9F@4.00	ea	83.20	226.00	189.00	498.20

Lightweight flatwork. Includes concrete, forms, and installation. Installation includes minor finish excavation but not general excavation, forming, pouring, and finishing concrete with control joints. Includes covering to protect concrete from the weather as needed.

	Craft@Hrs	Unit	Material	Labor	Equip.	Total
Lightweight concrete flatwork (no rebar), per cubic yard						
replace	9F@2.04	cy	201.00	115.00	—	316.00
remove	1D@2.35	cy	—	111.00	80.00	191.00
2" slab, per square foot						
replace	9F@.023	sf	1.25	1.30	—	2.55
remove	1D@.026	sf	—	1.23	1.02	2.25
4" slab, per square foot						
replace	9F@.024	sf	2.47	1.36	—	3.83
remove	1D@.026	sf	—	1.23	1.02	2.25
6" slab, per square foot						
replace	9F@.026	sf	3.79	1.47	—	5.26
remove	1D@.026	sf	—	1.23	1.02	2.25
Minimum charge	9F@3.00	ea	94.80	170.00	—	264.80

Flatwork. Includes concrete, forms, and installation. Installation includes minor finish excavation but not general excavation, forming, pouring, and finishing concrete with control joints. Includes covering to protect concrete from the weather as needed.

	Craft@Hrs	Unit	Material	Labor	Equip.	Total
Concrete flatwork (no rebar), per cubic yard						
replace	9F@2.04	cy	163.00	115.00	—	278.00
remove	1D@2.39	cy	—	113.00	80.70	193.70
4" slab, per square foot						
replace	9F@.035	sf	2.01	1.98	—	3.99
remove	1D@.027	sf	—	1.27	1.02	2.29
6" slab, per square foot						
replace	9F@.048	sf	3.02	2.71	—	5.73
remove	1D@.027	sf	—	1.27	1.02	2.29
8" slab, per square foot						
replace	9F@.048	sf	4.01	2.71	—	6.72
remove	1D@.027	sf	—	1.27	1.02	2.29
Minimum charge	9F@3.00	ea	77.30	170.00	—	247.30

Utility flatwork. Includes concrete, forms, and installation. Installation includes minor finish excavation but not general excavation, forming, pouring, and finishing concrete with control joints. Includes covering to protect concrete from the weather as needed.

	Craft@Hrs	Unit	Material	Labor	Equip.	Total
Concrete footing for chimney						
replace	9F@.044	sf	4.40	2.49	—	6.89
remove	1D@.030	sf	—	1.42	1.15	2.57
Concrete slab for exterior heat pump, condenser, or other equipment						
replace	9F@.044	sf	2.38	2.49	—	4.87
remove	1D@.027	sf	—	1.27	1.02	2.29

	Craft@Hrs	Unit	Material	Labor	Equip.	Total

Rebar for flatwork. Includes rebar, wire, and installation. For placement of rebar only does not include forming or grading. Wire mesh includes 6" x 6" #10 wire mesh, wire, and installation.

	Craft@Hrs	Unit	Material	Labor	Equip.	Total
Concrete slab						
replace, #4 24" on center	3F@.036	sf	.80	1.98	—	2.78
replace, #4 12" on center	3F@.080	sf	1.66	4.40	—	6.06
Wire mesh for concrete slab, 6" x 6" #10						
replace	3F@.015	sf	.27	.83	—	1.10

Flatwork base. Includes slab base, hauling and finish grading of base. Does not include tearout of existing slab, or rough grading.

	Craft@Hrs	Unit	Material	Labor	Equip.	Total
Slab base						
replace, 2" aggregate	2F@.017	sf	.05	.79	.19	1.03
replace, 4" aggregate	2F@.018	sf	.12	.83	.19	1.14
replace, 2" sand	2F@.014	sf	.05	.65	.19	.89
replace, 4" sand	2F@.016	sf	.12	.74	.19	1.05

Flatwork vapor barrier. Includes membrane and installation.

	Craft@Hrs	Unit	Material	Labor	Equip.	Total
Slab membrane						
replace, 6 mil	2F@.002	sf	.04	.09	—	.13
replace, 5 mil	2F@.002	sf	.04	.09	—	.13

Sidewalk. Sidewalk to 3' wide. Includes concrete, forms, and installation. Installation includes minor finish excavation but not general excavation, forming, pouring, and finishing concrete with control joints. Includes covering to protect concrete from the weather as needed.

	Craft@Hrs	Unit	Material	Labor	Equip.	Total
Concrete sidewalk, per cubic yard,						
3" to 6" thick						
replace	9F@4.17	cy	180.00	236.00	—	416.00
remove	1D@2.39	cy	—	113.00	80.70	193.70
Concrete sidewalk, per square foot						
3" thick						
replace	9F@.042	sf	1.50	2.37	—	3.87
remove	1D@.034	sf	—	1.60	1.02	2.62
4" thick						
replace	9F@.049	sf	2.01	2.77	—	4.78
remove	1D@.034	sf	—	1.60	1.02	2.62
6" thick						
replace	9F@.062	sf	3.02	3.50	—	6.52
remove	1D@.034	sf	—	1.60	1.02	2.62
Concrete sidewalk work, all thicknesses						
minimum charge	9F@3.00	ea	191.00	170.00	—	361.00

Exposed aggregate. Where required, add for disposal of slurry.

	Craft@Hrs	Unit	Material	Labor	Equip.	Total
Exposed aggregate finish, slab or sidewalk						
add	8F@.019	sf	.26	1.10	—	1.36

	Craft@Hrs	Unit	Material	Labor	Equip.	Total
Stamping concrete.						
Concrete slab or sidewalk stamping						
add	8F@.053	sf	—	3.07	—	3.07
add for paver-pattern	8F@.045	sf	—	2.61	—	2.61
add for paver w/ grouted joints	8F@.068	sf	.52	3.94	—	4.46

Concrete dye. For concrete quantities under 350 sf add **$37** per truck for wash-out. This charge does not apply to troweled-in dyes.

	Craft@Hrs	Unit	Material	Labor	Equip.	Total
For troweled-in concrete dye						
add	8F@.009	sf	2.32	.52	—	2.84
For concrete dye in flatwork,						
Add per 1" deep layer						
blacks and grays	—	sf	.08	—	—	.08
blues	—	sf	.49	—	—	.49
browns & terra cotta reds	—	sf	.48	—	—	.48
bright reds	—	sf	.49	—	—	.49
greens	—	sf	.97	—	—	.97

Curb & gutter. Includes concrete, forms, and installation. Installation includes minor finish excavation but not general excavation, forming, pouring, and finishing concrete with control joints. Includes barriers and covering to protect concrete from passersby and the weather.

	Craft@Hrs	Unit	Material	Labor	Equip.	Total
Concrete curb & gutter						
replace	9F@.274	lf	8.33	15.50	1.39	25.22
remove	1D@.145	lf	—	6.84	2.09	8.93
Minimum charge for curb & gutter work	9F@3.67	ea	61.50	207.00	125.00	393.50

Concrete step. Includes concrete, forms, rebar, and installation. Rebar reinforcement includes two horizontal pieces per step with vertical tie-ins. Installation includes minor finish excavation but not general excavation, forming, pouring, and finishing concrete with control joints. Includes covering to protect concrete from the weather as needed.

	Craft@Hrs	Unit	Material	Labor	Equip.	Total
Concrete step per cubic yard						
replace	9F@7.70	cy	450.00	435.00	—	885.00
remove	1D@3.12	cy	—	147.00	89.10	236.10
Concrete step per linear foot						
replace	9F@.232	lf	4.97	13.10	—	18.07
remove	1D@.047	lf	—	2.22	1.10	3.32
Concrete step per square foot of landing						
replace	9F@.068	sf	12.40	3.84	—	16.24
remove	1D@.046	sf	—	2.17	1.38	3.55
Minimum charge for a concrete step						
replace	9F@5.00	ea	119.00	283.00	—	402.00

Pargeting. Includes plaster and installation. Installation prepwork includes breaking of foundation ties as needed.

	Craft@Hrs	Unit	Material	Labor	Equip.	Total
Foundation wall pargeting, per square foot, replace	4P@.022	sf	.54	1.42	—	1.96
Foundation wall pargeting work minimum charge	4P@1.00	ea	39.20	64.70	—	103.90

	Craft@Hrs	Unit	Material	Labor	Equip.	Total
Concrete sawing. Includes blade wear.						
Saw concrete wall, per lf of cut, 1" deep						
with standard rebar package	20@.111	lf	—	6.96	2.56	9.52
with CEC rebar package	20@.125	lf	—	7.84	3.74	11.58
Saw concrete floor, per lf of cut, 1" deep						
no rebar	20@.050	lf	—	3.14	1.00	4.14
with rebar package	20@.054	lf	—	3.39	1.16	4.55
with wire mesh reinforcement	20@.052	lf	—	3.26	1.39	4.65
Saw expansion joint in "green" pad	20@.189	lf	—	11.90	3.98	15.88
Minimum charge for wall sawing	20@3.00	ea	—	188.00	165.00	353.00
Minimum charge for slab sawing	20@2.25	ea	—	141.00	144.00	285.00
Core drilling. Includes drill bit wear.						
Core drilling in concrete wall, per inch of depth						
2" thick wall	20@.066	li	—	4.14	1.13	5.27
4" thick wall	20@.073	li	—	4.58	1.31	5.89
Core drilling in concrete floor, per inch of depth						
2" thick floor	20@.064	li	—	4.01	1.04	5.05
4" thick floor	20@.067	li	—	4.20	1.20	5.40
Pump truck. When pumping concrete 50 feet or less and no more than three stories high.						
add for pumping footing concrete	—	lf	—	—	1.76	1.76
add for pumping found. wall concrete	—	sf	—	—	.45	.45
add for pumping pier concrete	—	lf	—	—	2.54	2.54
add for pumping grade-beam concrete	—	lf	—	—	1.39	1.39
add for pumping slab, sidewalk conc.	—	sf	—	—	.42	.42
minimum charge for pump truck	—	ea	7.85	—	622.00	629.85
Rebar. Tied and bent into place. Includes rebar, wire, and installation. Installation includes tying and bending rebar for typical concrete work such as slabs, footings, foundations, and so on.						
#3 rebar (3/8")	3F@.014	lf	.29	.77	—	1.06
#4 rebar (1/2")	3F@.014	lf	.38	.77	—	1.15
#5 rebar (5/8")	3F@.014	lf	.55	.77	—	1.32
#6 rebar (3/4")	3F@.014	lf	.88	.77	—	1.65
#7 rebar (7/8")	3F@.015	lf	1.18	.83	—	2.01
#8 rebar (1")	3F@.015	lf	1.52	.83	—	2.35
Foundation coating.						
Foundation coating per sf	2F@.017	sf	.05	.79	—	.84
Buttress foundation. Excavation, forming, pouring, and finishing of concrete buttress over existing foundation. Does not include shoring.						
Excavate exterior of foundation	10@.121	sf	—	10.80	2.28	13.08
Hand excavate interior of foundation in crawl space	2F@.457	sf	—	21.20	—	21.20
Hand excavate interior of footing in basement	2F@.204	sf	—	9.45	—	9.45

	Craft@Hrs	Unit	Material	Labor	Equip.	Total
Backfill foundation						
when work is complete	1O@.028	sf	—	2.49	.17	2.66
Buttress exterior of existing foundation	9F@.065	sf	6.31	3.67	—	9.98
Buttress interior of existing foundation						
with crawl space	9F@.091	sf	6.31	5.14	—	11.45
with basement	9F@.208	sf	6.31	11.80	—	18.11
Buttress existing foundation						
with concrete, minimum charge	9F@7.00	ea	306.00	396.00	—	702.00

Compaction grouting. Does not include excavation. (See excavation under concrete buttress above.) This process can form a "pier" for the foundation and compacts the destabilized soil. An engineer is required to supervise and dictate specifications.

	Craft@Hrs	Unit	Material	Labor	Equip.	Total
2" core drill footings for compaction grouting	2G@.379	ea	—	25.60	13.00	38.60
Insert 2" compaction grouting pipes into ground	2G@.143	lf	—	9.67	—	9.67
Pump pressurized concrete grout through pipes	2G@.251	lf	6.86	17.00	48.20	72.06

Asphalt graded base. Includes grading of driveway, hauling and placement of graded base. Does not include tear-out of existing driveway. Driveway is assumed to have already been rough-graded or to have had a previous driveway of concrete or asphalt over it.

	Craft@Hrs	Unit	Material	Labor	Equip.	Total
Graded base for asphalt						
replace 4"	2G@.001	sf	.04	.07	.17	.28
replace 6"	2G@.001	sf	.06	.07	.19	.32
Minimum grading charge						
replace	2G@1.66	ea	91.70	112.00	391.00	594.70

Asphalt driveway. Includes asphalt, equipment, hauling, and installation. Installed over a graded base. Does not include graded base.

	Craft@Hrs	Unit	Material	Labor	Equip.	Total
2" asphalt						
replace	2G@.010	sf	.88	.68	.74	2.30
remove	2G@.005	sf	—	.34	—	.34
3" asphalt						
replace	2G@.012	sf	1.29	.81	.83	2.93
remove	2G@.006	sf	—	.41	—	.41
4" asphalt						
replace	2G@.013	sf	1.72	.88	.93	3.53
remove	2G@.006	sf	—	.41	—	.41

Asphalt overlay.

	Craft@Hrs	Unit	Material	Labor	Equip.	Total
1" asphalt						
replace	2G@.012	sf	.45	.81	.78	2.04
remove	2G@.006	sf	—	.41	—	.41
2" asphalt						
replace	2G@.014	sf	.88	.95	.86	2.69
remove	2G@.006	sf	—	.41	—	.41
3" asphalt						
replace	2G@.016	sf	1.29	1.08	.95	3.32
remove	2G@.006	sf	—	.41	—	.41

	Craft@Hrs	Unit	Material	Labor	Equip.	Total
Fill pothole in driveway						
replace	2G@.379	ea	7.16	25.60	—	32.76
Seal crack in driveway						
replace	2G@.013	lf	.44	.88	—	1.32
Minimum charge for asphalt paving work						
replace	2G@4.00	ea	220.00	270.00	1,180.00	1,670.00
Seal asphalt						
replace	2G@.002	sf	.06	.14	—	.20
Seal and sand asphalt						
replace	2G@.003	sf	.12	.20	—	.32

Time & Material Charts (selected items)

Concrete Materials

	Craft@Hrs	Unit	Material	Labor	Total
Ready mix concrete delivered					
5 bag mix 2,500 psi, per cy, ($149.00), 6% waste	—	cy	158.00	—	158.00
5.5 bag mix 3,000 psi, per cy, ($149.00), 6% waste	—	cy	158.00	—	158.00
6 bag mix, 3,500 psi, per cy, ($151.00), 6% waste	—	cy	160.00	—	160.00
6.5 bag mix, 4,000 psi, per cy, ($152.00), 6% waste	—	cy	161.00	—	161.00
7 bag mix, 4,500 psi, per cy, ($155.00), 6% waste	—	cy	165.00	—	165.00
7.5 bag mix, 5,000 psi, per cy, ($158.00), 6% waste	—	cy	167.00	—	167.00
add for synthetic fiber reinforcing, 0% waste	—	cy	18.00	—	18.00
add for high early strength, 0% waste	—	cy	19.30	—	19.30
Concrete form materials					
form ties, box of 400 ($75.40), 5% waste	—	box	79.20	—	79.20
footing forms (forms 6 uses, key 3 uses) 0% waste	—	lf	.57	—	.57
Concrete footings					
with rebar per cy ($174.00 cy, 1 cy), 6% waste	—	cy	183.00	—	183.00
with CEC rebar per cy ($214.00 cy, 1 cy), 6% waste	—	cy	231.00	—	231.00
16" wide by 10" deep ($152.00 lf, 24.3 lf), 6% waste	—	lf	6.65	—	6.65
rebar package for 16" by 10" ($.60 pound, .7 lf), 6% waste	—	lf	.90	—	.90
CEC rebar package for 16" by 10" ($.60 pound, .2 lf), 6% waste	—	lf	3.19	—	3.19
20" wide by 10" deep ($152.00 lf, 19.44 lf), 6% waste	—	lf	8.30	—	8.30
rebar package for 20" by 10" ($.60 pound, .74 lf), 2% waste	—	lf	.82	—	.82
CEC rebar package for 20" by 10" ($.60 pound, .19 lf), 4% waste	—	lf	3.29	—	3.29

	Craft@Hrs	Unit	Material	Labor	Total
24" wide by 12" deep (\$152.00 lf, 13.5 lf), 6% waste	—	lf	12.00	—	12.00
rebar package for 24" by 12" (\$.60 pound, .5 lf), 2% waste	—	lf	1.22	—	1.22
CEC rebar package for 24" by 10" (\$.60 pound, .16 lf), 4% waste	—	lf	3.90	—	3.90
32" wide by 14" deep (\$152.00 lf, 8.68 lf), 6% waste	—	lf	18.60	—	18.60
rebar package for 32" by 14" (\$.60 pound, .5 lf), 2% waste	—	lf	1.22	—	1.22
CEC rebar package for 32" by 14" (\$.60 pound, .16 lf), 4% waste	—	lf	3.90	—	3.90
Foundation wall concrete					
including rebar package per cy (\$166.00 cy, 1 cy), 6% waste	—	cy	177.00	—	177.00
including CEC rebar package per cy (\$198.00 cy, 1 cy), 6% waste	—	cy	210.00	—	210.00
5" wide wall (\$152.00 cy, 65.06 ea), 6% waste	—	ea	2.49	—	2.49
rebar package for 5" wide (\$.60 pound, 1.71 sf), 2% waste	—	sf	.35	—	.35
CEC rebar package for 5" wide (\$.60 pound, .55 sf), 4% waste	—	sf	1.15	—	1.15
6" wide wall (\$152.00 cy, 54 ea), 6% waste	—	ea	3.00	—	3.00
rebar package for 6" wide wall (\$.60 pound, 1.71 sf), 2% waste	—	sf	.35	—	.35
CEC rebar package for 6" wide wall (\$.60 pound, .55 sf), 4% waste	—	sf	1.15	—	1.15
8" wide wall (\$152.00 cy, 40.5 ea), 6% waste	—	ea	4.00	—	4.00
rebar package for 8" wide wall (\$.60 sf, 1.71 sf), 2% waste	—	sf	.35	—	.35
CEC rebar package for 8" wide wall (\$.60 sf, .55 sf), 4% waste	—	sf	1.15	—	1.15
10" wide wall (\$152.00 cy, 32.4 ea), 6% waste	—	ea	4.97	—	4.97
rebar package for 10" wide wall (\$.60 sf, 1.71 sf), 2% waste	—	sf	.35	—	.35
CEC rebar package for 10" wide wall (\$.60 sf, .55 sf), 4% waste	—	sf	1.15	—	1.15
insulating foam stay-in-place foundation wall forms (\$2.20 sf, 1 sf), 0% waste	—	sf	2.20	—	2.20
Pier concrete					
concrete (\$152.00 cy, 8.6 lf), 6% waste	—	lf	18.80	—	18.80
fiber tube with bracing (\$147.00 10' tube, 3.33 ea), 2% waste	—	ea	44.90	—	44.90
prefabricated rebar cage (\$.60 pound, .01 lf), 0% waste	—	lf	60.00	—	60.00

	Craft@Hrs	Unit	Material	Labor	Total
Grade beam concrete					
concrete for 12" by 18" ($152.00 cy, 18 lf), 6% waste	—	lf	8.93	—	8.93
concrete for 12" by 24" ($152.00 cy, 13.5 lf), 6% waste	—	lf	12.00	—	12.00
concrete for 14" by 18" ($152.00 cy, 15.43 lf), 6% waste	—	lf	10.40	—	10.40
concrete for 14" by 24" ($152.00 cy, 11.57 lf), 6% waste	—	lf	14.00	—	14.00
concrete for 16" by 18" ($152.00 cy, 13.5 lf), 6% waste	—	lf	12.00	—	12.00
concrete for 16" by 24" ($152.00 cy, 10.13 lf), 6% waste	—	lf	15.90	—	15.90
concrete for 18" by 18" ($152.00 cy, 12 lf), 6% waste	—	lf	13.50	—	13.50
concrete for 18" by 24" ($152.00 cy, 9 lf), 6% waste	—	lf	18.00	—	18.00
rebar cage ($.55 pound, .01 lf), 0% waste	—	lf	55.40	—	55.40
Lightweight concrete flatwork					
per cy ($193.00 cy, 1 cy), 6% waste	—	cy	203.00	—	203.00
2" slab ($193.00 cy, 162 sf), 6% waste	—	sf	1.25	—	1.25
4" slab ($193.00 cy, 81 sf), 6% waste	—	sf	2.52	—	2.52
6" slab ($193.00 cy, 54 sf), 6% waste	—	sf	3.78	—	3.78
Concrete flatwork					
per cy ($152.00 cy, 1 cy), 6% waste	—	cy	161.00	—	161.00
4" slab ($152.00 cy, 81 sf), 6% waste	—	sf	1.99	—	1.99
6" slab ($152.00 cy, 54 sf), 6% waste	—	sf	3.00	—	3.00
8" slab ($152.00 cy, 40.5 sf), 6% waste	—	sf	4.00	—	4.00
rebar #4, 24" on center ($.60 pound, .75 sf), 2% waste	—	sf	.81	—	.81
rebar #4, 12" on center ($.60 pound, .37 sf), 2% waste	—	sf	1.66	—	1.66
wire mesh (6" x 6" #10) ($191.00 roll, 750 sf), 2% waste	—	sf	.26	—	.26
aggregate for 2" slab base ($11.10 cy, 162 sf), 0% waste	—	sf	.06	—	.06
aggregate for 4" slab base ($11.10 cy, 81 sf), 0% waste	—	sf	.13	—	.13
sand for 2" slab base ($11.10 cy, 162 sf), 0% waste	—	sf	.06	—	.06
sand for 4" slab base ($11.10 cy, 81 sf), 0% waste	—	sf	.13	—	.13
Sidewalk					
3" sidewalk ($152.00 cy, 108 sf), 6% waste	—	sf	1.49	—	1.49
4" sidewalk ($152.00 cy, 81 sf), 6% waste	—	sf	1.99	—	1.99
6" sidewalk ($152.00 cy, 54 sf), 6% waste	—	sf	3.00	—	3.00

	Craft@Hrs	Unit	Material	Labor	Total
Rebar					
#3 (3/8" .376 pound per lf) ($5.50 20' bar, 20 lf), 4% waste	—	lf	.29	—	.29
#4 (1/2" .668 pounds per lf) ($7.01 20' bar, 20 lf), 4% waste	—	lf	.35	—	.35
#5 (5/8" 1.043 pounds per lf) ($10.90 20' bar, 20 lf), 4% waste	—	lf	.57	—	.57
#6 (3/4" 1.502 pounds per lf) ($16.50 20' bar, 20 lf), 4% waste	—	lf	.86	—	.86
#7 (7/8" 2.044 pounds per lf) ($22.50 20' bar, 20 lf), 4% waste	—	lf	1.18	—	1.18
#8 (1" 2.670 pounds per lf) ($29.30 20' bar, 20 lf), 4% waste	—	lf	1.52	—	1.52

Concrete Rental Equipment

	Craft@Hrs	Unit	Material	Labor	Equip.	Total
Forms						
plywood wall forms (per sf of form) ($.32 per day, .25 sf)	—	sf	—	—	1.29	1.29
steel curb & gutter forms ($1.58 per day, 1 lf)	—	lf	—	—	1.58	1.58
Tractor-mounted auger rental						
hourly charge ($368.00 per hour, 90 lf)	—	lf	—	—	4.10	4.10
delivery and take-home charge	—	ea	—	—	794.00	794.00
Concrete pump truck rental						
foundation wall ($232.00 hour, 450 sf of wall)	—	sf	—	—	.51	.51
footings ($232.00 hour, 112 lf)	—	lf	—	—	2.07	2.07
flatwork ($232.00 hour, 475 sf of flat)	—	sf	—	—	.49	.49
piers ($232.00 hour, 78 lf of pier)	—	lf	—	—	2.98	2.98
grade beam ($232.00 hour, 146 lf of beam)	—	lf	—	—	1.58	1.58
pressure grout ($232.00 hour, 4 lf 2" pipe)	—	sf	—	—	58.00	58.00
Dump truck rental						
5 cy, average aggregate base haul ($113.00 day, 540 sf)	—	sf	—	—	.22	.22
Jackhammer with compressor rental						
foundation wall demo ($353.00 day, 110 sf)	—	sf	—	—	3.20	3.20
footing demo ($353.00 day, 66 lf)	—	lf	—	—	5.35	5.35
flatwork demo ($353.00 day, 295 sf)	—	sf	—	—	1.20	1.20
grade beam demo ($353.00 day, 46 lf)	—	lf	—	—	7.66	7.66
jacket grade beam ($353.00 day, 42 lf)	—	lf	—	—	8.38	8.38
jacket pier ($353.00 day, 5.6 each)	—	ea	—	—	62.70	62.70
curb & gutter ($353.00 day, 145 lf)	—	lf	—	—	2.43	2.43

	Craft@Hrs	Unit	Material	Labor	Equip.	Total
Portable concrete saw rental						
wall sawing ($234.00 day, 78 1" x 1')	—	lf	—	—	3.01	3.01
floor sawing ($182.00 day, 160 1" x 1')	—	lf	—	—	1.14	1.14
Concrete core drill rental						
per day ($163.00 day, 125 1" x 1')	—	lf	—	—	1.31	1.31
Crane rental						
concrete pier removal ($1,080.00 day, 98 day)	—	lf	—	—	11.00	11.00
Pressure pot rental						
for pressurized epoxy injection ($319.00 day, 120 day)	—	lf	—	—	2.56	2.56
Backhoe rental						
per day ($849.00 day, 322 day)	—	cy	—	—	2.63	2.63

Concrete Labor

Laborer	base wage	paid leave	true wage	taxes & ins.	total
Concrete form installer	$35.40	2.76	$38.16	25.44	$63.60
Concrete finisher	$35.30	2.75	$38.05	25.35	$63.40
Concrete finisher's helper	$28.50	2.22	$30.72	21.68	$52.40
Equipment operator	$52.10	4.06	$56.16	32.74	$88.90
Plasterer	$37.30	2.91	$40.21	24.49	$64.70
Concrete saw operator	$35.50	2.77	$38.27	24.43	$62.70
Compaction grouter	$51.00	3.98	$54.98	33.82	$88.80
Laborer	$24.70	1.93	$26.63	19.67	$46.30
Demolition laborer	$25.90	2.02	$27.92	19.28	$47.20

Paid leave is calculated based on two weeks paid vacation, one week sick leave, and seven paid holidays. Employer's matching portion of **FICA** is 7.65 percent. **FUTA** (Federal Unemployment) is .8 percent. **Worker's compensation** was calculated using a national average of 18.08 percent for the concrete trade; 15.32 percent for the equipment operator, 15.32 percent for the concrete saw operator; 13.28 percent for the plastering trade; 18.08 percent for the compaction grouting specialist; and 14.26 percent for the demolition laborer. **Unemployment insurance** was calculated using a national average of 8 percent. **Health insurance** was calculated based on a projected national average for 2020 of $1,280 per employee (and family when applicable) per month. Employer pays 80 percent for a per month cost of $1,024 per employee. **Retirement** is based on a 401(k) retirement program with employer matching of 50 percent. Employee contributions to the 401(k) plan are an average of 6 percent of the true wage. **Liability insurance** is based on a national average of 12.0 percent.

Concrete Labor Productivity

	Craft@Hrs	Unit	Material	Labor	Total
Demolition					
remove footings including rebar	1D@1.64	cy	—	77.40	77.40
remove footings including CEC rebar	1D@1.96	cy	—	92.50	92.50
remove footings with rebar	1D@.121	lf	—	5.71	5.71
remove footings with CEC rebar	1D@.147	lf	—	6.94	6.94
remove foundation including rebar	1D@2.44	cy	—	115.00	115.00
remove foundation including CEC rebar	1D@2.94	cy	—	139.00	139.00
remove 5" or 6" foundation with rebar	1D@.073	sf	—	3.45	3.45
remove 5" or 6" foundation with CEC rebar	1D@.088	sf	—	4.15	4.15

	Craft@Hrs	Unit	Material	Labor	Total
remove 8" or 10" foundation with rebar	1D@.076	sf	—	3.59	3.59
remove 8" or 10" foundation with CEC rebar	1D@.092	sf	—	4.34	4.34
remove footing & foundation with rebar	1D@1.96	cy	—	92.50	92.50
remove footing & foundation with CEC rebar	1D@2.33	cy	—	110.00	110.00
remove footing & 3' foundation	1D@.346	lf	—	16.30	16.30
remove footing & 3' foundation with CEC rebar	1D@.418	lf	—	19.70	19.70
remove footing & 4' foundation	1D@.422	lf	—	19.90	19.90
remove footing & 4' foundation with CEC rebar	1D@.510	lf	—	24.10	24.10
remove pier per cy	1D@.699	cy	—	33.00	33.00
remove pier	1D@.082	lf	—	3.87	3.87
remove grade beam per cy	1D@2.33	cy	—	110.00	110.00
remove grade beam	1D@.174	lf	—	8.21	8.21
remove lightweight slab	1D@.026	sf	—	1.23	1.23
remove lightweight slab with rebar	1D@.029	sf	—	1.37	1.37
remove lightweight slab with wire mesh	1D@.031	sf	—	1.46	1.46
remove slab	1D@.027	sf	—	1.27	1.27
remove slab with rebar	1D@.030	sf	—	1.42	1.42
remove slab with wire mesh	1D@.033	sf	—	1.56	1.56
remove sidewalk	1D@.034	sf	—	1.60	1.60
remove curb & gutter	1D@.145	lf	—	6.84	6.84
remove step	1D@.047	lf	—	2.22	2.22
remove step landing	1D@.046	sf	—	2.17	2.17

Forming and rebar setting crew

form and set rebar	conc. form installer	$63.60
form and set rebar	laborer	$46.30
form and set rebar	forming crew	$55.00

Pouring and finishing crew

pour and finish concrete	conc. finisher	$63.40
pour and finish concrete	conc. finisher's helper	$52.40
pour and finish concrete	finishing crew	$57.90

	Craft@Hrs	Unit	Material	Labor	Total
Concrete footings					
form, set rebar, pour, & finish per cy	9F@2.44	cy	—	138.00	138.00
form, set CEC rebar, pour, & finish per cy	9F@2.94	cy	—	166.00	166.00
set forms for concrete footings	3F@.041	lf	—	2.26	2.26
strip forms from concrete footings & clean	3F@.009	lf	—	.50	.50
set rebar in forms for concrete footings	3F@.036	lf	—	1.98	1.98
set CEC rebar in forms for concrete footings	3F@.067	lf	—	3.69	3.69
pour and finish 16" by 10" footings	8F@.065	lf	—	3.76	3.76
pour and finish 20" by 10" footings	8F@.074	lf	—	4.28	4.28
pour and finish 24" by 12" footings	8F@.081	lf	—	4.69	4.69
pour and finish 32" by 14" footings	8F@.091	lf	—	5.27	5.27

	Craft@Hrs	Unit	Material	Labor	Total
Concrete foundation wall					
form, set rebar, pour, & finish per cy	9F@2.70	cy	—	153.00	153.00
form, set CEC rebar, pour, & finish per cy	9F@3.13	cy	—	177.00	177.00
set forms	3F@.029	sf	—	1.60	1.60
strip forms & clean	3F@.006	sf	—	.33	.33
set rebar package	3F@.037	sf	—	2.04	2.04
set CEC rebar package	3F@.067	sf	—	3.69	3.69
pour and finish 5" wide wall	8F@.024	sf	—	1.39	1.39
pour and finish 6" wide wall	8F@.029	sf	—	1.68	1.68
pour and finish 8" wide wall	8F@.039	sf	—	2.26	2.26
pour and finish 10" wide wall	8F@.049	sf	—	2.84	2.84
Tractor-mounted auger (hole drilling for pier)					
delivery and take home charges (mobilization)	10@6.25	ea	—	556.00	556.00
drill pier hole	10@.083	lf	—	7.38	7.38
Concrete pier					
form, set rebar cage, pour and finish per cy	9F@14.9	cy	—	842.00	842.00
set and brace fiber tube cap	3F@.125	lf	—	6.88	6.88
pour concrete pier	8F@1.54	lf	—	89.20	89.20
set and brace fiber tube for above-grade pier	3F@.183	lf	—	10.10	10.10
strip fiber tube	3F@.056	lf	—	3.08	3.08
pour above-grade concrete pier	8F@.185	lf	—	10.70	10.70
set rebar cage for concrete pier	3F@.091	lf	—	5.01	5.01
Jacket pier					
jackhammer for jacket, drill for new rebar	1D@2.32	ea	—	110.00	110.00
set forms	3F@1.09	ea	—	60.00	60.00
set rebar, secure with epoxy	3F@.833	ea	—	45.80	45.80
pour concrete jacket	8F@1.30	ea	—	75.30	75.30
Grade beam					
form, set rebar cage, pour, and finish per cy	9F@4.00	cy	—	226.00	226.00
set forms	3F@.127	lf	—	6.99	6.99
set rebar cage	3F@.067	lf	—	3.69	3.69
pour and finish 12" by 18"	8F@.073	lf	—	4.23	4.23
pour and finish 12" by 24"	8F@.088	lf	—	5.10	5.10
pour and finish 14" by 18"	8F@.118	lf	—	6.83	6.83
pour and finish 14" by 24"	8F@.235	lf	—	13.60	13.60
pour and finish 16" by 18"	8F@.353	lf	—	20.40	20.40
pour and finish 16" by 24"	8F@.470	lf	—	27.20	27.20
pour and finish 18" by 18"	8F@.118	lf	—	6.83	6.83
pour and finish 18" by 24"	8F@.235	lf	—	13.60	13.60
Jacket grade beam					
jackhammer for jacket	1D@.088	lf	—	4.15	4.15
set forms	3F@.150	lf	—	8.25	8.25
set rebar, secure with epoxy	3F@.083	lf	—	4.57	4.57
pour concrete	8F@.240	lf	—	13.90	13.90

	Craft@Hrs	Unit	Material	Labor	Total
Lightweight concrete flatwork					
form, pour, and finish per cy	9F@2.04	cy	—	115.00	115.00
pour and finish 2" slab	8F@.012	sf	—	.69	.69
pour and finish 4" slab	8F@.013	sf	—	.75	.75
pour and finish 6" slab	8F@.015	sf	—	.87	.87
Concrete flatwork					
form, pour, and finish per cy	9F@2.04	cy	—	115.00	115.00
set forms for concrete slab	3F@.011	sf	—	.61	.61
set rebar 24" on center	3F@.036	sf	—	1.98	1.98
set rebar 12" on center	3F@.080	sf	—	4.40	4.40
set wire mesh	3F@.015	sf	—	.83	.83
pour and finish 4" slab	8F@.024	sf	—	1.39	1.39
pour and finish 6" slab	8F@.036	sf	—	2.08	2.08
pour and finish 8" slab	8F@.036	sf	—	2.08	2.08
haul aggregate for slab base to job site	10@.002	sf	—	.18	.18
spread 2" aggregate slab base	2F@.013	sf	—	.60	.60
spread 4" aggregate slab base	2F@.015	sf	—	.69	.69
haul sand for slab base to job site	10@.001	sf	—	.09	.09
spread 2" sand slab base	2F@.012	sf	—	.56	.56
spread 4" sand slab base	2F@.014	sf	—	.65	.65
Sidewalk flatwork					
form, pour, and finish per cy	9F@4.17	cy	—	236.00	236.00
set forms	3F@.023	sf	—	1.27	1.27
pour and finish 3" sidewalk	8F@.019	sf	—	1.10	1.10
pour and finish 4" sidewalk	8F@.026	sf	—	1.51	1.51
pour and finish 6" sidewalk	8F@.039	sf	—	2.26	2.26
Concrete dye					
trowel into wet concrete	8F@.009	sf	—	.52	.52
Stamp concrete					
slab or sidewalk	8F@.053	sf	—	3.07	3.07
with paver pattern	8F@.045	sf	—	2.61	2.61
grout stamped paver pattern joints	8F@.023	sf	—	1.33	1.33
Exposed aggregate					
Finish with exposed aggregate finish	8F@.019	sf	—	1.10	1.10
Curb & gutter work					
set forms	3F@.123	lf	—	6.77	6.77
pour and finish	8F@.150	lf	—	8.69	8.69
Concrete step					
form, set rebar, pour, and finish per cy	9F@7.69	cy	—	434.00	434.00
set forms and rebar per lf of step	3F@.133	lf	—	7.32	7.32
pour and finish per lf of step	8F@.100	lf	—	5.79	5.79
set forms and rebar for landing per cubic foot	3F@.011	sf	—	.61	.61
pour and finish landing per cubic foot	8F@.056	sf	—	3.24	3.24

	Craft@Hrs	Unit	Material	Labor	Total
Saw concrete					
wall with rebar per lf, 1" deep	20@.111	lf	—	6.96	6.96
wall with CEC rebar per lf, 1" deep	20@.125	lf	—	7.84	7.84
floor per lf, 1" deep	20@.050	lf	—	3.14	3.14
floor with rebar per lf, 1" deep	20@.054	lf	—	3.39	3.39
floor with wire mesh per lf, 1" deep	20@.052	lf	—	3.26	3.26
expansion joint in "green" slab	20@.189	lf	—	11.90	11.90
Core drill concrete					
2" wall, per inch of depth	20@.066	li	—	4.14	4.14
4" wall, per inch of depth	20@.073	li	—	4.58	4.58
2" floor, per inch of depth	20@.064	li	—	4.01	4.01
4" floor, per inch of depth	20@.067	li	—	4.20	4.20
Set rebar					
#3 (3/8")	3F@.014	lf	—	.77	.77
#4 (1/2")	3F@.014	lf	—	.77	.77
#5 (5/8")	3F@.014	lf	—	.77	.77
#6 (3/4")	3F@.014	lf	—	.77	.77
#7 (7/8")	3F@.015	lf	—	.83	.83
#8 (1")	3F@.015	lf	—	.83	.83
Buttress foundation					
machine excavate to 12" to 24" of ext. foundation	10@.027	sf	—	2.40	2.40
hand excavate exterior of foundation	2F@.161	sf	—	7.45	7.45
backfill buttressed foundation	2F@.048	sf	—	2.22	2.22
set forms for exterior buttress	3F@.021	sf	—	1.16	1.16
set rebar for exterior buttress	3F@.017	sf	—	.94	.94
pour and finish exterior buttress	8F@.027	sf	—	1.56	1.56
hand excavate interior crawl space for buttress	2F@.457	sf	—	21.20	21.20
set forms for interior buttress in crawl space	3F@.032	sf	—	1.76	1.76
set rebar for interior buttress in crawl space	3F@.023	sf	—	1.27	1.27
pour and finish interior buttress in crawl space	8F@.037	sf	—	2.14	2.14
hand excavate interior foundation in basement	2F@.204	sf	—	9.45	9.45
set forms for interior buttress in basement	3F@.025	sf	—	1.38	1.38
set rebar for interior buttress in basement	3F@.019	sf	—	1.05	1.05
pour and finish interior buttress in basement	8F@.032	sf	—	1.85	1.85

Compaction grouting crew

core drill, insert pipes, pump grout	compaction grouting specialist	$88.80
core drill, insert pipes, pump grout	laborer	$46.30
compaction grouting crew	compaction grouting crew	$67.60

	Craft@Hrs	Unit	Material	Labor	Total
Foundation & footing stabilization with pressurized grout					
2" core drill through footings	2G@.379	ea	—	25.60	25.60
insert 2" pipes into ground beneath footings	2G@.143	lf	—	9.67	9.67
pump pressurized grout through pipes	2G@.250	lf	—	16.90	16.90

	Craft@Hrs	Unit	Material	Labor	Equip.	Total

Demolition & Hauling

Minimum charge.

	Craft@Hrs	Unit	Material	Labor	Equip.	Total
Demolition work	1D@2.00	ea	—	94.40	154.00	248.40

Dump fee. Typical metro area landfill fees. Will be up to 60% higher in major metro areas and 30% to 60% lower in rural areas.

	Craft@Hrs	Unit	Material	Labor	Equip.	Total
Charge per cubic yard	—	cy	61.70	—	—	61.70
Charge per ton	—	tn	117.00	—	—	117.00

Dumpster. Includes up to two dumps per week.

	Craft@Hrs	Unit	Material	Labor	Equip.	Total
3 cy, rental per week	—	ea	—	—	170.00	170.00
5 cy, rental per week	—	ea	—	—	292.00	292.00
10 cy, rental per week	—	ea	—	—	370.00	370.00
30 cy, rental per week	—	ea	—	—	540.00	540.00
Minimum dumpster charge	—	ea	—	—	232.00	232.00

Refuse chute.

	Craft@Hrs	Unit	Material	Labor	Equip.	Total
Plywood	1D@1.63	lf	17.80	76.90	—	94.70
Minimum charge for plywood refuse chute	1D@3.67	ea	93.50	173.00	—	266.50
18" prefabricated circular steel	1D@1.18	lf	20.10	55.70	—	75.80
36" prefabricated circular steel	1D@1.60	lf	36.30	75.50	—	111.80
Minimum charge for circular steel refuse chute	1D@3.30	ea	106.00	156.00	—	262.00

Debris hauling.

	Craft@Hrs	Unit	Material	Labor	Equip.	Total
By trailer or dump truck						
per cubic yard	—	cy	—	—	33.40	33.40
per ton	—	tn	—	—	90.90	90.90
By pick-up truck						
per pick-up truck load	—	ea	—	—	34.20	34.20

Strip room. Remove all wall and floor coverings, doors, other trim, and underlayment. Stud walls and plywood subfloor remain. For rooms with contents add **4%**.

	Craft@Hrs	Unit	Material	Labor	Equip.	Total
Stripped to bare walls and subfloor, per square foot						
typical room	1D@.074	sf	—	3.49	—	3.49
bathroom	1D@.150	sf	—	7.08	—	7.08
kitchen	1D@.115	sf	—	5.43	—	5.43
utility room	1D@.086	sf	—	4.06	—	4.06
laundry room	1D@.095	sf	—	4.48	—	4.48

	Craft@Hrs	Unit	Material	Labor	Equip.	Total
Time & Material Charts (selected items)						
Demolition & Hauling Materials						
Refuse chute						
plywood	—	lf	17.70	—	—	17.70
18" prefabricated circular steel	—	lf	20.30	—	—	20.30
36" prefabricated circular steel	—	lf	36.80	—	—	36.80
Demolition & Hauling Rental Equipment						
Dumpster rental per week						
3 cy ($160.00 week)	—	ea	—	—	167.00	167.00
5 cy ($302.00 week)	—	ea	—	—	315.00	315.00
10 cy ($383.00 week)	—	ea	—	—	399.00	399.00
30 cy ($560.00 week)	—	ea	—	—	584.00	584.00
Truck rental with driver						
3 cy dump truck ($762.00 day, 22 cy)	—	cy	—	—	36.20	36.20
($762.00 day, 8.08 ton)	—	ton	—	—	98.30	98.30
pick-up truck ($320.00 day, 9 load)	—	ea	—	—	37.00	37.00

Demolition & Hauling Labor

Laborer	base wage	paid leave	true wage	taxes & ins.	total
Demolition laborer	$25.90	2.02	$27.92	19.28	$47.20

Paid leave is calculated based on two weeks paid vacation, one week sick leave, and seven paid holidays. Employer's matching portion of **FICA** is 7.65 percent. **FUTA** (Federal Unemployment) is .8 percent. **Worker's compensation** for Demolition was calculated using a national average of 14.26 percent. **Unemployment insurance** was calculated using a national average of 8 percent. **Health insurance** was calculated based on a projected national average for 2020 of $1,280 per employee (and family when applicable) per month. Employer pays 80 percent for a per month cost of $1,024 per employee. **Retirement** is based on a 401(k) retirement program with employer matching of 50 percent. Employee contributions to the 401(k) plan are an average of 6 percent of the true wage. **Liability insurance** is based on a national average of 12.0 percent.

	Craft@Hrs	Unit	Material	Labor	Equip.	Total
Demolition & Hauling Labor Productivity						
Build / install refuse chute						
plywood	1D@1.63	lf	—	76.90	—	76.90
18" prefabricated circular steel	1D@1.18	lf	—	55.70	—	55.70
36" prefabricated circular steel	1D@1.60	lf	—	75.50	—	75.50
Strip room to bare walls and subfloor						
typical room	1D@.074	sf	—	3.49	—	3.49
bathroom	1D@.150	sf	—	7.08	—	7.08
kitchen	1D@.115	sf	—	5.43	—	5.43
utility	1D@.086	sf	—	4.06	—	4.06
laundry	1D@.095	sf	—	4.48	—	4.48

Doors

Using Door Items. When replacing a door, select the door type, then the jamb type, and add the two together. Door prices are for the door slab only and include routing and drilling for standard hardware and hinges. Door tear-out is for the door slab only. Use this tear-out price when removing the slab and leaving the jamb and casing intact. Jamb and casing prices include the cost to prehang the door in the jamb and the cost of hinges (but not the door lockset). Tear-out includes removal of the entire door. Use this tear-out price when removing the door, jamb and casing.

	Craft@Hrs	Unit	Material	Labor	Total
Minimum charge.					
for door work	1C@2.50	ea	50.80	169.00	219.80

Folding Door. Folding doors usually used in closets. Includes folding door section with hardware (including track, pins, hinges and a basic pull as needed). Priced per section of folding door. Does not include jamb or casing. Folding doors are often called bi-fold doors.

	Craft@Hrs	Unit	Material	Labor	Total
Folding door (per section)					
replace, hardboard smooth or wood-textured	1C@.394	ea	42.20	26.60	68.80
replace, hardboard wood-textured and embossed	1C@.394	ea	48.30	26.60	74.90
replace, mahogany (lauan) or birch veneer	1C@.394	ea	45.00	26.60	71.60
replace, ash or oak veneer	1C@.394	ea	49.20	26.60	75.80
replace, walnut or cherry veneer	1C@.394	ea	56.40	26.60	83.00
replace, paint-grade pine panel	1C@.394	ea	73.50	26.60	100.10
replace, stain-grade pine panel	1C@.394	ea	90.50	26.60	117.10
replace, paint-grade pine full-louvered	1C@.394	ea	86.60	26.60	113.20
replace, stain-grade pine full-louvered	1C@.394	ea	101.00	26.60	127.60
replace, paint-grade pine half-louvered	1C@.394	ea	82.30	26.60	108.90
replace, stain-grade pine half-louvered	1C@.394	ea	95.60	26.60	122.20
replace, red oak folding panel	1C@.394	ea	110.00	26.60	136.60
replace, folding mirrored	1C@.394	ea	127.00	26.60	153.60
remove, folding door	1D@.114	ea	—	5.38	5.38
remove, folding doors for work, reinstall	1C@.148	ea	—	10.00	10.00

standard *panel* *full-louver* *half-louver*

Bypassing Door. Usually used in closets. Priced per door section. Includes track and hardware. For jamb & casing see below.

	Craft@Hrs	Unit	Material	Labor	Total
Bypassing door (per section)					
replace, hardboard smooth or wood textured	1C@.755	ea	83.10	51.00	134.10
replace, hardboard wood-textured and embossed	1C@.755	ea	93.30	51.00	144.30
replace, mahogany (lauan) or birch veneer	1C@.755	ea	90.50	51.00	141.50
replace, ash or oak veneer	1C@.755	ea	94.20	51.00	145.20
replace, walnut or cherry veneer	1C@.755	ea	117.00	51.00	168.00
replace, paint-grade pine panel	1C@.755	ea	143.00	51.00	194.00
replace, stain-grade pine panel	1C@.755	ea	175.00	51.00	226.00
replace, red oak panel door	1C@.755	ea	223.00	51.00	274.00
replace, paint-grade pine full-louvered	1C@.755	ea	167.00	51.00	218.00
replace, stain-grade pine full-louvered	1C@.755	ea	198.00	51.00	249.00

standard *panel* *full-louver*

	Craft@Hrs	Unit	Material	Labor	Total
replace, paint-grade pine half-louvered	1C@.755	ea	159.00	51.00	210.00
replace, stain-grade pine half-louvered	1C@.755	ea	189.00	51.00	240.00
replace, mirrored	1C@.755	ea	245.00	51.00	296.00
remove, bypassing door	1D@.139	ea	—	6.56	6.56
remove, bypassing sliding doors for work, then reinstall	1C@.220	ea	—	14.90	14.90

half-louver

Jamb & casing for closet door. Includes jamb, casing, finish nails, and installation in opening for a folding or bypassing door set. Includes casing strip to cover door track. Does not include door hardware including the track. Per lf around opening.

	Craft@Hrs	Unit	Material	Labor	Total
Jamb & casing for folding or bypassing door opening					
replace, paint grade	1C@.109	lf	1.48	7.37	8.85
replace, stain grade	1C@.109	lf	1.64	7.37	9.01
replace, birch	1C@.109	lf	1.59	7.37	8.96
replace, mahogany	1C@.109	lf	2.29	7.37	9.66
replace, ash or oak	1C@.109	lf	2.47	7.37	9.84
replace, walnut or cherry	1C@.109	lf	3.34	7.37	10.71
remove, jamb & casing for closet door	1D@.026	lf	—	1.23	1.23

Interior door. Price per door. Includes hollow-core interior door with hinges and installation in an existing jamb. Pre-drilled and routed for hardware and hinges. Does not include jamb, casing, or door hardware.

	Craft@Hrs	Unit	Material	Labor	Total
replace, hardboard smooth or wood-textured	1C@.215	ea	112.00	14.50	126.50
replace, hardboard wood-textured and embossed	1C@.215	ea	156.00	14.50	170.50
replace, mahogany or birch veneer	1C@.215	ea	125.00	14.50	139.50
replace, ash or oak veneer	1C@.215	ea	134.00	14.50	148.50
replace, walnut or cherry veneer	1C@.215	ea	178.00	14.50	192.50
replace, mirrored	1C@.215	ea	292.00	14.50	306.50
replace, plastic laminate face	1C@.215	ea	450.00	14.50	464.50
remove, interior door	1D@.217	ea	—	10.20	10.20
add for solid-core, 1 hr. fire-rated door	—	ea	159.00	—	159.00

French door. Includes interior French door with hinges and installation in an existing jamb. Pre-drilled and routed for hardware and hinges. Does not include jamb, casing, or door hardware.

	Craft@Hrs	Unit	Material	Labor	Total
replace, paint-grade wood	1C@.215	ea	292.00	14.50	306.50
replace, stain-grade fir	1C@.215	ea	349.00	14.50	363.50
replace, metal	1C@.215	ea	301.00	14.50	315.50
replace, full-lite door with simulated French grid	1C@.215	ea	261.00	14.50	275.50
remove, French door	1D@.217	ea	—	10.20	10.20
add for double-glazed insulated glass	—	ea	114.00	—	114.00
add for fire-rated glass	—	ea	130.00	—	130.00

Full-louvered door. Includes full-louvered door with hinges and installation in an existing jamb. Pre-drilled and routed for hardware and hinges. Does not include jamb, casing, or door hardware.

	Craft@Hrs	Unit	Material	Labor	Total
replace, paint-grade pine	1C@.215	ea	245.00	14.50	259.50
replace, stain-grade pine	1C@.215	ea	301.00	14.50	315.50
remove	1D@.217	ea	—	10.20	10.20

Half-louvered door. Includes half-louvered door with hinges and installation in an existing jamb. Pre-drilled and routed for hardware and hinges. Does not include jamb, casing, or door hardware.

	Craft@Hrs	Unit	Material	Labor	Total
replace, paint-grade pine	1C@.215	ea	221.00	14.50	235.50
replace, stain-grade pine	1C@.215	ea	291.00	14.50	305.50
remove	1D@.217	ea	—	10.20	10.20

Remove interior door for work and reinstall.

	Craft@Hrs	Unit	Material	Labor	Total
door slab	1C@.151	ea	—	10.20	10.20
door, jamb & casing	1C@1.63	ea	—	110.00	110.00
double door slabs	1C@.293	ea	—	19.80	19.80
double doors, jamb & casing for work	1C@2.67	ea	—	180.00	180.00

Jamb & casing for interior door. Price per door. Includes jamb and casing for interior door, shims, finish nails, and installation. Includes hanging door in jamb, installing jamb in rough opening, and casing jamb. Does not include door, hinges, or door hardware.

	Craft@Hrs	Unit	Material	Labor	Total
replace, paint-grade pine	1C@1.43	ea	26.50	96.70	123.20
replace, paint-grade pine for double door	1C@1.43	ea	49.50	96.70	146.20
replace, stain-grade pine	1C@1.43	ea	29.70	96.70	126.40
replace, stain-grade pine for double door	1C@1.43	ea	55.50	96.70	152.20
replace, birch	1C@1.43	ea	32.70	96.70	129.40
replace, birch for double door	1C@1.43	ea	60.10	96.70	156.80
replace, mahogany	1C@1.43	ea	39.10	96.70	135.80
replace, mahogany veneer	1C@1.43	ea	36.60	96.70	133.30
replace, mahogany for double door	1C@1.43	ea	75.40	96.70	172.10
replace, mahogany veneer for double door	1C@1.43	ea	74.40	96.70	171.10
replace, ash or oak	1C@1.43	ea	47.20	96.70	143.90
replace, ash or oak veneer	1C@1.43	ea	42.60	96.70	139.30
replace, ash or oak for double door	1C@1.43	ea	84.20	96.70	180.90
replace, ash or oak veneer for double door	1C@1.43	ea	80.80	96.70	177.50
replace, walnut or cherry	1C@1.43	ea	68.40	96.70	165.10
replace, walnut or cherry veneer	1C@1.43	ea	64.80	96.70	161.50
replace, walnut or cherry for double door	1C@1.43	ea	127.00	96.70	223.70
replace, walnut or cherry veneer for double door	1C@1.43	ea	120.00	96.70	216.70
replace, steel	1C@1.43	ea	181.00	96.70	277.70
replace, steel for double door	1C@1.43	ea	349.00	96.70	445.70
remove, jamb & casing for interior door	1D@.388	ea	—	18.30	18.30
remove, jamb & casing for steel double interior door	1D@.484	ea	—	22.80	22.80
add for installation for pocket door	1C@.322	ea	22.70	21.80	44.50
add for installation for double pocket door	1C@.871	ea	38.00	58.90	96.90

Rough framing for pocket door. Includes track, hardware, and side framing kit. Antique systems include framing for walls on each side of the pocket door.

	Craft@Hrs	Unit	Material	Labor	Total
replace, in 2" x 4" wall	1C@1.50	ea	37.10	101.00	138.10
replace, for double doors in 2" x 4" wall	1C@2.88	ea	59.50	195.00	254.50
replace, in 2" x 6" wall	1C@1.52	ea	40.20	103.00	143.20
replace, for double doors in 2" x 6" wall	1C@2.95	ea	71.40	199.00	270.40
replace, in 2" x 8" wall	1C@1.54	ea	43.30	104.00	147.30
replace, for double doors in 2" x 8" wall	1C@3.03	ea	76.40	205.00	281.40

	Craft@Hrs	Unit	Material	Labor	Total
replace, for antique pocket door (2" x 4" walls on both sides)	1C@2.45	ea	130.00	166.00	296.00
replace, for antique double pocket doors (2" x 4" walls on both sides)	1C@5.12	ea	234.00	346.00	580.00

Pocket door track.

	Craft@Hrs	Unit	Material	Labor	Total
replace, overhead for antique door system	1C@1.30	ea	76.40	87.90	164.30
replace, overhead for antique double door system	1C@2.34	ea	104.00	158.00	262.00
replace, clean & realign overhead double door track	1C@1.96	ea	—	132.00	132.00
replace, floor for antique door system	1C@1.21	ea	84.20	81.80	166.00
replace, floor for antique double door system	1C@2.20	ea	114.00	149.00	263.00
clean and realign floor double door track	1C@2.64	ea	—	178.00	178.00
remove, pocket door track	1D@.751	ea	—	35.40	35.40
remove, double pocket door system	1D@1.28	ea	—	60.40	60.40

Remove pocket door for work and reinstall.

	Craft@Hrs	Unit	Material	Labor	Total
door and trim	1C@1.15	ea	—	77.70	77.70
door, jamb & casing	1C@3.23	ea	—	218.00	218.00
double doors & trim	1C@1.58	ea	—	107.00	107.00
double doors, jamb & casing	1C@3.39	ea	—	229.00	229.00

Panel doors. Includes paint-grade panel door with hinges and installation in an existing jamb. Pre-drilled and routed for hardware and hinges. Does not include jamb, casing, or door hardware. Door contains from 6 to 9 panels. Each wood species includes line items for a new door, a reconditioned door, and a door that is custom milled to match a specific pattern. Reconditioned Door: Reconditioned panel doors have been removed from a historical structure and reconditioned by an architectural salvage company. These doors will have some scratches, dents, patching and may have been stripped of paint. "Graining" is a method of painting wood to make it look like the grain of another wood. This process was used extensively in the 19th century, most often to paint pine to appear like quartersawn oak. Custom-Milled Door: Custom-milled doors include the additional costs for panel doors that are milled to match an existing door pattern. For these doors it is necessary to cut custom shaper knives for the panels and custom molder (or shaper) knives for the stiles and rails. Add **60%** if milling two or fewer doors.

	Craft@Hrs	Unit	Material	Labor	Total
Paint grade doors					
replace	1C@.178	ea	745.00	12.00	757.00
add for panel door custom milled to match	—	%	67.0	—	—
replace, reconditioned antique	1C@.178	ea	621.00	12.00	633.00
replace, reconditioned grained antique	1C@.178	ea	905.00	12.00	917.00
Stain-grade fir doors					
replace	1C@.178	ea	1,020.00	12.00	1,032.00
add for panel door custom milled to match	—	%	35.0	—	—
replace, reconditioned antique	1C@.178	ea	936.00	12.00	948.00
Redwood doors					
replace	1C@.178	ea	1,100.00	12.00	1,112.00
add for panel door custom milled to match	—	%	26.0	—	—
replace, reconditioned antique	1C@.178	ea	1,060.00	12.00	1,072.00
Cypress doors					
replace	1C@.178	ea	1,040.00	12.00	1,052.00
add for panel door custom milled to match	—	%	17.0	—	—
replace, reconditioned antique	1C@.178	ea	976.00	12.00	988.00

	Craft@Hrs	Unit	Material	Labor	Total
Oak doors					
replace	1C@.178	ea	933.00	12.00	945.00
add for panel door custom milled to match	—	%	35.0	—	—
replace, reconditioned painted antique	1C@.178	ea	693.00	12.00	705.00
replace, reconditioned stained antique	1C@.178	ea	803.00	12.00	815.00
Mahogany doors					
replace	1C@.178	ea	1,080.00	12.00	1,092.00
add for panel door custom milled to match	—	%	29.0	—	—
replace, reconditioned antique door	1C@.178	ea	994.00	12.00	1,006.00
Walnut doors					
replace	1C@.178	ea	1,620.00	12.00	1,632.00
add for panel door custom milled to match	—	%	17.0	—	—
replace, reconditioned antique	1C@.178	ea	1,420.00	12.00	1,432.00
Remove panel door	1D@.217	ea	—	10.20	10.20

Additional panel door costs.

	Craft@Hrs	Unit	Material	Labor	Total
add for tempered glass lite	—	ea	366.00	—	366.00
add for beveled or edged glass lite	—	ea	686.00	—	686.00
add for leaded or colored glass lite	—	ea	796.00	—	796.00
add for panel door with round or elliptical top	1C@4.55	ea	933.00	308.00	1,241.00

Remove panel door for work and reinstall.

	Craft@Hrs	Unit	Material	Labor	Total
slab only	1C@.152	ea	—	10.30	10.30
door, jamb, & casing	1C@1.63	ea	—	110.00	110.00

Panel door repair. Not including custom knife charges.

	Craft@Hrs	Unit	Material	Labor	Total
Antique panel door repair					
replace door panel	1C@2.74	ea	156.00	185.00	341.00
replace door stile	1C@2.49	ea	114.00	168.00	282.00
Panel door minimum charges					
repair door with no custom milling	1C@1.50	ea	181.00	101.00	282.00
repair door when custom milling is required	1C@2.50	ea	621.00	169.00	790.00
Custom milling					
shaper or molder knife charge	—	ea	492.00	—	492.00
shaper or molder setup charge	—	ea	233.00	—	233.00
minimum shaper or molder charge	—	ea	1,030.00	—	1,030.00
Antique door repair					
repair cracks & dry rot in door with epoxy, tighten stiles	1C@4.49	ea	224.00	304.00	528.00
repair cracks & dry rot in jamb with epoxy	1C@3.40	ea	130.00	230.00	360.00
repair cracks & dry rot in casing with epoxy	1C@3.40	ea	101.00	230.00	331.00
remove casing, re-plumb door, & reinstall casing	1C@2.95	ea	—	199.00	199.00
add for antique panel door transom	—	%	38.0	—	—

	Craft@Hrs	Unit	Material	Labor	Total

Batten door. Built from vertical boards, usually tongue-and-groove, held in place by horizontal battens and diagonal braces. Standard grade, sometimes called ledged-and-braced doors, have boards held in place by diagonally braced horizontal battens. High grade have tongue-and-groove boards held in place by a stile frame and a single horizontal batten. The frame is joined with mortise and tenon joints, and diagonal braces. These are sometimes called framed ledged-and-braced doors.

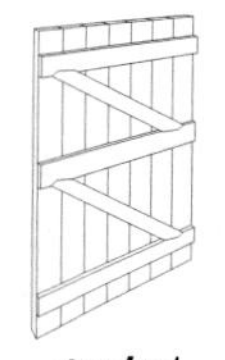
standard

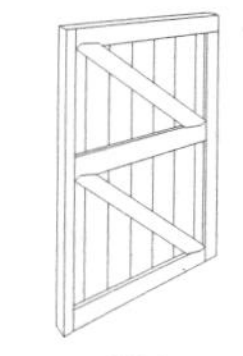
high

Batten door. Includes batten style door with hinges and installation in an existing jamb. Pre-drilled and routed for hardware and hinges. Does not include jamb, casing, or door hardware.

	Craft@Hrs	Unit	Material	Labor	Total
replace, paint-grade, standard grade	1C@.178	ea	582.00	12.00	594.00
replace, paint grade, high grade	1C@.178	ea	842.00	12.00	854.00
replace, stain-grade, standard grade	1C@.178	ea	1,060.00	12.00	1,072.00
replace, stain-grade, high grade	1C@.178	ea	1,480.00	12.00	1,492.00
remove	1D@.217	ea	—	10.20	10.20

Storm/screen door. Price per door. Includes storm/screen door set including jamb and hinges with fasteners. Complete unit is installed in jamb of existing door. Does not include furring or other modification to existing jamb.

	Craft@Hrs	Unit	Material	Labor	Total
replace, metal, economy grade	1C@1.21	ea	195.00	81.80	276.80
replace, metal, standard grade	1C@1.21	ea	291.00	81.80	372.80
replace, metal, high grade	1C@1.21	ea	420.00	81.80	501.80
replace, metal, deluxe grade	1C@1.21	ea	532.00	81.80	613.80
replace, wood, economy grade	1C@1.21	ea	223.00	81.80	304.80
replace, wood, standard grade	1C@1.21	ea	362.00	81.80	443.80
replace, wood, high grade	1C@1.21	ea	428.00	81.80	509.80
replace, wood, deluxe grade	1C@1.21	ea	532.00	81.80	613.80

economy *metal* *high* *deluxe*

	Craft@Hrs	Unit	Material	Labor	Total
replace, Victorian style wood, standard grade	1C@1.21	ea	582.00	81.80	663.80
replace, Victorian style wood, high grade	1C@1.21	ea	693.00	81.80	774.80
replace, Victorian style wood, deluxe grade	1C@1.21	ea	976.00	81.80	1,057.80
remove, storm door	1D@.227	ea	—	10.70	10.70
remove for work, then reinstall	1C@1.70	ea	—	115.00	115.00

standard *standard* *high* *deluxe*

	Craft@Hrs	Unit	Material	Labor	Total

Exterior French door. Includes exterior French door with hinges and installation in an existing jamb. Pre-drilled and routed for hardware and hinges. Does not include jamb, casing, or door hardware.

	Craft@Hrs	Unit	Material	Labor	Total
replace, paint-grade pine	1C@.212	ea	582.00	14.30	596.30
replace, stain-grade fir	1C@.212	ea	664.00	14.30	678.30
replace, metal	1C@.212	ea	646.00	14.30	660.30
replace, full-lite insulated with simulated grid	1C@.212	ea	555.00	14.30	569.30
remove, exterior French door	1D@.283	ea	—	13.40	13.40

Exterior veneer door. Price per door. Includes exterior veneer door with hinges and installation in an existing jamb. Pre-drilled and routed for hardware and hinges. Does not include jamb, casing, or door hardware.

	Craft@Hrs	Unit	Material	Labor	Total
replace, mahogany (lauan) or birch veneer	1C@.212	ea	274.00	14.30	288.30
replace, ash or red oak veneer	1C@.212	ea	301.00	14.30	315.30
replace, walnut or cherry veneer	1C@.212	ea	334.00	14.30	348.30
replace, insulated steel	1C@.212	ea	274.00	14.30	288.30
remove, exterior veneer door	1D@.283	ea	—	13.40	13.40

Additional exterior door costs.

	Craft@Hrs	Unit	Material	Labor	Total
add for full-lite	—	ea	354.00	—	354.00
add for leaded or colored glass full-lite	—	ea	789.00	—	789.00
add for half-lite	—	ea	152.00	—	152.00
add for leaded or colored glass half-lite	—	ea	441.00	—	441.00
add for 1' by 1' lite	—	ea	47.00	—	47.00

Bottom: *flush*, *raised panel*, *crossbuck*, *shelf*

Top: *full lite*, *four lite*, *six lite*, *nine lite*

Dutch door. Includes Dutch-style door with hinges and installation in an existing jamb. Pre-drilled and routed for hardware and hinges. Includes only one of the two door segments. Does not include jamb, casing, or door hardware.

	Craft@Hrs	Unit	Material	Labor	Total
replace, bottom section, flush	1C@.325	ea	273.00	22.00	295.00
replace, bottom section, raised panel	1C@.325	ea	409.00	22.00	431.00
replace, bottom section, cross buck	1C@.325	ea	528.00	22.00	550.00
replace, add for shelf on bottom section	—	ea	85.80	—	85.80
replace, top section with full-lite	1C@.325	ea	362.00	22.00	384.00
replace, top section with four lites	1C@.325	ea	435.00	22.00	457.00
replace, top section with six lites	1C@.325	ea	463.00	22.00	485.00
replace, top section with nine lites	1C@.325	ea	499.00	22.00	521.00
remove door, jamb, & casing for work, then reinstall	1C@2.74	ea	—	185.00	185.00
remove door slabs for work, then reinstall	1C@.393	ea	—	26.60	26.60
remove, Dutch door	1D@.321	ea	—	15.20	15.20

	Craft@Hrs	Unit	Material	Labor	Total

Entry doors. All entry doors are solid wood with occasional veneered wood components in the lower grades. Entry door prices do not include jamb and casing. See below for exterior jambs and quality indicators.

Entry door grades. Standard grade includes multiple panel doors (nine and more) with some embossed components. May have lites in place of panels. High grade is the same as Standard grade with some embossed and carved components. May also may have carved onlays. Deluxe grade includes door components that have moderate carvings. May have exotic wood components. Custom grade includes door components that have extensive carvings. May have exotic wood components, leaded glass, or beveled glass. Custom Deluxe grade is the same as Custom grade with oval or elliptical top windows.

Steel entry door. Price per door. Standard grade is flush face and high grade is embossed. Does not include jamb or casing.

	Craft@Hrs	Unit	Material	Labor	Total
replace, standard grade	1C@.212	ea	334.00	14.30	348.30
replace, high grade	1C@.212	ea	443.00	14.30	457.30
remove, steel entry door	1D@.283	ea	—	13.40	13.40

Steel entry door side-lite. Includes door side-lite, fasteners, and installation. Does not include wall framing, caulking, or finishing.

	Craft@Hrs	Unit	Material	Labor	Total
replace, standard grade	1C@1.14	ea	172.00	77.10	249.10
replace, high grade	1C@1.14	ea	221.00	77.10	298.10
remove, steel entry door side-lite	1D@.508	ea	—	24.00	24.00

Paint-grade wood entry door. Price per door. Includes paint-grade wood entry door with hinges and installation in an existing jamb. Pre-drilled and routed for hardware and hinges. Does not include jamb, casing, or door hardware.

	Craft@Hrs	Unit	Material	Labor	Total
replace, standard grade	1C@.212	ea	358.00	14.30	372.30
replace, high grade	1C@.212	ea	509.00	14.30	523.30
replace, deluxe grade	1C@.212	ea	689.00	14.30	703.30
replace, custom grade	1C@.212	ea	994.00	14.30	1,008.30
replace, custom deluxe grade	1C@.212	ea	1,150.00	14.30	1,164.30
remove, wood entry door	1D@.283	ea	—	13.40	13.40

Paint-grade wood entry door side-lite. Includes door side-lite, fasteners, and installation. Does not include wall framing, caulking, or finishing.

	Craft@Hrs	Unit	Material	Labor	Total
replace, standard grade	1C@1.14	ea	243.00	77.10	320.10
replace, high grade	1C@1.14	ea	335.00	77.10	412.10
replace, deluxe grade	1C@1.14	ea	486.00	77.10	563.10
replace, custom grade	1C@1.14	ea	670.00	77.10	747.10
replace, custom deluxe grade	1C@1.14	ea	809.00	77.10	886.10
remove, entry door side lite	1D@.508	ea	—	24.00	24.00

Fir entry door. Stain grade. Price per door. Does not include jamb or casing.

	Craft@Hrs	Unit	Material	Labor	Total
replace, standard grade	1C@.212	ea	794.00	14.30	808.30
replace, high grade	1C@.212	ea	965.00	14.30	979.30
replace, deluxe grade	1C@.212	ea	1,230.00	14.30	1,244.30
replace, custom grade	1C@.212	ea	1,820.00	14.30	1,834.30
replace, custom deluxe grade	1C@.212	ea	2,420.00	14.30	2,434.30
remove, entry door slab only	1D@.283	ea	—	13.40	13.40

	Craft@Hrs	Unit	Material	Labor	Total
Fir entry door side-lite. Stain grade. Includes door side-lite, fasteners, and installation. Does not include wall framing, caulking, or finishing.					
replace, standard grade	1C@1.14	ea	514.00	77.10	591.10
replace, high grade	1C@1.14	ea	664.00	77.10	741.10
replace, deluxe grade	1C@1.14	ea	823.00	77.10	900.10
replace, custom grade	1C@1.14	ea	1,260.00	77.10	1,337.10
replace, custom deluxe grade	1C@1.14	ea	1,670.00	77.10	1,747.10
remove, entry door side lite	1D@.508	ea	—	24.00	24.00
Mahogany entry door. Price per door. Does not include jamb or casing.					
replace, standard grade	1C@.411	ea	671.00	27.80	698.80
replace, high grade	1C@.411	ea	1,020.00	27.80	1,047.80
replace, deluxe grade	1C@.411	ea	1,580.00	27.80	1,607.80
replace, custom grade	1C@.411	ea	2,130.00	27.80	2,157.80
replace, custom deluxe grade	1C@.411	ea	2,660.00	27.80	2,687.80
remove, entry door	1D@.283	ea	—	13.40	13.40
Mahogany entry door side-lite. Includes door side-lite, fasteners, and installation. Does not include wall framing, caulking, or finishing.					
replace, standard grade	1C@1.14	ea	466.00	77.10	543.10
replace, high grade	1C@1.14	ea	699.00	77.10	776.10
replace, deluxe grade	1C@1.14	ea	1,100.00	77.10	1,177.10
replace, custom grade	1C@1.14	ea	1,480.00	77.10	1,557.10
replace, custom deluxe grade	1C@1.14	ea	1,900.00	77.10	1,977.10
remove, entry door side lite	1D@.508	ea	—	24.00	24.00

Wood embossing. Wood is often embossed to appear like it has been carved. Embossing stamps a pattern onto the wood. Embossing is far less expensive than carving and can only be used for relatively shallow patterns.

Wood onlays. Wood onlays are mass-produced carved wood components that are nailed and glued on doors to give a hand-carved look.

	Craft@Hrs	Unit	Material	Labor	Total
Ash or oak entry door. Price per door. Does not include jamb or casing.					
replace, standard grade	1C@.411	ea	643.00	27.80	670.80
replace, high grade	1C@.411	ea	994.00	27.80	1,021.80
replace, deluxe grade	1C@.411	ea	1,310.00	27.80	1,337.80
replace, custom grade	1C@.411	ea	2,230.00	27.80	2,257.80
replace, custom deluxe grade	1C@.411	ea	2,830.00	27.80	2,857.80
remove, entry door slab only	1D@.283	ea	—	13.40	13.40
Ash or oak entry door side-lite. Includes door side-lite, fasteners, and installation. Does not include wall framing, caulking, or finishing.					
replace, standard grade	1C@1.14	ea	435.00	77.10	512.10
replace, high grade	1C@1.14	ea	702.00	77.10	779.10
replace, deluxe grade	1C@1.14	ea	868.00	77.10	945.10
replace, custom grade	1C@1.14	ea	1,570.00	77.10	1,647.10
replace, custom deluxe grade	1C@1.14	ea	2,010.00	77.10	2,087.10
remove, entry door side lite	1D@.508	ea	—	24.00	24.00

	Craft@Hrs	Unit	Material	Labor	Total
Redwood entry door. Price per door. Does not include jamb or casing.					
replace, standard grade	1C@.411	ea	823.00	27.80	850.80
replace, high grade	1C@.411	ea	1,330.00	27.80	1,357.80
replace, deluxe grade	1C@.411	ea	1,720.00	27.80	1,747.80
replace, custom grade	1C@.411	ea	2,330.00	27.80	2,357.80
replace, custom deluxe grade	1C@.411	ea	2,650.00	27.80	2,677.80
remove, entry door slab only	1D@.283	ea	—	13.40	13.40
Redwood entry door side-lite. Includes door side-lite, fasteners, and installation. Does not include wall framing, caulking, or finishing.					
replace, standard grade	1C@1.14	ea	544.00	77.10	621.10
replace, high grade	1C@1.14	ea	933.00	77.10	1,010.10
replace, deluxe grade	1C@1.14	ea	1,200.00	77.10	1,277.10
replace, custom grade	1C@1.14	ea	1,630.00	77.10	1,707.10
replace, custom deluxe grade	1C@1.14	ea	1,890.00	77.10	1,967.10
remove, entry door side lite	1D@.508	ea	—	24.00	24.00
Cypress entry door. Price per door. Does not include jamb or casing.					
replace, standard grade	1C@.411	ea	750.00	27.80	777.80
replace, high grade	1C@.411	ea	1,010.00	27.80	1,037.80
replace, deluxe grade	1C@.411	ea	1,270.00	27.80	1,297.80
replace, custom grade	1C@.411	ea	1,830.00	27.80	1,857.80
replace, custom deluxe grade	1C@.411	ea	2,390.00	27.80	2,417.80
remove, entry door slab only	1D@.283	ea	—	13.40	13.40
Cypress entry door side-lite. Includes door side-lite, fasteners, and installation. Does not include wall framing, caulking, or finishing.					
replace, standard grade	1C@1.14	ea	512.00	77.10	589.10
replace, high grade	1C@1.14	ea	686.00	77.10	763.10
replace, deluxe grade	1C@1.14	ea	886.00	77.10	963.10
replace, custom grade	1C@1.14	ea	1,290.00	77.10	1,367.10
replace, custom deluxe grade	1C@1.14	ea	1,700.00	77.10	1,777.10
remove, entry door side lite	1D@.508	ea	—	24.00	24.00
Walnut or cherry entry door. Price per door. Does not include jamb or casing.					
replace, standard grade	1C@.411	ea	1,050.00	27.80	1,077.80
replace, high grade	1C@.411	ea	1,570.00	27.80	1,597.80
replace, deluxe grade	1C@.411	ea	1,980.00	27.80	2,007.80
replace, custom grade	1C@.411	ea	2,970.00	27.80	2,997.80
replace, custom deluxe grade	1C@.411	ea	3,820.00	27.80	3,847.80
remove, entry door slab only	1D@.283	ea	—	13.40	13.40
Walnut or cherry entry door side-lite. Includes door side-lite, fasteners, and installation. Does not include wall framing, caulking, or finishing.					
replace, standard grade	1C@1.14	ea	685.00	77.10	762.10
replace, high grade	1C@1.14	ea	1,080.00	77.10	1,157.10
replace, deluxe grade	1C@1.14	ea	1,350.00	77.10	1,427.10
replace, custom grade	1C@1.14	ea	2,130.00	77.10	2,207.10
replace, custom deluxe grade	1C@1.14	ea	2,730.00	77.10	2,807.10
remove, entry door side lite	1D@.508	ea	—	24.00	24.00

	Craft@Hrs	Unit	Material	Labor	Total
Remove exterior door for work and reinstall.					
door, jamb, & casing	1C@2.20	ea	—	149.00	149.00
door slab	1C@.292	ea	—	19.70	19.70
double door, jamb & casing	1C@3.12	ea	—	211.00	211.00
double door slab	1C@.366	ea	—	24.70	24.70
Exterior door jamb & casing. Price per door. Includes jamb and casing for exterior door, shims, finish nails, and installation. Includes hanging door in jamb, installing jamb in rough opening, and casing jamb. Does not include door, hinges, or door hardware.					
replace, paint-grade pine	1C@1.75	ea	68.40	118.00	186.40
replace, paint-grade pine for double door	1C@3.11	ea	122.00	210.00	332.00
replace, stain-grade pine	1C@1.75	ea	80.30	118.00	198.30
replace, stain-grade pine for double door	1C@3.11	ea	145.00	210.00	355.00
replace, birch	1C@1.75	ea	76.40	118.00	194.40
replace, birch for double door	1C@3.11	ea	134.00	210.00	344.00
replace, mahogany	1C@1.75	ea	90.50	118.00	208.50
replace, mahogany veneered jamb & solid casing	1C@1.75	ea	82.00	118.00	200.00
replace, mahogany for double door	1C@3.11	ea	162.00	210.00	372.00
replace, mahogany veneered jamb & solid casing for double door	1C@3.11	ea	149.00	210.00	359.00
replace, ash or oak	1C@1.75	ea	95.60	118.00	213.60
replace, ash or oak veneered jamb & solid casing	1C@1.75	ea	88.00	118.00	206.00
replace, ash or oak for double door	1C@3.11	ea	172.00	210.00	382.00
replace, ash or oak veneered jamb & solid casing for double door	1C@3.11	ea	159.00	210.00	369.00
replace, redwood	1C@1.75	ea	85.50	118.00	203.50
replace, redwood for double door	1C@3.11	ea	151.00	210.00	361.00
replace, cypress	1C@1.75	ea	79.40	118.00	197.40
replace, cypress double door	1C@3.11	ea	143.00	210.00	353.00
replace, walnut or cherry	1C@1.75	ea	112.00	118.00	230.00
replace, walnut or cherry veneered jamb & solid casing	1C@1.75	ea	104.00	118.00	222.00
replace, walnut or cherry for double door	1C@3.11	ea	193.00	210.00	403.00
replace, walnut or cherry veneered jamb & solid casing for double door	1C@3.11	ea	183.00	210.00	393.00
replace, steel	1C@1.75	ea	110.00	118.00	228.00
replace, steel for double door	1C@3.11	ea	159.00	210.00	369.00
remove, exterior door, jamb & casing	1D@.496	ea	—	23.40	23.40
Additional exterior jamb & casing costs.					
add for hanging Dutch door	1C@1.75	ea	—	118.00	118.00
add for round- or elliptical-top door	—	ea	236.00	—	236.00

	Craft@Hrs	Unit	Material	Labor	Total

Entry door fanlite. Includes fanlite, fasteners, and installation. Fanlites are semi-circular or half-elliptical windows that appear above doors. True fanlites have mullions that fan out from the bottom center of the lite. Lower grades will have a simulated fan grid. All fanlites include jamb and casing. Does not include wall framing, caulking, or finishing.

	Craft@Hrs	Unit	Material	Labor	Total
replace, single entry door, standard grade	1C@1.85	ea	287.00	125.00	412.00
replace, single entry door, high grade	1C@1.85	ea	483.00	125.00	608.00
replace, single entry door, deluxe grade	1C@1.85	ea	670.00	125.00	795.00
replace, single entry door, custom grade	1C@1.85	ea	927.00	125.00	1,052.00
replace, single entry door, custom deluxe grade	1C@1.85	ea	1,220.00	125.00	1,345.00
remove entry door fanlite	1D@.747	ea	—	35.30	35.30
remove fanlite & casing for work, reinstall	1C@3.13	ea	—	212.00	212.00

Additional fanlite sizes. Includes jamb and casing.

add 34% for fanlite for single door with side-lite
add 85% for fanlite for single door with two side-lites
add 99% for fanlite for double doors
add 147% for fanlite for double doors with two side-lites

Entry door fixed transom. Includes transom with jamb and casing, fasteners, and installation. Does not include wall framing. Transoms may have circular, semi-circular, or elliptical components, but the unit fits in a rectangular jamb.

	Craft@Hrs	Unit	Material	Labor	Total
replace, single entry door, standard grade	1C@1.00	ea	190.00	67.60	257.60
replace, high grade	1C@1.00	ea	322.00	67.60	389.60
replace, deluxe grade	1C@1.00	ea	535.00	67.60	602.60
replace, custom grade	1C@1.00	ea	780.00	67.60	847.60
replace, custom deluxe grade	1C@1.00	ea	1,100.00	67.60	1,167.60
remove fixed transom	1D@.924	ea	—	43.60	43.60
remove fixed transom for work, reinstall	1C@2.74	ea	—	185.00	185.00

Additional transom sizes. Includes jamb and casing.

replace, add 42% for fixed transom for single door with side lite
replace, add 65% for fixed transom for single door with two side lites
replace, add 89% for fixed transom for double doors
replace, add 106% for fixed transom for double doors with two side lites

Café (bar) doors. Includes hardware.

	Craft@Hrs	Unit	Material	Labor	Total
replace, louvered paint-grade	1C@.681	ea	138.00	46.00	184.00
replace, louvered stain-grade	1C@.681	ea	178.00	46.00	224.00
replace, Victorian spindle & raised-panel paint-grade	1C@.681	ea	181.00	46.00	227.00
replace, Victorian spindle & raised-panel stain-grade	1C@.681	ea	224.00	46.00	270.00
remove café door	1D@.254	ea	—	12.00	12.00
remove café doors for work, reinstall	1C@1.02	ea	—	69.00	69.00

louvered

Victorian spindle & raised-panel

Sliding patio door. Includes sliding patio door set with integrated jamb, roller assembly, hardware, and installation. Does not include caulking.

	Craft@Hrs	Unit	Material	Labor	Total
replace, 6' by 6'8" bronze finish single-glazed	1C@4.39	ea	684.00	297.00	981.00
replace, 6' by 6'8" mill finish double-glazed	1C@4.39	ea	924.00	297.00	1,221.00
replace, 8' by 6'8" bronze finish single-glazed	1C@4.39	ea	826.00	297.00	1,123.00
replace, 8' by 6'8" mill finish double-glazed	1C@4.39	ea	1,080.00	297.00	1,377.00
replace, 12' by 6'8" bronze finish single-glazed (3 lites)	1C@5.75	ea	1,230.00	389.00	1,619.00
replace, 12' by 6'8" mill finish double-glazed (3 lites)	1C@5.75	ea	1,750.00	389.00	2,139.00
replace, 6' by 6'8" wood single-glazed	1C@4.39	ea	1,450.00	297.00	1,747.00
replace, 6' by 6'8" wood double-glazed	1C@4.39	ea	1,630.00	297.00	1,927.00
replace, 8' by 6'8" wood single-glazed	1C@4.39	ea	1,760.00	297.00	2,057.00
replace, 8' by 6'8" wood double-glazed	1C@4.39	ea	1,950.00	297.00	2,247.00
replace, 12' by 6'8" wood single-glazed (3 lites)	1C@5.75	ea	2,650.00	389.00	3,039.00
replace, 12' by 6'8" wood double-glazed (3 lites)	1C@5.75	ea	2,920.00	389.00	3,309.00
replace, 6' by 6'8" wood single-glazed clad exterior	1C@4.39	ea	1,480.00	297.00	1,777.00
replace, 6' by 6'8" wood double-glazed clad exterior	1C@4.39	ea	1,810.00	297.00	2,107.00
replace, 8' by 6'8" wood single-glazed clad exterior	1C@5.75	ea	1,750.00	389.00	2,139.00
replace, 8' by 6'8" wood double-glazed clad exterior	1C@5.75	ea	2,140.00	389.00	2,529.00
replace, 12' by 6'8" wood single-glazed clad exterior (3 lites)	1C@5.75	ea	2,690.00	389.00	3,079.00
replace, 12' by 6'8" wood double-glazed clad exterior (3 lites)	1C@5.75	ea	3,510.00	389.00	3,899.00
remove, 6' or 8' wide sliding patio door	1D@1.87	ea	—	88.30	88.30
remove, 12' wide sliding patio door	1D@2.81	ea	—	133.00	133.00

two lite patio door

three lite patio door

Additional sliding patio door costs.

	Craft@Hrs	Unit	Material	Labor	Total
add for tinted glass (per lite)	—	ea	338.00	—	338.00
remove door, jamb, & casing for work, then reinstall	1C@5.55	ea	—	375.00	375.00
remove door lite for work, then reinstall	1C@1.65	ea	—	112.00	112.00
replace, minimum charge	1C@2.50	ea	39.80	169.00	208.80
recondition door, replace hardware	1C@2.33	ea	90.50	158.00	248.50

Sliding patio door screen. Includes screen door set to match existing sliding patio door and installation. Includes screen assembly with hardware and rollers.

	Craft@Hrs	Unit	Material	Labor	Total
replace, 36" door screen	1C@.125	ea	90.50	8.45	98.95
replace, 48" door screen	1C@.125	ea	112.00	8.45	120.45
remove, door screen	1D@.125	ea	—	5.90	5.90

Pet door. Includes lock, aluminum frame, and PVC door.

	Craft@Hrs	Unit	Material	Labor	Total
replace, for small dogs, cats	1C@.840	ea	80.30	56.80	137.10
replace, for average size dogs	1C@.840	ea	138.00	56.80	194.80
replace, for large dogs	1C@.844	ea	221.00	57.10	278.10
remove, pet door	1D@.291	ea	—	13.70	13.70
remove for work, then reinstall	1C@1.29	ea	10.00	87.20	97.20

Garage door with hardware. Includes garage door, track, hardware, fasteners, and installation. Does not include garage door opener, framing, or finishing. Economy grade includes overhead doors that are plastic-faced polystyrene or flush steel or uninsulated fiberglass. Single piece doors with flush surface or tongue-and-groove boards in herringbone or chevron patterns. Standard grade includes overhead doors that are textured or embossed steel, aluminum or fiberglass, usually insulated. Single-piece doors same as economy grade with lites, and paneled patterns. High grade overhead doors are same as standard grade with lites. Single-piece stain-grade wood doors with lites and carvings. Deluxe grade overhead doors are stain-grade wood doors, specialty patterns, designed carvings or embossing, fan-tops, and so forth.

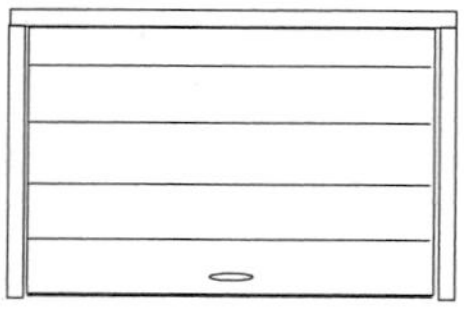

	Craft@Hrs	Unit	Material	Labor	Total
8' garage door with hardware					
replace, economy grade	1C@6.25	ea	571.00	423.00	994.00
replace, standard grade	1C@6.25	ea	664.00	423.00	1,087.00
replace, high grade	1C@6.25	ea	913.00	423.00	1,336.00
replace, deluxe grade	1C@6.25	ea	1,020.00	423.00	1,443.00
remove, garage door	1D@3.63	ea	—	171.00	171.00
9' garage door with hardware					
replace, economy grade	1C@6.25	ea	621.00	423.00	1,044.00
replace, standard grade	1C@6.25	ea	846.00	423.00	1,269.00
replace, high grade	1C@6.25	ea	994.00	423.00	1,417.00
replace, deluxe grade	1C@6.25	ea	1,120.00	423.00	1,543.00
remove, garage door	1D@3.63	ea	—	171.00	171.00
10' garage door with hardware					
replace, economy grade	1C@6.25	ea	712.00	423.00	1,135.00
replace, standard grade	1C@6.25	ea	886.00	423.00	1,309.00
replace, high grade	1C@6.25	ea	1,090.00	423.00	1,513.00
replace, deluxe grade	1C@6.25	ea	1,250.00	423.00	1,673.00
remove, garage door	1D@3.63	ea	—	171.00	171.00
12' garage door with hardware					
replace, economy grade	1C@6.25	ea	803.00	423.00	1,226.00
replace, standard grade	1C@6.25	ea	994.00	423.00	1,417.00
high grade	1C@6.25	ea	1,260.00	423.00	1,683.00
replace, deluxe grade	1C@6.25	ea	1,420.00	423.00	1,843.00
remove, garage door	1D@3.63	ea	—	171.00	171.00
14' garage door with hardware					
replace, economy grade	1C@6.25	ea	860.00	423.00	1,283.00
replace, standard grade	1C@6.25	ea	1,020.00	423.00	1,443.00
replace, high grade	1C@6.25	ea	1,260.00	423.00	1,683.00
replace, deluxe grade	1C@6.25	ea	1,510.00	423.00	1,933.00
remove, garage door	1D@3.63	ea	—	171.00	171.00
16' garage door with hardware					
replace, economy grade	1C@6.25	ea	933.00	423.00	1,356.00
replace, standard grade	1C@6.25	ea	1,180.00	423.00	1,603.00
replace, high grade	1C@6.25	ea	1,380.00	423.00	1,803.00
replace, deluxe grade	1C@6.25	ea	1,620.00	423.00	2,043.00
remove, garage door	1D@3.63	ea	—	171.00	171.00
18' garage door with hardware					
replace, economy grade	1C@6.25	ea	1,020.00	423.00	1,443.00
replace, standard grade	1C@6.25	ea	1,260.00	423.00	1,683.00
replace, high grade	1C@6.25	ea	1,450.00	423.00	1,873.00
replace, deluxe grade	1C@6.25	ea	1,740.00	423.00	2,163.00
remove, garage door	1D@3.63	ea	—	171.00	171.00
Replace, garage door spring	1C@2.00	ea	83.10	135.00	218.10
Remove single-car door for work, then reinstall	1C@7.70	ea	—	521.00	521.00
Remove two-car door for work, then reinstall	1C@9.25	ea	—	625.00	625.00

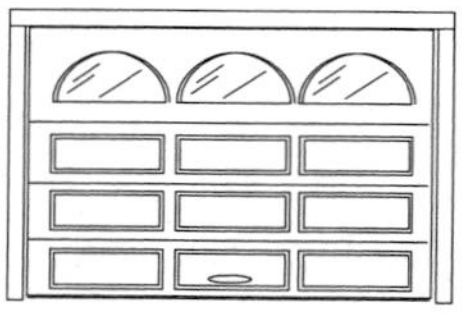

economy

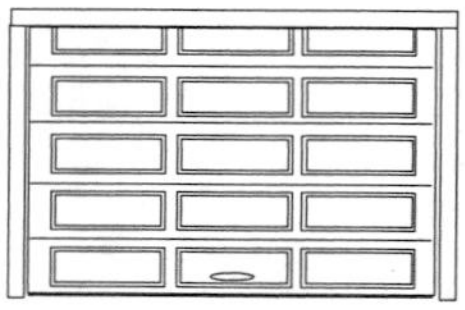

standard

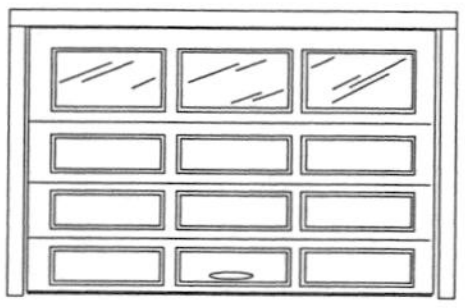

high

deluxe

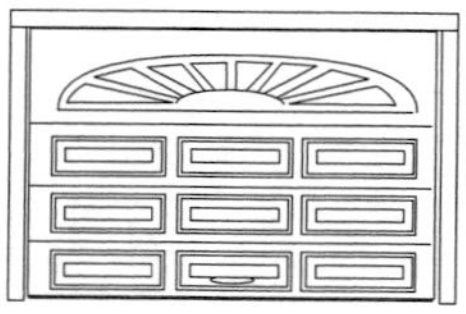

deluxe (fanlite)

	Craft@Hrs	Unit	Material	Labor	Total

Add for tall door. All door types.

Add 14% for 7' tall doors
Add 22% for 8' tall doors

Garage door opener. Includes garage door opener, hardware, fasteners, and installation. Does not include garage door, electrical wiring for the opener, or framing.

	Craft@Hrs	Unit	Material	Labor	Total
replace, standard grade	1C@2.45	ea	463.00	166.00	629.00
replace, high grade	1C@2.45	ea	501.00	166.00	667.00
replace, deluxe grade	1C@2.45	ea	553.00	166.00	719.00
remove, garage door opener	1D@.765	ea	—	36.10	36.10
replace, remote radio transmitter for opener	—	ea	57.80	—	57.80
remove garage door opener for work, then reinstall	1C@3.50	ea	—	237.00	237.00

Minimum charge.

	Craft@Hrs	Unit	Material	Labor	Total
for garage door work	1C@2.50	ea	50.80	169.00	219.80

Time & Material Charts (selected items)

Doors Materials

See Doors material prices with the line items above.

Doors Labor

Laborer	base wage	paid leave	true wage	taxes & ins.	total
Finish carpenter	$38.10	2.97	$41.07	26.53	$67.60
Demolition laborer	$25.90	2.02	$27.92	19.28	$47.20

Paid leave is calculated based on two weeks paid vacation, one week sick leave, and seven paid holidays. Employer's matching portion of **FICA** is 7.65 percent. **FUTA** (Federal Unemployment) is .8 percent. **Worker's compensation** for the doors trade was calculated using a national average of 17.21 percent. **Unemployment insurance** was calculated using a national average of 8 percent. **Health insurance** was calculated based on a projected national average for 2020 of $1,280 per employee (and family when applicable) per month. Employer pays 80 percent for a per month cost of $1,024 per employee. **Retirement** is based on a 401(k) retirement program with employer matching of 50 percent. Employee contributions to the 401(k) plan are an average of 6 percent of the true wage. **Liability insurance** is based on a national average of 12.0 percent.

Doors Labor Productivity

Demolition of doors	Craft@Hrs	Unit	Material	Labor	Total
remove, folding door	1D@.114	ea	—	5.38	5.38
remove, bypassing door	1D@.139	ea	—	6.56	6.56
remove, bypassing or folding door jamb & casing	1D@.026	lf	—	1.23	1.23
remove, pre-hung interior door, jamb, & casing	1D@.388	ea	—	18.30	18.30
remove, pre-hung interior door slab only	1D@.217	ea	—	10.20	10.20
remove, double pre-hung int. door, jamb, & casing	1D@.585	ea	—	27.60	27.60
remove, double pre-hung interior door slabs only	1D@.321	ea	—	15.20	15.20
remove, gar. door, overhead tracks, jamb, & casing	1D@3.65	ea	—	172.00	172.00
remove, garage door, hardware, jamb, & casing	1D@2.90	ea	—	137.00	137.00
remove, garage door only	1D@1.82	ea	—	85.90	85.90

	Craft@Hrs	Unit	Material	Labor	Total
remove, transom	1D@.451	ea	—	21.30	21.30
remove, entry door fan lite & casing	1D@.749	ea	—	35.40	35.40
remove, double entry door fan lite & casing	1D@1.71	ea	—	80.70	80.70
remove, storm door	1D@.227	ea	—	10.70	10.70
remove, entry door, jamb & casing	1D@.496	ea	—	23.40	23.40
remove, entry door slab only	1D@.283	ea	—	13.40	13.40
remove, double entry door, jamb, cas., & threshold	1D@.842	ea	—	39.70	39.70
remove, double entry door slabs only	1D@.492	ea	—	23.20	23.20
remove, entry door side lite	1D@.508	ea	—	24.00	24.00
remove, Dutch door slabs, jamb, & casing	1D@.883	ea	—	41.70	41.70
remove, Dutch door slabs only	1D@.321	ea	—	15.20	15.20
remove, cafe door	1D@.254	ea	—	12.00	12.00
remove, sliding glass patio door, per lite	1D@.427	ea	—	20.20	20.20
remove, pet door	1D@.290	ea	—	13.70	13.70
Install interior doors					
folding door per section	1C@.394	ea	—	26.60	26.60
bypassing door per section	1C@.755	ea	—	51.00	51.00
bypassing or folding door jamb & casing	1C@.109	lf	—	7.37	7.37
pre-hung interior door & casing	1C@1.43	ea	—	96.70	96.70
pre-hung interior door slab only	1C@.215	ea	—	14.50	14.50
double pre-hung interior door & casing	1C@2.45	ea	—	166.00	166.00
double pre-hung interior door slabs only	1C@.412	ea	—	27.90	27.90
Install garage door					
with overhead tracks, jamb, & casing	1C@6.25	ea	—	423.00	423.00
with hardware, jamb, & casing	1C@5.21	ea	—	352.00	352.00
door only	1C@3.15	ea	—	213.00	213.00
opener	1C@2.45	ea	—	166.00	166.00
Install pocket door					
door slabs	1C@.855	ea	—	57.80	57.80
door, jamb, trim, & casing	1C@1.35	ea	—	91.30	91.30
double doors	1C@2.34	ea	—	158.00	158.00
double door, jamb, trim, & casing	1C@3.12	ea	—	211.00	211.00
Install pocket door rough framing package					
Including track and hardware					
single door in 2" x 4" wall	1C@1.50	ea	—	101.00	101.00
double doors in 2" x 4" wall	1C@2.88	ea	—	195.00	195.00
single door in 2" x 6" wall	1C@1.52	ea	—	103.00	103.00
double doors in 2" x 6" wall	1C@2.95	ea	—	199.00	199.00
single door in 2" x 8" wall	1C@1.54	ea	—	104.00	104.00
double doors in 2" x 8" wall	1C@3.03	ea	—	205.00	205.00
rough framing for antique door	1C@2.45	ea	—	166.00	166.00
rough framing for antique double doors	1C@5.12	ea	—	346.00	346.00
Install pocket door track					
overhead track for antique door	1C@1.30	ea	—	87.90	87.90
overhead track for antique double door	1C@2.34	ea	—	158.00	158.00
clean & realign overhead ant. door track	1C@1.96	ea	—	132.00	132.00
floor track for antique door	1C@1.21	ea	—	81.80	81.80
floor track for antique double door	1C@2.20	ea	—	149.00	149.00
clean and realign floor antique doors	1C@2.64	ea	—	178.00	178.00

	Craft@Hrs	Unit	Material	Labor	Total
Install panel door					
hang door & install casing	1C@1.23	ea	—	83.10	83.10
slab only	1C@.178	ea	—	12.00	12.00
Install entry door, fanlite, transom					
fanlite & casing	1C@1.85	ea	—	125.00	125.00
transom & casing	1C@1.00	ea	—	67.60	67.60
hang entry door & install casing	1C@1.75	ea	—	118.00	118.00
slab only	1C@.212	ea	—	14.30	14.30
hang double-entry door & casing	1C@3.11	ea	—	210.00	210.00
double door slabs only	1C@.330	ea	—	22.30	22.30
side-lite	1C@1.14	ea	—	77.10	77.10
Install Dutch door					
hang door & install casing	1C@1.70	ea	—	115.00	115.00
slabs only	1C@.325	ea	—	22.00	22.00
Install door lite					
cut hole in door and install	1C@1.12	ea	—	75.70	75.70
Install cafe door					
per set	1C@.681	ea	—	46.00	46.00
Install sliding glass patio doors					
per lite	1C@4.39	ea	—	297.00	297.00
Install pet door					
per door	1C@.844	ea	—	57.10	57.10
Assemble & install storm door					
per door	1C@1.21	ea	—	81.80	81.80
Remove for work, then reinstall					
folding doors per section	1C@.148	ea	—	10.00	10.00
bypass sliding doors per section	1C@.220	ea	—	14.90	14.90
door slab	1C@.151	ea	—	10.20	10.20
door slab, jamb & casing	1C@1.63	ea	—	110.00	110.00
double door slabs	1C@.293	ea	—	19.80	19.80
double door slabs, jamb & casing	1C@2.67	ea	—	180.00	180.00
pocket door & casing	1C@1.15	ea	—	77.70	77.70
pocket door, jamb & casing	1C@3.23	ea	—	218.00	218.00
double pocket doors & casing	1C@1.58	ea	—	107.00	107.00
double pocket doors, jamb & casing	1C@3.39	ea	—	229.00	229.00
panel door slab	1C@.152	ea	—	10.30	10.30
panel door, jamb, & casing	1C@1.63	ea	—	110.00	110.00
antique door casing & re-plumb door	1C@2.95	ea	—	199.00	199.00
transom	1C@1.12	ea	—	75.70	75.70
storm door	1C@1.70	ea	—	115.00	115.00
Dutch door, jamb, & casing	1C@2.74	ea	—	185.00	185.00
Dutch door slabs	1C@.394	ea	—	26.60	26.60
exterior door, jamb, & casing	1C@2.20	ea	—	149.00	149.00
exterior door slab	1C@.292	ea	—	19.70	19.70
exterior double door, jamb, & casing	1C@3.12	ea	—	211.00	211.00
exterior double door slabs	1C@.366	ea	—	24.70	24.70
entry door fanlite & casing	1C@3.12	ea	—	211.00	211.00
entry door fixed transom & casing	1C@2.74	ea	—	185.00	185.00
cafe door	1C@1.02	ea	—	69.00	69.00
sliding patio door, jamb, & casing	1C@5.55	ea	—	375.00	375.00
sliding patio door lites	1C@1.65	ea	—	112.00	112.00

	Craft@Hrs	Unit	Material	Labor	Total
pet door	1C@1.29	ea	—	87.20	87.20
garage door opener	1C@3.50	ea	—	237.00	237.00
single-car garage door	1C@7.70	ea	—	521.00	521.00
two-car garage door	1C@9.25	ea	—	625.00	625.00
Door repair					
replace antique panel door panel (not including custom knife charges)	1C@2.74	ea	—	185.00	185.00
replace antique panel door stile (not including custom knife charges)	1C@2.49	ea	—	168.00	168.00
repair cracks & dry rot in antique door with epoxy, tighten stiles	1C@4.49	ea	—	304.00	304.00
repair cracks & dry rot in antique door jamb with epoxy	1C@3.40	ea	—	230.00	230.00
repair cracks & dry rot in antique door casing with epoxy	1C@3.40	ea	—	230.00	230.00
Additional door labor charges					
add for install. of full-lite in ext. door	1C@1.39	ea	—	94.00	94.00
add for install. of half-lite in ext. door	1C@1.07	ea	—	72.30	72.30
add for install. of 1' by 1' lite in ext. door	1C@.594	ea	—	40.20	40.20
recondition sliding door, replace hardware	1C@2.33	ea	—	158.00	158.00

Door Hardware

Hardware Quality. Some rules of thumb – Standard: Light gauge metal, chrome or brass plated with little or no pattern. High: Brass, chrome over brass, or nickel over brass with minimal detail, or plated with ornate detail. Deluxe: Brass, chrome over brass, or nickel over brass with moderate detail. Custom: Brass, chrome over brass, or nickel over brass with ornate detail.

Deadbolt.

	Craft@Hrs	Unit	Material	Labor	Total
replace, standard grade	1C@.678	ea	35.70	45.80	81.50
replace, high grade	1C@.678	ea	52.70	45.80	98.50
replace, deluxe grade	1C@.678	ea	65.70	45.80	111.50
remove, deadbolt	1D@.276	ea	—	13.00	13.00
remove for work, then reinstall	1C@1.10	ea	—	74.40	74.40

Exterior lockset.

	Craft@Hrs	Unit	Material	Labor	Total
replace, standard grade	1C@.461	ea	36.50	31.20	67.70
replace, high grade	1C@.461	ea	58.70	31.20	89.90
replace, deluxe grade	1C@.461	ea	78.90	31.20	110.10
remove, exterior lockset	1D@.314	ea	—	14.80	14.80
remove for work, then reinstall	1C@1.03	ea	—	69.60	69.60

	Craft@Hrs	Unit	Material	Labor	Total
Entry lockset.					
replace, standard grade	1C@1.45	ea	129.00	98.00	227.00
replace, high grade	1C@1.45	ea	342.00	98.00	440.00
replace, deluxe grade	1C@1.45	ea	572.00	98.00	670.00
replace, custom grade	1C@1.45	ea	795.00	98.00	893.00
remove, entry door lockset	1D@.696	ea	—	32.90	32.90
remove for work, then reinstall	1C@2.73	ea	—	185.00	185.00
Thumb latch.					
Antique style					
replace, standard grade	1C@1.73	ea	352.00	117.00	469.00
replace, high grade	1C@1.73	ea	427.00	117.00	544.00
remove, thumb latch	1D@.516	ea	—	24.40	24.40
Interior door lockset.					
replace, standard grade	1C@.405	ea	27.10	27.40	54.50
replace, high grade	1C@.405	ea	68.10	27.40	95.50
replace, deluxe grade	1C@.405	ea	108.00	27.40	135.40
remove, interior door lockset	1D@.281	ea	—	13.30	13.30
remove for work, then reinstall	1C@.710	ea	—	48.00	48.00
Door closer.					
replace, spring hinge style	1C@.257	ea	27.10	17.40	44.50
replace, standard grade	1C@.431	ea	66.90	29.10	96.00
replace, high grade	1C@.431	ea	118.00	29.10	147.10
remove, door closer	1D@.291	ea	—	13.70	13.70
remove for work, then reinstall	1C@.742	ea	—	50.20	50.20
Security chain.					
replace, typical	1C@.262	ea	20.00	17.70	37.70
remove, security chain	1D@.258	ea	—	12.20	12.20
remove for work, then reinstall	1C@.445	ea	—	30.10	30.10
Door hinges. Deluxe grade hinges include embossed antique patterns.					
replace, standard grade	1C@.438	ea	18.50	29.60	48.10
replace, high grade	1C@.438	ea	38.50	29.60	68.10
replace, deluxe grade	1C@.438	ea	59.70	29.60	89.30
replace, for batten door	1C@.495	ea	89.80	33.50	123.30
remove, door hinges	1D@.485	ea	—	22.90	22.90
remove for work, then reinstall	1C@.786	ea	—	53.10	53.10

deluxe grade

	Craft@Hrs	Unit	Material	Labor	Total
Door knocker.					
replace, standard grade	1C@.351	ea	27.10	23.70	50.80
replace, high grade	1C@.351	ea	61.30	23.70	85.00
replace, deluxe grade	1C@.351	ea	120.00	23.70	143.70
remove, door knocker	1D@.250	ea	—	11.80	11.80
remove for work, then reinstall	1C@.636	ea	—	43.00	43.00
Storm door hinges.					
replace, typical	1C@.342	ea	30.00	23.10	53.10
remove, screen door hinges	1D@.457	ea	—	21.60	21.60
remove for work, then reinstall	1C@.607	ea	—	41.00	41.00
Storm door lockset.					
replace, standard grade	1C@.387	ea	38.50	26.20	64.70
replace, high grade	1C@.387	ea	57.10	26.20	83.30
remove, screen door lockset	1D@.327	ea	—	15.40	15.40
remove for work, then reinstall	1C@.685	ea	—	46.30	46.30
Door stop.					
replace, hinge type	1C@.128	ea	7.12	8.65	15.77
replace, flexible spring baseboard	1C@.138	ea	9.26	9.33	18.59
replace, rigid baseboard	1C@.138	ea	10.70	9.33	20.03
remove, door stop	1D@.084	ea	—	3.96	3.96
remove for work, then reinstall	1C@.252	ea	—	17.00	17.00

rigid *flexible spring*

	Craft@Hrs	Unit	Material	Labor	Total
Door push plate.					
replace, standard grade	1C@.417	ea	31.00	28.20	59.20
replace, high grade	1C@.417	ea	58.60	28.20	86.80
remove, door push plate	1D@.160	ea	—	7.55	7.55
remove for work, then reinstall	1C@.742	ea	—	50.20	50.20

high grade

	Craft@Hrs	Unit	Material	Labor	Total
Door kick plate.					
replace, standard grade	1C@.445	ea	53.90	30.10	84.00
replace, high grade	1C@.445	ea	78.50	30.10	108.60
remove, door kick plate	1D@.213	ea	—	10.10	10.10
remove for work, then reinstall	1C@.809	ea	—	54.70	54.70
Letter plate.					
replace, standard grade	1C@.500	ea	44.50	33.80	78.30
replace, high grade	1C@.500	ea	87.10	33.80	120.90
remove, door letter plate	1D@.216	ea	—	10.20	10.20
remove for work, then reinstall	1C@.902	ea	—	61.00	61.00

	Craft@Hrs	Unit	Material	Labor	Total
Door peep hole.					
replace, typical	1C@.504	ea	28.50	34.10	62.60
remove, door peep hole	1D@.205	ea	—	9.68	9.68
remove for work, then reinstall	1C@.879	ea	—	59.40	59.40
Door sweep.					
replace, typical	1C@.431	ea	44.50	29.10	73.60
remove, door sweep	1D@.239	ea	—	11.30	11.30
remove for work, then reinstall	1C@.763	ea	—	51.60	51.60
Door threshold.					
replace, standard grade	1C@.477	ea	24.30	32.20	56.50
replace, high grade	1C@.477	ea	45.80	32.20	78.00
remove, door threshold	1D@.262	ea	—	12.40	12.40
remove for work, then reinstall	1C@.862	ea	—	58.30	58.30
Door weatherstripping.					
replace, typical	1C@.417	ea	30.00	28.20	58.20
remove door weatherstripping	1D@.195	ea	—	9.20	9.20
Garage door hardware.					
replace, standard grade	1C@2.30	ea	118.00	155.00	273.00
replace, high grade	1C@2.30	ea	162.00	155.00	317.00
remove, garage door hardware	1D@.762	ea	—	36.00	36.00
remove for work, then reinstall	1C@4.05	ea	—	274.00	274.00

Time & Material Charts (selected items)
Door Hardware Materials

See Door Hardware material prices with the line items above.

Doors Hardware Labor

Laborer	base wage	paid leave	true wage	taxes & ins.	total
Finish carpenter	$38.10	2.97	$41.07	26.53	$67.60
Demolition laborer	$25.90	2.02	$27.92	19.28	$47.20

Paid leave is calculated based on two weeks paid vacation, one week sick leave, and seven paid holidays. Employer's matching portion of **FICA** is 7.65 percent. **FUTA** (Federal Unemployment) is .8 percent. **Worker's compensation** for the door hardware trade was calculated using a national average of 17.21 percent. **Unemployment insurance** was calculated using a national average of 8 percent. **Health insurance** was calculated based on a projected national average for 2020 of $1,280 per employee (and family when applicable) per month. Employer pays 80 percent for a per month cost of $1,024 per employee. **Retirement** is based on a 401(k) retirement program with employer matching of 50 percent. Employee contributions to the 401(k) plan are an average of 6 percent of the true wage. **Liability insurance** is based on a national average of 12.0 percent.

Door Hardware Labor Productivity

	Craft@Hrs	Unit	Material	Labor	Total
Demolition of door hardware					
deadbolt	1D@.276	ea	—	13.00	13.00
exterior door keyed lockset	1D@.314	ea	—	14.80	14.80
entry door keyed lockset	1D@.696	ea	—	32.90	32.90
antique-style door thumb latch	1D@.516	ea	—	24.40	24.40
interior door lockset	1D@.281	ea	—	13.30	13.30
door closer	1D@.291	ea	—	13.70	13.70
security chain	1D@.258	ea	—	12.20	12.20
hinges	1D@.485	ea	—	22.90	22.90
door knocker	1D@.250	ea	—	11.80	11.80
storm/screen door hinges	1D@.457	ea	—	21.60	21.60
storm/screen door lockset	1D@.327	ea	—	15.40	15.40
door stop	1D@.084	ea	—	3.96	3.96
door push plate	1D@.160	ea	—	7.55	7.55
door kick plate	1D@.213	ea	—	10.10	10.10
door letter plate	1D@.216	ea	—	10.20	10.20
door peep hole	1D@.205	ea	—	9.68	9.68
door sweep	1D@.239	ea	—	11.30	11.30
door threshold	1D@.262	ea	—	12.40	12.40
door weatherstripping	1D@.195	ea	—	9.20	9.20
garage door hardware	1D@.762	ea	—	36.00	36.00
Install deadbolt					
install	1C@.678	ea	—	45.80	45.80
remove for work, then reinstall	1C@1.10	ea	—	74.40	74.40
Install exterior door keyed lockset					
install	1C@.461	ea	—	31.20	31.20
remove for work, then reinstall	1C@1.03	ea	—	69.60	69.60
Install entry door keyed lockset					
install	1C@1.45	ea	—	98.00	98.00
remove for work, then reinstall	1C@2.73	ea	—	185.00	185.00
Install antique-style door thumb latch					
install	1C@1.73	ea	—	117.00	117.00
Install interior door lockset					
install	1C@.405	ea	—	27.40	27.40
remove for work, then reinstall	1C@.710	ea	—	48.00	48.00
Install door closer					
spring hinge	1C@.257	ea	—	17.40	17.40
typical	1C@.431	ea	—	29.10	29.10
remove for work, then reinstall	1C@.742	ea	—	50.20	50.20
Install security chain					
install	1C@.262	ea	—	17.70	17.70
remove for work, then reinstall	1C@.445	ea	—	30.10	30.10
Install door hinges					
typical	1C@.438	ea	—	29.60	29.60
for batten door	1C@.495	ea	—	33.50	33.50
remove for work, then reinstall	1C@.786	ea	—	53.10	53.10

	Craft@Hrs	Unit	Material	Labor	Total
Install door knocker					
install	1C@.351	ea	—	23.70	23.70
remove for work, then reinstall	1C@.636	ea	—	43.00	43.00
Install screen door hinges					
install	1C@.342	ea	—	23.10	23.10
remove for work, then reinstall	1C@.607	ea	—	41.00	41.00
Install screen door lockset					
install	1C@.387	ea	—	26.20	26.20
remove for work, then reinstall	1C@.685	ea	—	46.30	46.30
Install door stop					
hinge	1C@.128	ea	—	8.65	8.65
baseboard	1C@.138	ea	—	9.33	9.33
remove for work, then reinstall	1C@.252	ea	—	17.00	17.00
Install door push plate					
install	1C@.417	ea	—	28.20	28.20
remove for work, then reinstall	1C@.742	ea	—	50.20	50.20
Install door kick plate					
install	1C@.445	ea	—	30.10	30.10
remove for work, then reinstall	1C@.809	ea	—	54.70	54.70
Install letter plate					
install	1C@.500	ea	—	33.80	33.80
remove for work, then reinstall	1C@.902	ea	—	61.00	61.00
Install door peep hole					
install	1C@.504	ea	—	34.10	34.10
remove for work, then reinstall	1C@.879	ea	—	59.40	59.40
Install door sweep					
install	1C@.431	ea	—	29.10	29.10
remove for work, then reinstall	1C@.763	ea	—	51.60	51.60
Install door threshold					
install	1C@.477	ea	—	32.20	32.20
remove for work, then reinstall	1C@.862	ea	—	58.30	58.30
Install door weatherstripping					
install	1C@.417	ea	—	28.20	28.20
Install garage door hardware					
install	1C@2.30	ea	—	155.00	155.00
remove for work, then reinstall	1C@4.05	ea	—	274.00	274.00

	Craft@Hrs	Unit	Material	Labor	Total

Drywall

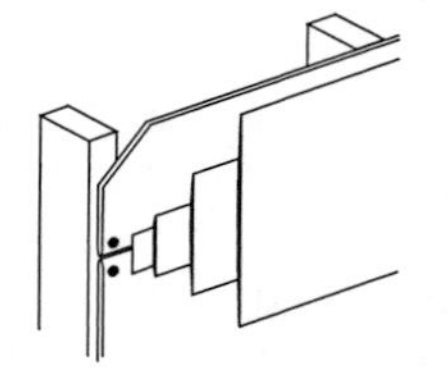

Drywall Installation. Drywall is hung and taped according to industry standards. Drywall is attached with nails when hung, then permanently affixed with screws. Perfatape and three coats of mud are applied plus texture. One lf of corner bead is calculated for every 24 sf of installed drywall. Productivity is based on repair installations.

Remove drywall. Tear-out and debris removal to a truck or dumpster on site. Does not include hauling, dumpster, or dump fees. No salvage value is assumed.

	Craft@Hrs	Unit	Material	Labor	Total
remove, drywall & prep walls	1D@.008	sf	—	.38	.38
scrape (remove) acoustic ceiling texture	1D@.011	sf	—	.52	.52
scrape (remove) painted acoustic ceiling texture	1D@.019	sf	—	.90	.90
remove, furring strips on wood, 16" oc	1D@.010	sf	—	.47	.47
remove, furring strips on concrete / masonry, 16" oc	1D@.011	sf	—	.52	.52
remove, furring strips on wood, 24" oc	1D@.009	sf	—	.42	.42
remove, furring strips on concrete / masonry, 24" oc	1D@.010	sf	—	.47	.47

3/8" drywall installed. Includes drywall board, drywall mud, drywall nails and/or screws, drywall tape, and corner bead.

	Craft@Hrs	Unit	Material	Labor	Total
replace, with machine texture	6D@.031	sf	1.01	1.79	2.80
replace, with knock-down machine texture	6D@.033	sf	1.01	1.90	2.91
replace, with light texture	6D@.034	sf	1.01	1.96	2.97
replace, with medium texture	6D@.035	sf	1.13	2.02	3.15
replace, with heavy texture	6D@.036	sf	1.22	2.07	3.29
replace, with smooth-wall finish	6D@.041	sf	1.06	2.36	3.42
replace, all coats, no texture	6D@.028	sf	.82	1.61	2.43
replace, hung & fire taped only	6D@.021	sf	.79	1.21	2.00
hung only (no tape, coating or texture)	6D@.011	sf	.73	.63	1.36
remove, drywall	1D@.008	sf	—	.38	.38
add for installation over existing plaster wall	6D@.003	sf	—	.17	.17
add for 3/8" foil-backed drywall	—	sf	.21	—	.21

1/2" drywall installed. Includes drywall board, drywall mud, drywall nails and/or screws, drywall tape, and corner bead.

	Craft@Hrs	Unit	Material	Labor	Total
replace, with machine texture	6D@.031	sf	.99	1.79	2.78
replace, with knock-down machine texture	6D@.033	sf	.99	1.90	2.89
replace, with light texture	6D@.034	sf	1.00	1.96	2.96
replace, with medium texture	6D@.035	sf	1.06	2.02	3.08
replace, with heavy texture	6D@.036	sf	1.18	2.07	3.25
replace, with smooth-wall finish	6D@.041	sf	1.02	2.36	3.38
replace, all coats, no texture	6D@.028	sf	.79	1.61	2.40
replace, hung & fire taped only	6D@.021	sf	.72	1.21	1.93
replace, hung only (no tape, coating or texture)	6D@.011	sf	.69	.63	1.32
remove, drywall	1D@.008	sf	—	.38	.38
add for installation over existing plaster wall	6D@.003	sf	—	.17	.17
add for 1/2" moisture-resistant drywall	—	sf	.25	—	.25
add for 1/2" type X fire-rated drywall	—	sf	.19	—	.19
add for 1/2" foil-backed drywall	—	sf	.29	—	.29
add for 1/2" type X fire-rated, foil-backed drywall	—	sf	.48	—	.48

	Craft@Hrs	Unit	Material	Labor	Total
5/8" drywall installed. Includes drywall board, drywall mud, drywall nails and/or screws, drywall tape, and corner bead.					
replace, with machine texture	6D@.031	sf	1.13	1.79	2.92
replace, with knock-down machine texture	6D@.033	sf	1.13	1.90	3.03
replace, with light texture	6D@.034	sf	1.14	1.96	3.10
replace, with medium texture	6D@.035	sf	1.24	2.02	3.26
replace, with heavy texture	6D@.036	sf	1.32	2.07	3.39
replace, with smooth-wall finish	6D@.041	sf	1.21	2.36	3.57
replace, all coats, no texture	6D@.028	sf	.91	1.61	2.52
replace, hung & fire taped only	6D@.021	sf	.88	1.21	2.09
replace, hung only (no tape, coating or texture)	6D@.011	sf	.86	.63	1.49
remove, drywall	1D@.008	sf	—	.38	.38
add for installation over existing plaster wall	6D@.003	sf	—	.17	.17
add for 5/8" moisture-resistant drywall	—	sf	.22	—	.22
add for 5/8" foil-backed drywall	—	sf	.40	—	.40
Acoustic ceiling texture (cottage cheese). Includes acoustic ceiling texture mix, optional glitter, and installation.					
replace, light coverage	5D@.008	sf	.30	.50	.80
replace, medium coverage	5D@.011	sf	.40	.69	1.09
replace, heavy coverage	5D@.016	sf	.43	1.01	1.44
remove, acoustic ceiling	1D@.011	sf	—	.52	.52
glitter application to acoustic ceiling texture	5D@.002	sf	.01	.13	.14
minimum charge for acoustic ceiling texture	5D@1.25	ea	216.00	78.90	294.90
Drywall patch.					
drywall patch (match existing texture or finish)	5D@1.42	ea	57.20	89.60	146.80
minimum charge for drywall repair	6D@6.50	ea	71.50	374.00	445.50
Texture drywall.					
with light hand texture	5D@.005	sf	.19	.32	.51
with medium hand texture	5D@.006	sf	.30	.38	.68
with heavy hand texture	5D@.006	sf	.40	.38	.78
with thin-coat smooth-wall	5D@.012	sf	.27	.76	1.03
machine applied	5D@.004	sf	.19	.25	.44
machine applied and knocked down with trowel	5D@.005	sf	.19	.32	.51
Metal furring channel installed on wood. Includes metal channel, nails and installation.					
Metal hat-shaped furring channel					
replace, 16" on center	2D@.011	sf	.28	.60	.88
replace, 24" on center	2D@.008	sf	.19	.44	.63
Metal sound-resistant furring channel					
replace, 16" on center	2D@.011	sf	.30	.60	.90
replace, 24" on center	2D@.008	sf	.24	.44	.68
Remove					
furring strips on wood, 16" oc	1D@.010	sf	—	.47	.47
furring strips on wood, 24" oc	1D@.009	sf	—	.42	.42

hat-shaped channel

	Craft@Hrs	Unit	Material	Labor	Total
Metal furring channel installed on concrete or masonry. Includes metal channel, case hardened or powder-actuated nails and installation.					
Metal hat-shaped furring channel					
replace, 16" on center	2D@.012	sf	.28	.66	.94
replace, 24" on center	2D@.008	sf	.19	.44	.63
Metal sound-resistant furring channel					
replace, 16" on center	2D@.012	sf	.30	.66	.96
replace, 24" on center	2D@.008	sf	.24	.44	.68
Remove					
furring strips on concrete or masonry, 16" oc	1D@.011	sf	—	.52	.52
furring strips on concrete or masonry, 24" oc	1D@.010	sf	—	.47	.47

sound-resistant channel

	Craft@Hrs	Unit	Material	Labor	Total
Metal Z channel. Includes metal Z channel, case hardened or powder-actuated nails, and installation on interior of masonry walls.					
Use with rigid foam insulation					
replace, 16" on center	2D@.012	sf	.19	.66	.85
replace, 24" on center	2D@.008	sf	.14	.44	.58

Z channel

Time & Material Charts (selected items)
Drywall Materials

	Craft@Hrs	Unit	Material	Labor	Total
3/8" drywall board					
standard ($659.00 1,000 sf), 8% waste	—	sf	.70	—	.70
foil-backed ($857.00 1,000 sf), 8% waste	—	sf	.91	—	.91
1/2" drywall board					
standard ($613.00 1,000 sf), 7% waste	—	sf	.66	—	.66
moisture-resistant ($853.00 1,000 sf), 7% waste	—	sf	.90	—	.90
foil-backed ($892.00 1,000 sf), 7% waste	—	sf	.96	—	.96
fire-rated type-X ($765.00 1,000 sf), 7% waste	—	sf	.82	—	.82
fire-rated type-X foil-backed ($1,080.00 1,000 sf), 7% waste	—	sf	1.15	—	1.15
5/8" drywall board					
type X ($765.00 1,000 sf), 7% waste	—	sf	.82	—	.82
moisture-resistant ($970.00 1,000 sf), 7% waste	—	sf	1.05	—	1.05
Fasteners					
drywall nails, used to secure when hanging ($75.50 per box, 32,000 sf), 4% waste	—	sf	.01	—	.01
drywall screws, used to attach after hung ($156.00 per box, 8,534 sf), 3% waste	—	sf	.01	—	.01
Drywall taping materials					
perfatape (per 500' roll) ($10.20 per roll, 1,111 sf), 6% waste	—	sf	.01	—	.01
corner bead ($1.45 per stick, 204 sf), 4% waste	—	sf	.01	—	.01
joint compound ($29.90 per 50 lb box, 965 sf), 16% waste	—	sf	.03	—	.03

	Craft@Hrs	Unit	Material	Labor	Total
Drywall texture compound					
machine-sprayed					
($32.40 per 50 lb box, 190 sf), 13% waste	—	sf	.20	—	.20
light hand texture					
($32.40 per 50 lb box, 175 sf), 13% waste	—	sf	.21	—	.21
medium hand texture					
($32.40 per 50 lb box, 120 sf), 13% waste	—	sf	.30	—	.30
heavy hand texture					
($32.40 per 50 lb box, 93 sf), 14% waste	—	sf	.40	—	.40
heavy trowel texture					
($32.40 per 50 lb box, 77 sf), 15% waste	—	sf	.48	—	.48
smooth wall					
($32.40 per 50 lb box, 142 sf), 14% waste	—	sf	.27	—	.27
Acoustic ceiling (cottage cheese) texture compound					
light texture					
($56.50 per 50 lb bag, 220 sf), 17% waste	—	sf	.30	—	.30
medium texture					
($56.50 per 50 lb box, 170 sf), 17% waste	—	sf	.39	—	.39
heavy texture					
($56.50 per 50 lb box, 153 sf), 17% waste	—	sf	.43	—	.43
Acoustic ceiling glitter (cottage cheese) texture compound (per pound), light to heavy application					
($5.28 per pound, 2,350 sf), 22% waste	—	sf	.01	—	.01
Metal furring strips					
hat-shaped channel installed 24" oc					
($4.14 per 12' stick, 21.34 sf), 3% waste	—	sf	.20	—	.20
sound-resistant channel installed 24" oc					
($4.56 per 12' stick, 21.34 sf), 3% waste	—	sf	.23	—	.23
Z channel installed 24" on center					
($2.16 per 8' stick, 15.11 sf), 3% waste	—	sf	.15	—	.15

	Craft@Hrs	Unit	Material	Labor	Equip.	Total
Drywall Rental Equipment						
Ames taping tools						
($16.70 per day, 416 sf per day)	—	sf	—	—	.03	.03
texture equipment (gun, hopper & compressor)						
($118.00 per day, 1,536 sf per day)	—	sf	—	—	.07	.07
glitter blower for acoustic ceilings						
($8.32 per day, 5,000 sf per day)	—	csf	—	—	.01	.01
drywall lifter (for one-person hanging)						
($33.40 per day, 102 sf per day)	—	sf	—	—	.33	.33
minimum charge	—	ea	—	—	83.30	83.30

Drywall Labor

Laborer	base wage	paid leave	true wage	taxes & ins.	total
Drywall hanger	$36.40	2.84	$39.24	23.36	$62.60
Drywall hanger's helper	$26.40	2.06	$28.46	18.74	$47.20
Drywall taper	$36.70	2.86	$39.56	23.54	$63.10
Demolition laborer	$25.90	2.02	$27.92	19.28	$47.20

Paid leave is calculated based on two weeks paid vacation, one week sick leave, and seven paid holidays. Employer's matching portion of **FICA** is 7.65 percent. **FUTA** (Federal Unemployment) is .8 percent. **Worker's compensation** for the drywall trade was calculated using a national average of 11.56 percent. **Unemployment insurance** was calculated using a national average of 8 percent. **Health insurance** was calculated based on a projected national average for 2020 of $1,280 per employee (and family when applicable) per month. Employer pays 80 percent for a per month cost of $1,024 per employee. **Retirement** is based on a 401(k) retirement program with employer matching of 50 percent. Employee contributions to the 401(k) plan are an average of 6 percent of the true wage. **Liability insurance** is based on a national average of 12.0 percent.

Drywall Labor Productivity

	Craft@Hrs	Unit	Material	Labor	Total
Demolition of drywall					
remove, drywall and prep walls	1D@.008	sf	—	.38	.38
scrape and prep acoustical ceiling	1D@.011	sf	—	.52	.52
scrape and prep painted acoustical ceiling	1D@.019	sf	—	.90	.90
remove, furring strips on wood, 16" oc	1D@.010	sf	—	.47	.47
remove, furring strips on conc. / masonry, 16" oc	1D@.011	sf	—	.52	.52
remove, furring strips on wood, 24" oc	1D@.009	sf	—	.42	.42
remove, furring strips on conc. / masonry, 24" oc	1D@.010	sf	—	.47	.47

Hang drywall

hang drywall	drywall hanger	$62.60
hang drywall	hanger's helper	$47.20
hang drywall	hanging crew	$54.90

	Craft@Hrs	Unit	Material	Labor	Total
Tape and coat drywall					
to finish (no texture)	5D@.017	sf	—	1.07	1.07
tape coat only	5D@.010	sf	—	.63	.63
tape coat and one additional coat only	5D@.012	sf	—	.76	.76
to finish, apply machine-sprayed texture	5D@.020	sf	—	1.26	1.26
to finish, apply machine-sprayed, knock-down	5D@.021	sf	—	1.33	1.33
to finish, apply light hand texture	5D@.022	sf	—	1.39	1.39
to finish, apply heavy hand texture	5D@.024	sf	—	1.51	1.51
to finish, apply smooth-wall texture	5D@.029	sf	—	1.83	1.83
Apply texture on drywall					
with compressor and spray gun	5D@.004	sf	—	.25	.25
with spray gun, then knock down with trowel	5D@.005	sf	—	.32	.32
light hand texture	5D@.005	sf	—	.32	.32
heavy hand texture	5D@.006	sf	—	.38	.38
thin coat smooth-wall finish	5D@.012	sf	—	.76	.76
Apply acoustical ceiling texture					
Light	5D@.008	sf	—	.50	.50
Heavy	5D@.016	sf	—	1.01	1.01
apply glitter to freshly sprayed ceiling	5D@.002	sf	—	.13	.13

Electrical

	Craft@Hrs	Unit	Material	Labor	Total
Minimum charge.					
for electrical work	9E@3.25	ea	31.80	205.00	236.80

Complete house electrical wiring, outlets, switches & light fixtures. Complete house wiring, boxes, switches, outlets and fixtures to code for a house that is not electrically heated. (Electrical hookups for furnace, AC unit, and/or heat pump included.) Does not include breaker panel, main disconnect, or service drop. Grades refer to quality of light fixtures.

	Craft@Hrs	Unit	Material	Labor	Total
replace, economy grade	9E@.053	sf	2.73	3.35	6.08
replace, standard grade	9E@.090	sf	3.11	5.69	8.80
replace, high grade	9E@.104	sf	3.59	6.57	10.16
replace, deluxe grade	9E@.109	sf	3.82	6.89	10.71
replace, low-voltage system	9E@.100	sf	3.44	6.32	9.76
add complete house wiring in conduit	9E@.055	sf	3.48	3.48	6.96

Complete house rough electrical (no light fixtures). Same as above but does not include fixtures.

	Craft@Hrs	Unit	Material	Labor	Total
replace, standard	9E@.070	sf	1.72	4.42	6.14
replace, in conduit	9E@.154	sf	3.83	9.73	13.56
replace, low-voltage system	9E@.075	sf	1.89	4.74	6.63

Complete house light fixtures. Fixtures only.

	Craft@Hrs	Unit	Material	Labor	Total
replace, economy grade	9E@.015	sf	1.77	.95	2.72
replace, standard grade	9E@.016	sf	2.27	1.01	3.28
replace, high grade	9E@.016	sf	3.19	1.01	4.20
replace, deluxe grade	9E@.017	sf	4.21	1.07	5.28
replace, low-voltage system	9E@.013	sf	3.06	.82	3.88

Electrical outlets and switches with wiring. Unless otherwise indicated, includes outlet or switch, box, and up to 26 lf wire. Voltage listed as "120 volt" applies to 110, 115, 120 and 125 volt applications. Voltage listed as "240 volt" applies to 220, 240 or 250 volt applications.

	Craft@Hrs	Unit	Material	Labor	Total
120 volt outlet with wiring for general-purpose use, average of 6 outlets per single-pole breaker (breaker not included)					
replace, up to 15 lf of #12/2 wire	9E@.977	ea	24.20	61.70	85.90
120 volt GFCI outlet with wiring for general-purpose use, average of 3 outlets per single-pole breaker					
replace, up to 26 lf of #12/2 wire	9E@.977	ea	79.50	61.70	141.20
120 volt exterior waterproof outlet with wiring for general-purpose use, average of 3 outlets per single-pole breaker (breaker not included)					
replace, up to 26 lf of #12/2 wire	9E@.977	ea	31.80	61.70	93.50
120 volt switch with wiring					
replace, up to 15 lf of #12/2 wire	9E@.977	ea	31.80	61.70	93.50
120 volt 3-way switch with wiring					
replace, up to 20 lf of #12/2 wire	9E@1.95	ea	55.80	123.00	178.80

	Craft@Hrs	Unit	Material	Labor	Total
120 volt exterior waterproof switch with wiring					
replace, up to 26 lf of #12/2 wire	9E@.977	ea	31.80	61.70	93.50
240 volt outlet with wiring					
replace, 30 amp, with up to 30 lf #10/3 wire and 30 amp double-pole breaker	9E@1.34	ea	111.00	84.70	195.70
replace, 40 amp, with up to 30 lf #8/3 wire and 40 amp double-pole breaker	9E@1.34	ea	128.00	84.70	212.70
replace, 50 amp, with up to 30 lf #6/3 wire and 50 amp double-pole breaker	9E@1.34	ea	149.00	84.70	233.70
240 volt switch with wiring					
replace, with up to 15 lf of #10/2 wire	9E@1.34	ea	128.00	84.70	212.70
Low-voltage switch wiring with up to 18 lf of #18/2 wire	9E@.977	ea	39.80	61.70	101.50

Electrical outlets & switches. Includes outlet or switch and cover only (no wire or box).

	Craft@Hrs	Unit	Material	Labor	Total
120 volt outlet with cover					
replace, outlet	9E@.226	ea	16.00	14.30	30.30
replace, GFCI outlet	9E@.226	ea	51.00	14.30	65.30
replace, exterior waterproof outlet	9E@.226	ea	25.50	14.30	39.80
remove	1D@.150	ea	—	7.08	7.08
120 volt switch with cover					
replace, switch	9E@.226	ea	10.40	14.30	24.70
replace, lighted switch	9E@.226	ea	14.90	14.30	29.20
replace, push-button switch	9E@.226	ea	31.80	14.30	46.10
replace, rotary dimmer switch	9E@.226	ea	41.60	14.30	55.90
replace, touch dimmer switch	9E@.226	ea	50.10	14.30	64.40
replace, exterior waterproof	9E@.226	ea	26.20	14.30	40.50
remove	1D@.150	ea	—	7.08	7.08
240 volt outlet or switch with cover					
replace, outlet	9E@.292	ea	24.20	18.50	42.70
replace, switch	9E@.292	ea	26.20	18.50	44.70
remove	1D@.200	ea	—	9.44	9.44
Low-voltage outlet or switch with cover					
replace, computer network outlet	9E@.335	ea	50.10	21.20	71.30
replace, sound system speaker outlet	9E@.256	ea	14.90	16.20	31.10
replace, TV outlet	9E@.256	ea	13.60	16.20	29.80
replace, phone outlet	9E@.256	ea	9.57	16.20	25.77
replace, switch	9E@.226	ea	18.60	14.30	32.90
remove	1D@.150	ea	—	7.08	7.08
add for high grade outlet or switch cover	—	ea	1.52	—	1.52
add for deluxe grade outlet or switch cover	—	ea	4.58	—	4.58

push-button switch

Outlet or switch covers.

	Craft@Hrs	Unit	Material	Labor	Total
replace, standard grade	9E@.020	ea	1.63	1.26	2.89
replace, high grade	9E@.020	ea	6.58	1.26	7.84
replace, deluxe grade	9E@.021	ea	11.40	1.33	12.73
remove	1D@.035	ea	—	1.65	1.65

120 volt wiring runs. Does not include breaker or finish electrical. Replace only.

	Craft@Hrs	Unit	Material	Labor	Total
120 volt dedicated appliance circuit with up to 26 lf #12/2 wire	9E@1.02	ea	71.90	64.50	136.40
120 volt wiring for through-wall AC and/or heater with up to 26 lf of #12/2 wire and box	9E@1.25	ea	55.20	79.00	134.20
120 volt wiring for bathroom fan with up to 20 lf of #12/2 wire	9E@.779	ea	28.50	49.20	77.70
120 volt wiring with GFCI for bathroom ceiling heater with up to 24 lf of #12/2 wire	9E@.820	ea	78.10	51.80	129.90
120 volt wiring for clothes washer, includes 1 to 2 outlets per circuit and up to 20 lf of #12/2 wire	9E@.779	ea	50.10	49.20	99.30
120 volt wiring for dishwasher on direct-wired run with up to 26 lf of #13/2 wire	9E@.779	ea	31.80	49.20	81.00
120 volt wiring for door bell or chime with up to 15 lf of #12/2 wire and up to 30 lf of #18 low-voltage wiring, and box	9E@.779	ea	49.30	49.20	98.50
120 volt wiring for electric baseboard heater 1 direct-wired run per 9 lf of baseboard heater or per location, with up to 30 lf of #12/2 wire	9E@.820	ea	34.20	51.80	86.00
120 volt wiring for electrical-resistance heating direct-wired run with a single-pole breaker, does not include electrical resistance wiring, includes up to 30 lf of #12/2 wire	9E@.896	ea	47.90	56.60	104.50
120 volt wiring for furnace, direct-wired run with up to 30 lf of #12/2 wire	9E@.779	ea	19.20	49.20	68.40
120 volt wiring for garage door opener with up to 30 lf of #12/2 wire	9E@.820	ea	27.20	51.80	79.00
120 volt wiring for garbage disposal, direct-wired run with outlet, switch and up to 30 lf of #12/2 wire	9E@1.00	ea	43.70	63.20	106.90
120 volt wiring for gas oven or range, light appliance circuit, 2 to 3 outlets per circuit with up to 20 lf of #12/2 wire	9E@.779	ea	28.50	49.20	77.70
120 volt wiring for humidifier, no more than 1 to 3 fixtures or outlets per run, with up to 30 lf of #12/2 wire	9E@.779	ea	27.20	49.20	76.40
120 volt wiring for in-sink hot water dispenser, direct-wired run with up to 30 lf of #12/2 wire	9E@.779	ea	27.20	49.20	76.40
120 volt wiring for intercom master station with up to 20 lf of #12/2 wire	9E@.896	ea	46.20	56.60	102.80
120 volt wiring for kitchen fan with up to 20 lf of #12/2 wire	9E@.779	ea	31.00	49.20	80.20
120 volt light fixture circuit, general-purpose, average of 6 outlets or lights per single-pole breaker with up to15 lf #12/2 wire and box	9E@.779	ea	45.40	49.20	94.60

	Craft@Hrs	Unit	Material	Labor	Total
120 volt wiring for range hood fan and light general-purpose circuit with up to 20 lf of #12/2 wire	9E@.779	ea	29.60	49.20	78.80
120 volt wiring for security system, general-purpose circuit with up to 15 lf of #12/2 wire and box	9E@.779	ea	50.10	49.20	99.30
120 volt wiring for smoke or carbon monoxide detector (or for direct-wired radon and carbon monoxide detectors), general-purpose circuit, with up to 15 lf of #12/2 wire and box	9E@.779	ea	41.60	49.20	90.80
120 volt wiring for sump / water pump, direct-wired run, includes up to 13 lf liquid-tight flexible conduit, remote switch with box, single-pole breaker, and up to 50 lf of #12/2 wire	9E@5.38	ea	31.80	340.00	371.80
120 volt wiring for trash compactor on garbage disposal run, and up to 15 lf of #12/2 wire	9E@.977	ea	41.60	61.70	103.30
120 volt wiring for built-in vacuum system with up to 30 lf of #12/2 wire	9E@.820	ea	27.20	51.80	79.00
120 volt wiring for typical outlet, general-purpose circuit with up to 15 lf #12/2 wire	9E@.802	ea	55.80	50.70	106.50
120 volt wiring for water heater, direct-wired run, single-pole breaker, and up to 30 lf of #12/2 wire	9E@2.11	ea	31.80	133.00	164.80
120 volt wiring for whole-house fan with remote switch with box, and up to 36 lf of #12/2 wire	9E@.820	ea	36.50	51.80	88.30

240 volt wiring runs. Unless indicated, does not include breaker or finish electrical. Replace only.

	Craft@Hrs	Unit	Material	Labor	Total
240 volt wiring for air conditioner, includes up to 36 lf of #6/3 wire, up to 32 lf of #3/2 wire, box, 60 amp disconnect switch, 40 amp 2 pole breaker, and up to 6 lf of liquid-tight flexible conduit	9E@2.99	ea	255.00	189.00	444.00
240 volt wiring, box & outlet for clothes dryer with a 30 amp double-pole breaker and up to 30 lf of #8/3 wire	9E@1.34	ea	115.00	84.70	199.70
240 volt wiring, box & outlet for clothes dryer with a 40 amp double-pole breaker and up to 30 lf of #8/3 wire	9E@1.34	ea	128.00	84.70	212.70
240 volt wiring for electric cook top, includes a 40 amp double-pole breaker and up to 38 lf of #8/3 wire	9E@1.34	ea	139.00	84.70	223.70
240 volt wiring for electric baseboard heater, direct-wired run per 9 lf of baseboard heater or per location, with up to 30 lf of #10/2 wire	9E@1.34	ea	41.60	84.70	126.30
240 volt wiring for electrical-resistance heating, direct-wired run and double-pole breaker, does not include electrical resistance wiring, includes up to 30 lf of #8/2 wire	9E@1.22	ea	638.00	77.10	715.10

	Craft@Hrs	Unit	Material	Labor	Total
240 volt wiring, box & outlet for electric oven or range, 40 amp double-pole breaker and up to 30 lf of #8/3 wire	9E@1.34	ea	115.00	84.70	199.70
240 volt wiring, box & outlet for electric oven or range, 50 amp double-pole breaker and up to 30 lf of #6/3 wire	9E@1.34	ea	146.00	84.70	230.70
240 volt wiring for heat pump, includes up to 42 lf of #6/3 wire, up to 36 lf of #3/2 wire, box, 60 amp disconnect switch, 100 amp 2 pole breaker, 40 amp 2 pole breaker, up to 6 lf of liquid-tight flexible conduit	9E@7.69	ea	558.00	486.00	1,044.00
240 volt wiring for water heater, direct-wired run with double-pole breaker and up to 30 lf of #10/2 wire	9E@1.34	ea	122.00	84.70	206.70

Low-voltage wiring runs. Does not include breaker or finish electrical. Replace only.

	Craft@Hrs	Unit	Material	Labor	Total
Low voltage computer network wiring (per outlet) with up to 28 lf computer network cable	9E@.358	ea	46.20	22.60	68.80
Low-voltage wiring for door bell or chime (per button) with up to 32 lf of #18/2 low-voltage wire	9E@.563	ea	5.72	35.60	41.32
Low-voltage wiring for door or gate latch release with up to 40 lf of #18/2 low-voltage wire	9E@.716	ea	24.20	45.30	69.50
Low-voltage wiring for intercom (per station) with up to 32 lf of #18/2 low-voltage wire	9E@.619	ea	22.90	39.10	62.00
Low voltage wiring for light fixture, average of 6 lights per circuit, with up to 18 lf #18/2 wire and box	9E@.779	ea	50.10	49.20	99.30
Low voltage wiring for phone wiring (per outlet) with up to 30 lf 4 or 8 bell wire	9E@.358	ea	8.76	22.60	31.36
Low voltage wiring for sound system speaker (per speaker) with up to 30 lf #20/2 wire	9E@.358	ea	12.80	22.60	35.40
Low-voltage wiring for thermostat with up to 28 lf of #18/2 low-voltage wire	9E@.538	ea	20.80	34.00	54.80
Low-voltage wiring for TV outlet (per outlet) with up to 30 lf of shielded coaxial television cable	9E@.358	ea	11.90	22.60	34.50

Conduit, average installation. Based on typical home or light commercial installation. Per sf of floor. Replace only.

	Craft@Hrs	Unit	Material	Labor	Total
Electric metallic tubing (EMT) of various diameters: 94% 1/2"; 3% 3/4"; 2% misc. flexible and larger IMC and RMC conduit	9E@.077	sf	1.29	4.87	6.16
Flexible metal conduit of various diameters: 72% 1/2"; 13% 3/4"; 15% misc. EMT, IMC and RMC conduit	9E@.041	sf	3.04	2.59	5.63
PVC conduit of various diameters: 94% 1/2"; 3% 3/4"; 2% misc. larger sizes.	9E@.045	sf	.82	2.84	3.66
Intermediate conduit (IMC) of various diameters: 44% 1/2"; 35% 3/4" or 1"; 22% misc. larger sizes	9E@.113	sf	2.85	7.14	9.99
Rigid metal conduit (RMC) of various diameters: 44% 1/2"; 35% 3/4" or 1"; 22% misc. larger sizes	9E@.067	sf	9.08	4.23	13.31

	Craft@Hrs	Unit	Material	Labor	Total
Wire. With ground non-metallic sheathed cable (Romex). Average installation and conditions for single family housing and light commercial construction. Per lf of installed wire. Replace only.					
#10 2	9E@.051	lf	1.04	3.22	4.26
#12 2	9E@.047	lf	.62	2.97	3.59
#14 2	9E@.040	lf	.49	2.53	3.02
Breaker panel. Does not include main disconnect breaker.					
replace, 100 amp exterior panel 20 breaker capacity, 5 breakers installed	7E@4.89	ea	589.00	354.00	943.00
replace, 100 amp interior panel 20 breaker capacity, 5 breakers installed	7E@4.30	ea	558.00	311.00	869.00
replace, 150 amp exterior panel 20 breaker capacity, 10 breakers installed	7E@7.68	ea	756.00	556.00	1,312.00
replace, 150 amp interior panel 20 breaker capacity, 10 breakers installed	7E@7.17	ea	719.00	519.00	1,238.00
replace, 200 amp exterior panel 40 breaker capacity, 15 breakers installed	7E@10.8	ea	974.00	782.00	1,756.00
replace, 200 amp interior panel 40 breaker capacity, 15 breakers installed	7E@9.77	ea	916.00	707.00	1,623.00
replace, 300 amp exterior panel 42 breaker capacity, 30 breakers installed	7E@14.3	ea	996.00	1,040.00	2,036.00
replace, 300 amp interior panel 42 breaker capacity, 30 breakers installed	7E@13.4	ea	933.00	970.00	1,903.00
replace, 40 amp interior sub-panel 10 breaker capacity, 2 breakers installed	7E@2.99	ea	255.00	216.00	471.00
replace, 50 amp interior sub-panel 10 breaker capacity, 3 breakers installed.	7E@3.16	ea	335.00	229.00	564.00
replace, 70 amp interior sub-panel 10 breaker capacity, 4 breakers installed.	7E@3.47	ea	398.00	251.00	649.00
remove	1D@6.09	ea	—	287.00	287.00
remove panel for work, then reinstall	7E@8.28	ea	—	599.00	599.00

breaker panel

sub-panel

Electrical service. Overhead installation. Add **13%** for underground supply — includes excavation, trench shaping, sand bed, backfill, up to 38 lf USE cable, and underground materials. Does not include breaker panel, see above.

	Craft@Hrs	Unit	Material	Labor	Total
100 amp cable service with up to 7 lf service entrance cable type SE, insulator, weather head, meter base, main disconnect breaker, up to 6 lf copper ground wire, grounding rod, clamps, fasteners, and waterproof connectors.					
replace	7E@11.9	ea	1,040.00	862.00	1,902.00
remove	1D@10.5	ea	—	496.00	496.00

cable service

	Craft@Hrs	Unit	Material	Labor	Total

150 amp cable service with up to 7 lf service entrance cable type SE, insulator, weather head, meter base, main disconnect breaker, up to 6 lf copper ground wire, grounding rod, clamps, fasteners, and waterproof connectors.

	Craft@Hrs	Unit	Material	Labor	Total
replace	7E@14.3	ea	1,400.00	1,040.00	2,440.00
remove	1D@10.5	ea	—	496.00	496.00

cable service

150 amp conduit service with up to 7 lf service entrance cable type SE, insulator, weather head for conduit, EMT conduit with liquid-tight couplings, conduit straps, conduit entrance ell, meter base, main disconnect breaker, up to 6 lf copper ground wire, grounding rod, clamps, fasteners, and waterproof connectors. Do not use this item if conduit extends through roof. See rigid steel mast installations.

	Craft@Hrs	Unit	Material	Labor	Total
replace	7E@15.4	ea	1,460.00	1,110.00	2,570.00
remove	1D@13.1	ea	—	618.00	618.00

conduit service

150 amp rigid steel mast service. Rigid steel conduit extends through roof. Up to 10 lf service entrance cable type SE, insulator (clamps to mast), weather head for conduit, 2" galvanized rigid steel conduit with liquid-tight connections, roof flashing, conduit supports, conduit offset fittings, conduit entrance ell, meter base, main disconnect breaker, up to 6 lf copper ground wire, grounding rod, clamps, fasteners, and waterproof connectors. Add **$141** if guy wire is needed to support mast.

	Craft@Hrs	Unit	Material	Labor	Total
replace	7E@15.4	ea	1,550.00	1,110.00	2,660.00
remove	1D@15.4	ea	—	727.00	727.00

rigid steel mast service

200 amp cable service with up to 7 lf service entrance cable type SE, insulator, weather head, meter base, main disconnect breaker, up to 6 lf copper ground wire, grounding rod, clamps, fasteners, and waterproof connectors.

	Craft@Hrs	Unit	Material	Labor	Total
replace	7E@18.5	ea	1,860.00	1,340.00	3,200.00
remove	1D@10.5	ea	—	496.00	496.00

200 amp conduit service with up to 7 lf service entrance cable type SE, insulator, weather head for conduit, EMT conduit with liquid-tight couplings, conduit straps, conduit entrance ell, meter base, main disconnect breaker, up to 6 lf copper ground wire, grounding rod, clamps, fasteners, and waterproof connectors. Do not use this item if conduit extends through roof. See rigid steel mast installations.

	Craft@Hrs	Unit	Material	Labor	Total
replace	7E@19.9	ea	1,920.00	1,440.00	3,360.00
remove	1D@13.1	ea	—	618.00	618.00

200 amp rigid steel mast service. Rigid steel conduit extends through roof. Up to 10 lf service entrance cable type SE, insulator (clamps to mast), weather head for conduit, 2" galvanized rigid steel conduit with liquid-tight connections, roof flashing, conduit supports, conduit offset fittings, conduit entrance ell, meter base, main disconnect breaker, up to 6 lf copper ground wire, grounding rod, clamps, fasteners, and waterproof connectors. Add **$141** if guy wire is needed to support mast.

	Craft@Hrs	Unit	Material	Labor	Total
replace	7E@19.9	ea	2,000.00	1,440.00	3,440.00
remove	1D@15.4	ea	—	727.00	727.00

	Craft@Hrs	Unit	Material	Labor	Total
Circuit breaker.					
replace, single pole (120 volt) circuit breaker	7E@.636	ea	31.80	46.00	77.80
replace, double pole (240 volt) circuit breaker	7E@.649	ea	62.10	47.00	109.10
replace, ground fault circuit interrupter breaker	7E@.689	ea	102.00	49.90	151.90
replace, main disconnect circuit breaker	7E@3.47	ea	589.00	251.00	840.00
remove	1D@.200	ea	—	9.44	9.44
remove breaker for work, then reinstall	7E@1.08	ea	—	78.20	78.20

single pole

double pole

Door bell or chime. Includes door bell or chime, fasteners, and installation. Door bell or chime unit includes interior unit and up to three buttons. Does not include door bell button wiring or electrical wiring for the main unit.

	Craft@Hrs	Unit	Material	Labor	Total
replace, economy grade, single tone, entry door only	9E@1.24	ea	92.50	78.40	170.90
replace, standard grade, separate tones for entry and secondary doors	9E@1.24	ea	165.00	78.40	243.40
replace, high grade, select from multiple tones	9E@1.24	ea	229.00	78.40	307.40
replace, deluxe grade, programmable, multiple tones	9E@1.24	ea	430.00	78.40	508.40
remove, door bell or chime	1D@.209	ea	—	9.86	9.86
replace, button only (no wire)	9E@.137	ea	10.50	8.66	19.16
remove, button only (no wire)	1D@.100	ea	—	4.72	4.72
remove chime for work, then reinstall	9E@1.90	ea	—	120.00	120.00

Bathroom exhaust fan. Includes up to 10 lf of 4" flexible vent duct and wall outlet.

	Craft@Hrs	Unit	Material	Labor	Total
replace, economy grade, 50 cfm	9E@.685	ea	57.20	43.30	100.50
replace, standard grade, 65 cfm, low noise	9E@.685	ea	95.70	43.30	139.00
replace, high grade, 110 cfm, low noise	9E@.685	ea	136.00	43.30	179.30
replace, deluxe grade, 110 cfm, low noise, variable speeds	9E@.685	ea	215.00	43.30	258.30
remove	1D@.506	ea	—	23.90	23.90
remove fan for work, then reinstall	9E@1.12	ea	—	70.80	70.80

Bathroom exhaust fan with heat lamp. Includes up to 10 lf of 4" flexible vent duct and wall outlet.

	Craft@Hrs	Unit	Material	Labor	Total
replace, standard grade, 75 cfm, single heat lamp in center, single switch	9E@.883	ea	119.00	55.80	174.80
replace, high grade, 110 cfm, double heat lamp in center, separate timer switch for heat lamp and wall outlet	9E@.883	ea	176.00	55.80	231.80
remove	1D@.726	ea	—	34.30	34.30

	Craft@Hrs	Unit	Material	Labor	Total

Bathroom exhaust fan with heater. Includes up to 10 lf of 4" flexible vent duct and wall outlet.

	Craft@Hrs	Unit	Material	Labor	Total
replace, standard grade, 110 cfm, 200 watt heater, separate timer switch for heater	9E@.952	ea	208.00	60.20	268.20
replace, high grade, 110 cfm, 450 watt heater, separate timer switch for heater	9E@.952	ea	304.00	60.20	364.20
remove	1D@.726	ea	—	34.30	34.30
remove fan for work, then reinstall	9E@1.44	ea	—	91.00	91.00

Kitchen exhaust fan. Standard and high grade include up to 10 lf of 4" flexible vent duct and wall outlet.

	Craft@Hrs	Unit	Material	Labor	Total
replace, economy grade, 110 cfm, no outside vent, charcoal filtration system	9E@1.12	ea	119.00	70.80	189.80
replace, standard grade, 200 cfm	9E@1.12	ea	208.00	70.80	278.80
replace, high grade, 300 cfm	9E@1.12	ea	391.00	70.80	461.80
remove	1D@.539	ea	—	25.40	25.40
remove fan for work, then reinstall	9E@1.61	ea	—	102.00	102.00

Whole-house exhaust fan. Includes whole house exhaust fan, fasteners, control switch, and installation, usually in attic. Does not include electrical wiring.

	Craft@Hrs	Unit	Material	Labor	Total
replace, standard grade, 4,500 cfm, two-speed with automatic shutters	9E@3.10	ea	983.00	196.00	1,179.00
replace, high grade, 6,500 cfm, variable speed with automatic shutters	9E@3.10	ea	1,140.00	196.00	1,336.00
remove	1D@1.52	ea	—	71.70	71.70
remove fan for work, then reinstall	9E@4.11	ea	—	260.00	260.00

Intercom system master station. Standard grade and higher have AM/FM radio, room monitor, hands-free answer, and privacy features.

	Craft@Hrs	Unit	Material	Labor	Total
replace, economy grade, push to talk, release to listen	9E@2.47	ea	447.00	156.00	603.00
replace, standard grade	9E@2.47	ea	901.00	156.00	1,057.00
replace, high grade, includes cassette player	9E@2.47	ea	1,160.00	156.00	1,316.00
replace, deluxe grade, includes cassette player, built-in telephone, telephone answering machine, gate or door latch release	9E@2.47	ea	1,860.00	156.00	2,016.00
replace, custom grade, includes cassette player, built-in telephone, telephone answering machine, gate or door latch release, door video monitor	9E@2.47	ea	2,790.00	156.00	2,946.00
remove	1D@.749	ea	—	35.40	35.40
remove master station for work, then reinstall	9E@3.25	ea	—	205.00	205.00

Intercom system remote station. Standard grade and higher have hands-free answer, and privacy features.

	Craft@Hrs	Unit	Material	Labor	Total
replace, economy grade, push to talk, release to listen	9E@.685	ea	111.00	43.30	154.30
replace, standard grade	9E@.685	ea	229.00	43.30	272.30
replace, high grade, with built-in telephone	9E@.685	ea	430.00	43.30	473.30
remove	1D@.300	ea	—	14.20	14.20
remove, remote station for work, then reinstall	9E@1.03	ea	—	65.10	65.10

	Craft@Hrs	Unit	Material	Labor	Total
Sound system speakers with grille.					
replace, economy grade, 3" to 5" diameter speaker, plastic grille	9E@.244	ea	31.80	15.40	47.20
replace, standard grade, 5" to 8" diameter elliptical speaker, wood-grain plastic grille	9E@.244	ea	47.90	15.40	63.30
replace, high grade, 5" to 8" diameter elliptical speaker with 1" to 2" "tweeter" speaker, wood-grain plastic grille	9E@.244	ea	63.80	15.40	79.20
replace, deluxe grade, two 5" to 8" diameter speakers with two or three 1" to 2" "tweeter" speakers, wood-grain plastic grille	9E@.244	ea	104.00	15.40	119.40
remove sound system speaker	1D@.235	ea	—	11.10	11.10
Check electrical circuits with megameter.					
in small home	7E@3.91	ea	—	283.00	283.00
in average size home	7E@5.66	ea	—	410.00	410.00
in large home	7E@7.79	ea	—	564.00	564.00
Smoke detector.					
replace, battery operated	9E@.355	ea	55.80	22.40	78.20
replace, direct wired	9E@.398	ea	71.90	25.20	97.10
remove	1D@.200	ea	—	9.44	9.44
remove for work, then reinstall	9E@.553	ea	—	34.90	34.90
Carbon monoxide detector.					
replace, battery operated	9E@.355	ea	63.80	22.40	86.20
replace, direct wired	9E@.398	ea	79.50	25.20	104.70
remove	1D@.200	ea	—	9.44	9.44
remove for work, then reinstall	9E@.553	ea	—	34.90	34.90
Radon detector.					
replace, battery operated	9E@.355	ea	71.90	22.40	94.30
replace, direct wired	9E@.398	ea	87.60	25.20	112.80
remove	1D@.200	ea	—	9.44	9.44
remove for work, then reinstall	9E@.553	ea	—	34.90	34.90
Thermostat. See HVAC for additional thermostats.					
replace, for electric-resistance heating system	9E@.183	ea	13.60	11.60	25.20
replace, for electric heating system	9E@.326	ea	28.50	20.60	49.10
remove	1D@.200	ea	—	9.44	9.44
Television antenna. Includes up to 50 lf coaxial cable. Add **$233** for ground wire and grounding rod driven 8' into ground. (Required by many local codes.)					
replace, economy grade, 55" boom	9E@1.24	ea	176.00	78.40	254.40
replace, standard grade, 100" boom	9E@1.24	ea	304.00	78.40	382.40
replace, high grade, 150" boom	9E@1.24	ea	454.00	78.40	532.40
replace, deluxe grade, 200" boom	9E@1.24	ea	651.00	78.40	729.40
remove	1D@.835	ea	—	39.40	39.40
remove for work, then reinstall	9E@1.82	ea	—	115.00	115.00

	Craft@Hrs	Unit	Material	Labor	Total

Electric space heater recessed in wall. Includes recessed space heater, fasteners, and installation. Heater includes a built-in thermostat. Does not include wall framing, electrical wiring, thermostat wiring, or a wall-mounted thermostat. Heater should fit between standard wall framing spacing (e.g. 16" on center).

	Craft@Hrs	Unit	Material	Labor	Total
replace, economy grade, 120 or 240 volt, 1,500 watt	9E@2.47	ea	348.00	156.00	504.00
replace, standard grade, 240 volt, 2,250 watt	9E@2.47	ea	398.00	156.00	554.00
replace, high grade 240 volt, 4,000 watt	9E@2.47	ea	479.00	156.00	635.00
remove	1D@.506	ea	—	23.90	23.90
remove heater for work, then reinstall	9E@4.11	ea	—	260.00	260.00

Electric-resistance heating cable. Ceiling installation: Copper inner resistance wire embedded in gypsum. Floor installation: Wire contains copper inner resistance core and is sheathed with aluminum and a cross-linked polyethylene outer layer. Per sf of exposed floor space. Do not measure floor space underneath bathtubs, cabinets or appliances. When resistance heating system installed in floor is not the only heat source in the house, floor sensors determine when floor is below desired temperature.

	Craft@Hrs	Unit	Material	Labor	Total
replace, ceiling installation	9E@.036	sf	4.62	2.28	6.90
replace, floor installation per sf of exposed floor space	9E@.060	sf	8.45	3.79	12.24
replace, floor sensor for electrical resistance heating system	9E@.632	ea	144.00	39.90	183.90

Electric baseboard heater. Includes electric baseboard heater, fasteners, and installation. Heater includes a built-in thermostat but is often connected to a wall-mounted thermostat. Does not include electrical wiring, thermostat wiring, or a wall-mounted thermostat. 3 LF minimum.

	Craft@Hrs	Unit	Material	Labor	Total
replace, economy grade, 3 lf minimum, 165 watts per lf, thermostat on heater	9E@.512	lf	28.50	32.40	60.90
replace, standard grade, 3 lf minimum, 188 watts per lf, includes up to 18 lf #18/2 thermostat wire, box and line thermostat	9E@.512	lf	30.40	32.40	62.80
remove	1D@.165	lf	—	7.79	7.79
remove heater for work, then reinstall	9E@.717	lf	—	45.30	45.30

Underground wiring for exterior post light fixture. Includes sand bed in trench for wire and up to 10 lf of galvanized conduit. Does not include excavation.

	Craft@Hrs	Unit	Material	Labor	Total
Up to 30 lf with up to 36 lf #12/2 UF wire	9E@1.22	ea	348.00	77.10	425.10
Up to 50 lf with up to 56 lf #12/2 UF wire	9E@1.30	ea	541.00	82.20	623.20

Lighting. The price of a light fixture can vary a great deal, often without any visible explanation. For example, a light fixture that is designed by a prominent designer may be far more expensive than a similar light fixture with similar materials and design.

Light fixture quality. Economy: Very light-gauge plated metal components, simple, thin glass shades. Chandeliers may contain 4 to 8 arms with turned wood center. Standard: Light-gauge plated metal components, simple glass shades. Chandeliers may contain 4 to 8 arms. High: Heavy-gauge plated metal components, glass shades with some intricate pattern or design, may have antique finish. Chandeliers may contain 4 to 8 arms. Deluxe: Heavy-gauge metal components, solid brass components, intricate glass shades, may have antique finish. Chandeliers may contain 6 to 12 arms. Custom: Heavy-gauge metal components, solid brass components, intricate glass shades, specialty antique finishes, polished brass, antiqued brass, polished nickel (or chrome) over brass. Chandeliers may contain 6 to 14 arms. Custom deluxe: Designed by prominent designer. Heavy-gauge metal components, solid brass components, intricate glass shades, specialty antique finishes, polished brass, antiqued brass, polished nickel (or chrome) over brass. Chandeliers usually contain 6 to 20 arms, often in two tiers.

	Craft@Hrs	Unit	Material	Labor	Total
Light fixture allowance. Contractor's allowance per light fixture in complete home.					
replace, economy grade	9E@.574	ea	24.20	36.30	60.50
replace, standard grade	9E@.574	ea	63.80	36.30	100.10
replace, high grade	9E@.574	ea	111.00	36.30	147.30
replace, deluxe grade	9E@.574	ea	200.00	36.30	236.30
remove	1D@.300	ea	—	14.20	14.20
remove light fixture for work, then reinstall	9E@1.02	ea	—	64.50	64.50
Bathroom light bar. Standard grade light bars contain decorative light bulbs, high grade light bars contain decorative globe covers.					
with 2 lights, replace, standard grade	9E@.612	ea	79.50	38.70	118.20
with 3 lights, replace, high grade	9E@.612	ea	104.00	38.70	142.70
with 3 to 4 lights, replace, standard grade	9E@.612	ea	111.00	38.70	149.70
with 3 to 4 lights, replace, high grade	9E@.612	ea	119.00	38.70	157.70
with 5 to 6 lights, replace, standard grade	9E@.612	ea	149.00	38.70	187.70
with 5 to 6 lights, replace, high grade	9E@.612	ea	186.00	38.70	224.70
with 7 to 8 lights, replace, standard grade	9E@.612	ea	215.00	38.70	253.70
with 7 to 8 lights, replace, high grade	9E@.612	ea	262.00	38.70	300.70
remove	1D@.315	ea	—	14.90	14.90
remove light bar for work, then reinstall	9E@1.08	ea	—	68.30	68.30
Chandelier. To 28" in diameter. Suitable for installations on 8' and taller ceilings.					
replace, economy grade	9E@1.67	ea	215.00	106.00	321.00
replace, standard grade	9E@1.67	ea	335.00	106.00	441.00
replace, high grade	9E@1.67	ea	479.00	106.00	585.00
replace, deluxe grade	9E@1.67	ea	685.00	106.00	791.00
replace, custom grade	9E@1.67	ea	884.00	106.00	990.00
replace, custom deluxe grade	9E@1.67	ea	1,130.00	106.00	1,236.00
remove	1D@.557	ea	—	26.30	26.30
Entrance chandelier. To 40" in diameter. Suitable for installations on 12' and taller ceilings.					
replace, economy grade	9E@3.35	ea	479.00	212.00	691.00
replace, standard grade	9E@3.35	ea	685.00	212.00	897.00
replace, high grade	9E@3.35	ea	901.00	212.00	1,113.00
replace, deluxe grade	9E@3.35	ea	1,150.00	212.00	1,362.00
replace, custom grade	9E@3.35	ea	1,470.00	212.00	1,682.00
replace, custom deluxe grade	9E@3.35	ea	1,860.00	212.00	2,072.00
remove	1D@.557	ea	—	26.30	26.30
Grand entrance chandelier. To 60" in diameter. Suitable for installations on 18' and taller ceilings.					
replace, economy grade	9E@4.59	ea	925.00	290.00	1,215.00
replace, standard grade	9E@4.59	ea	1,520.00	290.00	1,810.00
replace, high grade	9E@4.59	ea	2,080.00	290.00	2,370.00
replace, deluxe grade	9E@4.59	ea	2,620.00	290.00	2,910.00
replace, custom grade	9E@4.59	ea	3,300.00	290.00	3,590.00
replace, custom deluxe grade	9E@4.59	ea	5,340.00	290.00	5,630.00
remove	1D@.557	ea	—	26.30	26.30

Early American reproduction chandelier. Early American candle chandeliers converted to electrical. These fixtures are antique reproductions and are hand tooled from tin, distressed tin, pewter, brass, or copper. Joints are hand soldered. The electric "candles" may be coated with bees wax. Although many modern fixtures use a candle style, they should not be confused with early American reproduction fixtures. Look for hand tooling inconsistencies and hand-soldered joints.

	Craft@Hrs	Unit	Material	Labor	Total
replace, economy grade	9E@1.67	ea	479.00	106.00	585.00
replace, standard grade	9E@1.67	ea	651.00	106.00	757.00
replace, high grade	9E@1.67	ea	901.00	106.00	1,007.00
replace, deluxe grade	9E@1.67	ea	1,290.00	106.00	1,396.00
replace, custom grade	9E@1.67	ea	1,550.00	106.00	1,656.00
replace, custom deluxe grade	9E@1.67	ea	2,080.00	106.00	2,186.00

Early electric reproduction chandelier. Reproductions of chandeliers that were made when electricity was first introduced. Use these prices for true design reproductions.

	Craft@Hrs	Unit	Material	Labor	Total
replace, economy grade	9E@1.67	ea	558.00	106.00	664.00
replace, standard grade	9E@1.67	ea	719.00	106.00	825.00
replace, high grade	9E@1.67	ea	963.00	106.00	1,069.00
replace, deluxe grade	9E@1.67	ea	1,390.00	106.00	1,496.00
replace, custom grade	9E@1.67	ea	1,600.00	106.00	1,706.00
replace, custom deluxe grade	9E@1.67	ea	2,150.00	106.00	2,256.00
remove	1D@.557	ea	—	26.30	26.30

Courtesy of: Rejuvenation Lamp & Fixture Co.

Early gas reproduction chandelier. Electrified reproductions of chandeliers that were originally designed to burn coal gas. Can usually be identified by a fake gas valve on the chandelier arm. Because a flame burns upward, gas reproduction fixture shades always face upward. Also use these prices for electric reproductions of lantern style chandeliers. Modern designs which use a gas or lantern style usually fall in the economy and standard grades.

	Craft@Hrs	Unit	Material	Labor	Total
replace, economy grade	9E@1.67	ea	582.00	106.00	688.00
replace, standard grade	9E@1.67	ea	707.00	106.00	813.00
replace, high grade	9E@1.67	ea	996.00	106.00	1,102.00
replace, deluxe grade	9E@1.67	ea	1,350.00	106.00	1,456.00
replace, custom grade	9E@1.67	ea	1,600.00	106.00	1,706.00
replace, custom deluxe grade	9E@1.67	ea	2,210.00	106.00	2,316.00
remove	1D@.557	ea	—	26.30	26.30

Early electric / gas combination reproduction chandelier. When electricity was first introduced, builders were unsure whether it or gas was the wave of the future. Many builders played it safe by installing fixtures that used both. The original gas components can be identified by the fake valve on the arm. Gas components always face up. Electrical components often have a pull switch and usually, but not always, face down.

	Craft@Hrs	Unit	Material	Labor	Total
replace, economy grade	9E@1.67	ea	613.00	106.00	719.00
replace, standard grade	9E@1.67	ea	733.00	106.00	839.00
replace, high grade	9E@1.67	ea	1,000.00	106.00	1,106.00
replace, deluxe grade	9E@1.67	ea	1,390.00	106.00	1,496.00
replace, custom grade	9E@1.67	ea	1,620.00	106.00	1,726.00
replace, custom deluxe grade	9E@1.67	ea	2,150.00	106.00	2,256.00
remove	1D@.557	ea	—	26.30	26.30

Imitation crystal chandelier. Cut glass components.

	Craft@Hrs	Unit	Material	Labor	Total
replace, economy grade	9E@1.67	ea	359.00	106.00	465.00
replace, standard grade	9E@1.67	ea	486.00	106.00	592.00
remove	1D@.557	ea	—	26.30	26.30

	Craft@Hrs	Unit	Material	Labor	Total
Crystal chandelier. High-quality crystal components. Deduct **15%** for lower quality crystal. To 25" diameter. Suitable for installation on 8' or taller ceilings.					
replace, standard grade	9E@1.67	ea	1,160.00	106.00	1,266.00
replace, high grade	9E@1.67	ea	1,480.00	106.00	1,586.00
replace, deluxe grade	9E@1.67	ea	1,760.00	106.00	1,866.00
replace, custom grade	9E@1.67	ea	2,000.00	106.00	2,106.00
replace, custom deluxe grade	9E@1.67	ea	2,150.00	106.00	2,256.00
remove	1D@.557	ea	—	26.30	26.30
Entrance crystal chandelier. High-quality crystal components. Deduct **15%** for lower quality crystal. To 32" diameter. Suitable for installation on 14' or taller ceilings.					
replace, standard grade	9E@3.35	ea	1,760.00	212.00	1,972.00
replace, high grade	9E@3.35	ea	2,080.00	212.00	2,292.00
replace, deluxe grade	9E@3.35	ea	2,290.00	212.00	2,502.00
replace, custom grade	9E@3.35	ea	3,040.00	212.00	3,252.00
replace, custom deluxe grade	9E@3.35	ea	4,300.00	212.00	4,512.00
remove	1D@.557	ea	—	26.30	26.30
Grand entrance crystal chandelier. High-quality crystal components. Deduct **15%** for lower quality crystal. To 52" diameter. Suitable for installation on 20' or taller ceilings.					
replace, standard grade	9E@4.59	ea	2,550.00	290.00	2,840.00
replace, high grade	9E@4.59	ea	5,580.00	290.00	5,870.00
replace, deluxe grade	9E@4.59	ea	7,560.00	290.00	7,850.00
replace, custom grade	9E@4.59	ea	10,300.00	290.00	10,590.00
replace, custom deluxe grade	9E@4.59	ea	11,900.00	290.00	12,190.00
remove	1D@.557	ea	—	26.30	26.30
Crystal tier chandelier. High-quality crystal components. Deduct **15%** for lower quality crystal. Crystal tier chandeliers resemble an upside down wedding cake. They are usually 1.5 to 2 times taller than they are wide, so normally are installed on taller ceilings.					
replace, standard grade	9E@1.67	ea	1,390.00	106.00	1,496.00
replace, high grade	9E@4.59	ea	1,480.00	290.00	1,770.00
replace, deluxe grade	9E@4.59	ea	1,860.00	290.00	2,150.00
replace, custom grade	9E@4.59	ea	2,000.00	290.00	2,290.00
replace, custom deluxe grade	9E@4.59	ea	2,720.00	290.00	3,010.00
remove	1D@.557	ea	—	26.30	26.30
Crystal wall fixture. High-quality crystal components. Deduct **15%** for lower quality crystal.					
replace, standard grade	9E@.556	ea	416.00	35.10	451.10
replace, high grade	9E@.556	ea	638.00	35.10	673.10
replace, deluxe grade	9E@.556	ea	940.00	35.10	975.10
replace, custom grade	9E@.556	ea	1,390.00	35.10	1,425.10
remove	1D@.300	ea	—	14.20	14.20
remove fixture for work, then reinstall	9E@1.02	ea	—	64.50	64.50
Repair crystal chandelier.					
Minor repairs, replace 2 to 4 components	9E@1.84	ea	104.00	116.00	220.00

	Craft@Hrs	Unit	Material	Labor	Total
Significant repairs, replace 5 to 10 components	9E@3.68	ea	208.00	233.00	441.00
Major repairs, replace 11 to 16 components	9E@7.34	ea	348.00	464.00	812.00
Rebuild, dismantle all or most of fixture, replace 25% to 35% of components	9E@15.3	ea	678.00	967.00	1,645.00

Additional chandelier costs.

	Craft@Hrs	Unit	Material	Labor	Total
remove for work, then reinstall	9E@2.82	ea	—	178.00	178.00
remove vaulted ceiling chandelier for work, then reinstall	9E@6.12	ea	—	387.00	387.00
remove high vaulted-ceiling chandelier for work, then reinstall	9E@8.35	ea	—	528.00	528.00
add for hanging on vaulted ceiling					
12' to 14' tall	9E@.919	ea	—	58.10	58.10
14' to 16' tall	9E@1.67	ea	—	106.00	106.00
17' to 22' tall	9E@2.55	ea	—	161.00	161.00

Single pendant light fixture. Pendant fixtures hang from a chain or rod and connect directly to an electrical box in the ceiling.

	Craft@Hrs	Unit	Material	Labor	Total
replace, economy grade	9E@.574	ea	104.00	36.30	140.30
replace, standard grade	9E@.574	ea	128.00	36.30	164.30
replace, high grade	9E@.574	ea	160.00	36.30	196.30
replace, deluxe grade	9E@.574	ea	192.00	36.30	228.30
replace, custom grade	9E@.574	ea	221.00	36.30	257.30
replace, custom deluxe grade	9E@.574	ea	255.00	36.30	291.30
remove	1D@.300	ea	—	14.20	14.20

Double pendant light fixture. Pendant fixtures hang from a chain or rod and connect directly to an electrical box in the ceiling.

	Craft@Hrs	Unit	Material	Labor	Total
replace, economy grade	9E@.574	ea	176.00	36.30	212.30
replace, standard grade	9E@.574	ea	221.00	36.30	257.30
replace, high grade	9E@.574	ea	255.00	36.30	291.30
replace, deluxe grade	9E@.574	ea	296.00	36.30	332.30
replace, custom grade	9E@.574	ea	335.00	36.30	371.30
replace, custom deluxe grade	9E@.574	ea	391.00	36.30	427.30
remove	1D@.300	ea	—	14.20	14.20

"Shower" style light fixture. Shower style fixtures have a large base. Two to four (sometimes more) small lights with shades hang from the base on chains or rods.

	Craft@Hrs	Unit	Material	Labor	Total
replace, economy grade	9E@1.67	ea	255.00	106.00	361.00
replace, standard grade	9E@1.67	ea	330.00	106.00	436.00
replace, high grade	9E@1.67	ea	437.00	106.00	543.00
replace, deluxe grade	9E@1.67	ea	510.00	106.00	616.00
replace, custom grade	9E@1.67	ea	661.00	106.00	767.00
replace, custom deluxe grade	9E@1.67	ea	719.00	106.00	825.00
remove	1D@.557	ea	—	26.30	26.30

	Craft@Hrs	Unit	Material	Labor	Total
Small bowl-shade light fixture. Fixture with a globe or bowl shade. Small sizes are typically found in halls.					
replace, economy grade	9E@.574	ea	55.80	36.30	92.10
replace, standard grade	9E@.574	ea	71.90	36.30	108.20
replace, high grade	9E@.574	ea	95.70	36.30	132.00
replace, deluxe grade	9E@.574	ea	136.00	36.30	172.30
replace, custom grade	9E@.574	ea	149.00	36.30	185.30
replace, custom deluxe grade	9E@.574	ea	186.00	36.30	222.30
remove	1D@.300	ea	—	14.20	14.20
Bowl-shade light fixture. Fixture with a globe or bowl shade.					
replace, economy grade	9E@.574	ea	79.50	36.30	115.80
replace, standard grade	9E@.574	ea	144.00	36.30	180.30
replace, high grade	9E@.574	ea	176.00	36.30	212.30
replace, deluxe grade	9E@.574	ea	229.00	36.30	265.30
replace, custom grade	9E@.574	ea	279.00	36.30	315.30
replace, custom deluxe grade	9E@.574	ea	318.00	36.30	354.30
remove	1D@.300	ea	—	14.20	14.20
Suspended bowl-shade light fixture. Bowl-shade suspended on three rods or chains.					
replace, economy grade	9E@1.67	ea	304.00	106.00	410.00
replace, standard grade	9E@1.67	ea	359.00	106.00	465.00
replace, high grade	9E@1.67	ea	510.00	106.00	616.00
replace, deluxe grade	9E@1.67	ea	638.00	106.00	744.00
replace, custom grade	9E@1.67	ea	701.00	106.00	807.00
replace, custom deluxe grade	9E@1.67	ea	892.00	106.00	998.00
remove	1D@.557	ea	—	26.30	26.30
Billiard-table style light fixture.					
replace, economy grade	9E@1.53	ea	335.00	96.70	431.70
replace, standard grade	9E@1.53	ea	733.00	96.70	829.70
replace, high grade	9E@1.53	ea	1,130.00	96.70	1,226.70
replace, deluxe grade	9E@1.53	ea	1,600.00	96.70	1,696.70
replace, custom grade	9E@1.53	ea	2,080.00	96.70	2,176.70
replace, custom deluxe grade	9E@1.53	ea	3,180.00	96.70	3,276.70
remove	1D@.561	ea	—	26.50	26.50
remove fixture for work, then reinstall	9E@2.61	ea	—	165.00	165.00
Wall-mount light fixture.					
replace, economy grade	9E@.556	ea	31.80	35.10	66.90
replace, standard grade	9E@.556	ea	63.80	35.10	98.90
replace, high grade	9E@.556	ea	160.00	35.10	195.10
replace, deluxe grade	9E@.556	ea	272.00	35.10	307.10
replace, custom grade	9E@.556	ea	304.00	35.10	339.10
replace, custom deluxe grade	9E@.556	ea	447.00	35.10	482.10
remove	1D@.300	ea	—	14.20	14.20
remove fixture for work, then reinstall	9E@1.02	ea	—	64.50	64.50

Courtesy of: Rejuvenation Lamp & Fixture Co.

	Craft@Hrs	Unit	Material	Labor	Total
Surface-mounted one-tube fluorescent strip light.					
replace, 1'	9E@1.05	ea	31.80	66.40	98.20
replace, 1-1/2'	9E@1.05	ea	31.80	66.40	98.20
replace, 2'	9E@1.05	ea	39.80	66.40	106.20
replace, 4'	9E@1.05	ea	47.90	66.40	114.30
replace, 6'	9E@1.05	ea	71.90	66.40	138.30
replace, 8'	9E@1.05	ea	104.00	66.40	170.40
remove	1D@.500	ea	—	23.60	23.60
Surface-mounted one-tube fluorescent strip light with diffuser.					
replace, 2'	9E@1.05	ea	47.90	66.40	114.30
replace, 4'	9E@1.05	ea	55.80	66.40	122.20
replace, 6'	9E@1.05	ea	71.90	66.40	138.30
replace, 8'	9E@1.05	ea	111.00	66.40	177.40
remove	1D@.500	ea	—	23.60	23.60
2' two-tube surface-mounted fluorescent strip light.					
replace, exposed lights	9E@1.05	ea	55.80	66.40	122.20
replace, with diffuser	9E@1.05	ea	87.60	66.40	154.00
replace, with vinyl trim & diffuser	9E@1.05	ea	95.70	66.40	162.10
replace, with solid wood trim & diffuser	9E@1.05	ea	119.00	66.40	185.40
remove	1D@.500	ea	—	23.60	23.60
4' two-tube surface-mounted fluorescent strip light.					
replace, exposed lights	9E@1.05	ea	79.50	66.40	145.90
replace, with diffuser	9E@1.05	ea	104.00	66.40	170.40
replace, with vinyl trim & diffuser	9E@1.05	ea	111.00	66.40	177.40
replace, with solid wood trim & diffuser	9E@1.05	ea	149.00	66.40	215.40
remove	1D@.500	ea	—	23.60	23.60
6' two-tube surface-mounted fluorescent strip light.					
replace, exposed lights	9E@1.05	ea	87.60	66.40	154.00
replace, with diffuser	9E@1.05	ea	119.00	66.40	185.40
replace, with vinyl trim & diffuser	9E@1.05	ea	149.00	66.40	215.40
replace, with solid wood trim & diffuser	9E@1.05	ea	165.00	66.40	231.40
remove	1D@.500	ea	—	23.60	23.60
8' two-tube surface-mounted fluorescent strip light.					
replace, exposed lights	9E@1.05	ea	119.00	66.40	185.40
replace, with diffuser	9E@1.05	ea	144.00	66.40	210.40
replace, with vinyl trim & diffuser	9E@1.05	ea	186.00	66.40	252.40
replace, with solid wood trim & diffuser	9E@1.05	ea	215.00	66.40	281.40
remove	1D@.500	ea	—	23.60	23.60
4' four-tube surface-mounted fluorescent strip light.					
replace, exposed lights	9E@1.05	ea	128.00	66.40	194.40
replace, with diffuser	9E@1.05	ea	192.00	66.40	258.40
replace, with vinyl trim & diffuser	9E@1.05	ea	221.00	66.40	287.40
replace, with solid wood trim & diffuser	9E@1.05	ea	279.00	66.40	345.40
remove	1D@.500	ea	—	23.60	23.60

	Craft@Hrs	Unit	Material	Labor	Total
Fluorescent fixture for suspended ceiling. Add $17 for boxed-in fixture for fire-rated ceiling.					
replace, 2' by 2' drop-in two-tube	9E@2.12	ea	128.00	134.00	262.00
replace, 2' by 4' drop-in two-tube	9E@2.12	ea	160.00	134.00	294.00
replace, 2' by 4' drop-in four-tube	9E@2.12	ea	200.00	134.00	334.00
remove	1D@.500	ea	—	23.60	23.60
Additional fluorescent fixture costs.					
remove fixture for work, then reinstall	9E@1.84	ea	—	116.00	116.00
remove and replace fluorescent fixture ballast	9E@1.41	ea	104.00	89.10	193.10
Fluorescent circline fixture.					
replace, economy grade	9E@.967	ea	55.80	61.10	116.90
replace, standard grade	9E@.967	ea	87.60	61.10	148.70
replace, high grade	9E@.967	ea	144.00	61.10	205.10
replace, deluxe grade	9E@.967	ea	279.00	61.10	340.10
remove circline fixture	1D@.371	ea	—	17.50	17.50
remove circline fixture for work, then reinstall	9E@1.67	ea	—	106.00	106.00
remove and replace circline fixture ballast	9E@1.41	ea	95.70	89.10	184.80
Ceiling fan. Fan only, no light.					
replace, economy grade	9E@2.45	ea	104.00	155.00	259.00
replace, standard grade	9E@2.45	ea	165.00	155.00	320.00
replace, high grade	9E@2.45	ea	272.00	155.00	427.00
replace, deluxe grade	9E@2.45	ea	424.00	155.00	579.00
remove	1D@.500	ea	—	23.60	23.60
remove fan for work, then reinstall	9E@2.74	ea	—	173.00	173.00
Ceiling fan with light.					
replace, economy grade	9E@3.68	ea	272.00	233.00	505.00
replace, standard grade	9E@3.68	ea	517.00	233.00	750.00
replace, high grade	9E@3.68	ea	454.00	233.00	687.00
replace, deluxe grade	9E@3.68	ea	651.00	233.00	884.00
replace, custom grade	9E@3.68	ea	781.00	233.00	1,014.00
remove	1D@.500	ea	—	23.60	23.60
remove fan with light for work, then reinstall	9E@2.82	ea	—	178.00	178.00
Light for ceiling fan.					
replace, economy grade	9E@.525	ea	55.80	33.20	89.00
replace, standard grade	9E@.525	ea	111.00	33.20	144.20
replace, high grade	9E@.525	ea	160.00	33.20	193.20
replace, deluxe grade	9E@.525	ea	272.00	33.20	305.20
replace, custom grade	9E@.525	ea	424.00	33.20	457.20
remove	1D@.300	ea	—	14.20	14.20
remove light from fan for work, then reinstall	9E@.919	ea	—	58.10	58.10
Hanging light fixture. Hanging light fixtures usually hang from ceiling hooks and plug into a switched outlet.					
replace, economy grade	9E@.305	ea	63.80	19.30	83.10
replace, standard grade	9E@.305	ea	79.50	19.30	98.80
replace, high grade	9E@.305	ea	95.70	19.30	115.00
replace, deluxe grade	9E@.305	ea	149.00	19.30	168.30
remove	1D@.100	ea	—	4.72	4.72

	Craft@Hrs	Unit	Material	Labor	Total
Double hanging light fixture.					
replace, economy grade	9E@.305	ea	95.70	19.30	115.00
replace, standard grade	9E@.305	ea	136.00	19.30	155.30
replace, high grade	9E@.305	ea	186.00	19.30	205.30
replace, deluxe grade	9E@.305	ea	249.00	19.30	268.30
remove	1D@.100	ea	—	4.72	4.72
Minimum charge.					
for light fixture work	9E@1.90	ea	31.80	120.00	151.80
Porcelain light fixture.					
replace, standard	9E@.230	ea	7.19	14.50	21.69
replace, with pull chain	9E@.230	ea	9.57	14.50	24.07
remove	1D@.100	ea	—	4.72	4.72
remove porcelain fixture for work, then reinstall	9E@.368	ea	—	23.30	23.30
Recessed spot light fixture.					
replace, economy grade	9E@.835	ea	79.50	52.80	132.30
replace, standard grade	9E@.835	ea	111.00	52.80	163.80
replace, high grade	9E@.835	ea	144.00	52.80	196.80
replace, deluxe grade	9E@.835	ea	192.00	52.80	244.80
replace, custom grade	9E@.835	ea	255.00	52.80	307.80
remove	1D@.400	ea	—	18.90	18.90
remove spot fixture for work, then reinstall	9E@.766	ea	—	48.40	48.40
remove spot fixture & can for work, then reinstall	9E@1.41	ea	—	89.10	89.10
Electric strip for strip lighting. Surface-mounted electric strip channel. Does not include strip-mounted spot lights. 3 lf minimum.					
replace, electric strip for strip lights	9E@.261	lf	17.60	16.50	34.10
remove	1D@.110	lf	—	5.19	5.19
remove electric strip for work, then reinstall	9E@.459	lf	—	29.00	29.00
Strip light. Spot light for surface mounted electric strip channel.					
replace, economy grade	9E@.368	ea	39.80	23.30	63.10
replace, standard grade	9E@.368	ea	55.80	23.30	79.10
replace, high grade	9E@.368	ea	111.00	23.30	134.30
remove	1D@.150	ea	—	7.08	7.08
remove light for work, then reinstall	9E@.645	ea	—	40.80	40.80
Exterior single flood light fixture.					
replace, standard grade	9E@.591	ea	71.90	37.40	109.30
replace, high grade	9E@.591	ea	95.70	37.40	133.10
remove	1D@.300	ea	—	14.20	14.20
remove fixture for work, then reinstall	9E@1.08	ea	—	68.30	68.30

strip spot light

	Craft@Hrs	Unit	Material	Labor	Total
Exterior double flood light fixture.					
replace, standard grade	9E@.591	ea	87.60	37.40	125.00
replace, high grade	9E@.591	ea	119.00	37.40	156.40
remove	1D@.300	ea	—	14.20	14.20
remove fixture for work, then reinstall	9E@1.08	ea	—	68.30	68.30
Exterior triple flood light fixture.					
replace, standard grade	9E@.591	ea	111.00	37.40	148.40
replace, high grade	9E@.591	ea	136.00	37.40	173.40
remove	1D@.300	ea	—	14.20	14.20
remove fixture for work, then reinstall	9E@1.08	ea	—	68.30	68.30
Exterior wall-mount light fixture.					
replace, economy grade	9E@.574	ea	55.80	36.30	92.10
replace, standard grade	9E@.574	ea	87.60	36.30	123.90
replace, high grade	9E@.574	ea	221.00	36.30	257.30
replace, deluxe grade	9E@.574	ea	424.00	36.30	460.30
remove	1D@.250	ea	—	11.80	11.80
remove fixture for work, then reinstall	9E@1.08	ea	—	68.30	68.30
Exterior recessed light fixture.					
replace, economy grade	9E@.835	ea	95.70	52.80	148.50
replace, standard grade	9E@.835	ea	136.00	52.80	188.80
replace, high grade	9E@.835	ea	242.00	52.80	294.80
replace, deluxe grade	9E@.835	ea	335.00	52.80	387.80
remove	1D@.500	ea	—	23.60	23.60
remove fixture for work, then reinstall	9E@1.67	ea	—	106.00	106.00
Exterior light fixture allowance. Contractor's allowance per light fixture in complete home.					
replace, economy grade	9E@.591	ea	39.80	37.40	77.20
replace, standard grade	9E@.591	ea	79.50	37.40	116.90
replace, high grade	9E@.591	ea	149.00	37.40	186.40
replace, deluxe grade	9E@.591	ea	285.00	37.40	322.40
remove	1D@.315	ea	—	14.90	14.90
remove fixture for work, then reinstall	9E@1.08	ea	—	68.30	68.30

Courtesy of: Rejuvenation Lamp & Fixture Co.

Exterior post with light fixture. Post up to 7 feet tall. Includes post hole, concrete base and hole backfill. Does not include wiring. Economy: Redwood post with beveled corners or light turnings, or straight metal post. Plastic or light metal fixture, usually one light with polystyrene lens. Standard: Turned redwood post. Black metal fixture, usually one light with polystyrene lens. High: Turned redwood post. Metal fixture, one to two lights, clear plastic or glass lens. Deluxe: Aluminum post with J bolts. Metal fixture, one to two lights, clear plastic or glass lens. Custom: Cast-iron post with J bolts. Cast-iron, brass, copper or pewter fixture, often simulated gas light with one to four lights, clear plastic or glass lens.

	Craft@Hrs	Unit	Material	Labor	Total
replace, economy grade	9E@5.41	ea	279.00	342.00	621.00
replace, standard grade	9E@5.41	ea	404.00	342.00	746.00
replace, high grade	9E@5.41	ea	517.00	342.00	859.00
replace, deluxe grade	9E@5.41	ea	756.00	342.00	1,098.00
replace, custom grade	9E@5.41	ea	1,350.00	342.00	1,692.00
remove	1D@1.20	ea	—	56.60	56.60
remove for work, then reinstall	9E@10.2	ea	—	645.00	645.00

	Craft@Hrs	Unit	Material	Labor	Total

Exterior post light fixture (no post). Economy: Plastic or light metal fixture, one interior light with polystyrene lens. Standard: Metal fixture, two interior lights with clear polystyrene lens. High: Metal fixture, two to four interior lights with clear polystyrene lens. Deluxe: Copper, pewter or antique finish, two to four interior lights with clear glass lens. Custom: Ornate copper, pewter, brass or antique finish, two to four interior lights with clear glass lens

	Craft@Hrs	Unit	Material	Labor	Total
replace, economy grade	9E@1.15	ea	90.60	72.70	163.30
replace, standard grade	9E@1.15	ea	149.00	72.70	221.70
replace, high grade	9E@1.15	ea	215.00	72.70	287.70
replace, deluxe grade	9E@1.15	ea	335.00	72.70	407.70
replace, custom grade	9E@1.15	ea	486.00	72.70	558.70
remove	1D@.500	ea	—	23.60	23.60
remove for work, then reinstall	9E@1.93	ea	—	122.00	122.00

Redwood post for exterior post light fixture. To 7' tall. Diameter to 4".

	Craft@Hrs	Unit	Material	Labor	Total
replace, standard grade, beveled edges or light turnings	9E@4.29	ea	398.00	271.00	669.00
replace, high grade, turned post	9E@4.29	ea	558.00	271.00	829.00
remove	1D@1.20	ea	—	56.60	56.60

Metal post for exterior post light fixture.

	Craft@Hrs	Unit	Material	Labor	Total
replace, standard grade	9E@4.09	ea	255.00	258.00	513.00
replace, high grade	9E@4.09	ea	398.00	258.00	656.00
replace, deluxe grade	9E@4.09	ea	558.00	258.00	816.00
replace, custom grade	9E@4.09	ea	707.00	258.00	965.00
replace, custom deluxe grade	9E@4.09	ea	1,100.00	258.00	1,358.00
remove	1D@1.20	ea	—	56.60	56.60

Additional exterior fixture costs.

	Craft@Hrs	Unit	Material	Labor	Total
add for exterior light motion sensor	—	ea	107.00	—	107.00
add for exterior light infrared detector	—	ea	107.00	—	107.00

Light fixture repair.

	Craft@Hrs	Unit	Material	Labor	Total
replace, light fixture socket and wiring	9E@.482	ea	10.40	30.50	40.90
replace, chandelier socket and wiring, per arm	9E@.510	ea	13.60	32.20	45.80

Bowl shade or globe for light fixture.

	Craft@Hrs	Unit	Material	Labor	Total
replace, economy grade	9E@.062	ea	19.20	3.92	23.12
replace, standard grade	9E@.062	ea	31.80	3.92	35.72
replace, high grade	9E@.062	ea	55.20	3.92	59.12
replace, deluxe grade	9E@.062	ea	79.50	3.92	83.42
replace, custom grade	9E@.062	ea	97.40	3.92	101.32

Time & Material Charts (selected items)

Electrical Materials

See Electrical material prices with the line items above.

Electrical Labor

Laborer	base wage	paid leave	true wage	taxes & ins.	total
Electrician	$43.80	3.42	$47.22	25.18	$72.40
Electrician's helper	$31.50	2.46	$33.96	19.94	$53.90
Demolition laborer	$25.90	2.02	$27.92	19.28	$47.20

Paid leave is calculated based on two weeks paid vacation, one week sick leave, and seven paid holidays. Employer's matching portion of **FICA** is 7.65 percent. **FUTA** (Federal Unemployment) is .8 percent. **Worker's compensation** for the electrical trade was calculated using a national average of 8.08 percent. **Unemployment insurance** was calculated using a national average of 8 percent. **Health insurance** was calculated based on a projected national average for 2020 of $1,280 per employee (and family when applicable) per month. Employer pays 80 percent for a per month cost of $1,024 per employee. **Retirement** is based on a 401(k) retirement program with employer matching of 50 percent. Employee contributions to the 401(k) plan are an average of 6 percent of the true wage. **Liability insurance** is based on a national average of 12.0 percent.

Electrical Labor Productivity

	Craft@Hrs	Unit	Material	Labor	Total
Demolition of electrical work					
remove 120 volt switch or outlet	1D@.150	ea	—	7.08	7.08
remove 240 volt switch or outlet	1D@.200	ea	—	9.44	9.44
remove bathroom fan	1D@.506	ea	—	23.90	23.90
remove bathroom fan / light with heater	1D@.726	ea	—	34.30	34.30
remove breaker panel	1D@6.09	ea	—	287.00	287.00
remove cable service drop	1D@10.5	ea	—	496.00	496.00
remove conduit service drop	1D@13.1	ea	—	618.00	618.00
remove rigid steel mast service drop	1D@15.4	ea	—	727.00	727.00
remove circuit breaker	1D@.200	ea	—	9.44	9.44
remove door bell or chime	1D@.100	ea	—	4.72	4.72
remove kitchen exhaust fan	1D@.539	ea	—	25.40	25.40
remove whole-house exhaust fan	1D@1.52	ea	—	71.70	71.70
remove intercom system master station	1D@.749	ea	—	35.40	35.40
remove intercom system remote station	1D@.300	ea	—	14.20	14.20
remove low-voltage outlet	1D@.150	ea	—	7.08	7.08
remove detector	1D@.200	ea	—	9.44	9.44
remove thermostat	1D@.200	ea	—	9.44	9.44
remove television antenna	1D@.835	ea	—	39.40	39.40
remove recessed electric space heater	1D@.506	ea	—	23.90	23.90
remove electric baseboard heater	1D@.165	lf	—	7.79	7.79
remove light fixture	1D@.300	ea	—	14.20	14.20
remove bathroom light bar	1D@.315	ea	—	14.90	14.90
remove chandelier	1D@.557	ea	—	26.30	26.30
remove florescent light fixture	1D@.500	ea	—	23.60	23.60
remove florescent circline light fixture	1D@.371	ea	—	17.50	17.50
remove ceiling fan	1D@.500	ea	—	23.60	23.60
remove hanging light fixture	1D@.100	ea	—	4.72	4.72

	Craft@Hrs	Unit	Material	Labor	Total
remove porcelain light fixture	1D@.100	ea	—	4.72	4.72
remove recessed spot light fixture	1D@.400	ea	—	18.90	18.90
remove electric strip for electric strip lights	1D@.110	lf	—	5.19	5.19
remove electric strip light	1D@.150	ea	—	7.08	7.08
remove exterior flood light	1D@.300	ea	—	14.20	14.20
remove exterior post with light fixture	1D@1.20	ea	—	56.60	56.60
remove exterior post light fixture	1D@.500	ea	—	23.60	23.60

Electrical crew

install wiring, boxes, and fixtures	electrician	$72.40
install wiring, boxes, and fixtures	electrician's helper	$53.90
install wiring, boxes, and fixtures	electrical crew	$63.20

	Craft@Hrs	Unit	Material	Labor	Total
Install complete house electrical wiring, outlets, switches & light fixtures					
economy grade	9E@.053	sf	—	3.35	3.35
deluxe grade	9E@.109	sf	—	6.89	6.89
low-voltage system	9E@.100	sf	—	6.32	6.32
add for wiring installation in conduit	9E@.055	sf	—	3.48	3.48
Install complete house rough electrical (no light fixtures)					
standard	9E@.070	sf	—	4.42	4.42
in conduit	9E@.154	sf	—	9.73	9.73
low-voltage system	9E@.075	sf	—	4.74	4.74
Install complete house light fixtures					
economy grade	9E@.015	sf	—	.95	.95
deluxe grade	9E@.017	sf	—	1.07	1.07
for low-voltage system	9E@.013	sf	—	.82	.82
240 volt system					
install outlet or switch with wiring run & box	9E@1.34	ea	—	84.70	84.70
120 volt system					
install switch or outlet with wiring run & box	9E@.977	ea	—	61.70	61.70
install 3-way switch with wiring run & box	9E@1.95	ea	—	123.00	123.00
Electric wiring run					
for typical 120 volt outlet or light	9E@.802	ea	—	50.70	50.70
for 120 volt appliance	9E@1.02	ea	—	64.50	64.50
Install conduit					
electric metallic tubing (EMT), average installation	9E@.077	sf	—	4.87	4.87
flexible metal, average installation	9E@.041	sf	—	2.59	2.59
PVC, average installation	9E@.045	sf	—	2.84	2.84
intermediate (IMC), average installation	9E@.113	sf	—	7.14	7.14
rigid metal (RMC), average installation	9E@.067	sf	—	4.23	4.23
Install wiring					
#10 2 wire with ground Romex	9E@.051	lf	—	3.22	3.22
#12 2 wire with ground Romex	9E@.047	lf	—	2.97	2.97
#14 2 wire with ground Romex	9E@.040	lf	—	2.53	2.53
Install breaker panel					
100 amp ext. breaker panel (5 breakers)	7E@4.89	ea	—	354.00	354.00
100 amp int. breaker panel (5 breakers)	7E@4.30	ea	—	311.00	311.00
150 amp ext. breaker panel (10 breakers)	7E@7.68	ea	—	556.00	556.00
150 amp int. breaker panel (10 breakers)	7E@7.17	ea	—	519.00	519.00

	Craft@Hrs	Unit	Material	Labor	Total
200 amp ext. breaker panel (15 breakers)	7E@10.8	ea	—	782.00	782.00
200 amp int. breaker panel (15 breakers)	7E@9.77	ea	—	707.00	707.00
300 amp ext. breaker panel (30 breakers)	7E@14.3	ea	—	1,040.00	1,040.00
300 amp int. breaker panel (30 breakers)	7E@13.4	ea	—	970.00	970.00
40 amp int. breaker sub-panel (5 breakers)	7E@2.99	ea	—	216.00	216.00
50 amp int. breaker sub-panel	7E@3.16	ea	—	229.00	229.00
70 amp int. breaker sub-panel	7E@3.47	ea	—	251.00	251.00
Install electrical service					
100 amp cable service	7E@11.9	ea	—	862.00	862.00
150 amp cable service	7E@14.3	ea	—	1,040.00	1,040.00
150 amp conduit service	7E@15.4	ea	—	1,110.00	1,110.00
150 amp rigid steel mast service	7E@15.4	ea	—	1,110.00	1,110.00
200 amp cable service	7E@18.5	ea	—	1,340.00	1,340.00
200 amp conduit service	7E@19.9	ea	—	1,440.00	1,440.00
200 amp rigid steel mast service	7E@19.9	ea	—	1,440.00	1,440.00
Install breaker					
single-pole (120 volt) circuit breaker	7E@.636	ea	—	46.00	46.00
double-pole (240 volt) circuit breaker	7E@.649	ea	—	47.00	47.00
ground fault circuit interrupter breaker	7E@.689	ea	—	49.90	49.90
main disconnect circuit breaker	7E@3.47	ea	—	251.00	251.00
Install door bell or chime					
door bell or chime button	9E@.137	ea	—	8.66	8.66
door bell or chime	9E@1.24	ea	—	78.40	78.40
low-voltage wiring (per button)	9E@.563	ea	—	35.60	35.60
Install exhaust fan					
bathroom exhaust fan	9E@.685	ea	—	43.30	43.30
bathroom exhaust fan with heat lamp	9E@.883	ea	—	55.80	55.80
bathroom exhaust fan with heater	9E@.952	ea	—	60.20	60.20
kitchen exhaust fan	9E@1.12	ea	—	70.80	70.80
whole-house exhaust fan	9E@3.10	ea	—	196.00	196.00
Install intercom					
intercom system master station	9E@2.47	ea	—	156.00	156.00
intercom system remote station	9E@.685	ea	—	43.30	43.30
120 volt wiring for intercom master station	9E@.896	ea	—	56.60	56.60
intercom wiring (per station)	9E@.619	ea	—	39.10	39.10
Install low-voltage outlet					
install outlet	9E@.256	ea	—	16.20	16.20
wiring (per outlet)	9E@.358	ea	—	22.60	22.60
Install sound system					
speaker with grill	9E@.244	ea	—	15.40	15.40
Install detector					
battery operated	9E@.355	ea	—	22.40	22.40
direct-wired	9E@.398	ea	—	25.20	25.20
Check electrical circuits					
megameter test small home	7E@3.91	ea	—	283.00	283.00
megameter test average size home	7E@5.66	ea	—	410.00	410.00
megameter test large home	7E@7.79	ea	—	564.00	564.00
Install thermostat					
for electric resistance heating system	9E@.183	ea	—	11.60	11.60

	Craft@Hrs	Unit	Material	Labor	Total
electric heating system	9E@.326	ea	—	20.60	20.60
wiring (per thermostat)	9E@.538	ea	—	34.00	34.00
Install TV antenna					
television antenna	9E@1.24	ea	—	78.40	78.40
Install recessed electric space heater					
electric space heater recessed in wall	9E@2.47	ea	—	156.00	156.00
Install electric-resistance heating					
cable, ceiling installation	9E@.036	sf	—	2.28	2.28
cable, floor installation	9E@.060	sf	—	3.79	3.79
electric-resistance heating floor sensor	9E@.632	ea	—	39.90	39.90
Install electric baseboard heater					
electric baseboard heater	9E@.512	lf	—	32.40	32.40
remove for work, then reinstall	9E@.717	lf	—	45.30	45.30
Install underground wiring for exterior post light fixture					
up to 30 feet	9E@1.22	ea	—	77.10	77.10
up to 50 feet	9E@1.30	ea	—	82.20	82.20
Lighting					
Install interior light fixture					
interior incandescent light fixture	9E@.574	ea	—	36.30	36.30
Install bathroom bar light					
bathroom bar light	9E@.612	ea	—	38.70	38.70
Install chandelier					
chandelier	9E@1.67	ea	—	106.00	106.00
entrance chandelier	9E@3.35	ea	—	212.00	212.00
grand entrance chandelier	9E@4.59	ea	—	290.00	290.00
Install wall-mount fixture					
wall-mount light fixture	9E@.556	ea	—	35.10	35.10
Install fluorescent light fixture					
surface-mounted fluorescent light fixture	9E@1.05	ea	—	66.40	66.40
drop-in fluorescent fixture in suspended ceiling	9E@2.12	ea	—	134.00	134.00
fluorescent circline fixture	9E@.967	ea	—	61.10	61.10
Install ceiling fan					
assemble and install ceiling fan	9E@2.45	ea	—	155.00	155.00
remove ceiling fan for work, then reinstall	9E@2.74	ea	—	173.00	173.00
assemble and install ceiling fan with light	9E@3.68	ea	—	233.00	233.00
light for ceiling fan	9E@.525	ea	—	33.20	33.20
Install hanging light fixture					
hanging light fixture	9E@.305	ea	—	19.30	19.30
Install recessed spot fixture					
recessed spot light fixture	9E@.835	ea	—	52.80	52.80
Install exterior flood light					
exterior flood light fixture	9E@.591	ea	—	37.40	37.40
Install exterior post light fixture					
exterior post with light fixture	9E@5.41	ea	—	342.00	342.00
exterior post light fixture (no post)	9E@1.15	ea	—	72.70	72.70
wood post for exterior post light fixture	9E@4.29	ea	—	271.00	271.00
metal post for exterior post light fixture	9E@4.09	ea	—	258.00	258.00

Excavation

	Craft@Hrs	Unit	Material	Labor	Total
Mobilization charge.					
Delivery and take-home charge	6E@4.00	ea	—	266.00	266.00

	Craft@Hrs	Unit	Material	Labor	Equip.	Total
Excavate for slab-on-grade structure, per sf of floor. Under-slab excavation up to 1-1/2' deep. Footing excavation up to 3' deep. Does not include mobilization.						
Typical soil	6E@.002	sf	—	.13	.14	.27
Medium rocky or clay soil	6E@.003	sf	—	.20	.15	.35
Rocky or hardpan soil	6E@.004	sf	—	.27	.18	.45
Backfill	6E@.002	sf	—	.13	.05	.18
Excavate for structure with crawl space, per sf of floor. Crawl space excavation up to 3' deep. Footing excavation up to 4' deep. Does not include mobilization.						
Typical soil	6E@.004	sf	—	.27	.23	.50
Medium rocky or clay soil	6E@.005	sf	—	.33	.27	.60
Rocky or hardpan soil	6E@.005	sf	—	.33	.31	.64
Backfill	6E@.002	sf	—	.13	.10	.23
Excavate for structure with basement, per sf of floor. Basement excavation up to 7' deep. Footing excavation up to 8' deep. Does not include mobilization.						
Typical soil	6E@.011	sf	—	.73	.59	1.32
Medium rocky or clay soil	6E@.013	sf	—	.86	.73	1.59
Rocky or hardpan soil	6E@.015	sf	—	1.00	.81	1.81
Backfill	6E@.006	sf	—	.40	.27	.67
Excavate footings. Excavation slopes to 3' width at bottom of footing trench. Does not include mobilization.						
3' to 4' deep, typical soil	6E@.011	lf	—	.73	.58	1.31
3' to 4' deep, medium rocky or clay soil	6E@.013	lf	—	.86	.72	1.58
3' to 4' deep, rocky or hardpan soil	6E@.015	lf	—	1.00	.80	1.80
5' to 6' deep, typical soil	6E@.013	lf	—	.86	.77	1.63
5' to 6' deep, medium rocky or clay soil	6E@.015	lf	—	1.00	.86	1.86
5' to 6' deep, rocky or hardpan soil	6E@.019	lf	—	1.26	1.02	2.28
7' to 8' deep, typical soil	6E@.016	lf	—	1.06	.98	2.04
7' to 8' deep, medium rocky or clay soil	6E@.020	lf	—	1.33	1.10	2.43
7' to 8' deep, rocky or hardpan soil	6E@.023	lf	—	1.53	1.28	2.81
9' to 10' deep, typical soil	6E@.020	lf	—	1.33	1.13	2.46
9' to 10' deep, medium rocky or clay soil	6E@.024	lf	—	1.60	1.31	2.91
9' to 10' deep, rocky or hardpan soil	6E@.026	lf	—	1.73	1.52	3.25
Backfill against foundation wall. Typical backfill on flat lot or mild slope per sf of foundation wall. Does not include mobilization.						
Per sf of wall	6E@.007	lf	—	.47	.34	.81
Excavate utility trench, per lf of trench. Square trench, 2' wide. Does not include mobilization.						
1' to 2' deep, sandy loam soil	6E@.004	lf	—	.27	.16	.43
1' to 2' deep, typical soil	6E@.004	lf	—	.27	.20	.47
1' to 2' deep, medium rocky or clay soil	6E@.004	lf	—	.27	.23	.50
1' to 2' deep, rocky or hardpan soil	6E@.005	lf	—	.33	.27	.60
1' to 2' deep, backfill	6E@.004	lf	—	.27	.15	.42

	Craft@Hrs	Unit	Material	Labor	Equip.	Total
3' to 4' deep, sandy loam soil	6E@.005	lf	—	.33	.27	.60
3' to 4' deep, typical soil	6E@.005	lf	—	.33	.31	.64
3' to 4' deep, rocky or clay soil	6E@.007	lf	—	.47	.34	.81
3' to 4' deep, rocky or hardpan soil	6E@.007	lf	—	.47	.44	.91
3' to 4' deep, backfill	6E@.005	lf	—	.33	.31	.64
5' to 6' deep, sandy loam soil	6E@.007	lf	—	.47	.35	.82
5' to 6' deep, typical soil	6E@.006	lf	—	.40	.47	.87
5' to 6' deep, medium rocky or clay soil	6E@.008	lf	—	.53	.50	1.03
5' to 6' deep, rocky or hardpan soil	6E@.009	lf	—	.60	.58	1.18
5' to 6' deep, backfill	6E@.008	lf	—	.53	.47	1.00

Excavate water-line trench, per lf of trench. Trench slopes to 2' width at bottom for 2' to 3' depth, 3' width at bottom for all other depths. Does not include mobilization.

	Craft@Hrs	Unit	Material	Labor	Equip.	Total
2' to 3' deep, sandy loam soil	6E@.009	lf	—	.60	.35	.95
2' to 3' deep, typical soil	6E@.009	lf	—	.60	.44	1.04
2' to 3' deep, medium rocky or clay soil	6E@.010	lf	—	.67	.51	1.18
2' to 3' deep, rocky or hardpan soil	6E@.011	lf	—	.73	.59	1.32
2' to 3' deep, backfill	6E@.007	lf	—	.47	.33	.80
4' to 5' deep, sandy loam soil	6E@.011	lf	—	.73	.63	1.36
4' to 5' deep, typical soil	6E@.013	lf	—	.86	.75	1.61
4' to 5' deep, medium rocky or clay soil	6E@.014	lf	—	.93	.85	1.78
4' to 5' deep, rocky or hardpan soil	6E@.015	lf	—	1.00	1.01	2.01
4' to 5' deep, backfill	6E@.011	lf	—	.73	.58	1.31
6' to 7' deep, sandy loam soil	6E@.015	lf	—	1.00	.89	1.89
6' to 7' deep, typical soil	6E@.016	lf	—	1.06	1.05	2.11
6' to 7' deep, medium rocky or clay soil	6E@.018	lf	—	1.20	1.21	2.41
6' to 7' deep, rocky or hardpan soil	6E@.020	lf	—	1.33	1.35	2.68
6' to 7' deep, backfill	6E@.013	lf	—	.86	.84	1.70
8' to 9' deep, sandy loam soil	6E@.018	lf	—	1.20	1.13	2.33
8' to 9' deep, typical soil	6E@.020	lf	—	1.33	1.32	2.65
8' to 9' deep, medium rocky or clay soil	6E@.023	lf	—	1.53	1.56	3.09
8' to 9' deep, rocky or hardpan soil	6E@.025	lf	—	1.66	1.82	3.48
8' to 9' deep, backfill	6E@.016	lf	—	1.06	1.07	2.13

Excavate sewer-line trench, per lf of trench. Trench slopes to 3-1/2' width at bottom. Does not include mobilization.

	Craft@Hrs	Unit	Material	Labor	Equip.	Total
3' to 4' deep, sandy loam soil	6E@.011	lf	—	.73	.59	1.32
3' to 4' deep, typical soil	6E@.013	lf	—	.86	.72	1.58
3' to 4' deep, medium rocky soil	6E@.014	lf	—	.93	.80	1.73
3' to 4' deep, rocky or hardpan soil	6E@.015	lf	—	1.00	.97	1.97
3' to 4' deep, backfill	6E@.010	lf	—	.67	.55	1.22
5' to 6' deep, sandy loam soil	6E@.015	lf	—	1.00	.89	1.89
5' to 6' deep, typical soil	6E@.016	lf	—	1.06	1.05	2.11
5' to 6' deep, medium rocky soil	6E@.018	lf	—	1.20	1.21	2.41
5' to 6' deep, rocky or hardpan soil	6E@.020	lf	—	1.33	1.38	2.71
5' to 6' deep, backfill	6E@.013	lf	—	.86	.84	1.70
7' to 8' deep, sandy loam soil	6E@.018	lf	—	1.20	1.19	2.39
7' to 8' deep, typical soil	6E@.020	lf	—	1.33	1.35	2.68
7' to 8' deep, medium rocky soil	6E@.023	lf	—	1.53	1.58	3.11
7' to 8' deep, rocky or hardpan soil	6E@.025	lf	—	1.66	1.82	3.48
7' to 8' deep, backfill	6E@.011	lf	—	.73	1.11	1.84

	Craft@Hrs	Unit	Material	Labor	Equip.	Total
Additional excavation costs.						
Excavate hole for repairs to water, sewer, or utility lines, per lf of depth	6E@.044	lf	—	2.93	2.48	5.41
General foundation, slab, & footing excavation, per cubic yard	6E@.046	cy	—	3.06	2.61	5.67
Minimum charge for excavation work	6E@5.00	ea	—	333.00	538.00	871.00

Time & Material Charts (selected items)

Excavation Rental Equipment

	Craft@Hrs	Unit	Material	Labor	Equip.	Total
Backhoe / loader						
Excavate slab-on-grade structure						
typical soil ($832.00 per day, 5,740 sf per day)	—	sf	—	—	.15	.15
rocky or hardpan soil ($832.00 per day, 4,250 sf per day)	—	sf	—	—	.20	.20
backfill ($832.00 per day, 9,420 sf per day)	—	sf	—	—	.08	.08
Excavate for structure with crawl space						
typical soil ($832.00 per day, 3,530 sf per day)	—	sf	—	—	.24	.24
rocky or hardpan soil ($832.00 per day, 2,610 sf per day)	—	sf	—	—	.32	.32
backfill ($832.00 per day, 6,123 sf per day)	—	sf	—	—	.14	.14
Excavate for structure with basement						
typical soil ($832.00 per day, 1,400 sf per day)	—	sf	—	—	.59	.59
rocky or hardpan soil ($832.00 per day, 1,030 sf per day)	—	sf	—	—	.80	.80
backfill ($832.00 per day, 3,140 sf per day)	—	sf	—	—	.27	.27
Excavate for 3' to 4' deep footing						
typical soil ($832.00 per day, 1,420 lf per day)	—	lf	—	—	.58	.58
rocky or hardpan soil ($832.00 per day, 1,050 lf per day)	—	lf	—	—	.79	.79
Excavate for 5' to 6' deep footing						
typical soil ($832.00 per day, 1,108 lf per day)	—	lf	—	—	.75	.75
rocky or hardpan soil ($832.00 per day, 830 lf per day)	—	lf	—	—	1.00	1.00
Excavate for 7' to 8' deep footing						
typical soil ($832.00 per day, 880 lf per day)	—	lf	—	—	.95	.95
rocky or hardpan soil ($832.00 per day, 660 lf per day)	—	lf	—	—	1.26	1.26

	Craft@Hrs	Unit	Material	Labor	Equip.	Total
Excavate for 9' to 10' deep footing						
typical soil ($832.00 per day, 738 lf per day)	—	lf	—	—	1.13	1.13
rocky or hardpan soil ($832.00 per day, 560 lf per day)	—	lf	—	—	1.49	1.49
Backfill foundation wall						
backfill ($832.00 per day, 2,230 lf per day)	—	lf	—	—	.36	.36
Excavate for 1' to 2' deep utility trench						
sandy loam soil ($832.00 per day, 4,930 lf per day)	—	lf	—	—	.17	.17
rocky or hardpan soil ($832.00 per day, 3,150 lf per day)	—	lf	—	—	.27	.27
backfill ($832.00 per day, 5,300 lf per day)	—	lf	—	—	.16	.16
Excavate for 3' to 4' deep utility trench						
sandy loam soil ($832.00 per day, 3,057 lf per day)	—	lf	—	—	.28	.28
rocky or hardpan soil ($832.00 per day, 1,960 lf per day)	—	lf	—	—	.43	.43
backfill ($832.00 per day, 2,650 lf per day)	—	lf	—	—	.31	.31
Excavate for 5' to 6' deep utility trench						
sandy loam soil ($832.00 per day, 2,219 lf per day)	—	lf	—	—	.38	.38
rocky or hardpan soil ($832.00 per day, 1,410 lf per day)	—	lf	—	—	.58	.58
backfill ($832.00 per day, 1,760 lf per day)	—	lf	—	—	.47	.47
Excavate for 2' to 3' deep water-line trench						
sandy loam soil ($832.00 per day, 2,200 lf per day)	—	lf	—	—	.38	.38
rocky or hardpan soil ($832.00 per day, 1,400 lf per day)	—	lf	—	—	.59	.59
backfill ($832.00 per day, 2,350 lf per day)	—	lf	—	—	.35	.35
Excavate for 4' to 5' deep water-line trench						
sandy loam soil ($832.00 per day, 1,330 lf per day)	—	lf	—	—	.59	.59
typical soil ($832.00 per day, 1,140 lf per day)	—	lf	—	—	.73	.73
medium rocky or clay soil ($832.00 per day, 980 lf per day)	—	lf	—	—	.84	.84
rocky or hardpan soil ($832.00 per day, 850 lf per day)	—	lf	—	—	.98	.98
backfill ($832.00 per day, 1,410 lf per day)	—	lf	—	—	.58	.58

	Craft@Hrs	Unit	Material	Labor	Equip.	Total
Excavate for 6' to 7' deep water-line trench						
sandy loam soil ($832.00 per day, 940 lf per day)	—	lf	—	—	.88	.88
typical soil ($832.00 per day, 810 lf per day)	—	lf	—	—	1.02	1.02
medium rocky or clay soil ($832.00 per day, 700 lf per day)	—	lf	—	—	1.19	1.19
rocky or hardpan soil ($832.00 per day, 610 lf per day)	—	lf	—	—	1.37	1.37
backfill ($832.00 per day, 1,010 lf per day)	—	lf	—	—	.82	.82
Excavate for 8' to 9' deep water-line trench						
sandy loam soil ($832.00 per day, 730 lf per day)	—	lf	—	—	1.14	1.14
typical soil ($832.00 per day, 630 lf per day)	—	lf	—	—	1.32	1.32
medium rocky or clay soil ($832.00 per day, 540 lf per day)	—	lf	—	—	1.54	1.54
rocky or hardpan soil ($832.00 per day, 460 lf per day)	—	lf	—	—	1.80	1.80
backfill ($832.00 per day, 780 lf per day)	—	lf	—	—	1.07	1.07
Excavate for 3' to 4' deep sewer-line trench						
sandy loam soil ($832.00 per day, 1,420 lf per day)	—	lf	—	—	.58	.58
typical soil ($832.00 per day, 1,220 lf per day)	—	lf	—	—	.69	.69
medium rocky or clay soil ($832.00 per day, 1,050 lf per day)	—	lf	—	—	.79	.79
rocky or hardpan soil ($832.00 per day, 900 lf per day)	—	lf	—	—	.93	.93
backfill ($832.00 per day, 1,520 lf per day)	—	lf	—	—	.55	.55
Excavate for 5' to 6' deep sewer-line trench						
sandy loam soil ($832.00 per day, 940 lf per day)	—	lf	—	—	.88	.88
typical soil ($832.00 per day, 810 lf per day)	—	lf	—	—	1.02	1.02
medium rocky or clay soil ($832.00 per day, 700 lf per day)	—	lf	—	—	1.19	1.19
backfill ($832.00 per day, 1,010 lf per day)	—	lf	—	—	.82	.82

	Craft@Hrs	Unit	Material	Labor	Equip.	Total
Excavate for 7' to 8' deep sewer-line trench						
sandy loam soil ($832.00 per day, 710 lf per day)	—	lf	—	—	1.18	1.18
typical soil ($832.00 per day, 610 lf per day)	—	lf	—	—	1.37	1.37
medium rocky or clay soil ($832.00 per day, 530 lf per day)	—	lf	—	—	1.56	1.56
rocky or hardpan soil ($832.00 per day, 460 lf per day)	—	lf	—	—	1.80	1.80
backfill ($832.00 per day, 750 lf per day)	—	lf	—	—	1.10	1.10
Excavate for repairs						
excavate hole ($832.00 per day, 340 lf per day)	—	lf	—	—	2.45	2.45
General excavation						
excavation ($832.00 per day, 322 cy per day)	—	cy	—	—	2.58	2.58
minimum charge ($534.00 per day, 1 per day)	—	ea	—	—	534.00	534.00
Trenching						
Trenching machine ($600.00 per day, 1 per day)	—	day	—	—	600.00	600.00
minimum charge for trenching machine ($368.00 per day, 1 per day)	—	ea	—	—	368.00	368.00
Compacting						
"Jumping jack" compactor ($189.00 per day, 1 per day)	—	day	—	—	189.00	189.00
minimum charge for compactor ($151.00 per day, 1 per day)	—	ea	—	—	151.00	151.00

Excavation Labor

Laborer	base wage	paid leave	true wage	taxes & ins.	total
Equipment operator	$52.10	4.06	$56.16	32.74	$88.90
Excavation laborer	$23.70	1.85	$25.55	18.45	$44.00

Paid leave is calculated based on two weeks paid vacation, one week sick leave, and seven paid holidays. Employer's matching portion of **FICA** is 7.65 percent. **FUTA** (Federal Unemployment) is .8 percent. **Worker's compensation** for the excavation trade was calculated using a national average of 15.32 percent for the equipment operator and 15.37 percent for the excavation laborer. **Unemployment insurance** was calculated using a national average of 8 percent. **Health insurance** was calculated based on a projected national average for 2020 of $1,280 per employee (and family when applicable) per month. Employer pays 80 percent for a per month cost of $1,024 per employee. **Retirement** is based on a 401(k) retirement program with employer matching of 50 percent. Employee contributions to the 401(k) plan are an average of 6 percent of the true wage. **Liability insurance** is based on a national average of 12.0 percent.

Excavation Labor Productivity

	Craft@Hrs	Unit	Material	Labor	Total
Excavation equipment delivery and return labor charges					
backhoe mobilization	10@4.00	ea	—	356.00	356.00
Excavate slab-on-grade structure (per sf of floor)					
typical soil, machine	10@.001	sf	—	.09	.09
typical soil, hand finish	5E@.001	sf	—	.04	.04
rocky or hardpan soil, machine	10@.002	sf	—	.18	.18
rocky or hardpan soil, hand finish	5E@.002	sf	—	.09	.09
Backfill slab-on-grade structure (per sf of floor)					
all soil types, machine	10@.001	sf	—	.09	.09
all soil types, hand labor	5E@.001	sf	—	.04	.04
Excavate for structure with crawl space (per sf of floor)					
typical soil, machine	10@.002	sf	—	.18	.18
typical soil, hand finish	5E@.002	sf	—	.09	.09
rocky or hardpan soil, machine	10@.003	sf	—	.27	.27
rocky or hardpan soil, hand finish	5E@.002	sf	—	.09	.09
Backfill structure with crawl space (per sf of floor)					
all soil types, machine	10@.001	sf	—	.09	.09
all soil types, hand labor	5E@.001	sf	—	.04	.04
Excavate for structure with basement (per sf of floor)					
typical soil, machine	10@.006	sf	—	.53	.53
typical soil, hand finish	5E@.005	sf	—	.22	.22
rocky or hardpan soil, machine	10@.008	sf	—	.71	.71
rocky or hardpan soil, hand finish	5E@.006	sf	—	.26	.26
Backfill structure with basement (per sf of floor)					
all soil types, machine	10@.003	sf	—	.27	.27
all soil types, hand labor	5E@.003	sf	—	.13	.13
Excavate for 3' to 4' deep footings					
typical soil, machine	10@.006	lf	—	.53	.53
typical soil, hand finish	5E@.005	lf	—	.22	.22
rocky or hardpan soil, machine	10@.008	lf	—	.71	.71
rocky or hardpan soil, hand finish	5E@.006	lf	—	.26	.26
Excavate for 5' to 6' deep footings					
typical soil, machine	10@.007	lf	—	.62	.62
typical soil, hand finish	5E@.006	lf	—	.26	.26
rocky or hardpan soil, machine	10@.010	lf	—	.89	.89
rocky or hardpan soil, hand finish	5E@.008	lf	—	.35	.35
Excavate for 7' to 8' deep footing					
typical soil, machine	10@.009	lf	—	.80	.80
typical soil, hand finish	5E@.007	lf	—	.31	.31
hardpan or rocky soil, machine	10@.012	lf	—	1.07	1.07
hardpan or rocky soil, hand finish	5E@.010	lf	—	.44	.44
Excavate for 9' to 10' deep footing					
typical soil, machine	10@.011	lf	—	.98	.98
typical soil, hand finish	5E@.009	lf	—	.40	.40
hardpan or rocky soil, machine	10@.014	lf	—	1.24	1.24
hardpan or rocky soil, hand finish	5E@.011	lf	—	.48	.48
Backfill foundation wall (per sf of wall)					
machine	10@.004	sf	—	.36	.36
hand labor	5E@.003	sf	—	.13	.13

	Craft@Hrs	Unit	Material	Labor	Total
Machine excavate for 1' to 2' deep utility trench					
sandy loam soil	10@.002	lf	—	.18	.18
rocky or hardpan soil	10@.003	lf	—	.27	.27
backfill	10@.002	lf	—	.18	.18
Machine excavate for 3' to 4' deep utility trench					
sandy loam soil	10@.003	lf	—	.27	.27
rocky or hardpan soil	10@.004	lf	—	.36	.36
backfill	10@.003	lf	—	.27	.27
Machine excavate for 5' to 6' deep utility trench					
sandy loam soil	10@.004	lf	—	.36	.36
rocky or hardpan soil	10@.006	lf	—	.53	.53
backfill	10@.005	lf	—	.44	.44
Hand finish excavation for utility trench					
all soil types	5E@.002	lf	—	.09	.09
Excavate for 2' to 3' deep water-line trench					
sandy loam soil	10@.004	lf	—	.36	.36
rocky or hardpan soil	10@.006	lf	—	.53	.53
backfill	10@.003	lf	—	.27	.27
Excavate for 4' to 5' deep water-line trench					
sandy loam soil	10@.006	lf	—	.53	.53
rocky or hardpan soil	10@.009	lf	—	.80	.80
backfill	10@.006	lf	—	.53	.53
Excavate for 6' to 7' deep water-line trench					
sandy loam soil	10@.009	lf	—	.80	.80
rocky or hardpan soil	10@.013	lf	—	1.16	1.16
backfill	10@.008	lf	—	.71	.71
Excavate for 8' to 9' deep water-line trench					
sandy loam soil	10@.011	lf	—	.98	.98
rocky or hardpan soil	10@.017	lf	—	1.51	1.51
backfill	10@.010	lf	—	.89	.89
Hand finish excavation for water-line trench					
all soil types	5E@.005	lf	—	.22	.22
Excavate for 3' to 4' deep sewer-line trench					
sandy loam soil	10@.006	lf	—	.53	.53
rocky or hardpan soil	10@.009	lf	—	.80	.80
backfill	10@.005	lf	—	.44	.44
Excavate for 5' to 6' deep sewer-line trench					
sandy loam soil	10@.009	lf	—	.80	.80
rocky or hardpan soil	10@.013	lf	—	1.16	1.16
backfill	10@.008	lf	—	.71	.71
Excavate for 7' to 8' deep sewer-line trench					
sandy loam soil	10@.011	lf	—	.98	.98
rocky or hardpan soil	10@.017	lf	—	1.51	1.51
backfill	10@.006	lf	—	.53	.53
Hand finish excavation for sewer-line trench					
all soil types	5E@.005	lf	—	.22	.22
Machine excavate hole for repairs to water, sewer, or utility lines					
machine excavate per lf of depth	10@.024	lf	—	2.13	2.13
hand finish per lf of depth	5E@.019	lf	—	.84	.84
General foundation, slab, & footing machine excavation					
machine excavate per cy	10@.025	cy	—	2.22	2.22
hand finish per cy	5E@.020	cy	—	.88	.88

Fees

Building permit. Permit fees vary widely. These fees are rules of thumb for most areas of the country. Rural areas will be less and high-density areas will be more. Includes plan-check fees.

	Craft@Hrs	Unit	Material	Labor	Total
Minimum	—	ea	—	—	147.00
For job $2,001 to $5,000	—	ea	—	—	158.00
For job $5,001 to $10,000	—	ea	—	—	472.00
For job $10,001 to $20,000	—	ea	—	—	863.00
For job $20,001 to $30,000	—	ea	—	—	1,100.00
For job $30,001 to $40,000	—	ea	—	—	1,240.00
For job $40,001 to $50,000	—	ea	—	—	1,420.00
For job $50,001 to $60,000	—	ea	—	—	1,800.00
For job $60,001 to $70,000	—	ea	—	—	2,040.00
For job $70,001 to $80,000	—	ea	—	—	2,190.00
For job $80,001 to $90,000	—	ea	—	—	2,440.00
For job $90,001 to $100,000	—	ea	—	—	2,610.00
For job $100,001 to $150,000	—	ea	—	—	3,300.00
For job $150,001 to $200,000	—	ea	—	—	3,920.00
For job $200,001 to $250,000	—	ea	—	—	4,720.00
For job $250,001 to $300,000	—	ea	—	—	5,310.00
For job $300,001 to $400,000	—	ea	—	—	6,730.00
For job $400,001 to $500,000	—	ea	—	—	8,180.00

Sewer connection fee. Typical charges for labor and materials installed by government entity. Does not include any development upcharges added to the fee. Includes connect to a city sewer line up to 50 lf from the structure, mobilization, excavation, pipe and fittings, backfill, repair to street, curb and gutter, and temporary signs.

	Craft@Hrs	Unit	Material	Labor	Total
Typical charge	—	ea	—	—	3,600.00

Water connection fee. Typical charges for labor and materials installed by government entity. Does not include any development upcharges added to the fee. Includes connect to a city water line up to 50 lf from the structure, mobilization, excavation, pipe and fittings, water meter, backfill, repair to street, curb and gutter, and temporary signs.

	Craft@Hrs	Unit	Material	Labor	Total
Typical charge	—	ea	—	—	3,410.00

Estimating fees. Typical rates charged to estimate insurance repair damages. Rule of thumb for estimating insurance repair damages on losses under $200,000. Percentage rate drops to as low as .75% on large losses or multiple unit losses. This percentage is used across the country. Do not adjust with Area Modification Factor.

	Craft@Hrs	Unit	Material	Labor	Total
Per hour	—	hr	—	—	74.50
Per day	—	dy	—	—	714.00
As percentage of estimate	—	%	—	—	1.5
Minimum	—	ea	—	—	297.00

Soils engineer fee.

	Craft@Hrs	Unit	Material	Labor	Total
Per hour	—	hr	—	—	158.00

Structural engineer fee.

	Craft@Hrs	Unit	Material	Labor	Total
Per hour	—	hr	—	—	173.00

Plan drawing fee, per sf of floor.

	Craft@Hrs	Unit	Material	Labor	Total
Simple structure	—	sf	—	—	.33
Typical structure	—	sf	—	—	.39
Complex structure	—	sf	—	—	.57
Ornate structure	—	sf	—	—	.79
Minimum	—	ea	—	—	322.00

Fences

Wood fence post. All wood fence posts include digging of 2' deep hole and setting in concrete. Post height is above ground. For example, a 4' high post is made from a 6' length with 2' underground and 4' above ground.

	Craft@Hrs	Unit	Material	Labor	Total
Cedar fence post					
4' high cedar	4F@.907	ea	32.10	53.50	85.60
6' high cedar	4F@.907	ea	37.10	53.50	90.60
8' high cedar	4F@.907	ea	42.00	53.50	95.50
Redwood fence post					
4' high	4F@.907	ea	36.10	53.50	89.60
6' high	4F@.907	ea	41.30	53.50	94.80
8' high	4F@.907	ea	46.70	53.50	100.20
Treated pine fence post					
4' high	4F@.907	ea	33.60	53.50	87.10
6' high	4F@.907	ea	38.80	53.50	92.30
8' high	4F@.907	ea	43.70	53.50	97.20
Remove wood fence post	1D@.367	ea	—	17.30	17.30

Basketweave fence. Posts are set in concrete and spaced up to 8' on center with an average spacing of 7.2' apart. Fence boards are 1" x 6" to 1" x 10". Gates are 3' wide with a frame made from 2" x 4" lumber, cross braced with turn buckle and cable. Includes heavy-duty hinges, latch, and spring closer.

	Craft@Hrs	Unit	Material	Labor	Total
Cedar basketweave fence					
4' high	4F@.317	lf	19.00	18.70	37.70
6' high	4F@.336	lf	25.70	19.80	45.50
8' high	4F@.360	lf	33.50	21.20	54.70
Redwood basketweave fence					
4' high	4F@.317	lf	20.90	18.70	39.60
6' high	4F@.336	lf	28.20	19.80	48.00
8' high	4F@.360	lf	37.10	21.20	58.30
Treated pine basketweave fence					
4' high	4F@.317	lf	19.60	18.70	38.30
6' high	4F@.336	lf	26.40	19.80	46.20
8' high	4F@.360	lf	34.90	21.20	56.10
Cedar gate with hardware					
4' high	4F@2.70	ea	108.00	159.00	267.00
6' high	4F@2.86	ea	134.00	169.00	303.00
8' high	4F@3.04	ea	157.00	179.00	336.00
Redwood gate with hardware					
4' high	4F@2.70	ea	119.00	159.00	278.00
6' high	4F@2.86	ea	147.00	169.00	316.00
8' high	4F@3.04	ea	172.00	179.00	351.00
Treated pine gate with hardware					
4' high	4F@2.70	ea	112.00	159.00	271.00
6' high	4F@2.86	ea	139.00	169.00	308.00
8' high	4F@3.04	ea	162.00	179.00	341.00
Remove wood fence					
4' high wood fence	1D@.161	lf	—	7.60	7.60
6' high wood fence	1D@.165	lf	—	7.79	7.79
8' high wood fence	1D@.170	lf	—	8.02	8.02
4' high fence gate	1D@.324	ea	—	15.30	15.30
6' high fence gate	1D@.336	ea	—	15.90	15.90
8' high fence gate	1D@.346	ea	—	16.30	16.30

Board fence. Fence with boards installed vertically or horizontally. Posts are set in concrete and spaced up to 8' on center with an average spacing of 7.2' apart. Fence boards are 1" x 4" to 1" x 8". 8' tall fence contains three supporting rails. All other heights contain two. Gates are 3' wide with horizontal and diagonal rails, heavy-duty hinges, latch, and spring closer.

	Craft@Hrs	Unit	Material	Labor	Total
Cedar, replace fence					
4' high	4F@.267	lf	19.00	15.80	34.80
6' high	4F@.286	lf	25.70	16.90	42.60
8' high	4F@.307	lf	33.50	18.10	51.60
Redwood, replace fence					
4' high	4F@.267	lf	20.90	15.80	36.70
6' high	4F@.286	lf	28.20	16.90	45.10
8' high	4F@.307	lf	37.10	18.10	55.20
Treated pine, replace fence					
4' high	4F@.267	lf	19.60	15.80	35.40
6' high	4F@.286	lf	26.40	16.90	43.30
8' high	4F@.307	lf	34.90	18.10	53.00
Cedar gate with hardware					
4' high	4F@2.17	ea	108.00	128.00	236.00
6' high	4F@2.33	ea	134.00	137.00	271.00
8' high	4F@2.44	ea	157.00	144.00	301.00
Redwood gate with hardware					
4' high	4F@2.17	ea	119.00	128.00	247.00
6' high	4F@2.33	ea	147.00	137.00	284.00
8' high	4F@2.44	ea	172.00	144.00	316.00
Treated pine gate with hardware					
4' high	4F@2.17	ea	112.00	128.00	240.00
6' high	4F@2.33	ea	139.00	137.00	276.00
8' high	4F@2.44	ea	162.00	144.00	306.00
Remove wood fence					
4' high wood fence	1D@.161	lf	—	7.60	7.60
6' high wood fence	1D@.165	lf	—	7.79	7.79
8' high wood fence	1D@.170	lf	—	8.02	8.02
4' high wood gate	1D@.324	ea	—	15.30	15.30
6' high wood gate	1D@.336	ea	—	15.90	15.90
8' high wood gate	1D@.346	ea	—	16.30	16.30

Board fence with lattice cap. Fence with boards installed vertically or horizontally capped with 12" of lattice on top. Lattice is trimmed with 2" x 4" lattice rail on top, bottom and sides. Posts are set in concrete and spaced up to 8' on center with an average spacing of 7.2' apart. Fence boards are 1" x 4" to 1" x 8". Gates are 3' wide with a frame made from 2" x 4" lumber, cross braced with turn buckle and cable. Includes heavy-duty hinges, latch, and spring closer.

	Craft@Hrs	Unit	Material	Labor	Total
Cedar board fence with lattice cap					
6' high	4F@.331	lf	33.70	19.50	53.20
8' high	4F@.360	lf	41.80	21.20	63.00
Redwood board fence with lattice cap					
6' high	4F@.331	lf	37.50	19.50	57.00
8' high	4F@.360	lf	46.40	21.20	67.60
Treated pine board fence with lattice cap					
6' high	4F@.331	lf	35.30	19.50	54.80
8' high	4F@.360	lf	43.60	21.20	64.80

	Craft@Hrs	Unit	Material	Labor	Total
Cedar gate with hardware					
6' high	4F@2.94	ea	146.00	173.00	319.00
8' high	4F@3.44	ea	170.00	203.00	373.00
Redwood gate with hardware					
6' high	4F@2.94	ea	163.00	173.00	336.00
8' high	4F@3.44	ea	190.00	203.00	393.00
Treated pine gate with hardware					
6' high	4F@2.94	ea	152.00	173.00	325.00
8' high	4F@3.44	ea	176.00	203.00	379.00
Remove fence					
6' high wood fence	1D@.165	lf	—	7.79	7.79
8' high wood fence	1D@.170	lf	—	8.02	8.02
6' high wood gate	1D@.335	ea	—	15.80	15.80
8' high wood gate	1D@.344	ea	—	16.20	16.20

Board-on-board fence. Fence with overlapping boards installed vertically. Posts are set in concrete and spaced up to 8' on center with an average spacing of 7.2' apart. Fence boards are 1" x 4" to 1" x 8". 8' tall fence contains three supporting rails. All other heights contain two. Gates are 3' wide with horizontal and diagonal rails. Includes heavy-duty hinges, latch, and spring closer.

	Craft@Hrs	Unit	Material	Labor	Total
Cedar board-on-board fence					
4' high	4F@.296	lf	28.20	17.50	45.70
6' high	4F@.312	lf	39.70	18.40	58.10
8' high	4F@.331	lf	52.60	19.50	72.10
Redwood board-on-board fence					
4' high	4F@.296	lf	31.40	17.50	48.90
6' high	4F@.312	lf	44.20	18.40	62.60
8' high	4F@.331	lf	58.30	19.50	77.80
Treated pine board-on-board fence					
4' high	4F@.296	lf	29.70	17.50	47.20
6' high	4F@.312	lf	41.40	18.40	59.80
8' high	4F@.331	lf	55.00	19.50	74.50
Cedar gate with hardware					
4' high	4F@2.56	ea	136.00	151.00	287.00
6' high	4F@2.70	ea	176.00	159.00	335.00
8' high	4F@2.86	ea	213.00	169.00	382.00
Redwood gate with hardware					
4' high	4F@2.56	ea	151.00	151.00	302.00
6' high	4F@2.70	ea	196.00	159.00	355.00
8' high	4F@2.86	ea	235.00	169.00	404.00
Treated pine gate with hardware					
4' high	4F@2.56	ea	143.00	151.00	294.00
6' high	4F@2.70	ea	185.00	159.00	344.00
8' high	4F@2.86	ea	220.00	169.00	389.00
Remove fence					
4' high wood fence	1D@.161	lf	—	7.60	7.60
6' high wood fence	1D@.165	lf	—	7.79	7.79
8' high wood fence	1D@.170	lf	—	8.02	8.02
4' high wood gate	1D@.324	ea	—	15.30	15.30
6' high wood gate	1D@.336	ea	—	15.90	15.90
8' high wood gate	1D@.346	ea	—	16.30	16.30

Board-and-batten fence. Vertically installed 1" x 8" to 1" x 12" boards with 1" x 2" to 1" x 4" battens overlapping joints. Posts are set in concrete and spaced up to 8' on center with an average spacing of 7.2' apart. Gates are 3' wide with horizontal and diagonal rails. Includes heavy-duty hinges, latch, and spring closer.

	Craft@Hrs	Unit	Material	Labor	Total
Cedar board-and-batten fence					
4' high	4F@.302	lf	24.30	17.80	42.10
6' high	4F@.307	lf	33.50	18.10	51.60
8' high	4F@.323	lf	44.00	19.10	63.10
Redwood board-and-batten fence					
4' high	4F@.302	lf	26.80	17.80	44.60
6' high	4F@.307	lf	37.10	18.10	55.20
8' high	4F@.323	lf	49.20	19.10	68.30
Treated pine board-and-batten fence					
4' high	4F@.302	lf	25.20	17.80	43.00
6' high	4F@.307	lf	34.90	18.10	53.00
8' high	4F@.323	lf	45.90	19.10	65.00
Cedar gate with hardware					
4' high	4F@2.63	ea	123.00	155.00	278.00
6' high	4F@2.78	ea	157.00	164.00	321.00
8' high	4F@2.94	ea	189.00	173.00	362.00
Redwood gate with hardware					
4' high	4F@2.63	ea	137.00	155.00	292.00
6' high	4F@2.78	ea	174.00	164.00	338.00
8' high	4F@2.94	ea	208.00	173.00	381.00
Treated pine gate with hardware					
4' high	4F@2.63	ea	129.00	155.00	284.00
6' high	4F@2.78	ea	163.00	164.00	327.00
8' high	4F@2.94	ea	196.00	173.00	369.00
Remove fence					
4' high wood fence	1D@.161	lf	—	7.60	7.60
6' high wood fence	1D@.165	lf	—	7.79	7.79
8' high wood fence	1D@.170	lf	—	8.02	8.02
4' high wood gate	1D@.324	ea	—	15.30	15.30
6' high wood gate	1D@.336	ea	—	15.90	15.90
8' high wood gate	1D@.346	ea	—	16.30	16.30

Picket fence. Posts are set in concrete and spaced up to 8' on center with an average spacing of 7.2' apart. Pickets are factory pre-cut. Gates are 3' wide with horizontal and diagonal rails. Includes heavy-duty hinges, latch, and spring closer.

	Craft@Hrs	Unit	Material	Labor	Total
Cedar picket fence					
3' high	4F@.222	lf	11.10	13.10	24.20
5' high	4F@.238	lf	18.90	14.00	32.90
Redwood picket fence					
3' high	4F@.222	lf	12.20	13.10	25.30
5' high	4F@.238	lf	20.50	14.00	34.50
Treated pine picket fence					
3' high	4F@.222	lf	11.50	13.10	24.60
5' high	4F@.238	lf	19.40	14.00	33.40

	Craft@Hrs	Unit	Material	Labor	Total
Cedar gate with hardware					
3' high	4F@2.00	ea	84.00	118.00	202.00
5' high	4F@2.38	ea	110.00	140.00	250.00
Redwood gate with hardware					
3' high	4F@2.00	ea	93.30	118.00	211.30
5' high	4F@2.38	ea	121.00	140.00	261.00
Treated pine gate with hardware					
3' high	4F@2.00	ea	87.40	118.00	205.40
5' high	4F@2.38	ea	115.00	140.00	255.00
Remove fence					
3' high wood fence	1D@.161	lf	—	7.60	7.60
5' high wood fence	1D@.165	lf	—	7.79	7.79
3' high wood gate	1D@.324	ea	—	15.30	15.30
5' high wood gate	1D@.336	ea	—	15.90	15.90

Additional wood fence costs.

	Craft@Hrs	Unit	Material	Labor	Total
minimum charge	4F@1.75	ea	48.50	103.00	151.50
remove and reinstall wood fence gate	4F@.804	ea	—	47.40	47.40

Wood gate hardware.

	Craft@Hrs	Unit	Material	Labor	Total
replace, with self-closing hinges	4F@.754	ea	51.50	44.50	96.00
remove	1D@.197	ea	—	9.30	9.30
replace, heavy gauge with self-closing hinges	4F@.754	ea	88.60	44.50	133.10
remove	1D@.197	ea	—	9.30	9.30
replace, closing spring	4F@.331	ea	31.50	19.50	51.00
remove	1D@.100	ea	—	4.72	4.72

Chain-link fence. 11 gauge chain link. Line posts and terminal posts are set 2' deep in concrete and spaced up to 10' on center with an average spacing of 9.2' apart. Includes all post caps, tie wire, tension bands, nuts, bolts, and tension rods.

	Craft@Hrs	Unit	Material	Labor	Total
replace, 3' high, 1-5/8" line posts, 2" terminal posts	4F@.114	lf	11.20	6.73	17.93
remove	1D@.116	lf	—	5.48	5.48
replace, 4' high, 1-7/8" line posts, 2" terminal posts	4F@.145	lf	14.90	8.56	23.46
remove	1D@.121	lf	—	5.71	5.71
replace, 5' high, 1-7/8" line posts, 2" terminal posts	4F@.183	lf	18.80	10.80	29.60
remove	1D@.125	lf	—	5.90	5.90
replace, 6' high, 1-7/8" line posts, 2" terminal posts	4F@.233	lf	22.20	13.70	35.90
remove	1D@.131	lf	—	6.18	6.18
replace, 7' high, 1-7/8" line posts, 2" terminal posts	4F@.296	lf	25.90	17.50	43.40
remove	1D@.136	lf	—	6.42	6.42
replace, 8' high, 2" line posts, 3" terminal posts	4F@.376	lf	29.80	22.20	52.00
remove	1D@.143	lf	—	6.75	6.75
replace, 10' high, 2" line posts, 3" terminal posts	4F@.476	lf	36.80	28.10	64.90
remove	1D@.152	lf	—	7.17	7.17

	Craft@Hrs	Unit	Material	Labor	Total
Chain-link gate with hardware. 3' wide.					
replace, 3' high	4F@2.56	ea	61.60	151.00	212.60
replace, 4' high	4F@2.56	ea	79.50	151.00	230.50
replace, 5' high	4F@2.56	ea	95.90	151.00	246.90
replace, 6' high	4F@2.63	ea	110.00	155.00	265.00
replace, 7' high	4F@2.63	ea	123.00	155.00	278.00
replace, 8' high	4F@2.63	ea	136.00	155.00	291.00
replace, 10' high	4F@2.63	ea	164.00	155.00	319.00
remove gate	1D@.203	ea	—	9.58	9.58
Chain-link driveway gate with hardware. Prices per lf. Includes all hardware. Minimum of 6 lf.					
replace, 3' high	4F@.640	lf	18.40	37.80	56.20
replace, 4' high	4F@.640	lf	33.50	37.80	71.30
replace, 5' high	4F@.640	lf	54.00	37.80	91.80
replace, 6' high	4F@.659	lf	62.50	38.90	101.40
replace, 7' high	4F@.659	lf	70.70	38.90	109.60
replace, 8' high	4F@.659	lf	78.00	38.90	116.90
remove gate	1D@.043	lf	—	2.03	2.03
Additional chain-link fence costs. The chain-link fence prices in the previous tables allow for an average of 1 corner or terminal post for every 15 lf of fence. Use the additional corner price only when estimating a fence that contains more corner or terminal posts.					
add for additional corner in chain-link fence	4F@1.93	ea	75.90	114.00	189.90
add for vinyl privacy slats	4F@.006	sf	.90	.35	1.25
add for aluminum privacy slats	4F@.006	sf	1.00	.35	1.35
add 3% for chain-link fence with top rail					
add 28% for 9 gauge galvanized chain-link					
add 25% for 9 gauge vinyl coated chain-link and posts					
add 41% for 6 gauge chain-link					
remove chain-link fence gate for work, then reinstall	4F@.291	ea	—	17.20	17.20
remove chain-link fence for work, then reinstall	4F@.013	sf	—	.77	.77
minimum charge	4F@2.50	ea	85.50	148.00	233.50
Galvanized chain-link post. All chain-link fence posts include digging of 2' deep hole and setting in concrete. Post height is above ground. For example, a 4' high post is made from a 6' length with 2' underground and 4' above ground.					
Galvanized chain-link line posts					
replace, 3' high, 1-7/8" diameter post	4F@.918	ea	23.10	54.20	77.30
replace, 4' high, 1-7/8" diameter post	4F@.918	ea	25.00	54.20	79.20
replace, 5' high, 1-7/8" diameter post	4F@.918	ea	26.50	54.20	80.70
replace, 6' high, 1-7/8" diameter post	4F@.918	ea	28.20	54.20	82.40
replace, 7' high, 1-7/8" diameter post	4F@.918	ea	30.10	54.20	84.30
replace, 8' high, 2" diameter post	4F@.918	ea	32.00	54.20	86.20
replace, 10' high, 2" diameter post	4F@.918	ea	36.00	54.20	90.20
remove post	1D@.401	ea	—	18.90	18.90
Galvanized gate or corner post					
replace, 3' high, 2" diameter post	4F@.918	ea	22.20	54.20	76.40
replace, 4' high, 2" diameter post	4F@.918	ea	23.60	54.20	77.80
replace, 5' high, 2" diameter post	4F@.918	ea	25.10	54.20	79.30
replace, 6' high, 2-1/2" diameter post	4F@.918	ea	28.50	54.20	82.70
replace, 7' high, 2-1/2" diameter post	4F@.918	ea	30.60	54.20	84.80

	Craft@Hrs	Unit	Material	Labor	Total
replace, 8' high, 3" diameter post	4F@.918	ea	49.50	54.20	103.70
replace, 10' high, 3" diameter post	4F@.918	ea	57.20	54.20	111.40
remove post	1D@.401	ea	—	18.90	18.90
add 22% for aluminized steel post					

Vinyl picket fence. 36" and 48" fences are 2 rail fences, 60" and 72" have three. Pickets are 7/8" x 1-1/2" spaced 3-7/8" apart. Top and bottom rails are 1-1/2" x 3-1/2". Bottom rail includes galvanized steel channel. Posts are 4" x 4" and spaced up to 8' on center with an average spacing of 7.2 feet apart. Includes picket and post caps. Gates are priced per lf with a 3 lf minimum and a 5 lf maximum and include all hardware. For wider gates, measure the gate lf, round up to the next even number, multiply by the fence lf price, then add **$229**.

	Craft@Hrs	Unit	Material	Labor	Total
replace, 36" high	4F@.352	lf	20.80	20.80	41.60
replace, 48" high	4F@.357	lf	25.40	21.10	46.50
replace, 60" high	4F@.362	lf	30.20	21.40	51.60
replace, 72" high	4F@.365	lf	34.80	21.50	56.30
remove fence	1D@.139	lf	—	6.56	6.56
remove fence panels, then reinstall	4F@.540	lf	—	31.90	31.90
Vinyl picket fence gate					
replace, 36" high	4F@.646	lf	61.40	38.10	99.50
replace, 48" high	4F@.661	lf	64.70	39.00	103.70
replace, 60" high	4F@.680	lf	67.70	40.10	107.80
replace, 72" high	4F@.696	lf	70.90	41.10	112.00
remove gate	1D@.152	lf	—	7.17	7.17
remove gate, then reinstall	4F@.680	lf	—	40.10	40.10

Vinyl picket fence with large pickets. Same as picket fence, with large 7/8" x 3-1/2" pickets spaced 3-7/8" apart.

	Craft@Hrs	Unit	Material	Labor	Total
replace, 36" high	4F@.352	lf	24.20	20.80	45.00
replace, 48" high	4F@.357	lf	30.00	21.10	51.10
replace, 60" high	4F@.362	lf	35.20	21.40	56.60
remove fence	1D@.139	lf	—	6.56	6.56
remove fence panels, then reinstall	4F@.540	lf	—	31.90	31.90
Gate for vinyl picket fence with large pickets					
replace, 36" high	4F@.646	lf	63.90	38.10	102.00
replace, 48" high	4F@.661	lf	67.70	39.00	106.70
replace, 60" high	4F@.680	lf	71.90	40.10	112.00
remove gate	1D@.152	lf	—	7.17	7.17
remove gate, then reinstall	4F@.680	lf	—	40.10	40.10

Vinyl picket fence with scalloped pickets. Same as picket fence, with decorative scalloped pickets.

	Craft@Hrs	Unit	Material	Labor	Total
replace, 36" high	4F@.352	lf	22.60	20.80	43.40
replace, 48" high	4F@.357	lf	27.30	21.10	48.40
replace, 60" high	4F@.362	lf	32.00	21.40	53.40
remove fence, 36" or 48" high	1D@.139	lf	—	6.56	6.56
remove fence, 60" high	1D@.139	lf	—	6.56	6.56
remove fence panels, then reinstall	4F@.540	lf	—	31.90	31.90
Gate for vinyl picket fence with scalloped pickets					
replace, 36" high	4F@.646	lf	62.40	38.10	100.50
replace, 48" high	4F@.661	lf	66.00	39.00	105.00
replace, 60" high	4F@.680	lf	69.20	40.10	109.30
remove gate	1D@.152	lf	—	7.17	7.17
remove gate, then reinstall	4F@.680	lf	—	40.10	40.10

Vinyl three-rail fence. Rails are 1-1/2" x 5-1/2". Posts are 5" x 5" and spaced up to 8' apart with an average spacing of 7.2'. Gates priced per lf: 3 lf minimum, 5 lf maximum. For gates larger than 5 lf, use lf fence price, round to even number and add **$137**.

	Craft@Hrs	Unit	Material	Labor	Total
replace, 52" high	4F@.211	lf	13.90	12.40	26.30
remove fence	1D@.104	lf	—	4.91	4.91
remove for work, then reinstall (rails only)	4F@.373	lf	—	22.00	22.00
Gate for vinyl three-rail fence					
replace, 52" high	4F@.299	lf	53.50	17.60	71.10
remove gate	1D@.104	lf	—	4.91	4.91
remove for work, then reinstall	4F@.540	lf	—	31.90	31.90

Vinyl spaced slat fence. Contains two widths of slats. The narrow slat is 7/8" x 1-1/2" and the wide slat is 7/8" x 3-1/2". Slats are spaced 2-1/2" apart. Top and bottom rails are 1-1/2" x 3-1/2". Bottom rail includes galvanized steel channel. Posts are 4" x 4" and spaced up to 8' on center with an average spacing of 7.2 feet apart. Includes post caps. Gates are priced per lf with a 3 lf minimum and a 5 lf maximum and include all hardware. For wider gates, measure the gate lf, round up to the next even number, multiply by the fence lf price, then add **$229**.

	Craft@Hrs	Unit	Material	Labor	Total
replace, 48" high	4F@.357	lf	21.30	21.10	42.40
replace, 60" high	4F@.362	lf	26.10	21.40	47.50
replace, 72" high	4F@.365	lf	30.70	21.50	52.20
remove fence	1D@.139	lf	—	6.56	6.56
Gate for vinyl spaced slat fence					
replace, 48" high	4F@.661	lf	61.70	39.00	100.70
replace, 60" high	4F@.680	lf	65.10	40.10	105.20
replace, 72" high	4F@.696	lf	68.10	41.10	109.20
remove gate	1D@.152	lf	—	7.17	7.17

Vinyl solid slat fence. Slat is 7/8" x 5-1/2" with no space between slats. Top and bottom rails are 1-1/2" x 3-1/2". Bottom rail includes galvanized steel channel. Posts are 5" x 5" and spaced up to 8' on center with an average spacing of 7.2 feet apart. Includes post caps. Gates are priced per lf with a 3 lf minimum and a 5 lf maximum and include all hardware. For wider gates, measure the gate lf, round up to the next even number, multiply by the fence lf price, then add **$229**.

	Craft@Hrs	Unit	Material	Labor	Total
replace, 48" high	4F@.357	lf	22.20	21.10	43.30
replace, 60" high	4F@.362	lf	45.40	21.40	66.80
replace, 72" high	4F@.365	lf	32.00	21.50	53.50
remove fence	1D@.139	lf	—	6.56	6.56
Gate for vinyl solid slat fence					
replace, 48" high	4F@.661	lf	62.30	39.00	101.30
replace, 60" high	4F@.680	lf	66.00	40.10	106.10
replace, 72" high	4F@.696	lf	69.10	41.10	110.20
remove gate	1D@.152	lf	—	7.17	7.17

Vinyl solid slat fence with lattice top.

	Craft@Hrs	Unit	Material	Labor	Total
replace, 60" high	4F@.373	lf	37.00	22.00	59.00
replace, 72" high	4F@.378	lf	41.80	22.30	64.10
remove fence	1D@.139	lf	—	6.56	6.56
Gate for vinyl solid slat fence with lattice top					
replace, 60" high	4F@.725	lf	72.50	42.80	115.30
replace, 72" high	4F@.757	lf	75.70	44.70	120.40
remove gate	1D@.152	lf	—	7.17	7.17

Ornamental iron fence. On all grades, except fences with ornamental casting, steel pickets are 5/8" square and 3-7/8" apart with 1" x 1" top and bottom rails. Posts are 93-1/2" apart. Pickets with decorative scrolls are straight pickets with the decorative scrollwork riveted or screwed to the pickets. Ornamental iron fences with ornamental casting have pickets that are cast in decorative scroll and plant-like patterns. Gates are priced per lf with a 3 lf minimum and a 5 lf maximum. For wider gates, measure the gate lf, round up to the next even number, multiply by the fence lf price, then add **$293**.

	Craft@Hrs	Unit	Material	Labor	Total
Ornamental iron fence with straight pickets					
replace, 60" high	4F@.373	lf	19.60	22.00	41.60
replace, 72" high	4F@.378	lf	22.70	22.30	45.00
remove fence	1D@.194	lf	—	9.16	9.16
Ornamental iron gate with straight pickets					
replace, 60" high	4F@.725	lf	78.40	42.80	121.20
replace, 72" high	4F@.757	lf	83.40	44.70	128.10
remove gate	1D@.202	lf	—	9.53	9.53
Ornamental iron fence with twisted pickets					
replace, 60" high	4F@.373	lf	21.20	22.00	43.20
replace, 72" high	4F@.378	lf	24.80	22.30	47.10
remove fence	1D@.194	lf	—	9.16	9.16
Ornamental iron gate with twisted pickets					
replace, 60" high	4F@.725	lf	84.00	42.80	126.80
replace, 72" high	4F@.757	lf	89.60	44.70	134.30
remove gate	1D@.202	lf	—	9.53	9.53
Ornamental iron fence with decorative scrolls					
replace, 60" high	4F@.373	lf	30.10	22.00	52.10
replace, 72" high	4F@.378	lf	34.70	22.30	57.00
remove fence	1D@.194	lf	—	9.16	9.16
Ornamental iron gate with decorative scrolls					
replace, 60" high	4F@.725	lf	118.00	42.80	160.80
replace, 72" high	4F@.757	lf	125.00	44.70	169.70
remove gate	1D@.202	lf	—	9.53	9.53
Ornamental iron fence with ornamental casting					
replace, 60" high	4F@.373	lf	53.10	22.00	75.10
replace, 72" high	4F@.378	lf	61.40	22.30	83.70
remove fence	1D@.194	lf	—	9.16	9.16
Ornamental iron gate with ornamental casting					
replace, 60" high	4F@.725	lf	212.00	42.80	254.80
replace, 72" high	4F@.757	lf	223.00	44.70	267.70
remove gate	1D@.202	lf	—	9.53	9.53
Remove ornamental iron fence for work, then reinstall					
panels only	4F@.558	lf	—	32.90	32.90
gate	4F@.976	lf	—	57.60	57.60

Electric gate opener. For swinging or roll-type gates. Includes two remote transmitters, keypad on post by gate, all hardware and electrical wiring up to 45 lf.

	Craft@Hrs	Unit	Material	Labor	Total
replace, 1/2 horsepower	7E@3.29	ea	4,010.00	238.00	4,248.00
replace, 3/4 horsepower	7E@3.29	ea	5,120.00	238.00	5,358.00
replace, 1 horsepower	7E@3.29	ea	5,830.00	238.00	6,068.00
replace, 2 horsepower	7E@3.29	ea	6,880.00	238.00	7,118.00
replace, 5 horsepower	7E@3.29	ea	7,830.00	238.00	8,068.00
remove gate opener	1D@.765	ea	—	36.10	36.10
remove gate opener for work, then reinstall	7E@5.30	ea	—	384.00	384.00

	Craft@Hrs	Unit	Material	Labor	Total
Time & Material Charts (selected items)					
Fences Materials (Also see material prices with the line items above.)					
4" x 4" fence posts					
8' cedar, ($18.50 ea, 1 ea), 4% waste	—	ea	19.20	—	19.20
10' cedar, ($22.90 ea, 1 ea), 4% waste	—	ea	23.70	—	23.70
12' cedar, ($27.50 ea, 2 ea), 4% waste	—	ea	14.40	—	14.40
1" x 2" x 8' fence boards					
#2 cedar, ($5.68 ea, 8 lf), 4% waste	—	lf	.74	—	.74
B grade redwood, ($12.80 ea, 8 lf), 4% waste	—	lf	1.66	—	1.66
treated-pine, ($3.63 ea, 8 lf), 4% waste	—	lf	.48	—	.48
1" x 4" x 8' fence boards					
#2 cedar, ($7.09 ea, 8 lf), 4% waste	—	lf	.91	—	.91
B grade redwood, ($15.90 ea, 8 lf), 4% waste	—	lf	2.08	—	2.08
treated-pine, ($7.23 ea, 8 lf), 4% waste	—	lf	.94	—	.94
1" x 6" x 8' fence boards					
#2 cedar, ($10.70 ea, 8 lf), 4% waste	—	lf	1.40	—	1.40
B grade redwood, ($24.20 ea, 8 lf), 4% waste	—	lf	3.15	—	3.15
treated-pine, ($6.94 ea, 8 lf), 4% waste	—	lf	.89	—	.89
1" x 8" x 8' fence boards					
#2 cedar, ($22.60 ea, 8 lf), 4% waste	—	lf	2.95	—	2.95
B grade redwood, ($50.40 ea, 8 lf), 4% waste	—	lf	6.56	—	6.56
treated-pine, ($14.40 ea, 8 lf), 4% waste	—	lf	1.87	—	1.87
1" x 10" x 8' fence boards					
#2 cedar, ($27.70 ea, 8 lf), 4% waste	—	lf	3.60	—	3.60
B grade redwood, ($62.40 ea, 8 lf), 4% waste	—	lf	8.12	—	8.12
treated-pine, ($17.90 ea, 8 lf), 4% waste	—	lf	2.33	—	2.33
2" x 4" x 8' fence boards					
#2 cedar, ($8.66 ea, 8 lf), 4% waste	—	lf	1.12	—	1.12
B grade redwood, ($13.60 ea, 8 lf), 4% waste	—	lf	1.75	—	1.75
treated-pine, ($12.80 ea, 8 lf), 4% waste	—	lf	1.66	—	1.66
1" x 3" mill-cut fence picket					
3' cedar, ($2.08 ea, 1 ea), 4% waste	—	ea	2.16	—	2.16
5' cedar, ($5.41 ea, 1 ea), 4% waste	—	ea	5.63	—	5.63
Lattice					
cedar lattice, ($28.90 sh, 32 sf), 4% waste	—	sf	.94	—	.94
cedar cap mold, ($11.10 ea, 8 lf), 4% waste	—	lf	1.45	—	1.45
Chain-link fence materials					
11 gauge chain-link, ($2.72 sf, 1 sf), 3% waste	—	sf	2.80	—	2.80

Fences Labor

Laborer	base wage	paid leave	true wage	taxes & ins.	total
Carpenter	$38.10	2.97	$41.07	26.53	$67.60
Carpenter's helper	$27.40	2.14	$29.54	20.86	$50.40
Electrician	$43.80	3.42	$47.22	25.18	$72.40
Demolition laborer	$25.90	2.02	$27.92	19.28	$47.20

Paid leave is calculated based on two weeks paid vacation, one week sick leave, and seven paid holidays. Employer's matching portion of **FICA** is 7.65 percent. **FUTA** (Federal Unemployment) is .8 percent. **Worker's compensation** for the fences trade was calculated using a national average of 15.74 percent. **Unemployment insurance** was calculated using a national average of 8 percent. **Health insurance** was calculated based on a projected national average for 2020 of $1,280 per employee (and family when applicable) per month. Employer pays 80 percent for a per month cost of $1,024 per employee. **Retirement** is based on a 401(k) retirement program with employer matching of 50 percent. Employee contributions to the 401(k) plan are an average of 6 percent of the true wage. **Liability insurance** is based on a national average of 12.0 percent.

Fences Labor Productivity

	Craft@Hrs	Unit	Material	Labor	Total
Demolition of fence					
remove 4' high wood fence	1D@.161	lf	—	7.60	7.60
remove 6' high wood fence	1D@.165	lf	—	7.79	7.79
remove 8' high wood fence	1D@.170	lf	—	8.02	8.02
remove wood fence post	1D@.367	ea	—	17.30	17.30
remove 3' high chain-link fence	1D@.116	lf	—	5.48	5.48
remove 4' high chain-link fence	1D@.121	lf	—	5.71	5.71
remove 5' high chain-link fence	1D@.125	lf	—	5.90	5.90
remove 6' high chain-link fence	1D@.131	lf	—	6.18	6.18
remove 7' high chain-link fence	1D@.136	lf	—	6.42	6.42
remove 8' high chain-link fence	1D@.143	lf	—	6.75	6.75
remove 10' high chain-link fence	1D@.152	lf	—	7.17	7.17
remove chain-link fence gate	1D@.203	ea	—	9.58	9.58
remove chain-link fence driveway gate	1D@.043	lf	—	2.03	2.03
remove chain-link fence post	1D@.401	ea	—	18.90	18.90
remove vinyl fence	1D@.139	lf	—	6.56	6.56
remove vinyl rail fence	1D@.104	lf	—	4.91	4.91
remove vinyl fence gate	1D@.152	lf	—	7.17	7.17
remove ornamental iron fence	1D@.194	lf	—	9.16	9.16
remove ornamental iron fence gate	1D@.202	lf	—	9.53	9.53
remove electric gate opener	1D@.765	ea	—	36.10	36.10

Fencing crew

fencing installation	carpenter	$67.60
fencing installation	carpenter's helper	$50.40
fencing installation	fencing crew	$59.00

	Craft@Hrs	Unit	Material	Labor	Total
Build basketweave fence					
4' high	6C@.317	lf	—	18.70	18.70
8' high	6C@.360	lf	—	21.20	21.20
4' high gate	6C@2.70	ea	—	159.00	159.00
8' high gate	6C@3.04	ea	—	179.00	179.00
Build board fence					
4' high	6C@.267	lf	—	15.80	15.80
8' high	6C@.307	lf	—	18.10	18.10
4' high gate	6C@2.17	ea	—	128.00	128.00
8' high gate	6C@2.44	ea	—	144.00	144.00
Build board fence with lattice cap					
6' high	6C@.331	lf	—	19.50	19.50
8' high	6C@.360	lf	—	21.20	21.20
6' high gate	6C@2.94	ea	—	173.00	173.00
8' high gate	6C@3.44	ea	—	203.00	203.00
Build board-on-board fence					
4' high	6C@.296	lf	—	17.50	17.50
8' high	6C@.331	lf	—	19.50	19.50
4' high gate	6C@2.56	ea	—	151.00	151.00
8' high gate	6C@2.86	ea	—	169.00	169.00
Build board-and-batten fence					
4' high	6C@.302	lf	—	17.80	17.80
8' high	6C@.323	lf	—	19.10	19.10
4' high gate	6C@2.63	ea	—	155.00	155.00
8' high gate	6C@2.94	ea	—	173.00	173.00
Build picket fence					
3' high	6C@.222	lf	—	13.10	13.10
5' high	6C@.238	lf	—	14.00	14.00
3' high gate	6C@2.00	ea	—	118.00	118.00
5' high gate	6C@2.38	ea	—	140.00	140.00
Install chain-link fence					
3' high	6C@.114	lf	—	6.73	6.73
5' high	6C@.183	lf	—	10.80	10.80
7' high	6C@.296	lf	—	17.50	17.50
10' high	6C@.476	lf	—	28.10	28.10

	Craft@Hrs	Unit	Material	Labor	Total

Finish Carpentry

Minimum charge.

	Craft@Hrs	Unit	Material	Labor	Total
for finish carpentry work	1C@3.00	ea	35.10	203.00	238.10

Other wood species. For moldings made from these species of wood, factor the cost of stain-grade pine by the percentage listed. All grades are kiln-dried select or better.

	Craft@Hrs	Unit	Material	Labor	Total
add for walnut	—	%	70.0	—	—
add for teak	—	%	563.0	—	—
add for cherry	—	%	73.0	—	—
add for hard maple	—	%	31.0	—	—
add for cypress	—	%	24.0	—	—
add for redwood	—	%	17.0	—	—
add for cedar	—	%	0.0	—	—

Wood embossing. Wood moldings are often embossed to appear like they have been carved. Embossing stamps a pattern onto the wood. Embossing is far less expensive than carving and is most successful on softer woods, such as pine.

Clamshell base. Includes clamshell base, finish nails, and installation labor. Outside joints are mitered; inside joints are coped. Includes 7% waste.

	Craft@Hrs	Unit	Material	Labor	Total
Replace 1-1/2" clamshell base					
finger-joint pine	1C@.038	lf	1.32	2.57	3.89
stain-grade pine	1C@.038	lf	1.54	2.57	4.11
poplar	1C@.038	lf	1.62	2.57	4.19
mahogany	1C@.038	lf	1.93	2.57	4.50
red oak	1C@.038	lf	2.50	2.57	5.07
Replace 2-1/2" clamshell base					
finger-joint pine	1C@.038	lf	1.53	2.57	4.10
stain-grade pine	1C@.038	lf	1.74	2.57	4.31
poplar	1C@.038	lf	1.87	2.57	4.44
mahogany	1C@.038	lf	2.24	2.57	4.81
red oak	1C@.038	lf	2.88	2.57	5.45
Replace 3-1/2" clamshell base					
finger-joint pine	1C@.038	lf	2.17	2.57	4.74
stain-grade pine	1C@.038	lf	2.46	2.57	5.03
poplar	1C@.038	lf	2.67	2.57	5.24
mahogany	1C@.038	lf	3.15	2.57	5.72
red oak	1C@.038	lf	4.05	2.57	6.62
Replace 4-1/2" clamshell base					
add 22% to the cost of 3-1/2" clamshell base					
All sizes and wood species					
remove	1D@.026	lf	—	1.23	1.23
remove for work, then reinstall	6C@.065	lf	—	3.84	3.84

	Craft@Hrs	Unit	Material	Labor	Total

Pattern base. Includes pattern base, finish nails, and installation labor. Outside joints are mitered; inside joints are coped. Includes 4% waste. Pattern base is a generic term for a variety of standard base styles that are commonly available at home improvement centers throughout North America. In many regions this style of base is referred to as "Colonial" base.

	Craft@Hrs	Unit	Material	Labor	Total
Replace 1-1/2" pattern base					
finger-joint pine	1C@.038	lf	1.59	2.57	4.16
stain-grade pine	1C@.038	lf	1.86	2.57	4.43
poplar	1C@.038	lf	2.02	2.57	4.59
mahogany	1C@.038	lf	2.38	2.57	4.95
red oak	1C@.038	lf	3.02	2.57	5.59
Replace 2-1/2" pattern base					
finger-joint pine	1C@.038	lf	2.61	2.57	5.18
stain-grade pine	1C@.038	lf	2.99	2.57	5.56
poplar	1C@.038	lf	3.21	2.57	5.78
mahogany	1C@.038	lf	3.83	2.57	6.40
red oak	1C@.038	lf	4.95	2.57	7.52
Replace 3-1/2" pattern base					
finger-joint pine	1C@.038	lf	3.44	2.57	6.01
stain-grade pine	1C@.038	lf	3.96	2.57	6.53
poplar	1C@.038	lf	4.27	2.57	6.84
mahogany	1C@.038	lf	5.05	2.57	7.62
red oak	1C@.038	lf	6.47	2.57	9.04
Replace 5" pattern base					
add 10% to the cost of 3-1/2" pattern base					
Replace 8" pattern base					
add 31% to the cost of the 3-1/2" pattern base					
All sizes and wood species					
remove	1D@.026	lf	—	1.23	1.23
remove for work, then reinstall	6C@.065	lf	—	3.84	3.84

1-1/2" pattern base

Base shoe. Includes 3/4" base shoe, finish nails, and installation labor. Outside joints are mitered; inside joints are coped. Includes 7% waste.

	Craft@Hrs	Unit	Material	Labor	Total
Replace 3/4" base shoe					
finger-joint pine	1C@.038	lf	.82	2.57	3.39
stain-grade pine	1C@.038	lf	.93	2.57	3.50
poplar	1C@.038	lf	1.03	2.57	3.60
mahogany	1C@.038	lf	1.18	2.57	3.75
red oak	1C@.038	lf	1.54	2.57	4.11
All wood species					
remove	1D@.026	lf	—	1.23	1.23
remove for work, then reinstall	6C@.065	lf	—	3.84	3.84

Clamshell casing. Includes clamshell casing, finish nails, and installation labor. Outside corners are 45-degree miter joints; inside corners are cope joints.

	Craft@Hrs	Unit	Material	Labor	Total
Replace 1-1/2" clamshell casing					
paint-grade pine	1C@.041	lf	1.36	2.77	4.13
stain-grade pine	1C@.041	lf	1.55	2.77	4.32
poplar	1C@.041	lf	1.65	2.77	4.42
mahogany	1C@.041	lf	1.99	2.77	4.76
red oak	1C@.041	lf	2.56	2.77	5.33

1-1/2" clamshell casing

	Craft@Hrs	Unit	Material	Labor	Total
Replace 2-1/2" clamshell casing					
finger-joint pine	1C@.041	lf	1.55	2.77	4.32
stain-grade pine	1C@.041	lf	1.80	2.77	4.57
poplar	1C@.041	lf	1.90	2.77	4.67
mahogany	1C@.041	lf	2.27	2.77	5.04
red oak	1C@.041	lf	2.92	2.77	5.69
Replace 3-1/2" clamshell casing					
finger-joint pine	1C@.041	lf	2.20	2.77	4.97
stain-grade pine	1C@.041	lf	2.51	2.77	5.28
poplar	1C@.041	lf	2.71	2.77	5.48
mahogany	1C@.041	lf	3.18	2.77	5.95
red oak	1C@.041	lf	4.14	2.77	6.91
All sizes and wood species					
remove	1D@.032	lf	—	1.51	1.51
remove for work, then reinstall	6C@.083	lf	—	4.90	4.90

Pattern casing. Includes 2-1/2" pattern casing, finish nails, and installation labor. Outside corners are 45-degree miter joints; inside corners are cope joints. Laminated back, factory fabricated molding. Half-round and elliptical are priced per linear foot of casing. Casing for round windows is priced per linear foot of window diameter.

	Craft@Hrs	Unit	Material	Labor	Total
Replace 1-1/2" pattern casing					
paint-grade pine	1C@.041	lf	1.61	2.77	4.38
stain-grade pine	1C@.041	lf	1.89	2.77	4.66
poplar	1C@.041	lf	2.06	2.77	4.83
mahogany	1C@.041	lf	2.42	2.77	5.19
red oak	1C@.041	lf	3.07	2.77	5.84
Replace 1-1/2" curved casing					
for half-round top window or door	1C@.144	lf	11.40	9.73	21.13
for elliptical-top window or door	1C@.144	lf	11.90	9.73	21.63
for round window, per lf of window diameter	1C@.294	lf	26.90	19.90	46.80
Replace 2-1/2" pattern casing					
finger-joint pine	1C@.041	lf	2.64	2.77	5.41
stain-grade pine	1C@.041	lf	3.02	2.77	5.79
poplar	1C@.041	lf	3.27	2.77	6.04
mahogany	1C@.041	lf	3.90	2.77	6.67
red oak	1C@.041	lf	5.00	2.77	7.77
Replace 2-1/2" curved casing					
for half-round top window or door	1C@.144	lf	12.30	9.73	22.03
for elliptical-top window or door	1C@.144	lf	12.90	9.73	22.63
for round window, per lf of window diameter	1C@.294	lf	28.80	19.90	48.70
Replace 3-1/2" pattern casing					
finger-joint pine	1C@.041	lf	3.52	2.77	6.29
stain-grade pine	1C@.041	lf	4.03	2.77	6.80
poplar	1C@.041	lf	4.32	2.77	7.09
mahogany	1C@.041	lf	5.12	2.77	7.89
red oak	1C@.041	lf	6.60	2.77	9.37
Replace 3-1/2" curved casing					
for half-round top window or door	1C@.144	lf	12.40	9.73	22.13
for elliptical-top window or door	1C@.144	lf	13.20	9.73	22.93
for round window, per lf of window diameter	1C@.294	lf	29.20	19.90	49.10

	Craft@Hrs	Unit	Material	Labor	Total
Replace 5" pattern casing					
add 10% to the cost of 3-1/2" pattern casing					
Replace 5" curved casing					
add 10% to the cost of 3-1/2" curved casing					
All sizes, species, shapes and styles of pattern casing					
remove	1D@.032	lf	—	1.51	1.51
remove for work, then reinstall	6C@.083	lf	—	4.90	4.90

Wood key. Includes wood key, finish nails, and installation labor. Designed to be installed at the apex of curved casing. Does not include installation of casing to sides of the key.

	Craft@Hrs	Unit	Material	Labor	Total
Replace wood key					
standard wood key	1C@.251	ea	31.50	17.00	48.50
with light carvings	1C@.251	ea	50.00	17.00	67.00
with medium carvings	1C@.251	ea	61.80	17.00	78.80
with heavy carvings	1C@.251	ea	80.60	17.00	97.60
All types of wood key					
remove	1D@.036	ea	—	1.70	1.70
remove for work, then reinstall	6C@.323	ea	—	19.10	19.10

with rosette

with carvings

Base block.

	Craft@Hrs	Unit	Material	Labor	Total
Pine	1C@.063	ea	19.40	4.26	23.66
poplar	1C@.063	ea	20.80	4.26	25.06
mahogany	1C@.063	ea	22.70	4.26	26.96
red oak	1C@.063	ea	25.60	4.26	29.86
All types and wood species of base block					
remove	1D@.036	ea	—	1.70	1.70
remove for work, then reinstall	6C@.080	ea	—	4.72	4.72

with rosette

Corner block.

	Craft@Hrs	Unit	Material	Labor	Total
Replace corner block with rosette					
pine	1C@.063	ea	15.50	4.26	19.76
poplar	1C@.063	ea	18.00	4.26	22.26
mahogany	1C@.063	ea	20.30	4.26	24.56
red oak	1C@.063	ea	22.30	4.26	26.56
All types and wood species of corner block					
remove	1D@.036	ea	—	1.70	1.70
remove for work, then reinstall	6C@.080	ea	—	4.72	4.72
Add 30% for corner blocks with carvings					

with rosette

	Craft@Hrs	Unit	Material	Labor	Total

Head block. Most commonly found on historical structures, head block is used in corners at the top of door casing and other cased openings. In addition to providing a decorative element, it is designed to provide a flat bottom edge and flat side edges that are wide enough for casing to tie into. This often allows designers to use more decorative molding as casing. Trimwork that uses head block often uses base blocks at the base of openings. Includes manufactured head block, fasteners, glue as needed, and installation labor. Does not include paint prep, painting, or caulking.

	Craft@Hrs	Unit	Material	Labor	Total
Replace head block with rosette					
pine	1C@.063	ea	19.40	4.26	23.66
poplar	1C@.063	ea	22.70	4.26	26.96
mahogany	1C@.063	ea	25.60	4.26	29.86
red oak	1C@.063	ea	28.10	4.26	32.36
All types and wood species of head block					
remove	1D@.037	ea	—	1.75	1.75
remove for work, then reinstall	6C@.080	ea	—	4.72	4.72
Add 31% for head block with carvings					

with rosette

with carvings

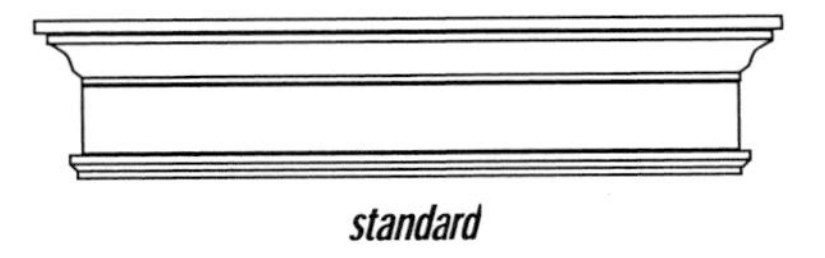

standard

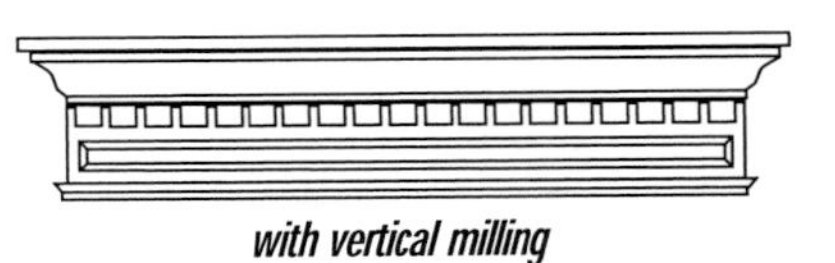

with vertical milling

with hand carvings

with heavy hand carvings

Overdoor molding. Includes overdoor moldings, finish board(s) as needed, finish nails, glue as needed, and installation labor. Does not include materials and labor to install side casing. Overdoor moldings can be manufactured to fit typical door sizes and installed on site or built on site. Site-built overdoor moldings are more common in the lower grades. Wherever they are built, overdoor moldings are usually a combination of trimwork such as crown, cove, half round, dentil molding, and rope molding which are often attached to a finish board. Moldings typically feature miter joints on each end that return the moldings into the wall.

	Craft@Hrs	Unit	Material	Labor	Total
Replace overdoor horizontal molding					
Pine	1C@.041	lf	18.50	2.77	21.27
poplar	1C@.041	lf	21.30	2.77	24.07
mahogany	1C@.041	lf	23.50	2.77	26.27
red oak	1C@.041	lf	26.90	2.77	29.67
All types and wood species of overdoor molding					
remove	1D@.028	lf	—	1.32	1.32
remove for work, then reinstall	6C@.057	lf	—	3.36	3.36
Add 72% for vertical milling in overdoor molding					
Add 227% for hand carving in overdoor molding					
Add 411% for heavy hand carving in overdoor molding					

	Craft@Hrs	Unit	Material	Labor	Total

Interior architrave. Includes moldings, trim blocks, finish nails, glue as needed, and installation labor. The architrave is the complete set of moldings that surround and finish the opening. In this price list architrave refers primarily to the style of trimwork found on many historical structures where the complete set of moldings includes base blocks, corner blocks, or head blocks at joints in the casing. Each of these components can also be estimated separately. Standard grade: paint-grade pine or poplar, or stain-grade pine with rosettes in head and base blocks. High grade: stain-grade pine or red oak with rosettes or light embossing in base blocks and overdoor with horizontal moldings at top and bottom. Deluxe grade: red oak, birch or similar hardwood with carving in base blocks and angled overdoor with horizontal moldings at top and bottom or with light carvings. Custom grade: red oak, birch or similar hardwood with carvings in base blocks and angled or curved overdoor with horizontal moldings at top and bottom or with heavy carvings.

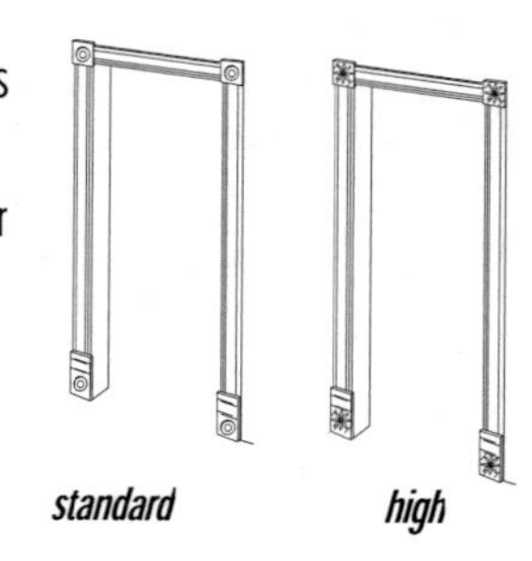

	Craft@Hrs	Unit	Material	Labor	Total
Replace interior architrave					
standard grade	1C@.062	lf	7.80	4.19	11.99
high grade	1C@.062	lf	10.20	4.19	14.39
deluxe grade	1C@.062	lf	14.30	4.19	18.49
custom grade	1C@.062	lf	20.10	4.19	24.29
All grades of interior architrave					
remove	1D@.032	lf	—	1.51	1.51
remove for work, then reinstall	6C@.103	lf	—	6.08	6.08

Exterior architrave. Includes moldings, trim blocks, finish nails, glue as needed, and installation labor. The architrave is the complete set of moldings that surround and finish the opening. In this book, architrave refers primarily to the style of trimwork found on many historical structures where the complete set of moldings includes base blocks, corner blocks, or head blocks at joints in the casing. Each of these components can also be estimated separately. All materials are back-primed. Economy grade: a standard width casing of paint- or stain-grade pine or poplar with rosettes in head and base blocks. Standard grade: wide casing (may be made from two moldings) of paint- or stain-grade pine or poplar. High grade: wide casing (may be made from two moldings) of stain-grade pine or red oak with rosettes or light embossing in base blocks, with horizontal moldings at top and bottom of overdoor. Deluxe and higher grades: wide casing (may be made from two moldings) of red oak, birch or similar hardwood, increasingly ornate embellishment of base blocks and overdoor.

	Craft@Hrs	Unit	Material	Labor	Total
Replace exterior architrave					
economy grade	1C@.080	lf	7.80	5.41	13.21
standard grade	1C@.080	lf	11.00	5.41	16.41
high grade	1C@.080	lf	15.60	5.41	21.01
deluxe grade	1C@.080	lf	22.30	5.41	27.71
custom grade	1C@.080	lf	28.30	5.41	33.71
custom deluxe grade	1C@.080	lf	39.50	5.41	44.91
All grades of exterior architrave					
remove	1D@.032	lf	—	1.51	1.51
remove for work, then reinstall	6C@.127	lf	—	7.49	7.49

	Craft@Hrs	Unit	Material	Labor	Total

economy *standard* *high* *deluxe* *custom* *custom deluxe*

Exterior door surround. Includes manufactured door surround to match door size, fasteners, and installation labor. Although door surrounds can be built on site from standard moldings, all labor productivities are based on installation of manufactured units. Material prices are based on classical door surrounds inspired by Greek and Roman designs. All grades standard are made from wood with increasingly intricate carved detail. Economy and standard grades are made of plastic or other synthetic materials. Illustrations show typical complexity for each grade, actual door styling varies widely.

	Craft@Hrs	Unit	Material	Labor	Total
Replace exterior door surround					
economy grade	1C@.234	lf	48.80	15.80	64.60
standard grade	1C@.257	lf	79.60	17.40	97.00
high grade	1C@.274	lf	141.00	18.50	159.50
deluxe grade	1C@.317	lf	271.00	21.40	292.40
custom grade	1C@.372	lf	423.00	25.10	448.10
custom deluxe grade	1C@.431	lf	703.00	29.10	732.10
All grades of exterior door					
remove	1D@.034	lf	—	1.60	1.60
remove for work, then reinstall	6C@.775	lf	—	45.70	45.70

economy *standard* *high* *deluxe* *custom* *custom deluxe*

Exterior window surround. Includes manufactured window surround to match window size, fasteners, and installation labor. Although window surrounds can be built on site from standard moldings, all labor productivities are based on installation of manufactured units. Material prices are based on classical window surrounds inspired by Greek and Roman designs.

	Craft@Hrs	Unit	Material	Labor	Total
Replace exterior window surround					
economy grade	1C@.207	lf	39.60	14.00	53.60
standard grade	1C@.227	lf	64.60	15.30	79.90
high grade	1C@.239	lf	113.00	16.20	129.20
deluxe grade	1C@.280	lf	223.00	18.90	241.90
custom grade	1C@.326	lf	344.00	22.00	366.00
custom deluxe grade	1C@.322	lf	572.00	21.80	593.80
All grades of exterior window surround					
remove	1D@.034	lf	—	1.60	1.60
remove for work, then reinstall	6C@.690	lf	—	40.70	40.70

	Craft@Hrs	Unit	Material	Labor	Total
Window stool. Includes 2-1/2" window stool, finish nails, silicone caulking as needed, and installation labor.					
Replace window stool					
paint-grade pine	1C@.058	lf	2.91	3.92	6.83
stain-grade pine	1C@.058	lf	3.41	3.92	7.33
poplar	1C@.058	lf	3.61	3.92	7.53
mahogany	1C@.058	lf	4.29	3.92	8.21
red oak	1C@.058	lf	4.64	3.92	8.56
All wood species of window stool					
remove	1D@.026	lf	—	1.23	1.23
remove for work, then reinstall	6C@.108	lf	—	6.37	6.37
Window apron. Includes 3-1/2" window apron, finish nails, silicone caulking as needed, and installation labor.					
Replace window apron					
paint-grade pine	1C@.049	lf	1.87	3.31	5.18
stain-grade pine	1C@.049	lf	2.23	3.31	5.54
poplar	1C@.049	lf	2.38	3.31	5.69
mahogany	1C@.049	lf	2.82	3.31	6.13
red oak	1C@.049	lf	3.01	3.31	6.32
All wood species of window apron					
remove	1D@.028	lf	—	1.32	1.32
remove for work, then reinstall	6C@.080	lf	—	4.72	4.72
Chair rail. Includes chair rail, finish nails, and installation labor. Outside joints are mitered; inside joints are coped. Includes 4% waste.					
Replace 2-1/2" chair rail					
paint-grade pine	1C@.038	lf	2.51	2.57	5.08
stain-grade pine	1C@.038	lf	2.92	2.57	5.49
poplar	1C@.038	lf	3.17	2.57	5.74
mahogany	1C@.038	lf	3.75	2.57	6.32
red oak	1C@.038	lf	4.00	2.57	6.57
Replace 3-1/2" chair rail					
paint-grade pine	1C@.038	lf	3.17	2.57	5.74
stain-grade pine	1C@.038	lf	3.68	2.57	6.25
poplar	1C@.038	lf	3.97	2.57	6.54
mahogany	1C@.038	lf	4.69	2.57	7.26
red oak	1C@.038	lf	5.12	2.57	7.69
All sizes and wood species of chair rail					
remove	1D@.026	lf	—	1.23	1.23
remove for work, then reinstall	6C@.065	lf	—	3.84	3.84
Drip cap. Includes drip cap, finish nails, silicone caulking as needed, and installation labor.					
Replace 1-5/8" drip cap					
paint-grade pine	1C@.038	lf	2.33	2.57	4.90
redwood	1C@.038	lf	2.69	2.57	5.26
cedar	1C@.038	lf	2.51	2.57	5.08

2-1/2" chair rail

	Craft@Hrs	Unit	Material	Labor	Total
Replace 2-1/4" drip cap					
paint-grade pine	1C@.038	lf	2.60	2.57	5.17
redwood	1C@.038	lf	3.00	2.57	5.57
cedar	1C@.038	lf	2.82	2.57	5.39
All sizes and wood species of drip cap					
remove	1D@.032	lf	—	1.51	1.51
remove for work, then reinstall	6C@.083	lf	—	4.90	4.90

Astragal. Includes astragal, finish nails or fasteners, and installation labor. Installation productivity is based on installation of a replacement astragal in previously installed doors.

	Craft@Hrs	Unit	Material	Labor	Total
Replace 1-3/4" astragal					
paint-grade pine	1C@.041	lf	4.85	2.77	7.62
stain-grade pine	1C@.041	lf	5.60	2.77	8.37
poplar	1C@.041	lf	6.04	2.77	8.81
mahogany	1C@.041	lf	7.14	2.77	9.91
red oak	1C@.041	lf	7.73	2.77	10.50
Replace 2-1/4" astragal					
paint-grade pine	1C@.041	lf	5.85	2.77	8.62
stain-grade pine	1C@.041	lf	6.77	2.77	9.54
poplar	1C@.041	lf	7.31	2.77	10.08
mahogany	1C@.041	lf	8.66	2.77	11.43
red oak	1C@.041	lf	9.38	2.77	12.15
All sizes and wood species of astragal					
remove	1D@.032	lf	—	1.51	1.51
remove for work, then reinstall	6C@.083	lf	—	4.90	4.90

Cove. Includes cove, finish nails, and installation labor. Outside joints are mitered; inside joints are coped. Includes 4% waste.

	Craft@Hrs	Unit	Material	Labor	Total
Replace 1/2" cove					
paint-grade pine	1C@.041	lf	.90	2.77	3.67
stain-grade pine	1C@.041	lf	1.06	2.77	3.83
poplar	1C@.041	lf	1.16	2.77	3.93
mahogany	1C@.041	lf	1.33	2.77	4.10
red oak	1C@.041	lf	1.43	2.77	4.20
Replace 3/4" cove					
paint-grade pine	1C@.041	lf	1.05	2.77	3.82
stain-grade pine	1C@.041	lf	1.20	2.77	3.97
poplar	1C@.041	lf	1.33	2.77	4.10
mahogany	1C@.041	lf	1.53	2.77	4.30
red oak	1C@.041	lf	1.61	2.77	4.38
Replace 1" cove					
paint-grade pine	1C@.041	lf	1.29	2.77	4.06
stain-grade pine	1C@.041	lf	1.48	2.77	4.25
poplar cove	1C@.041	lf	1.65	2.77	4.42
mahogany cove	1C@.041	lf	1.89	2.77	4.66
red oak cove	1C@.041	lf	2.03	2.77	4.80
All sizes and wood species of cove					
remove	1D@.032	lf	—	1.51	1.51
remove for work, then reinstall	6C@.065	lf	—	3.84	3.84

	Craft@Hrs	Unit	Material	Labor	Total

Ceiling cove. Includes ceiling cove (sometimes called "sprung cove"), finish nails, and installation labor. Outside joints are mitered; inside joints are coped. Includes 4% waste.

	Craft@Hrs	Unit	Material	Labor	Total
Replace 2-1/2" ceiling cove					
paint-grade pine	1C@.046	lf	2.27	3.11	5.38
stain-grade pine	1C@.046	lf	2.64	3.11	5.75
poplar	1C@.046	lf	2.83	3.11	5.94
mahogany	1C@.046	lf	3.42	3.11	6.53
red oak	1C@.046	lf	3.71	3.11	6.82
Replace 3-1/2" ceiling cove					
paint-grade pine	1C@.046	lf	2.92	3.11	6.03
stain-grade pine	1C@.046	lf	3.42	3.11	6.53
poplar	1C@.046	lf	3.68	3.11	6.79
mahogany	1C@.046	lf	4.41	3.11	7.52
red oak	1C@.046	lf	4.82	3.11	7.93
All sizes and wood species of cove					
remove	1D@.032	lf	—	1.51	1.51
remove for work, then reinstall	6C@.065	lf	—	3.84	3.84

"sprung" cove

Half-round. Includes half-round, finish nails, and installation labor. Outside joints are mitered; inside joints are coped. Includes 4% waste.

	Craft@Hrs	Unit	Material	Labor	Total
Replace 1" half-round					
paint-grade pine	1C@.041	lf	1.05	2.77	3.82
stain-grade pine	1C@.041	lf	1.23	2.77	4.00
poplar	1C@.041	lf	1.33	2.77	4.10
mahogany	1C@.041	lf	1.55	2.77	4.32
red oak	1C@.041	lf	1.67	2.77	4.44
Replace 2" half-round					
paint-grade pine	1C@.041	lf	1.47	2.77	4.24
stain-grade pine	1C@.041	lf	1.71	2.77	4.48
poplar	1C@.041	lf	1.87	2.77	4.64
mahogany	1C@.041	lf	2.21	2.77	4.98
red oak	1C@.041	lf	2.42	2.77	5.19
Replace 3" half-round					
add 19% to the cost of 2" half-round					
Replace 4" half-round					
add 40% to the cost of 2" half-round					
All sizes and wood species of half-round					
remove	1D@.032	lf	—	1.51	1.51
remove for work, then reinstall	6C@.083	lf	—	4.90	4.90

Quarter-round. Includes quarter-round, finish nails, and installation labor. Outside joints are mitered; inside joints are coped. Includes 4% waste.

	Craft@Hrs	Unit	Material	Labor	Total
Replace 1/2" quarter-round					
paint-grade pine	1C@.041	lf	.77	2.77	3.54
stain-grade pine	1C@.041	lf	.88	2.77	3.65
poplar	1C@.041	lf	1.01	2.77	3.78
mahogany	1C@.041	lf	1.17	2.77	3.94
red oak	1C@.041	lf	1.31	2.77	4.08

	Craft@Hrs	Unit	Material	Labor	Total
Replace 3/4" quarter-round					
paint-grade pine	1C@.041	lf	.90	2.77	3.67
stain-grade pine	1C@.041	lf	1.08	2.77	3.85
poplar	1C@.041	lf	1.16	2.77	3.93
mahogany	1C@.041	lf	1.40	2.77	4.17
red oak	1C@.041	lf	1.55	2.77	4.32
Replace 1" quarter-round					
paint-grade pine	1C@.041	lf	1.59	2.77	4.36
stain-grade pine	1C@.041	lf	1.90	2.77	4.67
poplar	1C@.041	lf	2.08	2.77	4.85
mahogany	1C@.041	lf	2.47	2.77	5.24
red oak	1C@.041	lf	2.75	2.77	5.52
All sizes and wood species of quarter-round					
remove	1D@.032	lf	—	1.51	1.51
remove for work, then reinstall	6C@.083	lf	—	4.90	4.90

Corner bead. Includes corner bead, finish nails or fasteners, and installation labor. Material costs for standard corner beads include a 4% allowance for waste. No waste included for turned corner beads with double finials.

	Craft@Hrs	Unit	Material	Labor	Total
Replace corner bead					
paint-grade pine	1C@.077	lf	1.77	5.21	6.98
stain-grade pine	1C@.077	lf	2.13	5.21	7.34
stain-grade pine, turned with double finials	1C@.077	lf	23.20	5.21	28.41
mahogany	1C@.077	lf	3.95	5.21	9.16
mahogany, turned with double finials	1C@.077	lf	28.50	5.21	33.71
red oak	1C@.077	lf	4.85	5.21	10.06
red oak, turned with double finials	1C@.077	lf	30.40	5.21	35.61
clear plastic	1C@.032	lf	1.62	2.16	3.78
All styles and material types of corner bead					
remove	1D@.043	lf	—	2.03	2.03
remove for work, then reinstall	6C@.113	lf	—	6.67	6.67

standard

turned with double finials

Bed mold. Includes 2-1/2" bed molding, finish nails, and installation labor. Outside joints are mitered; inside joints are coped. Includes 4% waste.

	Craft@Hrs	Unit	Material	Labor	Total
Replace bed mold					
paint-grade pine	1C@.038	lf	2.51	2.57	5.08
stain-grade pine	1C@.038	lf	2.93	2.57	5.50
poplar	1C@.038	lf	3.19	2.57	5.76
mahogany	1C@.038	lf	3.83	2.57	6.40
red oak	1C@.038	lf	4.14	2.57	6.71
All wood species of bed mold					
remove	1D@.026	lf	—	1.23	1.23
remove for work, then reinstall	6C@.065	lf	—	3.84	3.84

Crown mold. Includes crown molding, finish nails, and installation labor. Outside joints are mitered; inside joints are coped. Includes 4% waste.

	Craft@Hrs	Unit	Material	Labor	Total
Replace 3-1/2" bed mold					
paint-grade pine	1C@.046	lf	3.21	3.11	6.32
stain-grade pine	1C@.046	lf	3.77	3.11	6.88
poplar	1C@.046	lf	4.10	3.11	7.21
mahogany	1C@.046	lf	4.90	3.11	8.01
red oak	1C@.046	lf	5.31	3.11	8.42

	Craft@Hrs	Unit	Material	Labor	Total
Replace 4-1/2" crown mold					
paint-grade pine	1C@.046	lf	4.74	3.11	7.85
stain-grade pine	1C@.046	lf	5.56	3.11	8.67
poplar	1C@.046	lf	6.05	3.11	9.16
mahogany	1C@.046	lf	7.70	3.11	10.81
red oak	1C@.046	lf	8.34	3.11	11.45
All sizes and wood species of crown mold					
remove	1D@.026	lf	—	1.23	1.23
remove for work, then reinstall	6C@.065	lf	—	3.84	3.84

Hand rail. Includes wood hand rail, brackets with screws, and installation labor. Also includes 4% waste.

	Craft@Hrs	Unit	Material	Labor	Total
Replace hand rail					
paint-grade pine	1C@.098	lf	4.09	6.62	10.71
stain-grade pine	1C@.098	lf	5.58	6.62	12.20
mahogany	1C@.098	lf	7.95	6.62	14.57
red oak	1C@.098	lf	8.48	6.62	15.10
All wood species of hand rail					
remove	1D@.028	lf	—	1.32	1.32
remove for work, then reinstall	6C@.154	lf	—	9.09	9.09

Specialty molding. Includes specialty molding, finish nails, and installation labor. Outside joints are mitered; inside joints are coped. Includes 4% waste. Specialty molding is an all-purpose pricing item for all standard trimwork that has all vertical milling and is commonly available in home improvement centers throughout North America. This includes molding such as crown, chair rail, panel molds, cabinet trim and so forth.

	Craft@Hrs	Unit	Material	Labor	Total
Replace 3/4" specialty molding with vertical milling					
paint-grade pine	1C@.041	lf	3.80	2.77	6.57
stain-grade pine	1C@.041	lf	4.42	2.77	7.19
poplar	1C@.041	lf	4.76	2.77	7.53
mahogany	1C@.041	lf	5.95	2.77	8.72
red oak	1C@.041	lf	6.46	2.77	9.23
polyurethane	1C@.041	lf	5.67	2.77	8.44
stain-grade polyurethane	1C@.041	lf	6.09	2.77	8.86
Other 3/4" styles					
deduct for molding with embossed patterns (pine moldings only)	—	%	-41.0	—	—
add for molding with hand carvings	—	%	64.0	—	—
add for molding with heavy hand carvings	—	%	198.0	—	—
Replace 1-1/2" specialty molding with vertical milling					
paint-grade pine	1C@.041	lf	4.31	2.77	7.08
stain-grade pine	1C@.041	lf	5.06	2.77	7.83
poplar	1C@.041	lf	5.45	2.77	8.22
mahogany	1C@.041	lf	6.76	2.77	9.53
red oak	1C@.041	lf	7.37	2.77	10.14
polyurethane	1C@.041	lf	8.94	2.77	11.71
stain-grade polyurethane	1C@.041	lf	9.58	2.77	12.35
Replace other 1-1/2" styles of specialty molding					
deduct for molding with embossed patterns (pine moldings only)	—	%	-37.0	—	—
add for molding with hand carvings	—	%	66.0	—	—
add for molding with heavy hand carvings	—	%	205.0	—	—

dentil

bead and reel

bead

rope

	Craft@Hrs	Unit	Material	Labor	Total
Replace 2-1/2" specialty molding with vertical milling					
paint-grade pine	1C@.041	lf	4.88	2.77	7.65
stain-grade pine	1C@.041	lf	5.66	2.77	8.43
poplar	1C@.041	lf	6.12	2.77	8.89
mahogany	1C@.041	lf	7.57	2.77	10.34
red oak	1C@.041	lf	8.25	2.77	11.02
polyurethane	1C@.041	lf	10.00	2.77	12.77
stain-grade polyurethane	1C@.041	lf	10.70	2.77	13.47
Replace other 2-1/2" styles of specialty molding					
deduct for molding with embossed patterns (pine moldings only)	—	%	-37.0	—	—
add for molding with hand carvings	—	%	68.0	—	—
add for molding with heavy hand carvings	—	%	209.0	—	—
Replace 3-1/2" specialty molding with vertical milling					
paint-grade pine	1C@.041	lf	5.41	2.77	8.18
stain-grade pine	1C@.041	lf	6.31	2.77	9.08
poplar	1C@.041	lf	6.78	2.77	9.55
mahogany	1C@.041	lf	8.42	2.77	11.19
red oak	1C@.041	lf	9.16	2.77	11.93
polyurethane	1C@.041	lf	11.10	2.77	13.87
stain-grade polyurethane	1C@.041	lf	11.90	2.77	14.67
Replace other 3-1/2" styles of specialty molding					
deduct for molding with embossed patterns (pine moldings only)	—	%	-38.0	—	—
add for molding with hand carving	—	%	69.0	—	—
add for molding with heavy hand carving	—	%	213.0	—	—
Replace 5" specialty molding with vertical milling					
paint-grade pine	1C@.050	lf	7.79	3.38	11.17
stain-grade pine	1C@.050	lf	9.08	3.38	12.46
poplar	1C@.050	lf	9.78	3.38	13.16
mahogany	1C@.050	lf	12.00	3.38	15.38
red oak	1C@.050	lf	13.20	3.38	16.58
polyurethane	1C@.050	lf	15.90	3.38	19.28
stain-grade polyurethane	1C@.050	lf	17.00	3.38	20.38
Replace other 5" styles of vertical molding					
deduct for molding with embossed patterns (pine moldings only)	—	%	-39.0	—	—
add for molding with hand carving	—	%	71.0	—	—
add for molding with heavy hand carving	—	%	219.0	—	—
Replace 8" specialty molding with vertical milling					
paint-grade pine	1C@.055	lf	12.60	3.72	16.32
stain-grade pine	1C@.055	lf	14.80	3.72	18.52
poplar	1C@.055	lf	15.80	3.72	19.52
mahogany	1C@.055	lf	20.10	3.72	23.82
red oak	1C@.055	lf	21.80	3.72	25.52
polyurethane	1C@.055	lf	26.30	3.72	30.02
stain-grade polyurethane	1C@.055	lf	28.30	3.72	32.02

reed

acanthus

leaf and tongue

laurel and ribbon

shell

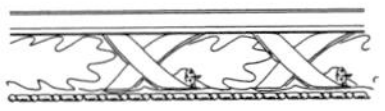

oak leaf and ribbon

meander

fret

egg and dart

	Craft@Hrs	Unit	Material	Labor	Total
Replace other 8" styles of specialty molding					
deduct for molding with embossed patterns (pine moldings only)	—	%	-41.0	—	—
add for molding with hand carving	—	%	75.0	—	—
add for molding with heavy hand carving	—	%	232.0	—	—
All widths and types of specialty moldings					
remove	1D@.032	lf	—	1.51	1.51
remove for work then reinstall					
wall molding	6C@.065	lf	—	3.84	3.84
window or door casing	6C@.083	lf	—	4.90	4.90

bundled reeds

ribbon

grape

Custom-milled moldings. Includes custom-milled molding, fasteners, and installation labor. Includes 4% waste. Custom milled to match existing pattern or a specially designed pattern. All milling is horizontal. Custom moldings are most common when repairing a historical structure that contains moldings that are no longer manufactured. Some custom homes also have custom moldings designed by an architect or designer. A custom molder knife is cut to match the existing molding and moldings are run to match. Linear foot prices do not include pattern charges, design charges, mill charges, molder setup charges, or minimum charges for small runs.

	Craft@Hrs	Unit	Material	Labor	Total
Setup charges, replace only					
molder setup charge	—	ea	278.00	—	278.00
molder custom knife charges	—	ea	619.00	—	619.00
Replace pine custom molding					
1" to 2"	1C@.041	lf	2.54	2.77	5.31
2" to 3"	1C@.041	lf	2.85	2.77	5.62
3" to 4"	1C@.050	lf	4.05	3.38	7.43
4" to 5"	1C@.050	lf	6.43	3.38	9.81
6" to 7"	1C@.055	lf	10.50	3.72	14.22
7" to 8"	1C@.055	lf	14.90	3.72	18.62
Replace other wood species of custom molding					
add for poplar molding	—	%	7.0	—	—
add for mahogany molding	—	%	21.0	—	—
add for red oak molding	—	%	30.0	—	—
All sizes and wood species of custom molding					
remove	1D@.032	lf	—	1.51	1.51

Panel molding corner radius. Includes panel corner radius, fasteners, and installation labor. Panel corner radius pieces are manufactured and often precut to match standard straight moldings. The corner radius is installed at corners of the molding to provide a radius instead of a 90-degree angle.

	Craft@Hrs	Unit	Material	Labor	Total
Replace					
standard	1C@.149	ea	2.90	10.10	13.00
with vertical milling	1C@.149	ea	6.82	10.10	16.92
with embossed pattern	1C@.149	ea	5.64	10.10	15.74
with hand carvings	1C@.149	ea	10.40	10.10	20.50
with heavy hand carvings	1C@.149	ea	14.80	10.10	24.90
All types					
remove	1D@.034	ea	—	1.60	1.60

	Craft@Hrs	Unit	Material	Labor	Total

Closet shelf brackets, shelving and rod. Includes shelf brackets, wood closet rod cut to length or metal telescoping closet rod, and bullnose shelf up to 16" wide, screws, and installation labor.

	Craft@Hrs	Unit	Material	Labor	Total
replace	1C@.332	lf	16.60	22.40	39.00
remove	1D@.063	lf	—	2.97	2.97
remove for work, then reinstall	6C@.500	lf	—	29.50	29.50

Closet organizer system. Includes closet shelves, side supports, rods, brackets, and — depending on quality — drawers. Installation labor includes some custom fitting to match walls. Closets are typically measured and a matching prefabricated unit installed. However, higher grades may include some custom work.

	Craft@Hrs	Unit	Material	Labor	Total
Replace					
standard grade	1C@.304	sf	15.60	20.60	36.20
high grade	1C@.414	sf	18.90	28.00	46.90
deluxe grade	1C@.526	sf	23.90	35.60	59.50
All grades					
remove	1D@.047	sf	—	2.22	2.22
remove for work, then reinstall	6C@.696	sf	—	41.10	41.10

Linen closet shelves. Includes bullnose paint-grade shelves up to 16" wide built on runners attached to walls and installation labor. Runners are made of pine or paint-grade casing. Shelves are spaced from 16" to 24" apart.

	Craft@Hrs	Unit	Material	Labor	Total
Replace					
standard grade	1C@.673	lf	12.40	45.50	57.90
high grade	1C@.673	lf	13.70	45.50	59.20
pull-out shelves	1C@.695	lf	15.70	47.00	62.70
All grades and styles					
remove	1D@.034	lf	—	1.60	1.60
remove for work, then reinstall	6C@.934	lf	—	55.10	55.10

Closet rod. Includes wood closet rod cut to length or metal expanding rod with installation labor. Includes screws to attach rod but does not include rod brackets.

	Craft@Hrs	Unit	Material	Labor	Total
Replace					
pine	1C@.080	lf	4.76	5.41	10.17
hardwood	1C@.080	lf	6.01	5.41	11.42
All closet rods					
remove	1D@.035	lf	—	1.65	1.65
remove for work, then reinstall	6C@.120	lf	—	7.08	7.08

Cedar closet lining. Includes tongue-and-groove aromatic cedar boards, finish nails, construction glue, and installation labor. Often installed on walls and ceilings of closet.

	Craft@Hrs	Unit	Material	Labor	Total
replace aromatic tongue-&-groove board	1C@.051	sf	.29	3.45	3.74
remove	1D@.028	sf	—	1.32	1.32
replace aromatic 1/4" veneer plywood	1C@.037	sf	4.44	2.50	6.94
remove	1D@.022	sf	—	1.04	1.04

	Craft@Hrs	Unit	Material	Labor	Total

Built-in bookcase. Includes finish boards, finish plywood, trimwork, brackets, fasteners, finish nails, and installation labor. Does not include lighting, caulking, finishing, or painting. Shelves may be manufactured or cut to width on site. Recessed and adjustable shelf hardware is also included but all shelves may not be adjustable. Shelves spaced approximately 12" on center. Measure square feet of bookcase face.

	Craft@Hrs	Unit	Material	Labor	Total
8" deep, replace					
paint-grade pine	1C@.269	sf	11.00	18.20	29.20
stain-grade pine	1C@.269	sf	12.90	18.20	31.10
mahogany	1C@.269	sf	14.00	18.20	32.20
oak	1C@.269	sf	16.20	18.20	34.40
walnut	1C@.269	sf	22.40	18.20	40.60
12" deep, replace					
paint-grade pine	1C@.304	sf	14.50	20.60	35.10
stain-grade pine	1C@.304	sf	16.60	20.60	37.20
mahogany	1C@.304	sf	18.10	20.60	38.70
oak	1C@.304	sf	21.60	20.60	42.20
walnut	1C@.304	sf	29.20	20.60	49.80
18" deep, replace					
paint-grade pine	1C@.317	sf	18.90	21.40	40.30
stain-grade pine	1C@.317	sf	22.30	21.40	43.70
mahogany	1C@.317	sf	23.90	21.40	45.30
oak	1C@.317	sf	28.20	21.40	49.60
walnut	1C@.317	sf	38.80	21.40	60.20
24" deep, replace					
paint-grade pine	1C@.332	sf	24.90	22.40	47.30
stain-grade pine	1C@.332	sf	29.10	22.40	51.50
mahogany	1C@.332	sf	31.30	22.40	53.70
oak	1C@.332	sf	37.10	22.40	59.50
walnut	1C@.332	sf	50.90	22.40	73.30
All bookcases, any wood species or depth					
remove	1D@.051	sf	—	2.41	2.41

Fireplace mantel beam. Includes mantel beam, fasteners, and installation labor. Some custom fitting to match is included. Does not include wall framing, masonry work, caulking, sanding, or painting.

	Craft@Hrs	Unit	Material	Labor	Total
Replace fireplace mantel beam					
rough sawn	1C@.539	lf	14.90	36.40	51.30
glue laminated	1C@.539	lf	42.70	36.40	79.10
All fireplace mantel beams, any type					
remove	1D@.080	lf	—	3.78	3.78
remove for work, then reinstall	6C@.799	lf	—	47.10	47.10

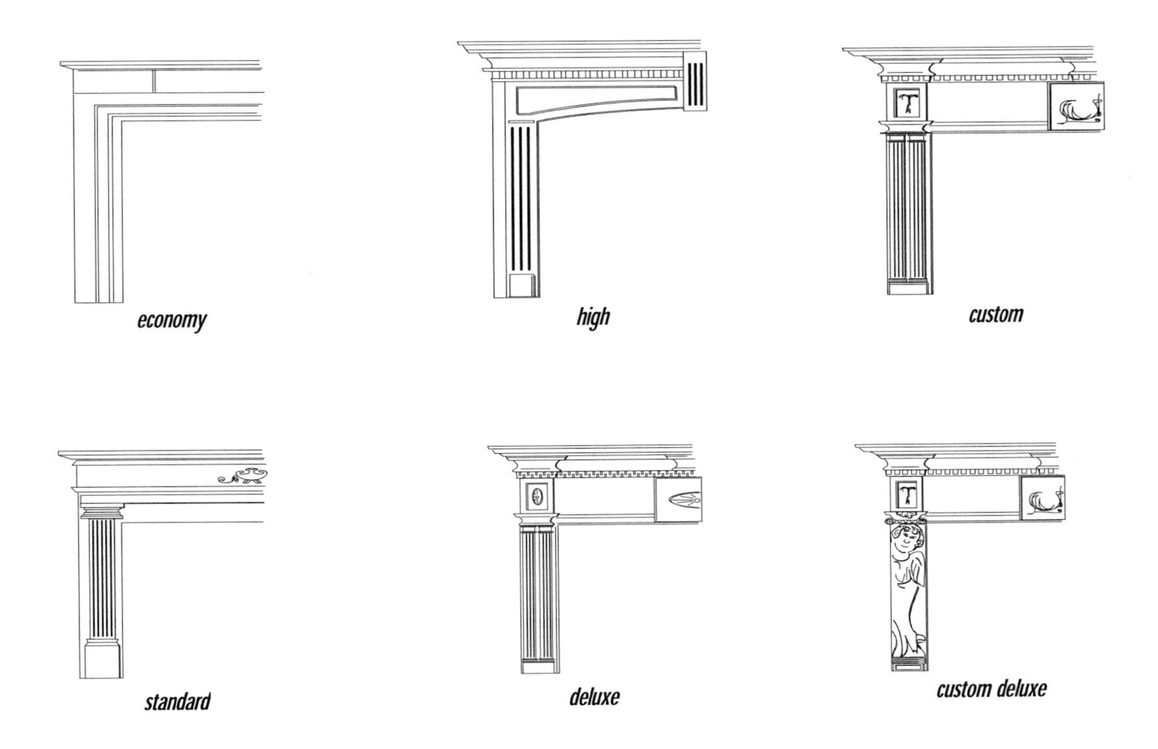

	Craft@Hrs	Unit	Material	Labor	Total

Fireplace mantel. Includes manufactured fireplace mantel to match fireplace opening, fasteners, masonry fasteners, and installation labor. Some custom fitting to match walls is included as is assembly of the mantel on site as needed. Does not include masonry work, caulking, sanding, or painting. Measure the LF of mantel width. Economy grade: paint-grade pine with simple pattern and moldings. Standard grade: paint- or stain-grade pine with dentil patterns and fluted pilasters. High grade: red oak, birch or stain-grade pine with a combination of some of the following: carved moldings, carved onlays, embossed patterns, fluted pilasters and face projections. Deluxe and higher grades: cherry or walnut, heart pine, red oak, birch or yellow pine with increasingly elaborate combinations of the elements listed for high grade, as well as the following: heavy hand carvings, carved or ornate (e.g. lion's head) corbels, circular patterns and moldings, round columns, curved pilasters, capitals (Ionic, Tuscan, Scamozzi, Corinthian, etc.), circular opening, paneling or mirror mantel.

	Craft@Hrs	Unit	Material	Labor	Total
Replace					
economy grade	1C@.653	lf	82.20	44.10	126.30
standard grade	1C@.653	lf	175.00	44.10	219.10
high grade	1C@.653	lf	380.00	44.10	424.10
deluxe grade	1C@.653	lf	853.00	44.10	897.10
custom grade	1C@.653	lf	1,700.00	44.10	1,744.10
custom deluxe grade	1C@.653	lf	2,620.00	44.10	2,664.10
All grades					
remove	1D@.094	lf	—	4.44	4.44
remove for work, then reinstall	6C@.828	lf	—	48.90	48.90

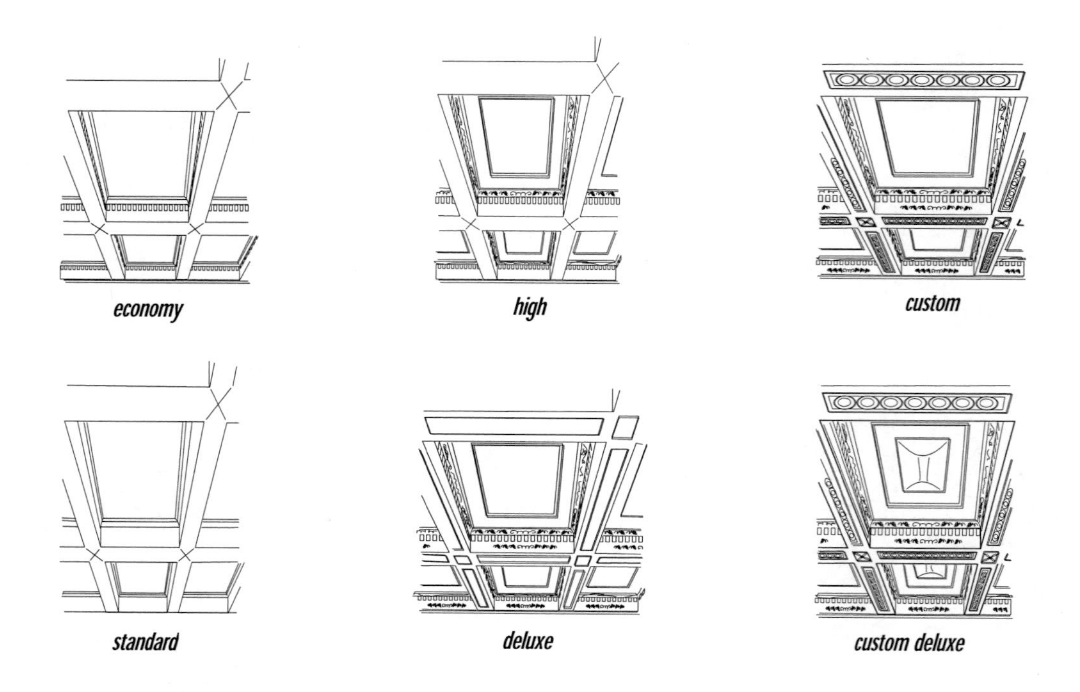

	Craft@Hrs	Unit	Material	Labor	Total

Coffered ceiling. Includes wood underframing and fire-taped drywall for coffers that measure between 42" to 48" on center with installation labor. Depending on grade, may also include finish drywall, millwork, finish hardware, finish plywood, onlays, corbels, and labor to install them. Does not include electrical work or fixtures, prepwork, painting or staining and varnishing. All inside molding joints are coped. Economy grade: coffers are finished drywall with corner bead, inside corners trimmed with 3-1/2" crown mold or similar. Standard grade: coffers and beams finished drywall, inside corners trimmed with vertical pattern moldings. High grade: covered with wood paneling, inside corners trimmed with vertical pattern or embossed moldings and coffers contain panel moldings. Deluxe grade: covered with hardwood plywood, inside corners trimmed with hand-carved moldings. Custom grade: covered with wood paneling, inside corners trimmed with heavy hand-carved moldings, wood onlays at each corner, recessed panels on beams or in coffers, corbels at wall corners. Custom deluxe grade: covered with wood paneling, inside corners trimmed with heavy hand-carved moldings, ornate wood onlays at each corner, recessed panels on beams or in coffers, ornately carved corbels (e.g. lion's head) at wall corners.

	Craft@Hrs	Unit	Material	Labor	Total
Replace					
economy grade	1C@.248	sf	15.90	16.80	32.70
standard grade	1C@.276	sf	22.40	18.70	41.10
high grade	1C@.308	sf	26.30	20.80	47.10
deluxe grade	1C@.347	sf	39.10	23.50	62.60
custom grade	1C@.430	sf	55.60	29.10	84.70
custom deluxe grade	1C@.567	sf	82.70	38.30	121.00
All grades					
remove	1D@.047	sf	—	2.22	2.22

	Craft@Hrs	Unit	Material	Labor	Total

Niche with casing and shelf bracket. Includes prefabricated niche with shelf bracket and hardware necessary to install in wall, and installation labor. Does not include wall framing. Made of high density polyurethane. Add **17%** to materials cost for niches made from fiber reinforced plaster.

	Craft@Hrs	Unit	Material	Labor	Total
Replace					
standard niche	1C@1.31	ea	283.00	88.60	371.60
with clamshell top	1C@1.31	ea	498.00	88.60	586.60
with clamshell top and clamshell shelf bracket	1C@1.31	ea	644.00	88.60	732.60
All types					
remove	1D@.317	ea	—	15.00	15.00
remove for work, then reinstall	6C@2.13	ea	—	126.00	126.00

standard

clamshell top

fleur-sawn *picket-sawn* *ball-and-dowel* *spindle*

Gingerbread running trim. Includes manufactured and assembled gingerbread running trim, fasteners, glue as needed, and installation labor. Does not include paint prep, painting, or caulking. Running trim maintains a consistent width and is used in a variety of applications. Paint-grade poplar. Add **30%** to materials cost for red oak.

	Craft@Hrs	Unit	Material	Labor	Total
Fleur-sawn, replace	1C@.160	lf	43.50	10.80	54.30
Picket-sawn, replace	1C@.160	lf	56.40	10.80	67.20
Ball-and-dowel, replace					
2" wide	1C@.160	lf	13.30	10.80	24.10
4" wide	1C@.160	lf	20.30	10.80	31.10
6" wide	1C@.160	lf	23.00	10.80	33.80
8" wide	1C@.160	lf	27.40	10.80	38.20
10" wide	1C@.160	lf	34.70	10.80	45.50
12" wide	1C@.160	lf	41.10	10.80	51.90
14" wide	1C@.160	lf	49.80	10.80	60.60
Spindle, replace					
2" wide	1C@.160	lf	12.50	10.80	23.30
4" wide	1C@.160	lf	15.70	10.80	26.50
6" wide	1C@.160	lf	20.30	10.80	31.10
8" wide	1C@.160	lf	24.50	10.80	35.30
10" wide	1C@.160	lf	33.40	10.80	44.20
12" wide	1C@.160	lf	40.00	10.80	50.80
14" wide	1C@.160	lf	48.80	10.80	59.60
All styles and sizes					
remove	1D@.038	lf	—	1.79	1.79
remove for work, then reinstall	6C@.224	lf	—	13.20	13.20

	Craft@Hrs	Unit	Material	Labor	Total

Gingerbread bracket. Includes manufactured and assembled bracket, fasteners, glue as needed, and installation labor. Does not include paint prep, painting, or caulking. All grades made from paint-grade poplar. Add **30%** to materials cost for red oak.

	Craft@Hrs	Unit	Material	Labor	Total
Replace					
standard grade, longest leg up to 12" long	1C@.348	ea	46.50	23.50	70.00
high grade, longest leg up to 23" long	1C@.348	ea	56.40	23.50	79.90
deluxe grade, longest leg up to 30" long	1C@.348	ea	62.10	23.50	85.60
custom grade, longest leg up to 30" long, with very ornate sawn pattern	1C@.348	ea	70.30	23.50	93.80
All grades					
remove	1D@.154	ea	—	7.27	7.27

standard
high
deluxe

Gingerbread fan bracket. Includes manufactured and assembled gingerbread fan bracket, fasteners, glue as needed, and installation labor. Does not include paint prep, painting, or caulking. Arched spandrels have arched end sections that typically contain either sawn woodwork or a spindle fan. The higher the grade the longer the leg against the wall and the more ornate the sawn woodwork. Highest grades may include post drops in the center section. All grades made from paint-grade poplar. Add **30%** to materials cost for red oak.

	Craft@Hrs	Unit	Material	Labor	Total
Replace					
standard grade, longest leg up to 14" long	1C@.348	ea	55.20	23.50	78.70
high grade, longest leg up to 25" long	1C@.348	ea	61.90	23.50	85.40
deluxe grade, longest leg up to 32" long	1C@.348	ea	76.70	23.50	100.20
custom grade, longest leg up to 32" long, with very ornate sawn pattern	1C@.348	ea	98.10	23.50	121.60
All grades					
remove	1D@.154	ea	—	7.27	7.27

Gingerbread post bracket. Includes turned or sawn elements in post bracket with fasteners, glue as needed, and installation labor. Does not include paint prep, painting, or caulking. Made from paint-grade poplar. Add **30%** to materials cost for red oak. The higher the grade the more ornate the pattern.

	Craft@Hrs	Unit	Material	Labor	Total
Replace					
standard grade	1C@.348	ea	71.30	23.50	94.80
high grade	1C@.348	ea	88.80	23.50	112.30
deluxe grade	1C@.348	ea	104.00	23.50	127.50
All grades					
remove	1D@.154	ea	—	7.27	7.27
remove for work, then reinstall	6C@.431	ea	—	25.40	25.40

	Craft@Hrs	Unit	Material	Labor	Total

Gingerbread corbel. Includes manufactured and assembled corbel, fasteners, glue as needed, and installation labor. Does not include paint prep, painting, or caulking. All grades made from paint-grade poplar. Add **30%** to materials cost for red oak. Simple sawn patterns for standard and high grades, other grades more ornate.

	Craft@Hrs	Unit	Material	Labor	Total
Replace					
standard grade, longest leg up to 12" long	1C@.359	ea	50.90	24.30	75.20
high grade, longest leg up to 16" long	1C@.359	ea	74.20	24.30	98.50
deluxe grade, longest leg up to 20" long	1C@.359	ea	103.00	24.30	127.30
custom grade, longest leg up to 20" long, with very ornate sawn pattern	1C@.359	ea	148.00	24.30	172.30
All grades					
remove	1D@.160	ea	—	7.55	7.55
remove for work, then reinstall	6C@.434	ea	—	25.60	25.60

Gingerbread door or window header. Includes manufactured and assembled gingerbread door or window header, fasteners, glue as needed, and installation labor. Does not include paint prep, painting, or caulking. All grades made from paint-grade poplar. Add **30%** to materials cost for red oak.

	Craft@Hrs	Unit	Material	Labor	Total
Replace					
standard grade, up to 14" long	1C@.232	lf	75.90	15.70	91.60
high grade, up to 25" long	1C@.232	lf	106.00	15.70	121.70
deluxe grade, up to 36" long	1C@.232	lf	143.00	15.70	158.70
All grades					
remove	1D@.037	lf	—	1.75	1.75
remove for work, then reinstall	6C@.333	lf	—	19.60	19.60

Gingerbread post drop. Includes turned or sawn post drop with fasteners, glue as needed, and installation labor. Does not include paint prep, painting, or caulking. Paint-grade poplar. Add **30%** to materials cost for red oak.

	Craft@Hrs	Unit	Material	Labor	Total
Replace					
10" long	1C@.322	ea	11.20	21.80	33.00
12" long	1C@.322	ea	14.10	21.80	35.90
18" long	1C@.322	ea	16.70	21.80	38.50
24" long	1C@.322	ea	19.90	21.80	41.70
All sizes					
remove	1D@.154	ea	—	7.27	7.27
remove for work, then reinstall	6C@.410	ea	—	24.20	24.20

	Craft@Hrs	Unit	Material	Labor	Total

Gingerbread spandrel. Includes manufactured and assembled gingerbread spandrel, fasteners, glue as needed, and installation labor. Spandrels come in typical opening sizes or can be custom-ordered to fit. Does not include paint prep, painting, or caulking. Spandrels are horizontal decorative elements for exterior and interior openings. Typically both corners and the center contain sawn elements. The remaining space is filled with either ball-and-dowel or spindle elements. All grades made from paint-grade poplar. Add **30%** to materials cost for red oak.

	Craft@Hrs	Unit	Material	Labor	Total
Replace					
standard grade, up to 6" deep in center, legs up to 14" long	1C@.237	lf	36.10	16.00	52.10
high grade, up to 9" deep in center, legs up to 17" long	1C@.237	lf	45.80	16.00	61.80
custom grade, up to 13" deep in center, legs up to 21" long	1C@.237	lf	59.70	16.00	75.70
deluxe grade, up to 16" deep in center, legs up to 24" long	1C@.237	lf	81.00	16.00	97.00
custom deluxe grade, up to 22" deep in center, legs up to 31" long	6C@.272	lf	102.00	16.00	118.00
All grades					
remove	1D@.038	lf	—	1.79	1.79
remove for work, then reinstall	6C@.341	lf	—	20.10	20.10

high

deluxe

Gingerbread arch spandrel. Includes manufactured to size and assembled arch spandrel, fasteners, glue as needed, and installation labor. Does not include paint prep, painting, or caulking. Also does not include extra work or trim that may be required when installing in out-of-plumb structures. Arched spandrels have arched end sections that typically contain either sawn woodwork or a spindle fan. The higher the grade the longer the leg against the wall and the more ornate the sawn woodwork. Highest grades may include post drops in the center section. All grades made from paint-grade poplar. Add **30%** to materials cost for red oak.

	Craft@Hrs	Unit	Material	Labor	Total
Replace					
standard grade, legs up to 12" long	1C@.237	lf	69.20	16.00	85.20
high grade, legs up to 20" long	1C@.237	lf	86.20	16.00	102.20
custom grade, legs up to 22" long	1C@.237	lf	99.50	16.00	115.50
deluxe grade, legs up to 26" long	1C@.237	lf	111.00	16.00	127.00
Replace custom deluxe grade, legs up to 28" long	1C@.237	lf	131.00	16.00	147.00
All grades					
remove	1D@.038	lf	—	1.79	1.79
remove for work, then reinstall	6C@.341	lf	—	20.10	20.10

custom

custom deluxe

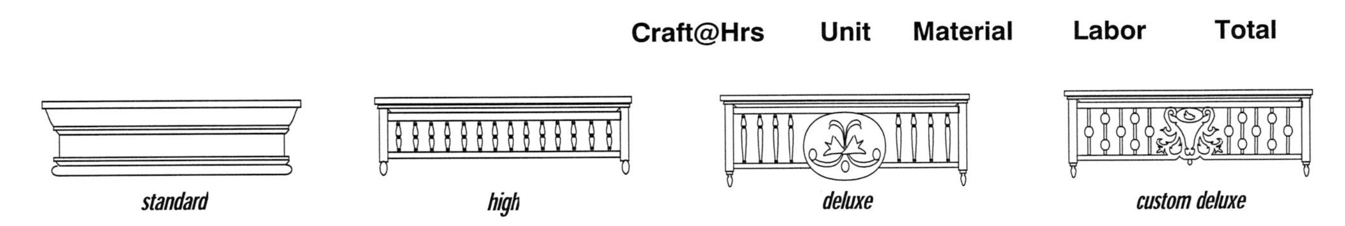

Gingerbread window cornice. Includes manufactured and assembled gingerbread window cornice, fasteners, glue as needed, and installation labor. Window cornices come in typical opening sizes or can be custom-ordered to fit. Does not include paint prep, painting, or caulking. All grades made from paint-grade poplar. Add **30%** to materials cost for red oak. Standard grade: made from finish board with bed mold on top and simple panel mold on bottom. High grade: side brackets with sawn drop, front cornice of 2" to 4" ball-and-dowel running trim. Deluxe grade: side brackets with sawn drop and spindles, front cornice of 6" to 8" spindle running trim with sawn work at center. Custom grade: side brackets with sawn drop and spindles, front cornice of 6" to 8" spindle running trim with very ornately sawn woodwork at center.

	Craft@Hrs	Unit	Material	Labor	Total
Replace					
standard grade, 5" deep and 6" high	1C@.242	lf	34.10	16.40	50.50
high grade, 5-1/2" deep and 8-1/2" high	1C@.242	lf	42.40	16.40	58.80
deluxe grade, 8" deep and 9" high	1C@.242	lf	53.60	16.40	70.00
custom grade, 8" deep and 9" high	1C@.242	lf	55.60	16.40	72.00
custom deluxe grade, 8" deep and 9" high	1C@.242	lf	64.80	16.40	81.20
All grades					
remove	1D@.041	lf	—	1.94	1.94
remove for work, then reinstall	6C@.352	lf	—	20.80	20.80

Gingerbread gable ornament. Includes manufactured and assembled gable ornament built to match the slope of the roof, fasteners, glue as needed, and installation labor. Does not include paint prep, painting, or caulking. Gable ornaments can be ordered for roofs of any slope. They are typically found on roofs with a slope of 6-in-12 or more. Ornaments are placed in the gables at the peak of the roof. All grades are made from paint-grade poplar. Add **30%** to materials cost for red oak.

	Craft@Hrs	Unit	Material	Labor	Total
Replace					
standard grade, longest leg up to 32"	1C@.582	ea	194.00	39.30	233.30
high grade, longest leg up to 36"	1C@.582	ea	334.00	39.30	373.30
deluxe grade, longest leg up to 45"	1C@.582	ea	507.00	39.30	546.30
custom grade, longest leg up to 48"	1C@.582	ea	897.00	39.30	936.30
custom deluxe grade, longest leg up to 60"	1C@.582	ea	1,090.00	39.30	1,129.30
All grades					
remove	1D@.254	ea	—	12.00	12.00
remove for work, then reinstall	6C@1.08	ea	—	63.70	63.70

standard

custom deluxe

Gingerbread gable finial. Includes turned or sawn gable finial with fasteners, glue as needed, and installation labor. Does not include paint prep, painting, or caulking. Finials are placed at the peak of the roof on each end. All grades are made from cedar.

	Craft@Hrs	Unit	Material	Labor	Total
Replace					
standard grade, with 28" post	1C@.560	ea	119.00	37.90	156.90
high grade, with 34" post	1C@.560	ea	161.00	37.90	198.90
All grades					
remove	1D@.239	ea	—	11.30	11.30
remove for work, then reinstall	6C@1.03	ea	—	60.80	60.80

Porch post. Includes post, hardware, and installation labor. Does not include bracing. For stain-grade pine add **16%** to material costs. Non-laminated posts made from clear, select Douglas fir or clear all-heart redwood. Standard grade posts may be made from cedar. Square posts are hollow and include a structural post inside. Standard grade: turned post, sometimes called colonist style. High grade: turned post with flutes, or square post with flutes trimmed with molding at base, 1/3 up the shaft and 12" to 16" from the top. Deluxe grade: turned post with spiral pattern or square post with flutes trimmed with panels 1/3 up the shaft, capped at base and top of panels with moldings and molding trim about 12" to 16" from top.

	Craft@Hrs	Unit	Material	Labor	Total
Replace 4" x 4" porch post					
standard grade	1C@.135	lf	14.00	9.13	23.13
high grade	1C@.135	lf	27.80	9.13	36.93
deluxe grade	1C@.135	lf	78.70	9.13	87.83
Replace 6" x 6" porch post					
standard grade	1C@.135	lf	20.80	9.13	29.93
high grade	1C@.135	lf	41.60	9.13	50.73
deluxe grade	1C@.135	lf	52.20	9.13	61.33
Replace 8" x 8" porch post					
standard grade	1C@.135	lf	23.50	9.13	32.63
high grade	1C@.135	lf	47.30	9.13	56.43
deluxe grade	1C@.135	lf	59.70	9.13	68.83
All sizes and grades of porch post					
remove	1D@.127	lf	—	5.99	5.99
remove for work, then reinstall	6C@.216	lf	—	12.70	12.70

Carved wood onlay. Includes wood onlay, finish nails or brads, glue as needed, and installation labor. Does not include sanding, caulking, or finishing. Onlays are available in a wide variety of styles and prices. Use these listings as indicators of complexity and price range. Made from maple. Add **25%** to material costs for red oak.

	Craft@Hrs	Unit	Material	Labor	Total
Replace small carved wood onlay					
acanthus	1C@.519	ea	54.10	35.10	89.20
ribbon	1C@.519	ea	56.70	35.10	91.80
grape	1C@.519	ea	90.90	35.10	126.00
shell	1C@.519	ea	79.50	35.10	114.60
floral	1C@.519	ea	83.10	35.10	118.20
Replace medium carved wood onlay					
acanthus	1C@.519	ea	84.20	35.10	119.30
ribbon	1C@.519	ea	88.10	35.10	123.20
grape	1C@.519	ea	141.00	35.10	176.10
shell	1C@.519	ea	123.00	35.10	158.10
floral	1C@.519	ea	129.00	35.10	164.10

	Craft@Hrs	Unit	Material	Labor	Total
Replace large carved wood onlay					
acanthus	1C@.519	ea	107.00	35.10	142.10
ribbon	1C@.519	ea	113.00	35.10	148.10
grape	1C@.519	ea	175.00	35.10	210.10
shell	1C@.519	ea	157.00	35.10	192.10
floral	1C@.519	ea	161.00	35.10	196.10
All grades and styles of onlay					
remove	1D@.062	ea	—	2.93	2.93
remove for work, then reinstall	6C@.743	ea	—	43.80	43.80

Finish board. Includes finish nails and installation. S4S = Smooth four sides. Includes 4% waste.

	Craft@Hrs	Unit	Material	Labor	Total
Replace S4S select pine finish board					
1" x 2"	1C@.049	lf	.97	3.31	4.28
1" x 4"	1C@.049	lf	1.87	3.31	5.18
1" x 6"	1C@.049	lf	2.84	3.31	6.15
1" x 8"	1C@.049	lf	3.79	3.31	7.10
1" x 10"	1C@.049	lf	4.71	3.31	8.02
1" x 12"	1C@.049	lf	5.67	3.31	8.98
Replace other wood species S4S finish board					
add for S4S poplar	—	%	3.0	—	—
add for S4S mahogany	—	%	13.0	—	—
add for S4S red oak	—	%	17.0	—	—
Other finish board costs, replace					
add for bullnose on 1 x boards	—	lf	.87	—	.87
All sizes and wood species of S4S finish board					
remove	1D@.026	lf	—	1.23	1.23
remove for work, then reinstall	6C@.080	lf	—	4.72	4.72

1/4" finish veneer plywood. Includes finish nails and installation.

	Craft@Hrs	Unit	Material	Labor	Total
Replace					
aromatic cedar	1C@.028	sf	4.44	1.89	6.33
birch	1C@.028	sf	4.29	1.89	6.18
cherry	1C@.028	sf	4.80	1.89	6.69
chestnut	1C@.028	sf	7.57	1.89	9.46
knotty pine	1C@.028	sf	3.37	1.89	5.26
lauan mahogany	1C@.028	sf	1.70	1.89	3.59
mahogany	1C@.028	sf	4.85	1.89	6.74
pecan	1C@.028	sf	4.97	1.89	6.86
red oak	1C@.028	sf	4.34	1.89	6.23
rosewood	1C@.028	sf	6.71	1.89	8.60
teak	1C@.028	sf	8.89	1.89	10.78
walnut	1C@.028	sf	6.43	1.89	8.32
All wood species of 1/4" finish veneer plywood					
remove	1D@.022	sf	—	1.04	1.04
remove for work, then reinstall	6C@.042	sf	—	2.48	2.48

	Craft@Hrs	Unit	Material	Labor	Total
Time & Material Charts (selected items)					
Finish Carpentry Materials					
Clamshell base					
1-1/2" stain-grade pine ($1.43 lf, 1 lf), 7% waste	—	lf	1.54	—	1.54
3-1/2" stain-grade pine ($2.28 lf, 1 lf), 7% waste	—	lf	2.46	—	2.46
Pattern base					
1-1/2" stain-grade pine ($1.70 lf, 1 lf), 7% waste	—	lf	1.82	—	1.82
2-1/2" stain-grade pine ($2.77 lf, 1 lf), 7% waste	—	lf	2.97	—	2.97
3-1/2" stain-grade pine ($3.66 lf, 1 lf), 7% waste	—	lf	4.06	—	4.06
Base shoe					
3/4" stain-grade pine ($.86 lf, 1 lf), 7% waste	—	lf	.92	—	.92
Clamshell casing					
1-1/2" stain-grade pine ($1.48 lf, 1 lf), 5% waste	—	lf	1.56	—	1.56
2-1/2" stain-grade pine ($1.67 lf, 1 lf), 5% waste	—	lf	1.75	—	1.75
3-1/2" stain-grade pine ($2.39 lf, 1 lf), 5% waste	—	lf	2.50	—	2.50
Pattern casing					
1-1/2" stain-grade pine ($1.81 lf, 1 lf), 5% waste	—	lf	1.90	—	1.90
2-1/2" stain-grade pine ($2.89 lf, 1 lf), 5% waste	—	lf	3.02	—	3.02
3-1/2" stain-grade pine ($3.83 lf, 1 lf), 5% waste	—	lf	4.02	—	4.02
Window moldings					
2-1/2" stain-grade pine stool ($3.23 lf, 1 lf), 4% waste	—	lf	3.36	—	3.36
3-1/2" stain-grade pine apron ($2.13 lf, 1 lf), 4% waste	—	lf	2.20	—	2.20
Chair rail					
2-1/2" stain-grade pine ($2.81 lf, 1 lf), 4% waste	—	lf	2.92	—	2.92
3-1/2" stain-grade pine ($3.54 lf, 1 lf), 4% waste	—	lf	3.68	—	3.68
Drip cap					
1-5/8" redwood ($2.61 lf, 1 lf), 4% waste	—	lf	2.70	—	2.70
2-1/4" redwood ($2.90 lf, 1 lf), 4% waste	—	lf	3.01	—	3.01

	Craft@Hrs	Unit	Material	Labor	Total
Astragal					
1-3/4" stain-grade pine ($5.37 lf, 1 lf), 4% waste	—	lf	5.56	—	5.56
2-1/4" stain-grade pine ($6.52 lf, 1 lf), 4% waste	—	lf	6.77	—	6.77
Cove					
1/2" stain-grade pine ($1.03 lf, 1 lf), 4% waste	—	lf	1.07	—	1.07
3/4" stain-grade pine ($1.13 lf, 1 lf), 4% waste	—	lf	1.18	—	1.18
1" stain-grade pine ($1.42 lf, 1 lf), 4% waste	—	lf	1.47	—	1.47
2-1/2" stain-grade pine ($2.52 lf, 1 lf), 4% waste	—	lf	2.63	—	2.63
3-1/2" stain-grade pine ($3.25 lf, 1 lf), 4% waste	—	lf	3.38	—	3.38
Half-round					
1" stain-grade pine ($1.18 lf, 1 lf), 4% waste	—	lf	1.23	—	1.23
2" stain-grade pine ($1.65 lf, 1 lf), 4% waste	—	lf	1.72	—	1.72
Quarter-round					
1/2" stain-grade pine ($.86 lf, 1 lf), 4% waste	—	lf	.89	—	.89
3/4" stain-grade pine ($1.05 lf, 1 lf), 4% waste	—	lf	1.09	—	1.09
1" stain-grade pine ($1.84 lf, 1 lf), 4% waste	—	lf	1.91	—	1.91
Corner bead					
stain-grade pine ($2.02 lf, 1 lf), 4% waste	—	lf	2.11	—	2.11
turned stain-grade pine with finials ($22.30 lf, 1 lf), 4% waste	—	lf	23.30	—	23.30
clear plastic ($1.57 lf, 1 lf), 4% waste	—	lf	1.62	—	1.62
Bed mold					
2-1/2" stain-grade pine ($2.83 lf, 1 lf), 4% waste	—	lf	2.95	—	2.95
Crown					
3-1/2" stain-grade pine ($3.61 lf, 1 lf), 4% waste	—	lf	3.75	—	3.75
4-1/2" stain-grade pine ($5.34 lf, 1 lf), 4% waste	—	lf	5.54	—	5.54
Hand rail					
stain-grade pine ($5.34 lf, 1 lf), 4% waste	—	lf	5.54	—	5.54

	Craft@Hrs	Unit	Material	Labor	Total
Specialty moldings, 3/4"					
stain-grade pine with vertical milling ($4.26 lf, 1 lf), 4% waste	—	lf	4.43	—	4.43
stain-grade pine embossed ($2.50 lf, 1 lf), 4% waste	—	lf	2.61	—	2.61
stain-grade pine with hand carving ($7.85 lf, 1 lf), 4% waste	—	lf	8.16	—	8.16
stain-grade pine with heavy hand carving ($15.40 lf, 1 lf), 4% waste	—	lf	15.90	—	15.90
Specialty moldings, 1-1/2"					
stain-grade pine with vertical milling ($4.85 lf, 1 lf), 4% waste	—	lf	5.05	—	5.05
stain-grade pine embossed ($2.88 lf, 1 lf), 4% waste	—	lf	2.99	—	2.99
stain-grade pine with hand carving ($8.96 lf, 1 lf), 4% waste	—	lf	9.31	—	9.31
stain-grade pine with heavy hand carving ($17.10 lf, 1 lf), 4% waste	—	lf	17.90	—	17.90
Specialty moldings, 2-1/2"					
stain-grade pine with vertical milling ($5.47 lf, 1 lf), 4% waste	—	lf	5.69	—	5.69
stain-grade pine embossed ($3.21 lf, 1 lf), 4% waste	—	lf	3.34	—	3.34
stain-grade pine with hand carving ($10.00 lf, 1 lf), 4% waste	—	lf	10.40	—	10.40
stain-grade pine with heavy hand carving ($19.40 lf, 1 lf), 4% waste	—	lf	20.20	—	20.20
Specialty moldings, 3-1/2"					
stain-grade pine with vertical milling ($6.04 lf, 1 lf), 4% waste	—	lf	6.27	—	6.27
stain-grade pine embossed ($3.57 lf, 1 lf), 4% waste	—	lf	3.71	—	3.71
stain-grade pine with hand carving ($11.10 lf, 1 lf), 4% waste	—	lf	11.60	—	11.60
stain-grade pine with heavy hand carving ($21.90 lf, 1 lf), 4% waste	—	lf	22.80	—	22.80
Specialty moldings, 5"					
stain-grade pine with vertical milling ($8.71 lf, 1 lf), 4% waste	—	lf	9.06	—	9.06
stain-grade pine embossed ($5.14 lf, 1 lf), 4% waste	—	lf	5.36	—	5.36
stain-grade pine with hand carving ($15.90 lf, 1 lf), 4% waste	—	lf	16.50	—	16.50
stain-grade pine with heavy hand carving ($31.50 lf, 1 lf), 4% waste	—	lf	32.60	—	32.60

	Craft@Hrs	Unit	Material	Labor	Total
Specialty moldings, 8"					
stain-grade pine with vertical milling ($14.20 lf, 1 lf), 4% waste	—	lf	14.70	—	14.70
stain-grade pine embossed ($8.42 lf, 1 lf), 4% waste	—	lf	8.75	—	8.75
stain-grade pine with hand carving ($26.30 lf, 1 lf), 4% waste	—	lf	27.20	—	27.20
stain-grade pine with heavy hand carving ($51.40 lf, 1 lf), 4% waste	—	lf	53.60	—	53.60
Custom-milled moldings					
molder setup charge, ($278.00 ea)	—	ea	278.00	—	278.00
molder custom knife charges, ($604.00 ea)	—	ea	604.00	—	604.00
2" to 3" pine, ($2.74 lf, 1 lf), 4% waste	—	lf	2.85	—	2.85
2" to 3" poplar, ($3.02 lf, 1 lf), 4% waste	—	lf	3.15	—	3.15
2" to 3" mahogany, ($3.61 lf, 1 lf), 4% waste	—	lf	3.75	—	3.75
2" to 3" red oak, ($3.96 lf, 1 lf), 4% waste	—	lf	4.11	—	4.11
Inside corner radius for panel molding					
standard ($2.75 ea, 1 ea), 4% waste	—	ea	2.87	—	2.87
with heavy carving ($14.20 ea, 1 ea), 4% waste	—	ea	14.70	—	14.70
Closet work					
shelf brackets, shelving, & rod ($16.00 lf, 1 lf), 4% waste	—	lf	16.70	—	16.70
organizer system, standard grade ($14.80 sf, 1 sf), 4% waste	—	sf	15.50	—	15.50
organizer system, deluxe grade ($23.00 sf, 1 sf), 4% waste	—	sf	24.00	—	24.00
linen closet shelves, standard grade ($11.90 lf, 1 lf), 4% waste	—	lf	12.40	—	12.40
linen closet pull-out shelves ($15.30 ea, 1 ea), 4% waste	—	ea	15.80	—	15.80
closet rod, pine ($4.57 lf, 1 lf), 4% waste	—	lf	4.74	—	4.74
Built-in bookcase					
8" deep stain-grade pine ($12.10 sf, 1 sf), 4% waste	—	sf	12.60	—	12.60
12" deep stain-grade pine ($16.00 sf, 1 sf), 4% waste	—	sf	16.70	—	16.70
18" deep stain-grade pine ($21.30 sf, 1 sf), 4% waste	—	sf	22.00	—	22.00
24" deep stain-grade pine ($27.80 sf, 1 sf), 4% waste	—	sf	29.00	—	29.00
Fireplace mantel beam					
rough sawn, ($14.30 lf, 1 lf), 4% waste	—	lf	14.80	—	14.80
glue laminated, ($40.70 lf, 1 lf), 4% waste	—	lf	42.40	—	42.40
Fireplace mantel					
economy grade ($82.20 lf, 1 lf)	—	lf	82.20	—	82.20
custom deluxe grade ($2,610.00 lf, 1 lf)	—	lf	2,610.00	—	2,610.00

	Craft@Hrs	Unit	Material	Labor	Total
Coffered ceiling					
economy grade, ($13.80 sf, 1 sf), 12% waste	—	sf	15.40	—	15.40
custom deluxe grade, ($71.70 sf, 1 sf), 12% waste	—	sf	80.40	—	80.40
Niche with casing and shelf bracket					
standard	—	ea	283.00	—	283.00
with clamshell top	—	ea	498.00	—	498.00
clamshell top and clamshell shelf	—	ea	644.00	—	644.00
Gingerbread running trim					
fleur-sawn, ($42.80 lf, 1 lf), 2% waste	—	lf	43.60	—	43.60
picket-sawn, ($55.20 lf, 1 lf), 2% waste	—	lf	56.40	—	56.40
2" ball-and-dowel, ($13.10 lf, 1 lf), 2% waste	—	lf	13.30	—	13.30
14" ball-and-dowel, ($48.50 lf, 1 lf), 2% waste	—	lf	49.50	—	49.50
2" spindle, ($12.30 lf, 1 lf), 2% waste	—	lf	12.50	—	12.50
14" spindle, ($47.60 lf, 1 lf), 2% waste	—	lf	48.50	—	48.50
Gingerbread bracket					
standard grade, ($46.50 ea)	—	ea	46.50	—	46.50
custom grade, ($70.30 ea)	—	ea	70.30	—	70.30
Gingerbread fan bracket					
standard grade, ($55.20 ea)	—	ea	55.20	—	55.20
custom grade, ($98.10 ea)	—	ea	98.10	—	98.10
Gingerbread post bracket					
standard grade, ($71.30 ea)	—	ea	71.30	—	71.30
deluxe grade, ($104.00 ea)	—	ea	104.00	—	104.00
Gingerbread corbel					
standard grade, ($50.90 ea)	—	ea	50.90	—	50.90
custom grade, ($148.00 ea)	—	ea	148.00	—	148.00
Gingerbread door or window header					
standard grade, ($75.90 ea)	—	ea	75.90	—	75.90
deluxe grade, ($143.00 ea)	—	ea	143.00	—	143.00
Gingerbread post drop					
10", ($11.20 ea)	—	ea	11.20	—	11.20
24", ($19.80 ea)	—	ea	19.80	—	19.80
Gingerbread spandrel					
standard grade, ($36.10 lf)	—	lf	36.10	—	36.10
custom deluxe grade, ($102.00 lf)	—	lf	102.00	—	102.00
Gingerbread arch spandrel					
standard grade, ($69.20 lf)	—	lf	69.20	—	69.20
custom deluxe grade, ($131.00 lf)	—	lf	131.00	—	131.00
Gingerbread window cornice					
standard grade, ($34.10 lf)	—	lf	34.10	—	34.10
custom deluxe grade, ($64.80 lf)	—	lf	64.80	—	64.80
Gingerbread gable ornament					
standard grade, ($194.00 ea)	—	ea	194.00	—	194.00
custom deluxe grade, ($1,090.00 ea)	—	ea	1,090.00	—	1,090.00
Gingerbread gable finial					
standard grade, ($119.00 ea)	—	ea	119.00	—	119.00
high grade, ($161.00 ea)	—	ea	161.00	—	161.00

	Craft@Hrs	Unit	Material	Labor	Total
Porch post					
4" x 4", standard grade ($111.00 8' post, 8 lf), 0% waste	—	lf	14.00	—	14.00
4" x 4", deluxe grade ($630.00 8' post, 8 lf), 0% waste	—	lf	78.90	—	78.90
6" x 6", standard grade ($162.00 8' post, 8 lf), 0% waste	—	lf	20.30	—	20.30
6" x 6", deluxe grade ($417.00 8' post, 8 lf), 0% waste	—	lf	52.20	—	52.20
8" x 8", standard grade ($187.00 8' post, 8 lf), 0% waste	—	lf	23.40	—	23.40
8" x 8", deluxe grade ($476.00 8' post, 8 lf), 0% waste	—	lf	59.40	—	59.40
Carved wood onlay					
medium acanthus ($84.20 ea)	—	ea	84.20	—	84.20
medium ribbon ($88.10 ea)	—	ea	88.10	—	88.10
medium grape ($141.00 ea)	—	ea	141.00	—	141.00
medium shell ($123.00 ea)	—	ea	123.00	—	123.00
medium floral ($129.00 ea)	—	ea	129.00	—	129.00
Finish board					
1" x 12" S4S select pine ($5.43 lf, 1 lf), 4% waste	—	lf	5.66	—	5.66
Hardwood plywood					
1/4" aromatic cedar veneer ($137.00 sheet, 32 sf), 4% waste	—	sf	4.43	—	4.43
1/4" birch veneer ($132.00 sheet, 32 sf), 4% waste	—	sf	4.27	—	4.27
1/4" cherry veneer ($147.00 sheet, 32 sf), 4% waste	—	sf	4.80	—	4.80
1/4" chestnut veneer ($234.00 sheet, 32 sf), 4% waste	—	sf	7.60	—	7.60
1/4" knotty pine veneer ($103.00 sheet, 32 sf), 4% waste	—	sf	3.35	—	3.35
1/4" lauan mahogany veneer ($53.00 sheet, 32 sf), 4% waste	—	sf	1.71	—	1.71
1/4" mahogany veneer ($149.00 sheet, 32 sf), 4% waste	—	sf	4.87	—	4.87
1/4" pecan veneer ($153.00 sheet, 32 sf), 4% waste	—	sf	4.95	—	4.95
1/4" red oak veneer ($133.00 sheet, 32 sf), 4% waste	—	sf	4.30	—	4.30
1/4" rosewood veneer ($208.00 sheet, 32 sf), 4% waste	—	sf	6.72	—	6.72
1/4" teak veneer ($272.00 sheet, 32 sf), 4% waste	—	sf	8.87	—	8.87
1/4" walnut veneer ($194.00 sheet, 32 sf), 4% waste	—	sf	6.30	—	6.30

Finish Carpentry Labor

Laborer	base wage	paid leave	true wage	taxes & ins.	total
Carpenter	$38.10	2.97	$41.07	26.53	$67.60
Carpenter's helper	$27.40	2.14	$29.54	20.86	$50.40
Demolition laborer	$25.90	2.02	$27.92	19.28	$47.20

Paid leave is calculated based on two weeks paid vacation, one week sick leave, and seven paid holidays. Employer's matching portion of **FICA** is 7.65 percent. **FUTA** (Federal Unemployment) is .8 percent. **Worker's compensation** for the finish carpentry trade was calculated using a national average of 17.21 percent. **Unemployment insurance** was calculated using a national average of 8 percent. **Health insurance** was calculated based on a projected national average for 2020 of $1,280 per employee (and family when applicable) per month. Employer pays 80 percent for a per month cost of $1,024 per employee. **Retirement** is based on a 401(k) retirement program with employer matching of 50 percent. Employee contributions to the 401(k) plan are an average of 6 percent of the true wage. **Liability insurance** is based on a national average of 12.0 percent.

Finish Carpentry Labor Productivity

	Craft@Hrs	Unit	Material	Labor	Total
Demolition of finish carpentry					
remove wall molding (base, chair rail, crown)	1D@.026	lf	—	1.23	1.23
remove door and window molding	1D@.032	lf	—	1.51	1.51
remove base or corner block	1D@.036	ea	—	1.70	1.70
remove head block	1D@.037	ea	—	1.75	1.75
remove overdoor molding	1D@.028	lf	—	1.32	1.32
remove window stool	1D@.026	lf	—	1.23	1.23
remove window apron	1D@.028	lf	—	1.32	1.32
remove corner bead	1D@.043	lf	—	2.03	2.03
remove hand rail	1D@.028	lf	—	1.32	1.32
remove shelf brackets, shelving, and rod	1D@.063	lf	—	2.97	2.97
remove closet organizer system	1D@.047	sf	—	2.22	2.22
remove linen closet shelves	1D@.034	lf	—	1.60	1.60
remove closet rod	1D@.035	lf	—	1.65	1.65
remove tongue-and-groove cedar closet lining	1D@.028	sf	—	1.32	1.32
remove cedar veneer plywood closet lining	1D@.022	sf	—	1.04	1.04
remove built-in bookcase	1D@.051	sf	—	2.41	2.41
remove fireplace mantel beam	1D@.080	lf	—	3.78	3.78
remove fireplace mantel	1D@.094	lf	—	4.44	4.44
remove coffered ceiling	1D@.047	sf	—	2.22	2.22
remove niche	1D@.317	ea	—	15.00	15.00
remove gingerbread running trim	1D@.038	lf	—	1.79	1.79
remove gingerbread bracket	1D@.154	ea	—	7.27	7.27
remove gingerbread corbel	1D@.160	ea	—	7.55	7.55
remove gingerbread door or window header	1D@.037	lf	—	1.75	1.75
remove gingerbread post drop	1D@.154	ea	—	7.27	7.27
remove gingerbread spandrel	1D@.038	lf	—	1.79	1.79
remove gingerbread window cornice	1D@.041	lf	—	1.94	1.94
remove gingerbread gable ornament	1D@.254	ea	—	12.00	12.00
remove gingerbread gable finial	1D@.239	ea	—	11.30	11.30
remove porch post	1D@.127	lf	—	5.99	5.99
remove onlay	1D@.062	ea	—	2.93	2.93
remove finish trim board	1D@.026	lf	—	1.23	1.23
remove finish plywood	1D@.022	sf	—	1.04	1.04

	Craft@Hrs	Unit	Material	Labor	Total
Finish carpentry crew					
finish carpentry · finish carpenter · $67.60					
finish carpentry · carpenter's helper · $50.40					
finish carpentry · finish crew · $59.00					
Install molding					
base, chair rail	1C@.038	lf	—	2.57	2.57
window or door casing or panel moldings	1C@.041	lf	—	2.77	2.77
crown	1C@.046	lf	—	3.11	3.11
Install casing on curved window or door					
half-round top window or door per lf	1C@.144	lf	—	9.73	9.73
elliptical top window or door per lf	1C@.144	lf	—	9.73	9.73
round window per lf of diameter	1C@.294	lf	—	19.90	19.90
Install wood key in curved molding					
install	1C@.251	ea	—	17.00	17.00
Install base, corner, or head block					
install	1C@.063	ea	—	4.26	4.26
Install overdoor molding					
install	1C@.041	lf	—	2.77	2.77
Install door architrave					
interior	1C@.062	lf	—	4.19	4.19
exterior	1C@.080	lf	—	5.41	5.41
Install exterior door surround					
economy grade	1C@.234	lf	—	15.80	15.80
custom deluxe grade	1C@.431	lf	—	29.10	29.10
Install exterior window surround					
economy grade	1C@.207	lf	—	14.00	14.00
custom deluxe grade	1C@.322	lf	—	21.80	21.80
Install window trim					
stool	1C@.058	lf	—	3.92	3.92
apron	1C@.049	lf	—	3.31	3.31
Install corner bead					
wood	1C@.077	lf	—	5.21	5.21
plastic	1C@.032	lf	—	2.16	2.16
Install hand rail					
install	1C@.098	lf	—	6.62	6.62
Install panel molding inside corner radius					
install	1C@.149	ea	—	10.10	10.10
Install closet shelf brackets, shelving, & rod					
install	1C@.332	lf	—	22.40	22.40
Install closet organizer system					
standard grade	1C@.304	sf	—	20.60	20.60
deluxe grade	1C@.526	sf	—	35.60	35.60
Install linen closet shelves					
typical	1C@.673	lf	—	45.50	45.50
with pull-out shelves	1C@.695	lf	—	47.00	47.00
Install closet rod					
install	1C@.080	lf	—	5.41	5.41

	Craft@Hrs	Unit	Material	Labor	Total
Install closet lining					
tongue-&-groove cedar board	1C@.051	sf	—	3.45	3.45
1/4" cedar veneer plywood	1C@.037	sf	—	2.50	2.50
Install built-in bookcase					
8" deep	1C@.269	sf	—	18.20	18.20
12" deep	1C@.304	sf	—	20.60	20.60
18" deep	1C@.317	sf	—	21.40	21.40
24" deep	1C@.332	sf	—	22.40	22.40
Install fireplace mantel beam					
install	1C@.539	lf	—	36.40	36.40
Install fireplace mantel					
install	1C@.653	lf	—	44.10	44.10
Install coffered ceiling					
economy grade	1C@.248	sf	—	16.80	16.80
custom deluxe grade	1C@.567	sf	—	38.30	38.30
Install niche					
with casing and shelf bracket	1C@1.31	ea	—	88.60	88.60
Install gingerbread					
running trim	1C@.160	lf	—	10.80	10.80
post bracket	1C@.348	ea	—	23.50	23.50
corbel	1C@.359	ea	—	24.30	24.30
door or window header	1C@.232	lf	—	15.70	15.70
post drop	1C@.322	ea	—	21.80	21.80
spandrel	1C@.237	lf	—	16.00	16.00
window cornice	1C@.242	lf	—	16.40	16.40
gable ornament	1C@.582	ea	—	39.30	39.30
gable finial	1C@.560	ea	—	37.90	37.90
Install porch post					
install	1C@.135	lf	—	9.13	9.13
Install carved wood onlay					
install	1C@.519	ea	—	35.10	35.10
Install finish board trim					
install	1C@.049	lf	—	3.31	3.31
Install 1/4" hardwood plywood					
install	1C@.028	sf	—	1.89	1.89

Fireplaces

Fireplace. Includes concrete reinforced hearth, ash drop, fire brick, damper, throat, smoke shelf and smoke chamber. Does not include foundation, flue or chimney, face, or finish hearth.

	Craft@Hrs	Unit	Material	Labor	Total
Open front fireplace					
30" wide by 16" deep by 29" high	1M@16.1	ea	629.00	1,140.00	1,769.00
36" wide by 16" deep by 29" high	1M@16.9	ea	756.00	1,200.00	1,956.00
40" wide by 16" deep by 29" high	1M@17.5	ea	841.00	1,240.00	2,081.00
48" wide by 18" deep by 32" high	1M@18.5	ea	1,010.00	1,320.00	2,330.00
open front					
Open front and one side fireplace					
32" wide by 16" deep by 26" high	1M@16.7	ea	592.00	1,190.00	1,782.00
40" wide by 16" deep by 29" high	1M@17.5	ea	822.00	1,240.00	2,062.00
48" wide by 20" deep by 29" high	1M@18.9	ea	986.00	1,340.00	2,326.00
open front and side					
Open two faces fireplace					
32" wide by 28" deep by 29" high	1M@18.5	ea	579.00	1,320.00	1,899.00
36" wide by 28" deep by 29" high	1M@18.9	ea	652.00	1,340.00	1,992.00
40" wide by 28" deep by 29" high	1M@19.2	ea	724.00	1,370.00	2,094.00
"see through"					
Open two faces and one side fireplace					
36" wide by 32" deep by 27" high	1M@19.2	ea	707.00	1,370.00	2,077.00
36" wide by 36" deep by 27" high	1M@20.0	ea	797.00	1,420.00	2,217.00
44" wide by 40" deep by 27" high	1M@20.4	ea	1,060.00	1,450.00	2,510.00
open two faces and side					
Remove fireplace	1D@3.85	ea	—	182.00	182.00

Prefabricated fireplace. Zero clearance box. Does not include flue. Actual sizes and style available vary from manufacturer to manufacturer. Choose the size and style that most closely matches.

	Craft@Hrs	Unit	Material	Labor	Total
Open front prefabricated fireplace					
replace, 30" wide by 16" deep by 29" high	1M@4.55	ea	714.00	324.00	1,038.00
replace, 36" wide by 16" deep by 29" high	1M@4.55	ea	864.00	324.00	1,188.00
replace, 40" wide by 16" deep by 29" high	1M@4.55	ea	1,430.00	324.00	1,754.00
replace, 48" wide by 18" deep by 32" high	1M@4.55	ea	1,740.00	324.00	2,064.00
open front					
Open front convection prefabricated fireplace					
replace, 30" wide by 16" deep by 29" high	1M@4.76	ea	819.00	338.00	1,157.00
replace, 36" wide by 16" deep by 29" high	1M@4.76	ea	994.00	338.00	1,332.00
replace, 40" wide by 16" deep by 29" high	1M@4.76	ea	1,100.00	338.00	1,438.00
replace, 48" wide by 18" deep by 32" high	1M@4.76	ea	1,320.00	338.00	1,658.00
Open front forced air prefabricated fireplace					
replace, 30" wide by 16" deep by 29" high	1M@5.26	ea	891.00	374.00	1,265.00
replace, 36" wide by 16" deep by 29" high	1M@5.26	ea	1,080.00	374.00	1,454.00
replace, 40" wide by 16" deep by 29" high	1M@5.26	ea	1,200.00	374.00	1,574.00
replace, 48" wide by 18" deep by 32" high	1M@5.26	ea	1,420.00	374.00	1,794.00
Open front and one side prefabricated fireplace					
replace, 32" wide by 16" deep by 26" high	1M@4.55	ea	1,200.00	324.00	1,524.00
replace, 40" wide by 16" deep by 29" high	1M@4.76	ea	1,420.00	338.00	1,758.00
replace, 48" wide by 20" deep by 29" high	1M@5.26	ea	1,720.00	374.00	2,094.00
open front and side					

	Craft@Hrs	Unit	Material	Labor	Total
Open two faces (see through) prefabricated fireplace					
replace, 32" wide by 28" deep by 29" high	1M@4.55	ea	1,430.00	324.00	1,754.00
replace, 36" wide by 28" deep by 29" high	1M@4.76	ea	1,740.00	338.00	2,078.00
replace, 40" wide by 28" deep by 29" high	1M@5.26	ea	2,100.00	374.00	2,474.00
"see through"					
Open two faces and one side prefabricated fireplace					
replace, 36" wide by 32" deep by 27" high	1M@4.55	ea	1,720.00	324.00	2,044.00
replace, 36" wide by 36" deep by 27" high	1M@4.76	ea	2,060.00	338.00	2,398.00
replace, 44" wide by 40" deep by 27" high	1M@5.26	ea	2,470.00	374.00	2,844.00
open two faces and side					
Remove, prefabricated fireplace	1D@2.44	ea	—	115.00	115.00

Fireplace form. Does not include face, hearth, flue or chimney. Actual sizes and style available vary from manufacturer to manufacturer. Choose the size and style that most closely matches.

	Craft@Hrs	Unit	Material	Labor	Total
Open front fireplace form					
replace, 30" wide by 16" deep by 29" high	1M@3.03	ea	819.00	215.00	1,034.00
replace, 36" wide by 16" deep by 29" high	1M@3.03	ea	985.00	215.00	1,200.00
replace, 40" wide by 16" deep by 29" high	1M@3.03	ea	1,090.00	215.00	1,305.00
replace, 48" wide by 18" deep by 32" high	1M@3.03	ea	1,330.00	215.00	1,545.00
open front					
Open front and one side fireplace form					
replace, 32" wide by 16" deep by 26" high	1M@3.03	ea	1,520.00	215.00	1,735.00
replace, 40" wide by 16" deep by 29" high	1M@3.03	ea	1,820.00	215.00	2,035.00
replace, 48" wide by 20" deep by 29" high	1M@3.03	ea	3,360.00	215.00	3,575.00
open front and side					
Open two faces fireplace form					
replace, 32" wide by 28" deep by 29" high	1M@3.03	ea	1,840.00	215.00	2,055.00
replace, 36" wide by 28" deep by 29" high	1M@3.03	ea	2,080.00	215.00	2,295.00
replace, 40" wide by 28" deep by 29" high	1M@3.03	ea	2,310.00	215.00	2,525.00
"see through"					
Open two faces and one side fireplace form					
replace, 36" wide by 32" deep by 27" high	1M@3.03	ea	2,280.00	215.00	2,495.00
replace, 36" wide by 36" deep by 27" high	1M@3.03	ea	2,560.00	215.00	2,775.00
replace, 44" wide by 40" deep by 27" high	1M@3.03	ea	2,850.00	215.00	3,065.00
open two faces and side					
Remove, fireplace form	1D@2.78	ea	—	131.00	131.00

Fireplace furnace.

	Craft@Hrs	Unit	Material	Labor	Total
replace, typical	1M@34.5	ea	7,550.00	2,450.00	10,000.00
remove	1D@3.85	ea	—	182.00	182.00

Fire brick. Installation of fire brick with fire clay only, no surrounding masonry.

	Craft@Hrs	Unit	Material	Labor	Total
replace, complete fireplace	1M@3.03	ea	200.00	215.00	415.00
replace, concrete simulated firebrick (complete fireplace)	1M@1.85	ea	151.00	132.00	283.00
remove, fire brick	1D@.952	ea	—	44.90	44.90
remove, concrete simulated fire brick	1D@.529	ea	—	25.00	25.00

	Craft@Hrs	Unit	Material	Labor	Total
Fireplace chimney. All brick types. Does not include foundation.					
replace, 16" by 16" with one 8" x 8" flue	5M@2.21	lf	53.20	151.00	204.20
replace, 16" by 24" with two 8" x 8" flues	5M@3.09	lf	70.20	211.00	281.20
replace, 16" by 20" with one 8" x 12" flue	5M@2.24	lf	63.20	153.00	216.20
replace, 20" by 24" with two 8" x 12" flues	5M@3.15	lf	81.10	215.00	296.10
replace, 20" by 20" with one 12" x 12" flue	5M@2.27	lf	67.00	155.00	222.00
replace, 20" by 32" with two 12" x 12" flues	5M@3.24	lf	92.80	221.00	313.80
replace, 20" by 24" with one 12" x 16" flue	5M@2.31	lf	70.90	158.00	228.90
replace, 20" by 40" with two 12" x 16" flues	5M@3.30	lf	119.00	225.00	344.00
replace, 24" by 24" with one 16" x 16" flue	5M@2.35	lf	74.90	160.00	234.90
replace, 24" by 40" with two 16" x 16" flues	5M@3.36	lf	132.00	229.00	361.00
remove, one flue	1D@.346	lf	—	16.30	16.30
remove, two flues	1D@.472	lf	—	22.30	22.30
add for bend in one flue	5M@3.33	ea	35.30	227.00	262.30
add for bend in two flues	5M@4.05	ea	39.70	276.00	315.70
add 14% for stone facing on chimney					
Reline chimney using grout and inflatable tube method. With a technique designed and licensed by Ahrens Chimney Technique. Replace only.					
reline	1M@.604	lf	9.28	42.90	52.18
add per bend in chimney	1M@3.08	ea	27.40	219.00	246.40
minimum charge	1M@4.00	ea	55.20	284.00	339.20
2" concrete chimney cap. Poured in place.					
replace, single flue	1M@1.00	ea	24.20	71.10	95.30
replace, double flue	1M@1.49	ea	38.60	106.00	144.60
remove, single flue	1D@.303	ea	—	14.30	14.30
remove, double flue	1D@.362	ea	—	17.10	17.10
Cast stone chimney cap. Manufactured cast stone cap with decorative patterns. Grades have progressively more complex molding patterns on edge.					
replace, standard grade	1M@1.28	ea	232.00	91.00	323.00
replace, high grade	1M@1.28	ea	276.00	91.00	367.00
replace, deluxe grade	1M@1.28	ea	367.00	91.00	458.00
remove	1D@.362	ea	—	17.10	17.10
Galvanized steel chimney cap. For use on metal or masonry flues, with spark arrestor.					
replace, cap	2M@.298	ea	184.00	19.40	203.40
remove	1D@.177	ea	—	8.35	8.35
remove for work, then reinstall	2M@.498	ea	—	32.50	32.50
Chimney bird screen. Replace only.					
for single flue	2M@.437	ea	71.40	28.50	99.90
for double flue	2M@.476	ea	87.40	31.00	118.40

	Craft@Hrs	Unit	Material	Labor	Total
Chimney pot. Grades have progressively more ornate designs on pot. Deluxe and above grades also have patterns around top edge.					
replace, standard grade	1M@1.18	ea	297.00	83.90	380.90
replace, high grade	1M@1.18	ea	435.00	83.90	518.90
replace, deluxe grade	1M@1.18	ea	547.00	83.90	630.90
replace, custom grade	1M@1.18	ea	673.00	83.90	756.90
replace, custom deluxe grade	1M@1.18	ea	859.00	83.90	942.90
remove	1D@.469	ea	—	22.10	22.10
remove for work, then reinstall	1M@2.04	ea	—	145.00	145.00

standard

high

custom deluxe

	Craft@Hrs	Unit	Material	Labor	Total
Fireplace grate.					
replace, 22" x 16"	2M@.119	ea	72.70	7.76	80.46
replace, 28" x 16"	2M@.119	ea	92.80	7.76	100.56
replace, 32" x 16"	2M@.119	ea	106.00	7.76	113.76
replace, 38" x 16"	2M@.119	ea	121.00	7.76	128.76
replace, 38" x 20"	2M@.119	ea	127.00	7.76	134.76
replace, 32" x 28"	2M@.119	ea	138.00	7.76	145.76
replace, 36" x 28"	2M@.119	ea	145.00	7.76	152.76
replace, 40" x 28"	2M@.119	ea	153.00	7.76	160.76
replace, 36" x 32"	2M@.119	ea	161.00	7.76	168.76
replace, 36" x 36"	2M@.119	ea	179.00	7.76	186.76
replace, 44" x 40"	2M@.119	ea	245.00	7.76	252.76
remove	1D@.030	ea	—	1.42	1.42
remove for work, then reinstall	2M@.147	ea	—	9.58	9.58
Fireplace screen.					
replace, standard grade	2M@.794	ea	161.00	51.80	212.80
replace, high grade	2M@.794	ea	247.00	51.80	298.80
replace, deluxe grade	2M@.794	ea	336.00	51.80	387.80
remove	1D@.185	ea	—	8.73	8.73
remove for work, then reinstall	2M@1.19	ea	—	77.60	77.60

	Craft@Hrs	Unit	Material	Labor	Total
Fireplace door. Standard grade has black face, higher grades have brass face.					
replace, standard grade	2M@1.00	ea	274.00	65.20	339.20
replace, high grade	2M@1.00	ea	418.00	65.20	483.20
replace, deluxe grade	2M@1.00	ea	544.00	65.20	609.20
replace, custom grade	2M@1.00	ea	724.00	65.20	789.20
remove	1D@.253	ea	—	11.90	11.90
remove for work, then reinstall	2M@1.56	ea	—	102.00	102.00
Fireplace clean-out. Cast iron door and jamb.					
replace, 8" x 8"	1M@.625	ea	30.40	44.40	74.80
replace, 12" x 12"	1M@.625	ea	63.70	44.40	108.10
replace, 18" x 24"	1M@.625	ea	190.00	44.40	234.40
remove	1D@.069	ea	—	3.26	3.26
salvage, then reinstall	1M@.971	ea	—	69.00	69.00
Damper. Replace only.					
Rotary controlled damper					
for 30" wide fireplace	1M@1.11	ea	127.00	78.90	205.90
for 36" wide fireplace	1M@1.11	ea	136.00	78.90	214.90
for 40" wide fireplace	1M@1.11	ea	171.00	78.90	249.90
for 48" wide fireplace	1M@1.11	ea	200.00	78.90	278.90
Poker controlled damper					
for 30" wide fireplace	1M@.901	ea	100.00	64.10	164.10
for 36" wide fireplace	1M@.901	ea	108.00	64.10	172.10
for 40" wide fireplace	1M@.901	ea	131.00	64.10	195.10
for 48" wide fireplace	1M@.901	ea	158.00	64.10	222.10
Chimney pipe. Stainless steel.					
Double wall chimney pipe					
replace, 8" diameter	2M@.578	lf	43.40	37.70	81.10
replace, 10" diameter	2M@.578	lf	62.60	37.70	100.30
replace, 12" diameter	2M@.578	lf	82.80	37.70	120.50
replace, 14" diameter	2M@.578	lf	103.00	37.70	140.70
Triple wall chimney pipe					
replace, 8" diameter	2M@.578	lf	55.20	37.70	92.90
replace, 10" diameter	2M@.578	lf	79.90	37.70	117.60
replace, 12" diameter	2M@.578	lf	106.00	37.70	143.70
replace, 14" diameter	2M@.578	lf	131.00	37.70	168.70
Remove	1D@.021	lf	—	.99	.99
Additional chimney pipe costs					
add through interior ceiling	2M@1.11	ea	75.60	72.40	148.00
add through roof	2M@2.94	ea	210.00	192.00	402.00
remove for work, then reinstall	2M@.884	lf	—	57.60	57.60

	Craft@Hrs	Unit	Material	Labor	Total
Gas log lighter. Log lighter, valve, key and up to 24 lf copper gas piping.					
replace, typical	2M@1.18	ea	141.00	76.90	217.90
remove	1D@.227	ea	—	10.70	10.70
remove for work, then reinstall	2M@1.96	ea	—	128.00	128.00
Gas fireplace kit.					
replace, typical	2M@1.25	ea	872.00	81.50	953.50
remove	1D@.251	ea	—	11.80	11.80
remove for work, then reinstall	2M@2.13	ea	—	139.00	139.00

Marble fireplace face. Economy grade: Marble tiles laid around inside of mantel. Standard grade: Single or large pieces of marble up to 5' tall and 4' wide, polished or natural finish. High grade: Same as standard grade, but with simple designs around the opening and at corner (cove or similar). Pieces may be up to 8' tall and 6' wide. Deluxe grade: Same as high grade, but with ornate designs around opening and at corner (crown or similar).

	Craft@Hrs	Unit	Material	Labor	Total
replace, economy grade	1M@3.57	ea	142.00	254.00	396.00
remove, economy	1D@1.66	ea	—	78.40	78.40
replace, standard grade	1M@7.63	ea	435.00	542.00	977.00
replace, high grade	1M@7.81	ea	714.00	555.00	1,269.00
replace, deluxe grade	1M@8.00	ea	1,060.00	569.00	1,629.00
remove	1D@3.03	ea	—	143.00	143.00

Brick fireplace face. Economy grade: Brick face in stacked or running bond, up to 5' tall and 5' wide. Standard grade: Brick face in specialty bond (see Masonry), up to 5' tall and 5' wide. May include built-in shelf brackets or small patterns in face. Or, may be same as economy, except 8' tall and 6' wide. High grade: Same as standard grade, but with arch over opening, or may be up to 11' high and 7' wide. Deluxe grade: Brick face in specialty bond, such as herringbone or basket weave mixed with other patterns (see Masonry for brick bonds), 5' tall and 5' wide. May have arched openings or curving sides. May be same as high grade except up to 8' tall and 6' wide; same as standard except up to 11' tall and 7' wide; or same as economy except up to 15' tall and 10' wide.

	Craft@Hrs	Unit	Material	Labor	Total
replace, economy grade	1M@16.1	ea	404.00	1,140.00	1,544.00
replace, standard grade	1M@20.4	ea	450.00	1,450.00	1,900.00
replace, high grade	1M@25.0	ea	546.00	1,780.00	2,326.00
replace, deluxe grade	1M@32.3	ea	661.00	2,300.00	2,961.00
remove	1D@3.23	ea	—	152.00	152.00

Rubble stone fireplace face. Economy grade: Flagstone or equivalent stone, up to 5' tall and 5' wide. Standard grade: Sandstone, limestone or similar, up to 5' tall and 5' wide. Or, same as economy, except up to 8' tall and 6' wide. High grade: Higher priced stones or stones with high-priced finishes, up to 5' tall and 5' wide. Or, same as standard, except up to 8' tall and 6' wide; or same as economy except up to 10' tall and 7' wide. Deluxe grade: Specialty stones or stones with high-priced finishes, up to 5' tall and 5' wide, with arch over opening or other specialty components. May be same as high, except up to 8' tall and 6' wide; same as standard, except 11' tall and 7' wide; or same as economy, except up to 14' tall and 8' wide.

	Craft@Hrs	Unit	Material	Labor	Total
replace, economy grade	1M@17.5	ea	689.00	1,240.00	1,929.00
replace, standard grade	1M@22.3	ea	765.00	1,590.00	2,355.00
replace, high grade	1M@27.1	ea	926.00	1,930.00	2,856.00
replace, deluxe grade	1M@34.5	ea	1,130.00	2,450.00	3,580.00
remove	1D@3.45	ea	—	163.00	163.00

Ashlar stone fireplace face. Economy grade: Lowest priced stone in area, up to 5' tall and 5' wide. Standard grade: Sandstone, limestone or similar stone, up to 5' tall and 5' wide. Or, same as economy, except up to 8' tall and 6' wide. High grade: Specialty high-priced stones or stones with high-priced finishes, up to 5' tall and 5' wide. Or, same as standard, except up to 8' tall and 6' wide; or same as economy except, up to 10' tall and 7' wide. Deluxe grade: Specialty high-priced stones or stones with high-priced finishes, up to 5' tall and 5' wide, with arch over opening or other specialty components or designs. May be same as high, except up to 8' tall and 6' wide; same as standard, except 11' tall and 7' wide; or same as economy, except up to 14' tall and 8' wide.

	Craft@Hrs	Unit	Material	Labor	Total
replace, economy grade	1M@17.2	ea	799.00	1,220.00	2,019.00
replace, standard grade	1M@21.7	ea	890.00	1,540.00	2,430.00
replace, high grade	1M@26.3	ea	1,080.00	1,870.00	2,950.00
replace, deluxe grade	1M@33.3	ea	1,300.00	2,370.00	3,670.00
remove	1D@3.45	ea	—	163.00	163.00

Tile fireplace face. Economy grade: Standard priced tiles laid around inside of mantel. Standard grade: High priced tiles. High grade: Specialty antique tiles, including designer hand-painted designs and raised pattern designs. May include tiles in fireback. Deluxe grade: Same as high grade with specialty trim pieces and patterns that match antique Victorian styles.

	Craft@Hrs	Unit	Material	Labor	Total
replace, economy grade	1M@3.70	ea	111.00	263.00	374.00
remove, economy	1D@1.82	ea	—	85.90	85.90
replace, standard grade	1M@7.94	ea	231.00	565.00	796.00
replace, high grade	1M@8.06	ea	300.00	573.00	873.00
replace, deluxe grade	1M@8.26	ea	639.00	587.00	1,226.00
remove	1D@2.94	ea	—	139.00	139.00

Fireplace hearth. Economy grade is flat hearth. All other grades are raised hearths. See fireplace face items for more information about quality.

	Craft@Hrs	Unit	Material	Labor	Total
Marble fireplace hearth					
replace, economy grade	1M@2.16	ea	218.00	154.00	372.00
remove, economy	1D@.775	ea	—	36.60	36.60
replace, standard grade	1M@4.60	ea	355.00	327.00	682.00
replace, high grade	1M@4.73	ea	572.00	336.00	908.00
replace, deluxe grade	1M@4.83	ea	855.00	343.00	1,198.00
remove	1D@1.45	ea	—	68.40	68.40
Brick fireplace hearth					
replace, economy grade	1M@7.04	ea	269.00	501.00	770.00
replace, standard grade	1M@8.85	ea	298.00	629.00	927.00
replace, high grade	1M@9.44	ea	364.00	671.00	1,035.00
replace, deluxe grade	1M@11.1	ea	440.00	789.00	1,229.00
remove	1D@1.52	ea	—	71.70	71.70
Rubble stone fireplace hearth					
replace, economy grade	1M@7.63	ea	458.00	542.00	1,000.00
replace, standard grade	1M@9.63	ea	509.00	685.00	1,194.00
replace, high grade	1M@10.2	ea	618.00	725.00	1,343.00
replace, deluxe grade	1M@12.0	ea	749.00	853.00	1,602.00
remove	1D@1.56	ea	—	73.60	73.60
Ashlar stone fireplace hearth					
replace, economy grade	1M@7.46	ea	532.00	530.00	1,062.00
replace, standard grade	1M@9.44	ea	594.00	671.00	1,265.00
replace, high grade	1M@10.0	ea	720.00	711.00	1,431.00
replace, deluxe grade	1M@11.8	ea	873.00	839.00	1,712.00
remove	1D@1.56	ea	—	73.60	73.60

	Craft@Hrs	Unit	Material	Labor	Total
Tile fireplace hearth					
replace, economy grade	1M@2.24	ea	170.00	159.00	329.00
remove, economy	1D@.875	ea	—	41.30	41.30
replace, standard grade	1M@4.77	ea	231.00	339.00	570.00
replace, high grade	1M@4.88	ea	300.00	347.00	647.00
replace, deluxe grade	1M@5.00	ea	639.00	356.00	995.00
remove	1D@1.43	ea	—	67.50	67.50

Time & Material Charts (selected items)

Fireplace Materials (Also see material prices with the line items above.)

	Craft@Hrs	Unit	Material	Labor	Total
Fireplace					
30" wide by 16" deep by 29" high, open front	—	ea	629.00	—	629.00
32" wide by 16" deep by 26" high, open front and one side	—	ea	592.00	—	592.00
32" wide by 28" deep by 29" high, open two sides	—	ea	579.00	—	579.00
36" wide by 32" deep by 27" high, open two faces and one side	—	ea	707.00	—	707.00
Prefabricated fireplace					
30" wide by 16" deep by 29" high, open front	—	ea	714.00	—	714.00
32" wide by 16" deep by 26" high, open front and one side	—	ea	1,200.00	—	1,200.00
32" wide by 28" deep by 29" high, open two faces (see through)	—	ea	1,430.00	—	1,430.00
36" wide by 32" deep by 27" high, open two faces and one side	—	ea	1,720.00	—	1,720.00
Fireplace form					
30" wide by 16" deep by 29" high, open front	—	ea	819.00	—	819.00
32" wide by 16" deep by 26" high, open front and one side	—	ea	1,520.00	—	1,520.00
32" wide by 28" deep by 29" high, open two faces (see through)	—	ea	1,840.00	—	1,840.00
36" wide by 32" deep by 27" high, open two faces and one side	—	ea	2,280.00	—	2,280.00
Fireplace chimney					
16" by 16" with one 8" x 8" flue	—	lf	53.20	—	53.20
16" by 24" with two 8" x 8" flues	—	lf	70.20	—	70.20
20" by 32" with two 12" x 12" flues	—	lf	92.80	—	92.80
20" by 40" with two 12" x 16" flues	—	lf	119.00	—	119.00
24" by 40" with two 16" x 16" flues	—	lf	132.00	—	132.00

	Craft@Hrs	Unit	Material	Labor	Total
Marble fireplace facing					
economy grade	—	ea	142.00	—	142.00
deluxe grade	—	ea	1,060.00	—	1,060.00
Brick fireplace facing					
economy grade	—	ea	404.00	—	404.00
deluxe grade	—	ea	661.00	—	661.00
Rubble stone fireplace facing					
economy grade	—	ea	689.00	—	689.00
deluxe grade	—	ea	1,130.00	—	1,130.00
Ashlar stone fireplace facing					
economy grade	—	ea	799.00	—	799.00
deluxe grade	—	ea	1,300.00	—	1,300.00
Tile fireplace facing					
economy grade	—	ea	111.00	—	111.00
deluxe grade	—	ea	639.00	—	639.00
Marble fireplace hearth					
economy grade	—	ea	218.00	—	218.00
deluxe grade	—	ea	855.00	—	855.00
Brick fireplace hearth					
economy grade	—	ea	269.00	—	269.00
deluxe grade	—	ea	440.00	—	440.00
Rubble stone fireplace hearth					
economy grade	—	ea	458.00	—	458.00
deluxe grade	—	ea	749.00	—	749.00
Ashlar stone fireplace hearth					
economy grade	—	ea	532.00	—	532.00
deluxe grade	—	ea	873.00	—	873.00
Tile fireplace hearth					
economy grade	—	ea	170.00	—	170.00
deluxe grade	—	ea	639.00	—	639.00

Fireplace Labor

Laborer	base wage	paid leave	true wage	taxes & ins.	total
Mason	$39.70	3.10	$42.80	28.30	$71.10
Mason's helper	$36.10	2.82	$38.92	26.28	$65.20
Demolition laborer	$25.90	2.02	$27.92	19.28	$47.20

Paid leave is calculated based on two weeks paid vacation, one week sick leave, and seven paid holidays. Employer's matching portion of **FICA** is 7.65 percent. **FUTA** (Federal Unemployment) is .8 percent. **Worker's compensation** for the masonry (fireplaces) trade was calculated using a national average of 19.41 percent. **Unemployment insurance** was calculated using a national average of 8 percent. **Health insurance** was calculated based on a projected national average for 2020 of $1,280 per employee (and family when applicable) per month. Employer pays 80 percent for a per month cost of $1,024 per employee. **Retirement** is based on a 401(k) retirement program with employer matching of 50 percent. Employee contributions to the 401(k) plan are an average of 6 percent of the true wage. **Liability insurance** is based on a national average of 12.0 percent.

	Craft@Hrs	Unit	Material	Labor	Total
Fireplace Labor Productivity					
Demolition of fireplaces					
remove fireplace	1D@3.85	ea	—	182.00	182.00
remove prefabricated fireplace	1D@2.44	ea	—	115.00	115.00
remove fire brick	1D@.952	ea	—	44.90	44.90
remove concrete simulated firebrick	1D@.529	ea	—	25.00	25.00
remove chimney with single flue	1D@.346	lf	—	16.30	16.30
remove chimney with double flue	1D@.472	lf	—	22.30	22.30
remove 2" concrete single flue cap	1D@.303	ea	—	14.30	14.30
remove 2" concrete double flue cap	1D@.362	ea	—	17.10	17.10
remove galvanized steel cap	1D@.177	ea	—	8.35	8.35
remove chimney pot	1D@.469	ea	—	22.10	22.10
remove fireplace grate	1D@.030	ea	—	1.42	1.42
remove fireplace screen	1D@.185	ea	—	8.73	8.73
remove fireplace door	1D@.253	ea	—	11.90	11.90
remove fireplace clean-out	1D@.069	ea	—	3.26	3.26
remove chimney pipe	1D@.021	lf	—	.99	.99
remove gas log lighter	1D@.227	ea	—	10.70	10.70
remove gas fireplace kit	1D@.251	ea	—	11.80	11.80
remove marble fireplace face	1D@3.03	ea	—	143.00	143.00
remove brick fireplace face	1D@3.23	ea	—	152.00	152.00
remove stone fireplace face	1D@3.45	ea	—	163.00	163.00
remove tile fireplace face	1D@2.94	ea	—	139.00	139.00
remove marble fireplace hearth	1D@1.45	ea	—	68.40	68.40
remove brick fireplace hearth	1D@1.52	ea	—	71.70	71.70
remove stone fireplace hearth	1D@1.56	ea	—	73.60	73.60
remove tile fireplace hearth	1D@1.43	ea	—	67.50	67.50

Fireplace crew

build fireplace	mason	$71.10
build fireplace	mason's helper	$65.20
build fireplace	mason & helper	$68.20

	Craft@Hrs	Unit	Material	Labor	Total
Build open front fireplace					
30" wide by 16" by 29"	1M@16.1	ea	—	1,140.00	1,140.00
36" wide by 16" by 29"	1M@16.9	ea	—	1,200.00	1,200.00
40" wide by 16" by 29"	1M@17.5	ea	—	1,240.00	1,240.00
48" wide by 18" by 32"	1M@18.5	ea	—	1,320.00	1,320.00
Build open front and one side fireplace					
32" wide by 16" by 26"	1M@16.7	ea	—	1,190.00	1,190.00
40" wide by 16" by 29"	1M@17.5	ea	—	1,240.00	1,240.00
48" wide by 20" by 29"	1M@18.9	ea	—	1,340.00	1,340.00
Build open two faces (see through) fireplace					
32" wide by 28" by 29"	1M@18.5	ea	—	1,320.00	1,320.00
36" wide by 28" by 29"	1M@18.9	ea	—	1,340.00	1,340.00
40" wide by 28" by 29"	1M@19.2	ea	—	1,370.00	1,370.00

	Craft@Hrs	Unit	Material	Labor	Total
Build open two faces and one side fireplace					
36" wide by 32" by 27"	1M@19.2	ea	—	1,370.00	1,370.00
36" wide by 36" by 27"	1M@20.0	ea	—	1,420.00	1,420.00
44" wide by 40" by 27"	1M@20.4	ea	—	1,450.00	1,450.00
Install prefabricated fireplace					
radiation	1M@4.55	ea	—	324.00	324.00
convection	1M@4.76	ea	—	338.00	338.00
forced-air	1M@5.26	ea	—	374.00	374.00
Install fireplace form					
form	1M@3.03	ea	—	215.00	215.00
Install fireplace furnace					
typical	1M@34.5	ea	—	2,450.00	2,450.00
Install fire brick					
complete fireplace	1M@3.03	ea	—	215.00	215.00
concrete simulated firebrick (complete fireplace)	1M@1.85	ea	—	132.00	132.00
Build brick chimney					
16" by 16" with one 8" x 8" flue	5M@2.21	lf	—	151.00	151.00
16" by 24" with two 8" x 8" flues	5M@3.09	lf	—	211.00	211.00
16" by 20" with one 8" x 12" flue	5M@2.24	lf	—	153.00	153.00
20" by 24" with two 8" x 12" flues	5M@3.15	lf	—	215.00	215.00
20" by 20" with one 12" x 12" flue	5M@2.27	lf	—	155.00	155.00
20" by 32" with two 12" x 12" flues	5M@3.24	lf	—	221.00	221.00
20" by 24" with one 12" x 16" flue	5M@2.31	lf	—	158.00	158.00
20" by 40" with two 12" x 16" flues	5M@3.30	lf	—	225.00	225.00
24" by 24" with one 16" x 16" flue	5M@2.35	lf	—	160.00	160.00
24" by 40" with two 16" x 16" flues	5M@3.36	lf	—	229.00	229.00
add for bend in one flue	5M@3.33	ea	—	227.00	227.00
add for bend in two flues	5M@4.05	ea	—	276.00	276.00
Pour 2" thick concrete chimney cap					
single flue	1M@1.00	ea	—	71.10	71.10
double flue	1M@1.49	ea	—	106.00	106.00
Install cast stone chimney cap					
typical	1M@1.28	ea	—	91.00	91.00
Install galvanized steel chimney cap					
with spark arrestor	2M@.298	ea	—	19.40	19.40
Install chimney pot					
typical	1M@1.18	ea	—	83.90	83.90
Install chimney bird screen					
for single flue	2M@.437	ea	—	28.50	28.50
for double flue	2M@.476	ea	—	31.00	31.00
Install fireplace grate					
typical	2M@.119	ea	—	7.76	7.76
Install fireplace screen					
typical	2M@.794	ea	—	51.80	51.80

	Craft@Hrs	Unit	Material	Labor	Total
Install fireplace door					
typical	2M@1.00	ea	—	65.20	65.20
Install fireplace clean-out					
typical	1M@.625	ea	—	44.40	44.40
Install fireplace damper					
rotary control	1M@1.11	ea	—	78.90	78.90
poker control	1M@.901	ea	—	64.10	64.10
Install chimney pipe					
typical	2M@.578	lf	—	37.70	37.70
add through interior ceiling	2M@1.11	ea	—	72.40	72.40
add through roof	2M@2.94	ea	—	192.00	192.00
Install gas log lighter					
typical	2M@1.18	ea	—	76.90	76.90
Install gas fireplace kit					
typical	2M@1.25	ea	—	81.50	81.50
Install fireplace face					
marble economy grade	1M@3.57	ea	—	254.00	254.00
marble deluxe grade	1M@8.00	ea	—	569.00	569.00
brick economy grade	1M@16.1	ea	—	1,140.00	1,140.00
brick deluxe grade	1M@32.3	ea	—	2,300.00	2,300.00
rubble stone economy grade	1M@17.5	ea	—	1,240.00	1,240.00
rubble stone deluxe grade	1M@34.5	ea	—	2,450.00	2,450.00
ashlar stone economy grade	1M@17.2	ea	—	1,220.00	1,220.00
ashlar stone deluxe grade	1M@33.3	ea	—	2,370.00	2,370.00
ashlar stone economy grade	1M@3.70	ea	—	263.00	263.00
ashlar stone deluxe grade	1M@8.26	ea	—	587.00	587.00
Install fireplace hearth					
marble economy grade	1M@2.16	ea	—	154.00	154.00
marble deluxe grade	1M@4.83	ea	—	343.00	343.00
brick economy grade	1M@7.04	ea	—	501.00	501.00
brick deluxe grade	1M@11.1	ea	—	789.00	789.00
rubble stone economy grade	1M@7.63	ea	—	542.00	542.00
rubble stone deluxe grade	1M@12.0	ea	—	853.00	853.00
ashlar stone economy grade	1M@7.46	ea	—	530.00	530.00
ashlar stone deluxe grade	1M@11.8	ea	—	839.00	839.00
tile economy grade	1M@2.24	ea	—	159.00	159.00
tile deluxe grade	1M@5.00	ea	—	356.00	356.00

	Craft@Hrs	Unit	Material	Labor	Total

Flooring

Carpet. Includes carpet, tackless strips, seaming tape, and installation. Includes 12% waste. Does not include pad. Carpet is available in virtually every price imaginable. These prices should be considered allowances for typical grades found in residential and light commercial structures. Lower grades have thin pile that can be pulled aside to expose the backing. Top grades include wool carpets, pattern carpets, and 52-ounce cut pile carpets.

	Craft@Hrs	Unit	Material	Labor	Total
replace, economy grade	5I@.138	sy	26.30	7.82	34.12
replace, standard grade	5I@.138	sy	31.90	7.82	39.72
replace, high grade	5I@.138	sy	42.30	7.82	50.12
replace, deluxe grade	5I@.138	sy	50.70	7.82	58.52
replace, custom grade	5I@.138	sy	61.30	7.82	69.12
remove	1D@.062	sy	—	2.93	2.93
remove for work, then re-lay	5I@.211	sy	—	12.00	12.00

Glue-down carpet. Includes carpet, glue, seaming material, and installation. Includes 12% waste. Pad may be integrated into carpet. These prices should be considered allowances for typical grades found in residential and light commercial structures.

	Craft@Hrs	Unit	Material	Labor	Total
replace, economy grade	5I@.167	sy	19.80	9.47	29.27
replace, standard grade	5I@.167	sy	23.20	9.47	32.67
replace, high grade	5I@.167	sy	30.50	9.47	39.97
replace, deluxe grade	5I@.167	sy	36.90	9.47	46.37
remove	1D@.155	sy	—	7.32	7.32

Indoor-outdoor carpet. Includes carpet, glue, seaming material, and installation. Includes 12% waste. Pad may be integrated into carpet. These prices should be considered allowances for typical grades found in residential and light commercial structures. Higher grades are similar to commercial quality glue-down carpet, lowest grade is plastic imitation grass.

	Craft@Hrs	Unit	Material	Labor	Total
replace, economy grade	5I@.167	sy	14.60	9.47	24.07
replace, standard grade	5I@.167	sy	20.80	9.47	30.27
replace, high grade	5I@.167	sy	30.50	9.47	39.97
replace, deluxe grade	5I@.167	sy	46.50	9.47	55.97
remove	1D@.153	sy	—	7.22	7.22

Wool carpet. Includes carpet, tackless strips, seaming tape, and installation. Includes 12% waste. Does not include pad. These prices should be considered allowances for typical grades of wool carpet found in residential and light commercial structures. Lower grades have thin pile that can be pulled aside to expose the backing. Top grades include patterns and thick piles. Many synthetic fiber carpets mimic the look of wool and can be hard to differentiate. If on a fire job, wool carpet has a noticeable "burnt-hair" smell.

	Craft@Hrs	Unit	Material	Labor	Total
replace, standard grade	5I@.139	sy	39.40	7.88	47.28
replace, high grade	5I@.139	sy	49.40	7.88	57.28
replace, deluxe grade	5I@.139	sy	65.00	7.88	72.88
replace, custom grade	5I@.139	sy	82.40	7.88	90.28
remove	1D@.062	sy	—	2.93	2.93

Carpet installation on stairs. Includes extra labor, waste and tackless strips needed to install carpet on steps. Does not include the carpet or pad. Installation can be either waterfall or tuck style.

	Craft@Hrs	Unit	Material	Labor	Total
per step	5I@.180	ea	2.18	10.20	12.38

	Craft@Hrs	Unit	Material	Labor	Total
Carpet pad. Includes carpet pad, glue, and installation.					
replace, urethane rebound	5I@.017	sy	7.19	.96	8.15
replace, urethane	5I@.017	sy	4.81	.96	5.77
replace, rubber waffle	5I@.017	sy	5.48	.96	6.44
replace, jute	5I@.017	sy	5.02	.96	5.98
remove	1D@.030	sy	—	1.42	1.42
Add for carpet cove. Carpet wrapped up the wall up to 8" high with metal cap. Grades refer to allowances for quality of carpet.					
replace, economy grade	5I@.135	lf	3.32	7.65	10.97
replace, standard grade	5I@.135	lf	4.14	7.65	11.79
replace, high grade	5I@.135	lf	5.48	7.65	13.13
replace, deluxe grade	5I@.135	lf	6.63	7.65	14.28
replace, custom grade	5I@.135	lf	7.80	7.65	15.45
remove	1D@.022	lf	—	1.04	1.04
remove for work, then re-lay	5I@.211	lf	—	12.00	12.00
Carpet tile. Includes carpet tile, glue when needed, and installation. Includes 12% waste.					
replace, economy grade	5I@.107	sy	42.80	6.07	48.87
replace, standard grade	5I@.107	sy	51.50	6.07	57.57
replace, high grade	5I@.107	sy	65.70	6.07	71.77
remove	1D@.155	sy	—	7.32	7.32
Minimum charge.					
for carpeting work	5I@2.50	ea	94.70	142.00	236.70
Stone floor. Grades vary by region. Standard grade includes flagstone, Chattahoochee, and some sandstone. High grade includes better sandstone and lower to medium grade limestone. Deluxe grade includes higher grade limestone, sandstone and granite.					
replace, standard grade	5I@.191	sf	9.69	10.80	20.49
replace, high grade	5I@.191	sf	12.80	10.80	23.60
replace, deluxe grade	5I@.191	sf	15.50	10.80	26.30
replace, salvage, then reinstall	5I@.500	sf	1.82	28.40	30.22
remove	1D@.082	sf	—	3.87	3.87
regrout	5I@.038	sf	.78	2.15	2.93
minimum charge	5I@3.00	ea	—	170.00	170.00
Marble floor. Includes marble tile, grout, mortar, and installation. Allowance is for typical grades of marble used in residential and light commercial structures.					
replace, standard grade	5I@.189	sf	16.20	10.70	26.90
replace, high grade	5I@.189	sf	33.90	10.70	44.60
replace, deluxe grade	5I@.189	sf	52.80	10.70	63.50
remove	1D@.080	sf	—	3.78	3.78
regrout	5I@.034	sf	.60	1.93	2.53
minimum charge	5I@3.00	ea	117.00	170.00	287.00

	Craft@Hrs	Unit	Material	Labor	Total

Slate floor. Allowances Includes slate tile, grout, mortar, and installation. Allowances for typical grades of slate used in residential and light commercial structures.

	Craft@Hrs	Unit	Material	Labor	Total
replace, standard grade	5I@.189	sf	12.60	10.70	23.30
replace, high grade	5I@.189	sf	15.60	10.70	26.30
replace, deluxe grade	5I@.189	sf	18.70	10.70	29.40
remove	1D@.080	sf	—	3.78	3.78
regrout	5I@.034	sf	.60	1.93	2.53
salvage, then reinstall	5I@.490	sf	1.15	27.80	28.95
minimum charge	5I@3.00	ea	113.00	170.00	283.00

Tile floor. Includes tile, grout, mortar, and installation. Allowances for typical grades of tile used in residential and light commercial structures. Tile quality can be hard to judge. Higher quality tile has distinct patterns. Blurry or fuzzy patterns are a sign of lower quality tile. Sharp edges and clear patterns are usually a sign of high quality tile. Higher quality tile is also free from blemishes, has consistently straight edges, is uniform in size, lies flat, and is free of glaze cracks (sometimes called alligatoring or lizard tails). Although color will vary in all tiles, even in the same dye lot, higher grade tile is more consistent in color. Some high grade tile will violate some of these indicators but in most cases they are good rules of thumb.

	Craft@Hrs	Unit	Material	Labor	Total
replace, standard grade	5I@.180	sf	8.14	10.20	18.34
replace, high grade	5I@.180	sf	10.80	10.20	21.00
replace, deluxe grade	5I@.180	sf	13.40	10.20	23.60
remove	1D@.080	sf	—	3.78	3.78
regrout tile floor	5I@.035	sf	.96	1.98	2.94
minimum charge	5I@3.00	ea	94.70	170.00	264.70

Quarry tile floor. Includes quarry tile, grout, mortar, and installation. Allowances for typical grades of quarry tile used in residential and light commercial structures. Quarry tile is clay-based and unglazed and typically comes in earth tones. Quarry tile quality can be hard to judge. Higher quality tile is free from blemishes, has consistently straight edges, is uniform in size, lays flat, and is free of defects. Although color will vary in all tile even in the same dye lot, higher grade tile is more consistent in color. Some high grade tile will violate some of these indicators but in most cases they are good rules of thumb. Does not include sealer.

	Craft@Hrs	Unit	Material	Labor	Total
replace, standard grade	5I@.143	sf	10.70	8.11	18.81
replace, high grade	5I@.143	sf	15.80	8.11	23.91
replace, deluxe grade	5I@.143	sf	21.30	8.11	29.41
remove	1D@.078	sf	—	3.68	3.68

Tile base. Allowances for typical grades of tile used in residential and light commercial structures.

	Craft@Hrs	Unit	Material	Labor	Total
replace, standard grade	5I@.128	lf	6.84	7.26	14.10
replace, high grade	5I@.128	lf	9.14	7.26	16.40
replace, deluxe grade	5I@.128	lf	11.30	7.26	18.56
remove	1D@.061	lf	—	2.88	2.88

Precast terrazzo floor tiles. Includes precast 12" x 12" terrazzo tile, self-leveling compound as needed, mortar, and installation. Cast in gray cement. Add **2%** for white cement.

	Craft@Hrs	Unit	Material	Labor	Total
replace, standard grade	5I@.116	sf	20.40	6.58	26.98
replace, high grade	5I@.116	sf	22.80	6.58	29.38
remove	1D@.081	sf	—	3.82	3.82
minimum charge	5I@3.00	ea	94.70	170.00	264.70

Precast terrazzo base. Includes precast terrazzo base, mortar, and installation. Cast in gray cement.

	Craft@Hrs	Unit	Material	Labor	Total
replace, 6" high	5I@.256	sf	13.40	14.50	27.90
replace, 8" high	5I@.256	sf	14.60	14.50	29.10
remove	1D@.061	sf	—	2.88	2.88

	Craft@Hrs	Unit	Material	Labor	Total

Cast-in-place thinset terrazzo floor. Includes terrazzo, zinc divider strips, equipment, and installation. Cast in gray cement. Add **3%** for brass divider strips. Standard grade is 1-1/2" deep, high grade is up to 3" thick.

	Craft@Hrs	Unit	Material	Labor	Total
replace, standard grade	5I@.128	sf	4.36	7.26	11.62
replace, high grade	5I@.128	sf	6.53	7.26	13.79
remove	1D@.082	sf	—	3.87	3.87
minimum charge	5I@5.00	ea	137.00	284.00	421.00

Vinyl floor. Includes sheet vinyl in either 6' or 12' widths, vinyl glue, and installation. Allowances for typical grades of vinyl used in residential and light commercial structures. Higher quality vinyl is thicker at 25-30 mils than lower quality vinyl which can be as thin as 10 to 15 mils. Blurry or fuzzy patterns are a sign of lower quality vinyl. Sharp edges, bright colors, and clear patterns are usually a sign of high quality vinyl. Higher quality vinyl is also free from blemishes. Does not include vinyl cove. Estimate vinyl cove separately.

	Craft@Hrs	Unit	Material	Labor	Total
replace, economy grade	5I@.183	sy	23.80	10.40	34.20
replace, standard grade	5I@.183	sy	31.20	10.40	41.60
replace, high grade	5I@.183	sy	38.90	10.40	49.30
replace, deluxe grade	5I@.183	sy	54.70	10.40	65.10
remove	1D@.175	sy	—	8.26	8.26
minimum charge	5I@3.00	ea	80.30	170.00	250.30

Vinyl tile floor. Includes vinyl tile with either a "peel-and-stick" or dry back, vinyl glue as needed, and installation. Allowances for typical grades of vinyl tile used in residential and light commercial structures. Higher quality vinyl tile is thicker at 25-30 mils than lower quality vinyl tile which can be as thin as 15 mils. Blurry or fuzzy patterns are a sign of lower quality vinyl. Sharp edges and clear patterns are usually a sign of high quality vinyl tile. Higher quality vinyl tile is also free from blemishes.

	Craft@Hrs	Unit	Material	Labor	Total
replace, economy grade	5I@.029	sf	1.66	1.64	3.30
replace, standard grade	5I@.029	sf	2.66	1.64	4.30
replace, high grade	5I@.029	sf	2.98	1.64	4.62
replace, deluxe grade	5I@.029	sf	4.70	1.64	6.34
remove	1D@.024	sf	—	1.13	1.13
minimum charge	5I@2.00	ea	51.00	113.00	164.00

Resilient tile floor. Includes resilient tile with either a "peel-and-stick" or dry back, vinyl glue as needed, and installation. Allowances for typical grades of resilient tile used in residential and light commercial structures. Higher quality resilient tile is thicker at 25-30 mils than lower quality tile which can be as thin as 15 mils. Blurry or fuzzy patterns are a sign of lower quality tile. Sharp edges and clear patterns are usually a sign of high quality resilient tile. Higher quality resilient tile is also free from blemishes.

	Craft@Hrs	Unit	Material	Labor	Total
replace, economy grade	5I@.029	sf	1.83	1.64	3.47
replace, standard grade	5I@.029	sf	2.32	1.64	3.96
replace, high grade	5I@.029	sf	2.66	1.64	4.30
replace, deluxe grade	5I@.029	sf	3.87	1.64	5.51
remove	1D@.024	sf	—	1.13	1.13
minimum charge	5I@2.00	ea	51.00	113.00	164.00

Antique style linoleum. Includes linoleum flooring, linoleum glue, and installation. Not to be confused with vinyl flooring. Antique style linoleum is no longer made in the USA and must be imported from Europe. In recent years linoleum has become much more common in commercial and residential applications. Tiles are normally available in limited styles and colors.

	Craft@Hrs	Unit	Material	Labor	Total
replace, plain and marbleized	5I@.189	sy	54.60	10.70	65.30
replace, inlaid & molded	5I@.189	sy	70.90	10.70	81.60
replace, battleship	5I@.189	sy	50.00	10.70	60.70
replace, tile	5I@.032	sf	7.47	1.81	9.28
remove	1D@.175	sy	—	8.26	8.26
minimum charge	5I@3.50	ea	155.00	198.00	353.00

	Craft@Hrs	Unit	Material	Labor	Total

Vinyl cove. Includes the additional labor and material cost to wrap vinyl up a wall with a metal cap. Add to the cost of vinyl floor covering with the same quality.

	Craft@Hrs	Unit	Material	Labor	Total
replace, 4" high	5I@.091	lf	3.32	5.16	8.48
replace, 6" high	5I@.091	lf	3.51	5.16	8.67
replace, 8" high	5I@.091	lf	3.82	5.16	8.98
remove	1D@.036	lf	—	1.70	1.70
minimum charge	5I@1.25	ea	43.70	70.90	114.60

Rubber base. Includes rubber base, glue, and installation. Economy grade is black or brown 2-1/2" high. Standard grade is 4" high standard colors. High grade is 6" high. Deluxe grade is 6" high, specialty colors.

	Craft@Hrs	Unit	Material	Labor	Total
replace, economy grade	5I@.020	lf	2.36	1.13	3.49
replace, standard grade	5I@.020	lf	2.98	1.13	4.11
replace, high grade	5I@.020	lf	3.99	1.13	5.12
replace, deluxe grade	5I@.020	lf	4.70	1.13	5.83
remove	1D@.011	lf	—	.52	.52
minimum charge	5I@.750	ea	29.10	42.50	71.60

Maple strip flooring. Includes maple strip flooring boards, nails, and installation. Strip flooring is up to 3/4" thick and up to 3-1/4" wide. Grading information based on rules established by the Maple Flooring Manufacturers Association (MFMA). All wood is kiln-dried, tongue-and-groove, end-matched and hollow or scratch backed. Does not include vapor barrier or sleepers.

	Craft@Hrs	Unit	Material	Labor	Total
replace, first grade	5I@.088	sf	12.70	4.99	17.69
replace, second grade	5I@.088	sf	9.78	4.99	14.77
replace, third grade	5I@.088	sf	7.47	4.99	12.46
remove	1D@.020	sf	—	.94	.94

Red oak strip flooring. Strip flooring is up to 3/4" thick and up to 3-1/4" wide. Grading information based on rules established by the National Oak Flooring Manufacturers Association (NOFMA). All wood is kiln-dried, tongue-and-groove, end-matched and hollow or scratch backed.

	Craft@Hrs	Unit	Material	Labor	Total
replace, select & better grade	5I@.088	sf	11.60	4.99	16.59
replace, #1 common grade	5I@.088	sf	9.66	4.99	14.65
replace, #2 common grade	5I@.088	sf	6.83	4.99	11.82
remove	1D@.020	sf	—	.94	.94

Red oak strip flooring, quartersawn.

	Craft@Hrs	Unit	Material	Labor	Total
replace, clear grade	5I@.088	sf	27.90	4.99	32.89
replace, select grade	5I@.088	sf	23.20	4.99	28.19
remove	1D@.020	sf	—	.94	.94

Prefinished red oak strip flooring. Includes pre-finished red oak strip flooring boards, nails, and installation. Strip flooring is up to 3/4" thick and up to 3-1/4" wide. Grading information based on rules established by the National Oak Flooring Manufacturers Association (NOFMA). All wood is kiln-dried, tongue-and-groove, end-matched and hollow or scratch backed. Does not include vapor barrier or sleepers.

	Craft@Hrs	Unit	Material	Labor	Total
replace, prime grade	5I@.091	sf	13.10	5.16	18.26
replace, standard & better grade	5I@.091	sf	11.10	5.16	16.26
replace, standard grade	5I@.091	sf	18.80	5.16	23.96
replace, tavern & better grade	5I@.091	sf	31.50	5.16	36.66
replace, tavern grade	5I@.091	sf	53.30	5.16	58.46
remove	1D@.020	sf	—	.94	.94

	Craft@Hrs	Unit	Material	Labor	Total
Red oak plank flooring. Includes red oak planks, nails, and installation. Plank flooring is up to 3/4" thick and over 3-1/4" wide. Floors that contain a mix of plank and strip floor widths are usually considered plank floors. Grading information based on rules established by the National Oak Flooring Manufacturers Association (NOFMA). Does not include vapor barrier or sleepers.					
replace, select & better grade	5I@.079	sf	25.90	4.48	30.38
replace, #1 common grade	5I@.079	sf	21.30	4.48	25.78
replace, #2 common grade	5I@.079	sf	15.30	4.48	19.78
remove	1D@.019	sf	—	.90	.90
Maple parquet or block flooring. Clear grade. Includes parquet blocks, adhesive, and installation. Face pieces are practically clear, may have some small areas of bright sap. For select grade deduct **8%**. For #1 common grade deduct **15%**.					
replace, unit block	5I@.092	sf	9.41	5.22	14.63
replace, laminated block	5I@.092	sf	11.00	5.22	16.22
replace, slat block	5I@.092	sf	12.00	5.22	17.22
remove	1D@.018	sf	—	.85	.85
Prefinished maple parquet or block flooring. Prime grade. Face pieces are almost flawless with very small character marks. (Sap and color variations are not flaws.) For standard grade deduct **10%**. For tavern grade deduct **14%**.					
replace, unit block	5I@.094	sf	10.60	5.33	15.93
replace, laminated block	5I@.094	sf	12.30	5.33	17.63
replace, slat block	5I@.094	sf	13.10	5.33	18.43
remove	1D@.018	sf	—	.85	.85
Red oak parquet or block flooring. Clear grade. Face pieces are practically clear, may have some small areas of bright sap. For select grade deduct **8%**. For #1 common grade deduct **15%**.					
replace, unit block	5I@.092	sf	10.90	5.22	16.12
replace, laminated block	5I@.092	sf	12.90	5.22	18.12
replace, slat block	5I@.092	sf	13.70	5.22	18.92
remove	1D@.018	sf	—	.85	.85
Prefinished red oak parquet or block flooring. Prime grade. Face pieces are almost flawless with very small character marks. (Sap and color variations are not flaws.) For standard grade deduct **10%**. For Tavern grade deduct **14%**.					
replace, unit block	5I@.094	sf	12.50	5.33	17.83
replace, laminated block	5I@.094	sf	14.50	5.33	19.83
replace, slat block	5I@.094	sf	15.50	5.33	20.83
remove	1D@.018	sf	—	.85	.85
Additional cost for parquet. Pickets are mitered wood trim around edges of block.					
Add for pickets around unit block	—	%	8.0	—	—
Additional wood species. Add to the costs of red oak for these wood species.					
deduct for beech flooring	—	%	-7.0	—	—
deduct for birch flooring	—	%	-7.0	—	—
deduct for ash flooring	—	%	-6.0	—	—
add for cherry flooring	—	%	46.0	—	—
add for Brazilian cherry	—	%	45.0	—	—
deduct for hickory flooring	—	%	-5.0	—	—
add for teak flooring	—	%	182.0	—	—
add for walnut flooring	—	%	70.0	—	—
deduct for white oak	—	%	-7.0	—	—
add for bamboo	—	%	37.0	—	—

	Craft@Hrs	Unit	Material	Labor	Total

Southern pine strip flooring. Strip flooring is up to 3/4" thick and up to 3-1/4" wide. Grading information based on rules established by the Southern Pine Inspection Bureau (SPIB). All wood is kiln-dried, tongue-&-groove, end-matched and hollow or scratch backed.

	Craft@Hrs	Unit	Material	Labor	Total
replace, B&B grade	5I@.088	sf	4.55	4.99	9.54
replace, C & better grade	5I@.088	sf	4.19	4.99	9.18
replace, C grade	5I@.088	sf	3.71	4.99	8.70
remove	1D@.020	sf	—	.94	.94

Southern pine plank flooring. Plank flooring is up to 3/4" thick and over 3-1/4" wide. Floors that contain a mix of plank and strip floor widths are usually considered plank floors. See southern pine strip flooring above for grading information.

	Craft@Hrs	Unit	Material	Labor	Total
replace, B&B grade	5I@.079	sf	5.48	4.48	9.96
replace, C & better grade	5I@.079	sf	4.97	4.48	9.45
replace, C grade	5I@.079	sf	4.50	4.48	8.98
remove	1D@.019	sf	—	.90	.90

Douglas fir strip flooring. Douglas fir or western hemlock, or white fir, or Sitka spruce. Strip flooring is up to 3/4" thick and up to 3-1/4" wide. Grading information based on rules established by the West Coast Lumbermen's Inspection Bureau (WCLIB). All wood is kiln-dried, tongue-&-groove, end-matched and hollow or scratch backed.

	Craft@Hrs	Unit	Material	Labor	Total
replace, C & better grade	5I@.088	sf	4.88	4.99	9.87
replace, D grade	5I@.088	sf	4.50	4.99	9.49
replace, E grade	5I@.088	sf	4.04	4.99	9.03
remove	1D@.020	sf	—	.94	.94

Douglas fir plank flooring. Plank flooring is up to 3/4" thick and over 3-1/4" wide. Floors that contain a mix of plank and strip floor widths are usually considered plank floors. See Douglas fir strip flooring for grading information.

	Craft@Hrs	Unit	Material	Labor	Total
replace, C & better grade	5I@.079	sf	5.94	4.48	10.42
replace, D grade	5I@.079	sf	5.45	4.48	9.93
replace, E grade	5I@.079	sf	4.81	4.48	9.29
remove	1D@.019	sf	—	.90	.90

Reclaimed antique longleaf pine flooring. Includes reclaimed antique longleaf pine flooring boards, nails, and installation. Strip or plank flooring. Although longleaf pine was once common in Southeastern America, new lumber is no longer available. Longleaf pine must be salvaged from existing structures. (Although Goodwin Heart Pine Company of Micanopy, Florida does salvage logs that sank to the bottom of rivers.) Grading is based on grades marketed by Mountain Lumber Company, Inc. of Ruckersville, Virginia. Does not include vapor barrier or sleepers.

	Craft@Hrs	Unit	Material	Labor	Total
replace, crown grade	5I@.088	sf	22.80	4.99	27.79
replace, select prime grade	5I@.088	sf	17.90	4.99	22.89
replace, prime grade	5I@.088	sf	17.20	4.99	22.19
replace, naily grade	5I@.088	sf	15.30	4.99	20.29
replace, cabin grade	5I@.088	sf	15.40	4.99	20.39
replace, distressed grade	5I@.088	sf	27.30	4.99	32.29
remove	1D@.020	sf	—	.94	.94

Sand, edge, and fill wood floor. Replace only.

	Craft@Hrs	Unit	Material	Labor	Total
new	5I@.026	sf	.43	1.47	1.90
light sand	5I@.030	sf	.36	1.70	2.06
medium sand	5I@.033	sf	.51	1.87	2.38
heavy sand	5I@.041	sf	.57	2.32	2.89

	Craft@Hrs	Unit	Material	Labor	Total
Sleepers on concrete floor, 24" on center. Glued and nailed with case-hardened nails to concrete floor.					
replace, 1" x 2"	5I@.008	sf	.33	.45	.78
replace, 1" x 3"	5I@.009	sf	.53	.51	1.04
replace, 2" x 4"	5I@.012	sf	1.29	.68	1.97
replace, 2" x 6"	5I@.013	sf	1.93	.74	2.67
remove	1D@.007	sf	—	.33	.33

Floor underlayment. Includes underlayment board, construction adhesive, staples or nails, and installation. When removing heavily glued and stapled underlayment, double the removal cost. Also includes 5% waste.

	Craft@Hrs	Unit	Material	Labor	Total
Particleboard underlayment					
replace, 3/8"	5I@.167	sy	6.47	9.47	15.94
replace, 1/2"	5I@.168	sy	6.70	9.53	16.23
replace, 5/8"	5I@.168	sy	7.44	9.53	16.97
replace, 3/4"	5I@.169	sy	8.60	9.58	18.18
Hardboard underlayment					
replace, 1/4"	5I@.174	sy	8.81	9.87	18.68
Plywood underlayment					
replace, 1/2"	5I@.168	sy	11.10	9.53	20.63
remove	1D@.070	sy	—	3.30	3.30
Minimum charge	5I@1.50	ea	58.40	85.10	143.50

Time & Material Charts (selected items)

Flooring Materials

	Craft@Hrs	Unit	Material	Labor	Total
Carpet					
economy grade, ($23.10 sy, 1 sy), 12% waste	—	sy	25.90	—	25.90
custom grade, ($53.80 sy, 1 sy), 12% waste	—	sy	60.20	—	60.20
Glue-down carpet					
economy grade, ($17.30 sy, 1 sy), 12% waste	—	sy	19.40	—	19.40
deluxe grade, ($32.50 sy, 1 sy), 12% waste	—	sy	36.50	—	36.50
Indoor-outdoor carpet					
economy grade, ($12.80 sy, 1 sy), 12% waste	—	sy	14.30	—	14.30
deluxe grade, ($40.90 sy, 1 sy), 12% waste	—	sy	46.00	—	46.00
Wool carpet					
standard grade, ($35.50 sy, 1 sy), 12% waste	—	sy	39.80	—	39.80
custom grade, ($72.50 sy, 1 sy), 12% waste	—	sy	81.30	—	81.30
Simulated wool berber carpet					
standard grade, ($31.50 sy, 1 sy), 12% waste	—	sy	35.20	—	35.20
custom grade, ($50.90 sy, 1 sy), 12% waste	—	sy	57.00	—	57.00
Carpet pad					
urethane rebound, ($6.32 sy, 1 sy), 12% waste	—	sy	7.08	—	7.08
jute, ($4.41 sy, 1 sy), 12% waste	—	sy	4.95	—	4.95
Carpet tile					
standard grade, ($37.80 sy, 1 sy), 12% waste	—	sy	42.30	—	42.30
custom grade, ($57.70 sy, 1 sy), 12% waste	—	sy	64.50	—	64.50
Stone floor					
standard grade, ($9.27 sf, 1 sf), 4% waste	—	sf	9.66	—	9.66
deluxe grade, ($14.70 sf, 1 sf), 4% waste	—	sf	15.40	—	15.40

	Craft@Hrs	Unit	Material	Labor	Total
Marble floor					
standard grade, ($15.60 sf, 1 sf), 4% waste	—	sf	16.10	—	16.10
deluxe grade, ($50.60 sf, 1 sf), 4% waste	—	sf	52.60	—	52.60
Slate floor					
standard grade, ($12.00 sf, 1 sf), 4% waste	—	sf	12.60	—	12.60
deluxe grade, ($18.00 sf, 1 sf), 4% waste	—	sf	18.70	—	18.70
Tile floor					
standard grade, ($7.81 sf, 1 sf), 4% waste	—	sf	8.13	—	8.13
deluxe grade, ($12.90 sf, 1 sf), 4% waste	—	sf	13.40	—	13.40
Quarry tile floor					
standard grade, ($10.30 sf, 1 sf), 4% waste	—	sf	10.70	—	10.70
deluxe grade, ($20.60 sf, 1 sf), 4% waste	—	sf	21.40	—	21.40
Precast terrazzo floor tiles					
standard grade, ($19.30 sf, 1 sf), 4% waste	—	sf	20.00	—	20.00
high grade, ($21.80 sf, 1 sf), 4% waste	—	sf	22.70	—	22.70
Cast-in-place thinset terrazzo floor					
standard grade, ($3.84 sf, 1 sf), 12% waste	—	sf	4.29	—	4.29
high grade, ($5.74 sf, 1 sf), 12% waste	—	sf	6.43	—	6.43
Vinyl floor					
economy grade, ($21.00 sy, 1 sy), 12% waste	—	sy	23.50	—	23.50
deluxe grade, ($48.10 sy, 1 sy), 12% waste	—	sy	53.80	—	53.80
Vinyl tile floor					
economy grade, ($1.56 sf, 1 sf), 12% waste	—	sf	1.73	—	1.73
deluxe grade, ($4.11 sf, 1 sf), 12% waste	—	sf	4.62	—	4.62
Resilient tile floor					
economy grade, ($1.62 sf, 1 sf), 12% waste	—	sf	1.82	—	1.82
deluxe grade, ($3.41 sf, 1 sf), 12% waste	—	sf	3.82	—	3.82
Antique style linoleum					
battleship, ($43.90 sy, 1 sy), 12% waste	—	sy	49.30	—	49.30
Rubber base					
economy grade, ($2.08 lf, 1 lf), 12% waste	—	lf	2.33	—	2.33
deluxe grade, ($4.11 lf, 1 lf), 12% waste	—	lf	4.62	—	4.62
Strip flooring					
maple, first grade, ($11.90 sf, 1 sf), 6% waste	—	sf	12.70	—	12.70
red oak, select & better grade, ($10.90 sf, 1 sf), 6% waste	—	sf	11.60	—	11.60
red oak, quarter sawn, clear grade, ($26.30 sf, 1 sf), 6% waste	—	sf	27.80	—	27.80
prefinished red oak, prime grade, ($12.50 sf, 1 sf), 6% waste	—	sf	13.10	—	13.10
Plank flooring					
red oak, select & better grade, ($24.10 sf, 1 sf), 6% waste	—	sf	25.50	—	25.50
Parquet or block flooring					
maple, slat block, clear grade, ($11.60 sf, 1 sf), 4% waste	—	sf	12.00	—	12.00
prefinished maple, slat block, prime grade, ($10.30 sf, 1 sf), 4% waste	—	sf	10.70	—	10.70

	Craft@Hrs	Unit	Material	Labor	Total
red oak, slat block, clear grade, ($10.60 sf, 1 sf), 4% waste	—	sf	11.00	—	11.00
prefinished red oak, slat block, prime grade, ($12.00 sf, 1 sf), 4% waste	—	sf	12.60	—	12.60
Softwood flooring					
Southern pine strip, B&B grade, ($4.27 sf, 1 sf), 6% waste	—	sf	4.53	—	4.53
Southern pine plank, B&B grade, ($5.16 sf, 1 sf), 6% waste	—	sf	5.48	—	5.48
Douglas fir strip, C & better grade, ($4.59 sf, 1 sf), 6% waste	—	sf	4.85	—	4.85
Douglas fir plank, C & better grade, ($5.60 sf, 1 sf), 6% waste	—	sf	5.93	—	5.93
reclaimed antique longleaf pine, crown grade, ($21.30 sf, 1 sf), 6% waste	—	sf	22.60	—	22.60
Sleepers on concrete floor, 24" on center					
1" x 2"	—	sf	.32	—	.32
2" x 4"	—	sf	1.26	—	1.26
Floor underlayment					
3/8" particleboard, ($21.70 sheet, 3.56 sy), 5% waste	—	sy	6.38	—	6.38
1/2" particleboard, ($22.70 sheet, 3.56 sy), 5% waste	—	sy	6.69	—	6.69
5/8" particleboard, ($25.20 sheet, 3.56 sy), 5% waste	—	sy	7.47	—	7.47
3/4" particleboard, ($28.90 sheet, 3.56 sy), 5% waste	—	sy	8.55	—	8.55
1/4" hardboard, ($15.20 sheet, 1.78 sy), 5% waste	—	sy	8.90	—	8.90
1/2" plywood, ($37.80 sheet, 3.56 sy), 5% waste	—	sy	11.10	—	11.10

Flooring Labor

Laborer	base wage	paid leave	true wage	taxes & ins.	total
Flooring installer	$37.90	2.96	$40.86	24.14	$65.00
Flooring installer's helper	$27.10	2.11	$29.21	19.09	$48.30
Demolition laborer	$25.90	2.02	$27.92	19.28	$47.20

Paid leave is calculated based on two weeks paid vacation, one week sick leave, and seven paid holidays. Employer's matching portion of **FICA** is 7.65 percent. **FUTA** (Federal Unemployment) is .8 percent. **Worker's compensation** for the flooring trade was calculated using a national average of 11.74 percent. **Unemployment insurance** was calculated using a national average of 8 percent. **Health insurance** was calculated based on a projected national average for 2020 of $1,280 per employee (and family when applicable) per month. Employer pays 80 percent for a per month cost of $1,024 per employee. **Retirement** is based on a 401(k) retirement program with employer matching of 50 percent. Employee contributions to the 401(k) plan are an average of 6 percent of the true wage. **Liability insurance** is based on a national average of 12.0 percent.

	Craft@Hrs	Unit	Material	Labor	Total
Flooring Labor Productivity					
Demolition of flooring					
remove carpet	1D@.062	sy	—	2.93	2.93
remove glue-down carpet	1D@.155	sy	—	7.32	7.32
remove indoor-outdoor carpet	1D@.153	sy	—	7.22	7.22
remove carpet pad	1D@.030	sy	—	1.42	1.42
remove carpet cove	1D@.022	lf	—	1.04	1.04
remove carpet tile	1D@.155	sy	—	7.32	7.32
remove stone floor	1D@.082	sf	—	3.87	3.87
remove marble floor	1D@.080	sf	—	3.78	3.78
remove slate floor	1D@.080	sf	—	3.78	3.78
remove tile floor	1D@.080	sf	—	3.78	3.78
remove quarry tile floor	1D@.078	sf	—	3.68	3.68
remove tile base	1D@.061	lf	—	2.88	2.88
remove precast terrazzo floor tiles	1D@.081	sf	—	3.82	3.82
remove cast-in-place terrazzo floor	1D@.082	sf	—	3.87	3.87
remove vinyl floor	1D@.175	sy	—	8.26	8.26
remove vinyl tile floor	1D@.024	sf	—	1.13	1.13
remove resilient tile floor	1D@.024	sf	—	1.13	1.13
remove antique style linoleum floor	1D@.175	sy	—	8.26	8.26
remove vinyl cove	1D@.036	lf	—	1.70	1.70
remove rubber base	1D@.011	lf	—	.52	.52
remove wood strip flooring	1D@.020	sf	—	.94	.94
remove wood plank flooring	1D@.019	sf	—	.90	.90
remove wood unit block flooring	1D@.018	sf	—	.85	.85
remove floor sleepers	1D@.007	sf	—	.33	.33
remove floor underlayment	1D@.070	sy	—	3.30	3.30

Flooring crew

install flooring	flooring installer	$65.00
install flooring	flooring installer's helper	$48.30
install flooring	flooring crew	$56.70

	Craft@Hrs	Unit	Material	Labor	Total
Install carpet					
typical on tackless	5I@.138	sy	—	7.82	7.82
cove	5I@.135	lf	—	7.65	7.65
remove for work, then re-lay	5I@.211	sy	—	12.00	12.00
glue-down	5I@.167	sy	—	9.47	9.47
wool or simulated wool	5I@.139	sy	—	7.88	7.88
add per step for installation on stairs	5I@.180	ea	—	10.20	10.20
pad	5I@.017	sy	—	.96	.96
tile	5I@.107	sy	—	6.07	6.07
Install stone floor					
typical	5I@.191	sf	—	10.80	10.80
salvage, then reinstall	5I@.500	sf	—	28.40	28.40
regrout	5I@.038	sf	—	2.15	2.15
Install marble floor					
typical	5I@.189	sf	—	10.70	10.70
regrout	5I@.034	sf	—	1.93	1.93

	Craft@Hrs	Unit	Material	Labor	Total
Install slate floor					
typical	5I@.189	sf	—	10.70	10.70
regrout	5I@.034	sf	—	1.93	1.93
salvage, then reinstall	5I@.490	sf	—	27.80	27.80
Install tile floor					
typical	5I@.180	sf	—	10.20	10.20
regrout	5I@.035	sf	—	1.98	1.98
base	5I@.128	lf	—	7.26	7.26
Install quarry tile floor					
typical	5I@.143	sf	—	8.11	8.11
Install terrazzo floor					
precast tiles	5I@.116	sf	—	6.58	6.58
precast base	5I@.256	sf	—	14.50	14.50
cast-in-place thinset	5I@.128	sf	—	7.26	7.26
Install vinyl floor					
sheet goods	5I@.183	sy	—	10.40	10.40
cove	5I@.091	lf	—	5.16	5.16
tile	5I@.029	sf	—	1.64	1.64
resilient tile	5I@.029	sf	—	1.64	1.64
Install antique-style linoleum					
typical	5I@.189	sy	—	10.70	10.70
tile	5I@.032	sf	—	1.81	1.81
Install rubber base					
typical	5I@.020	lf	—	1.13	1.13
Install wood floor					
hardwood strip	5I@.088	sf	—	4.99	4.99
prefinished hardwood strip	5I@.091	sf	—	5.16	5.16
hardwood plank	5I@.079	sf	—	4.48	4.48
parquet or block	5I@.092	sf	—	5.22	5.22
prefinished hardwood block or parquet	5I@.094	sf	—	5.33	5.33
softwood strip	5I@.088	sf	—	4.99	4.99
softwood plank	5I@.079	sf	—	4.48	4.48
Sand, edge, and fill wood floor					
new	5I@.026	sf	—	1.47	1.47
light sand	5I@.030	sf	—	1.70	1.70
medium	5I@.033	sf	—	1.87	1.87
heavy sand	5I@.041	sf	—	2.32	2.32
Install sleepers on concrete floor, 24" on center					
1" x 2"	5I@.008	sf	—	.45	.45
1" x 3"	5I@.009	sf	—	.51	.51
2" x 4"	5I@.012	sf	—	.68	.68
2" x 6"	5I@.013	sf	—	.74	.74
Install floor underlayment					
3/8" particleboard	5I@.167	sy	—	9.47	9.47
1/2" particleboard	5I@.168	sy	—	9.53	9.53
5/8" particleboard	5I@.168	sy	—	9.53	9.53
3/4" particleboard	5I@.169	sy	—	9.58	9.58
1/4" hardboard	5I@.174	sy	—	9.87	9.87
1/2" plywood	5I@.168	sy	—	9.53	9.53

Hazardous Materials

	Craft@Hrs	Unit	Material	Labor	Equip.	Total
Minimum charge.						
asbestos removal	1H@16.0	ea	205.00	918.00	289.00	1,412.00
Asbestos analysis. Includes up to four hours, up to 12 analyzed samples, and a written report indicating severity of asbestos risk for a typical home or small commercial structure.						
analysis	1H@16.7	ea	—	959.00	—	959.00
Seal area for asbestos work. Seal area where asbestos exists. Includes tape along all seams, caulk or foam insulation in any cracks, and negative air vent system.						
plastic cover attached to ceiling	1H@.044	sf	1.97	2.53	—	4.50
plastic cover attached to walls	1H@.042	sf	2.04	2.41	—	4.45
plastic cover and plywood over floor (2 layers)	1H@.102	sf	3.21	5.85	—	9.06
temporary containment walls with plastic cover	1H@.179	sf	3.85	10.30	—	14.15
Prefabricated decontamination unit. Four day minimum. Includes decontamination equipment, shower, and clean room.						
per day	—	dy	—	—	123.00	123.00
Encapsulate asbestos-based acoustical ceiling. Penetrating sealant sprayed on ceiling with airless sprayer.						
with sealant	1H@.014	sf	.49	.80	.18	1.47
Remove asbestos-based materials. Includes full Tyvek suits for workers changed four times per eight hour shift, respirators (add **20%** for respirators with air supply), air monitoring, supervision by qualified hygiene professionals, final seal of scraped or demo'ed walls, final seal and disposal of plastic cover materials, disposal in fiber drums and hauling to dump. Does not include dump fees (typically about $12.00 per drum).						
scrape acoustical ceiling	1H@.034	sf	.09	1.95	.27	2.31
insulation over 1/2" to 3/4" pipe	1H@.026	lf	.22	1.49	.15	1.86
insulation over 1" to 3" pipe	1H@.080	lf	.46	4.59	.30	5.35
asbestos-based siding	1H@.179	sf	1.31	10.30	.31	11.92
asbestos-based plaster	1H@.137	sf	.89	7.86	.31	9.06

	Craft@Hrs	Unit	Material	Labor	Total
Time & Material Charts (selected items) **Materials for Hazardous Materials Removal**					
polyethylene (plastic), ($.32 sf)	—	sf	.32	—	.32
Tyvek whole body suit, ($23.40 ea)	—	ea	23.40	—	23.40
respirator cartridge, ($6.66 ea)	—	ea	6.66	—	6.66
glove bag, 7 mil., 50" x 64", ($12.80 ea)	—	ea	12.80	—	12.80
glove bag, 10 mil., 44" x 60", ($13.90 ea)	—	ea	13.90	—	13.90
3 cf disposable polyethylene bags, 6 mil, ($1.66 ea)	—	ea	1.66	—	1.66
3 cf disposable fiber drums, ($14.00 ea)	—	ea	14.00	—	14.00
caution labels, ($.35 ea)	—	ea	.35	—	.35
encapsulation quality sealant, ($120.00 gal)	—	gal	120.00	—	120.00
vent fan filters, ($11.50 ea)	—	ea	11.50	—	11.50
respirator, ($34.40 ea)	—	ea	34.40	—	34.40

	Craft@Hrs	Unit	Material	Labor	Equip.	Total
Hazardous Materials Rental Equipment						
negative air vent system, two fans	—	day	—	—	93.60	93.60
HEPA vacuum cleaner, 16 gallon wet/dry	—	day	—	—	120.00	120.00
airless sprayer unit	—	day	—	—	84.90	84.90
light stand	—	day	—	—	34.00	34.00
prefabricated decontamination unit	—	day	—	—	127.00	127.00

Hazardous Materials Labor

Laborer	base wage	paid leave	true wage	taxes & ins.	total
Haz. mat. laborer	$29.60	2.31	$31.91	25.49	$57.40

Paid leave is calculated based on two weeks paid vacation, one week sick leave, and seven paid holidays. Employer's matching portion of **FICA** is 7.65 percent. **FUTA** (Federal Unemployment) is .8 percent. **Worker's compensation** for the hazardous materials trade was calculated using a national average of 29.01 percent. **Unemployment insurance** was calculated using a national average of 8 percent. **Health insurance** was calculated based on a projected national average for 2020 of $1,280 per employee (and family when applicable) per month. Employer pays 80 percent for a per month cost of $1,024 per employee. **Retirement** is based on a 401(k) retirement program with employer matching of 50 percent. Employee contributions to the 401(k) plan are an average of 6 percent of the true wage. **Liability insurance** is based on a national average of 12.0 percent.

Hazardous Materials Labor Productivity	Craft@Hrs	Unit	Material	Labor	Total
Minimum charge for asbestos removal	1H@16.0	ea	—	918.00	918.00
Asbestos analysis typical home	1H@16.7	ea	—	959.00	959.00
Seal area for work					
attach plastic cover to ceiling	1H@.044	sf	—	2.53	2.53
attach plastic cover to walls	1H@.042	sf	—	2.41	2.41
attach plastic cover & plywood to floor	1H@.102	sf	—	5.85	5.85
build walls and attach plastic cover	1H@.179	sf	—	10.30	10.30
Remove asbestos-based materials					
scrape acoustical ceiling	1H@.034	sf	—	1.95	1.95
encapsulate acoustical ceiling with sealant	1H@.014	sf	—	.80	.80
insulation over 1/2" to 3/4" pipe	1H@.026	lf	—	1.49	1.49
insulation over 1" to 3" pipe	1H@.080	lf	—	4.59	4.59
siding	1H@.179	sf	—	10.30	10.30
plaster	1H@.137	sf	—	7.86	7.86

HVAC

	Craft@Hrs	Unit	Material	Labor	Total
Duct work. Replace only.					
per duct	2H@5.00	ea	243.00	353.00	596.00
per sf of floor	2H@.032	sf	.75	2.26	3.01
for home to 1,200 sf	2H@35.7	ea	729.00	2,520.00	3,249.00
for home 1,200 to 1,900 sf	2H@43.8	ea	1,130.00	3,090.00	4,220.00
for home 1,900 to 2,400 sf	2H@53.7	ea	1,460.00	3,790.00	5,250.00
for home 2,400 to 2,900 sf	2H@70.0	ea	1,740.00	4,940.00	6,680.00
for home 2,900 to 3,400 sf	2H@100	ea	2,070.00	7,050.00	9,120.00
for home 3,400 to 3,900 sf	2H@130	ea	2,360.00	9,170.00	11,530.00
for home 3,900 to 4,500 sf	2H@163	ea	2,740.00	11,500.00	14,240.00

duct work

	Craft@Hrs	Unit	Material	Labor	Total
Electric forced-air furnace. Includes connection only. Does not include wiring run, plenums, ducts, or flues.					
replace, 10,200 btu	2H@10.8	ea	642.00	761.00	1,403.00
replace, 17,100 btu	2H@11.4	ea	680.00	804.00	1,484.00
replace, 27,300 btu	2H@12.6	ea	755.00	888.00	1,643.00
replace, 34,100 btu	2H@13.8	ea	832.00	973.00	1,805.00
remove	1D@1.89	ea	—	89.20	89.20
remove for work, then reinstall	2H@5.21	ea	—	367.00	367.00
Gas forced-air furnace. Includes connection only. Does not include wiring run, plenums, ducts, or flues.					
replace, 45,000 btu	2H@11.3	ea	680.00	797.00	1,477.00
replace, 60,000 btu	2H@12.3	ea	755.00	867.00	1,622.00
replace, 75,000 btu	2H@13.4	ea	832.00	945.00	1,777.00
replace, 100,000 btu	2H@15.0	ea	907.00	1,060.00	1,967.00
replace, 125,000 btu	2H@16.4	ea	1,060.00	1,160.00	2,220.00
replace, 150,000 btu	2H@17.3	ea	1,300.00	1,220.00	2,520.00
remove	1D@1.82	ea	—	85.90	85.90
remove for work, then reinstall	2H@5.08	ea	—	358.00	358.00
Oil forced-air furnace. Includes connection only. Does not include wiring run, plenums, ducts, or flues.					
replace, 56,000 btu	2H@12.4	ea	1,490.00	874.00	2,364.00
replace, 84,000 btu	2H@12.8	ea	1,810.00	902.00	2,712.00
replace, 95,000 btu	2H@13.2	ea	1,900.00	931.00	2,831.00
replace, 134,000 btu	2H@17.8	ea	2,270.00	1,250.00	3,520.00
remove	1D@2.04	ea	—	96.30	96.30
remove for work, then reinstall	2H@5.45	ea	—	384.00	384.00
Service furnace. Replace only.					
typical	2H@1.71	ea	14.70	121.00	135.70
Heat pump with supplementary heat coil. Includes heat pump with supplementary coil and installation. Includes some fitting and materials to connect to existing work but does not include electrical wiring runs, or interior pipe.					
replace, 2 ton	2H@19.5	ea	3,790.00	1,370.00	5,160.00
replace, 4 ton	2H@29.1	ea	6,800.00	2,050.00	8,850.00
replace, 5 ton	2H@34.9	ea	7,550.00	2,460.00	10,010.00
remove	1D@2.33	ea	—	110.00	110.00
remove for work, then reinstall	2H@21.9	ea	—	1,540.00	1,540.00

	Craft@Hrs	Unit	Material	Labor	Total
Humidifier. Includes humidifier and installation. Installed in existing ductwork. Centrifugal atomizing.					
replace, 5 pounds per hour	2H@3.83	ea	2,270.00	270.00	2,540.00
replace, 10 pounds per hour	2H@4.44	ea	2,640.00	313.00	2,953.00
remove	1D@1.69	ea	—	79.80	79.80
remove for work, then reinstall	2H@2.98	ea	—	210.00	210.00
Furnace vent pipe, double wall (all fuels). Includes double wall flue, sheet-metal screws, metallic duct tape, and installation.					
replace, 3" diameter	2H@.283	lf	6.06	20.00	26.06
replace, 4" diameter	2H@.318	lf	7.40	22.40	29.80
replace, 5" diameter	2H@.351	lf	8.77	24.70	33.47
replace, 6" diameter	2H@.390	lf	10.30	27.50	37.80
replace, 7" diameter	2H@.486	lf	15.10	34.30	49.40
replace, 8" diameter	2H@.533	lf	16.90	37.60	54.50
remove	1D@.040	lf	—	1.89	1.89
remove for work, then reinstall	2H@.283	lf	—	20.00	20.00
Central air conditioning system. Includes condensing unit, coil, copper lines and installation.					
replace, complete, 1 ton	2H@6.12	ea	1,210.00	431.00	1,641.00
replace, 1-1/2 ton	2H@7.08	ea	1,360.00	499.00	1,859.00
replace, 2 ton	2H@8.52	ea	1,480.00	601.00	2,081.00
replace, 3 ton	2H@7.33	ea	2,430.00	517.00	2,947.00
replace, 4 ton	2H@16.2	ea	3,000.00	1,140.00	4,140.00
replace, 5 ton	2H@23.3	ea	3,720.00	1,640.00	5,360.00
replace, air conditioning unit service	1D@2.93	ea	68.00	138.00	206.00
Recharge AC system with refrigerant. Includes refrigerant and labor to recover, evacuate and recharge AC system.					
replace, 5 lbs	2H@1.50	ea	173.00	106.00	279.00
replace, 10 lbs	2H@1.65	ea	349.00	116.00	465.00
replace, 14 lbs	2H@1.80	ea	490.00	127.00	617.00
Through-wall AC unit. Includes through-wall AC unit with built-in thermostat and installation. Does not include wall framing or electrical wiring runs.					
replace, 5,000 btu	2H@8.83	ea	1,250.00	623.00	1,873.00
replace, 8,000 btu	2H@9.11	ea	1,290.00	642.00	1,932.00
replace, 12,000 btu	2H@10.6	ea	1,490.00	747.00	2,237.00
replace, 18,000 btu	2H@11.8	ea	1,900.00	832.00	2,732.00
remove	1D@1.49	ea	—	70.30	70.30
remove for work, then reinstall	2H@4.98	ea	—	351.00	351.00
Through-wall combination AC and heat unit.					
replace, 6,000 btu AC and 4,040 btu heat unit	2H@10.2	ea	1,900.00	719.00	2,619.00
replace, 9,500 btu AC and 4,040 btu heat unit	2H@10.5	ea	2,170.00	740.00	2,910.00
replace, 11,300 btu AC and 9,200 btu heat unit	2H@12.2	ea	2,640.00	860.00	3,500.00
remove	1D@1.56	ea	—	73.60	73.60
remove for work, then reinstall	2H@5.35	ea	—	377.00	377.00

	Craft@Hrs	Unit	Material	Labor	Total
Evaporative cooler. Roof mount. Includes up to 36 lf of copper water supply tubing and electrical hookup.					
replace, 2,000 cfm	2H@7.49	ea	1,180.00	528.00	1,708.00
replace, 4,300 cfm	2H@9.55	ea	1,590.00	673.00	2,263.00
replace, 4,700 cfm	2H@10.1	ea	1,650.00	712.00	2,362.00
replace, 5,600 cfm	2H@10.9	ea	1,900.00	768.00	2,668.00
remove	1D@1.64	ea	—	77.40	77.40
replace, grille	2H@.680	ea	106.00	47.90	153.90
remove, grille	1D@.167	ea	—	7.88	7.88
remove for work, then reinstall	2H@9.42	ea	—	664.00	664.00
Cold-air grille.					
replace, cold-air return	2H@.952	ea	45.30	67.10	112.40
remove	1D@.167	ea	—	7.88	7.88
remove cold-air return for work, then reinstall	2H@.248	ea	—	17.50	17.50
Heat register.					
replace, heat register	2H@.630	ea	30.10	44.40	74.50
remove	1D@.135	ea	—	6.37	6.37
remove heat register for work, then reinstall	2H@.160	ea	—	11.30	11.30
Thermostat. Includes thermostat and installation. Does not include thermostat wire.					
replace, heat only	2H@1.53	ea	68.00	108.00	176.00
replace, heat and air conditioning	2H@2.38	ea	121.00	168.00	289.00
replace, programmable	2H@3.49	ea	136.00	246.00	382.00
replace, programmable with zone control	2H@4.29	ea	165.00	302.00	467.00
remove	1D@.233	ea	—	11.00	11.00
remove for work, then reinstall	2H@.622	ea	—	43.90	43.90

Time & Material Charts (selected items)

HVAC Materials

	Craft@Hrs	Unit	Material	Labor	Total
Ductwork materials					
per sf of floor	—	sf	.75	—	.75
Electric forced-air furnace					
10,200 btu	—	ea	642.00	—	642.00
34,100 btu	—	ea	832.00	—	832.00
Gas forced-air furnace					
45,000 btu	—	ea	680.00	—	680.00
100,000 btu	—	ea	1,300.00	—	1,300.00
Oil forced-air furnace					
56,000 btu	—	ea	1,490.00	—	1,490.00
134,000 btu	—	ea	2,270.00	—	2,270.00

HVAC Labor

Laborer	base wage	paid leave	true wage	taxes & ins.	total
HVAC installer	$42.10	3.28	$45.38	25.12	$70.50
Demolition laborer	$25.90	2.02	$27.92	19.28	$47.20

Paid leave is calculated based on two weeks paid vacation, one week sick leave, and seven paid holidays. Employer's matching portion of **FICA** is 7.65 percent. **FUTA** (Federal Unemployment) is .8 percent. **Worker's compensation** for the HVAC trade was calculated using a national average of 9.57 percent. **Unemployment insurance** was calculated using a national average of 8 percent. **Health insurance** was calculated based on a projected national average for 2020 of $1,280 per employee (and family when applicable) per month. Employer pays 80 percent for a per month cost of $1,024 per employee. **Retirement** is based on a 401(k) retirement program with employer matching of 50 percent. Employee contributions to the 401(k) plan are an average of 6 percent of the true wage. **Liability insurance** is based on a national average of 12.0 percent.

HVAC Labor Productivity

	Craft@Hrs	Unit	Material	Labor	Total
Demolition of HVAC equipment					
remove electric forced-air furnace	1D@1.89	ea	—	89.20	89.20
remove gas forced-air furnace	1D@1.82	ea	—	85.90	85.90
remove oil forced-air furnace	1D@2.04	ea	—	96.30	96.30
remove heat pump	1D@2.33	ea	—	110.00	110.00
remove humidifier	1D@1.69	ea	—	79.80	79.80
remove through-wall AC unit	1D@1.49	ea	—	70.30	70.30
remove combination AC and heat unit	1D@1.56	ea	—	73.60	73.60
remove evaporative cooler	1D@1.64	ea	—	77.40	77.40
remove heat register	1D@.135	ea	—	6.37	6.37
remove thermostat	1D@.233	ea	—	11.00	11.00
Install electric forced-air furnace					
10,200 btu	2H@10.8	ea	—	761.00	761.00
34,100 btu	2H@13.8	ea	—	973.00	973.00
Install gas forced-air furnace					
45,000 btu	2H@11.3	ea	—	797.00	797.00
150,000 btu	2H@17.3	ea	—	1,220.00	1,220.00
Install oil forced-air furnace					
56,000 btu	2H@12.4	ea	—	874.00	874.00
134,000 btu	2H@17.8	ea	—	1,250.00	1,250.00
Install heat pump with supplementary heat coil					
2 ton	2H@19.5	ea	—	1,370.00	1,370.00
Install humidifier					
5 pounds per hour	2H@3.83	ea	—	270.00	270.00
Double wall furnace flue (all fuels)					
4" diameter	2H@.318	lf	—	22.40	22.40
8" diameter	2H@.533	lf	—	37.60	37.60
Through-wall AC unit					
5,000 btu	2H@8.83	ea	—	623.00	623.00
18,000 btu	2H@11.8	ea	—	832.00	832.00
Through-wall combination AC and heat unit					
6,000 btu AC and 4,040 btu heat unit	2H@10.2	ea	—	719.00	719.00
11,300 btu AC and 9,200 btu heat unit	2H@12.2	ea	—	860.00	860.00
Evaporative cooler					
2,000 cfm	2H@7.49	ea	—	528.00	528.00
5,600 cfm	2H@10.9	ea	—	768.00	768.00
Thermostat					
heat only	2H@1.53	ea	—	108.00	108.00
heat and air conditioning	2H@2.38	ea	—	168.00	168.00
programmable	2H@3.49	ea	—	246.00	246.00
programmable, with zone control	2H@4.29	ea	—	302.00	302.00

Insulation

	Craft@Hrs	Unit	Material	Labor	Total
Minimum charge.					
for insulation work	1I@3.10	ea	57.70	152.00	209.70
Vermiculite attic insulation. Poured by hand from 3 cf bags.					
replace, 3" deep	1I@.008	sf	1.59	.39	1.98
replace, 4" deep	1I@.009	sf	2.15	.44	2.59
replace, 5" deep	1I@.009	sf	2.69	.44	3.13
replace, 6" deep	1I@.010	sf	3.23	.49	3.72
remove, 3"	1D@.014	sf	—	.66	.66
remove, 4"	1D@.017	sf	—	.80	.80
remove, 5"	1D@.020	sf	—	.94	.94
remove, 6"	1D@.025	sf	—	1.18	1.18
Blown mineral wool attic insulation.					
R11 (3-1/2" deep)	1I@.010	sf	.84	.49	1.33
R19 (6" deep)	1I@.011	sf	1.26	.54	1.80
R30 (10" deep)	1I@.013	sf	2.04	.64	2.68
R38 (12" deep)	1I@.016	sf	2.79	.78	3.57
remove, R11	1D@.012	sf	—	.57	.57
remove, R19	1D@.013	sf	—	.61	.61
remove, R30	1D@.019	sf	—	.90	.90
remove, R38	1D@.022	sf	—	1.04	1.04
Blown cellulose attic insulation.					
R12 (3-1/2" deep)	1I@.010	sf	.82	.49	1.31
R21 (6" deep)	1I@.012	sf	1.23	.59	1.82
R33 (10" deep)	1I@.013	sf	1.99	.64	2.63
R42 (12" deep)	1I@.016	sf	2.68	.78	3.46
remove, R12	1D@.014	sf	—	.66	.66
remove, R21	1D@.015	sf	—	.71	.71
remove, R33	1D@.023	sf	—	1.09	1.09
remove, R42	1D@.026	sf	—	1.23	1.23
Blown fiberglass attic insulation.					
R11 (5" deep)	1I@.010	sf	.75	.49	1.24
R19 (8" deep)	1I@.011	sf	1.07	.54	1.61
R30 (13" deep)	1I@.013	sf	1.74	.64	2.38
R38 (16" deep)	1I@.016	sf	2.37	.78	3.15
remove, R11	1D@.012	sf	—	.57	.57
remove, R19	1D@.013	sf	—	.61	.61
remove, R30	1D@.019	sf	—	.90	.90
remove, R38	1D@.022	sf	—	1.04	1.04
Add to blow insulation into existing wall.					
add for wall with stucco	—	%	—	285.0	—
add for wall with siding	—	%	—	233.0	—
add for wall with masonry veneer	—	%	—	318.0	—
add for wall with interior drywall	—	%	—	216.0	—

	Craft@Hrs	Unit	Material	Labor	Total
Fiberglass batt insulation, installed in ceiling, attic, or floor.					
R11 (3-1/2" deep)	1I@.009	sf	.60	.44	1.04
R19 (6" deep)	1I@.011	sf	.90	.54	1.44
R30 (9-1/2" deep)	1I@.012	sf	1.48	.59	2.07
R38 (12" deep)	1I@.015	sf	1.99	.74	2.73
remove, R11	1D@.009	sf	—	.42	.42
remove, R19	1D@.013	sf	—	.61	.61
remove, R30	1D@.015	sf	—	.71	.71
remove, R38	1D@.018	sf	—	.85	.85
Fiberglass batt insulation, installed in wall.					
R6 (1-3/4" deep) (between furring strips)	1I@.010	sf	.34	.49	.83
R11 (3-1/2" deep)	1I@.011	sf	.60	.54	1.14
R19 (6" deep)	1I@.013	sf	.90	.64	1.54
R19 (8" deep)	1I@.015	sf	1.18	.74	1.92
remove, R6	1D@.007	sf	—	.33	.33
remove, R11	1D@.009	sf	—	.42	.42
remove, R19 (6" deep)	1D@.013	sf	—	.61	.61
remove, R19 (8" deep)	1D@.015	sf	—	.71	.71
Add for batt face.					
add for Kraft face	—	%	8.0	—	—
add for foil face	—	%	10.0	—	—
add for plastic vapor barrier	—	%	14.0	—	—
Rigid foam insulation board. Includes rigid foam board, glue and/or nails, and installation on walls or roof.					
1/2"	1I@.014	sf	.77	.69	1.46
3/4"	1I@.014	sf	.90	.69	1.59
1"	1I@.014	sf	1.18	.69	1.87
2"	1I@.014	sf	1.57	.69	2.26
remove	1D@.007	sf	—	.33	.33

Time & Material Charts (selected items)
Insulation Materials

	Craft@Hrs	Unit	Material	Labor	Total
Vermiculite attic insulation					
3" deep, ($1.54 sf, 1 sf), 4% waste	—	sf	1.59	—	1.59
4" deep, ($2.05 sf, 1 sf), 4% waste	—	sf	2.14	—	2.14
5" deep, ($2.58 sf, 1 sf), 4% waste	—	sf	2.68	—	2.68
6" deep, ($3.11 sf, 1 sf), 4% waste	—	sf	3.24	—	3.24
Blown mineral wool insulation					
R11 (3-1/2" deep), ($.82 sf, 1 sf), 4% waste	—	sf	.85	—	.85
R19 (6" deep), ($1.21 sf, 1 sf), 4% waste	—	sf	1.25	—	1.25
R30 (10" deep), ($1.98 sf, 1 sf), 4% waste	—	sf	2.05	—	2.05
R38 (12" deep), ($2.66 sf, 1 sf), 4% waste	—	sf	2.75	—	2.75

	Craft@Hrs	Unit	Material	Labor	Total
Blown cellulose insulation					
R12 blown (3-1/2" deep), ($.80 sf, 1 sf), 4% waste	—	sf	.83	—	.83
R21 blown (6" deep), ($1.18 sf, 1 sf), 4% waste	—	sf	1.22	—	1.22
R33 blown (10" deep), ($1.91 sf, 1 sf), 4% waste	—	sf	1.99	—	1.99
R42 blown (12" deep), ($2.57 sf, 1 sf), 4% waste	—	sf	2.67	—	2.67
Blown fiberglass insulation					
R11 (5" deep), ($.72 sf, 1 sf), 4% waste	—	sf	.76	—	.76
R19 (8" deep), ($1.02 sf, 1 sf), 4% waste	—	sf	1.07	—	1.07
R30 (13" deep), ($1.67 sf, 1 sf), 4% waste	—	sf	1.74	—	1.74
R38 (16" deep), ($2.28 sf, 1 sf), 4% waste	—	sf	2.37	—	2.37
Fiberglass batt insulation in ceiling or floor					
R11 (3-1/2" deep), ($.55 sf, 1 sf), 4% waste	—	sf	.58	—	.58
R19 (6" deep), ($.86 sf, 1 sf), 4% waste	—	sf	.90	—	.90
R30 (9-1/2" deep), ($1.43 sf, 1 sf), 4% waste	—	sf	1.48	—	1.48
R38 (12" deep), ($1.91 sf, 1 sf), 4% waste	—	sf	1.99	—	1.99
Fiberglass batt insulation in wall					
R6 (1-3/4" deep) (between furring strips), ($.33 sf, 1 sf), 4% waste	—	sf	.34	—	.34
R11 (3-1/2" deep), ($.55 sf, 1 sf), 4% waste	—	sf	.58	—	.58
R19 (6" deep), ($.86 sf, 1 sf), 4% waste	—	sf	.90	—	.90
R25 (8" deep), ($1.12 sf, 1 sf), 4% waste	—	sf	1.17	—	1.17
Rigid foam insulation board					
1/2", ($.74 sf, 1 sf), 4% waste	—	sf	.77	—	.77
3/4", ($.86 sf, 1 sf), 4% waste	—	sf	.90	—	.90
1", ($1.12 sf, 1 sf), 4% waste	—	sf	1.17	—	1.17
2", ($1.52 sf, 1 sf), 4% waste	—	sf	1.57	—	1.57

Insulation Labor

Laborer	base wage	paid leave	true wage	taxes & ins.	total
Insulation installer	$25.60	2.00	$27.60	21.40	$49.00
Demolition laborer	$25.90	2.02	$27.92	19.28	$47.20

Paid leave is calculated based on two weeks paid vacation, one week sick leave, and seven paid holidays. Employer's matching portion of **FICA** is 7.65 percent. **FUTA** (Federal Unemployment) is .8 percent. **Worker's compensation** for the insulation trade was calculated using a national average of 22.68 percent. **Unemployment insurance** was calculated using a national average of 8 percent. **Health insurance** was calculated based on a projected national average for 2020 of $1,280 per employee (and family when applicable) per month. Employer pays 80 percent for a per month cost of $1,024 per employee. **Retirement** is based on a 401(k) retirement program with employer matching of 50 percent. Employee contributions to the 401(k) plan are an average of 6 percent of the true wage. **Liability insurance** is based on a national average of 12.0 percent.

Insulation Labor Productivity

	Craft@Hrs	Unit	Material	Labor	Total
Pour vermiculite attic insulation					
3" deep	1I@.008	sf	—	.39	.39
4" deep	1I@.009	sf	—	.44	.44
5" deep	1I@.009	sf	—	.44	.44
6" deep	1I@.010	sf	—	.49	.49
Blow in mineral wool insulation					
R11 (3-1/2" deep)	1I@.010	sf	—	.49	.49
R19 (6" deep)	1I@.011	sf	—	.54	.54
R30 (10" deep)	1I@.013	sf	—	.64	.64
R38 (12" deep)	1I@.016	sf	—	.78	.78
Blow in cellulose insulation					
R12 (3-1/2" deep)	1I@.010	sf	—	.49	.49
R21 (6" deep)	1I@.012	sf	—	.59	.59
R33 (10" deep)	1I@.013	sf	—	.64	.64
R42 (12" deep)	1I@.016	sf	—	.78	.78
Blow in fiberglass insulation					
R11 (5" deep)	1I@.010	sf	—	.49	.49
R19 (8" deep)	1I@.011	sf	—	.54	.54
R30 (13" deep)	1I@.013	sf	—	.64	.64
R38 (16" deep)	1I@.016	sf	—	.78	.78
Install fiberglass batt insulation in ceiling or floor					
R11 (3-1/2" deep)	1I@.009	sf	—	.44	.44
R19 (6" deep)	1I@.011	sf	—	.54	.54
R30 (9-1/2" deep)	1I@.012	sf	—	.59	.59
R38 (12" deep)	1I@.015	sf	—	.74	.74
Install fiberglass batt insulation in wall					
R6 (1-3/4" deep) (between furring strips)	1I@.010	sf	—	.49	.49
R11 (3-1/2" deep)	1I@.011	sf	—	.54	.54
R19 (6" deep)	1I@.013	sf	—	.64	.64
R25 (8" deep)	1I@.015	sf	—	.74	.74
Install rigid foam insulation board					
all thicknesses	1I@.014	sf	—	.69	.69

Manufactured Housing

Mini price list. This chapter contains some of the most common items found in manufactured housing.

Appliances

Clothes line. Steel posts, umbrella-style supports with vinyl-coated rayon lines.

	Craft@Hrs	Unit	Material	Labor	Total
replace, 30-line, exterior	7M@1.06	ea	205.00	51.50	256.50
remove	1D@.264	ea	—	12.50	12.50
replace, 7-line, exterior	7M@1.02	ea	121.00	49.60	170.60
remove	1D@.251	ea	—	11.80	11.80
remove, then reinstall	7M@.348	ea	—	16.90	16.90

Sidewall exhaust ventilator. Louvered inside grille, insulated outside door, approximately 9" x 9".

	Craft@Hrs	Unit	Material	Labor	Total
replace, 2" to 3" thick walls	7M@1.96	ea	83.50	95.30	178.80
replace, 2-3/4" to 4" thick walls	7M@1.96	ea	86.90	95.30	182.20
remove	1D@.374	ea	—	17.70	17.70
remove, then reinstall	7M@3.04	ea	3.38	148.00	151.38

Cabinets

Mobile-home type cabinets, average grade.

	Craft@Hrs	Unit	Material	Labor	Total
replace, lower units	7M@.205	lf	78.80	9.96	88.76
replace, upper units	7M@.169	lf	53.50	8.21	61.71
replace, lower island units	7M@.256	lf	110.00	12.40	122.40
replace, upper island units	7M@.184	lf	95.40	8.94	104.34
remove	1D@.116	lf	—	5.48	5.48
remove, then reinstall	7M@.236	lf	—	11.50	11.50

Doors

Access door.

	Craft@Hrs	Unit	Material	Labor	Total
replace, water heater or furnace	7M@.820	ea	72.30	39.90	112.20
remove	1D@.554	ea	—	26.10	26.10
remove, then reinstall	7M@1.02	ea	—	49.60	49.60

Mobile-home type entry door, fiberglass.

	Craft@Hrs	Unit	Material	Labor	Total
replace, flush	7M@.771	ea	159.00	37.50	196.50
replace, flush with fixed lite	7M@.771	ea	183.00	37.50	220.50
replace, flush with sliding window	7M@.771	ea	199.00	37.50	236.50
remove	1D@.581	ea	—	27.40	27.40
remove, then reinstall	7M@1.22	ea	—	59.30	59.30

	Craft@Hrs	Unit	Material	Labor	Total
Mobile-home type combination entry door (with storm door).					
replace, flush	7M@.875	ea	208.00	42.50	250.50
replace, flush with fixed lite	7M@.875	ea	238.00	42.50	280.50
replace, flush with sliding window	7M@.875	ea	263.00	42.50	305.50
remove	1D@.696	ea	—	32.90	32.90
remove, then reinstall	7M@1.09	ea	—	53.00	53.00

Door Hardware

	Craft@Hrs	Unit	Material	Labor	Total
Mobile-home type lockset.					
replace, interior passage	7M@.398	ea	20.40	19.30	39.70
replace, interior privacy	7M@.398	ea	23.50	19.30	42.80
remove	1D@.240	ea	—	11.30	11.30
remove, then reinstall	7M@.462	ea	—	22.50	22.50
Mobile-home type entrance lockset.					
replace, standard	7M@.569	ea	46.20	27.70	73.90
remove	1D@.297	ea	—	14.00	14.00
remove, then reinstall	7M@.755	ea	—	36.70	36.70
Mobile-home type deadbolt.					
replace, standard	7M@.651	ea	30.10	31.60	61.70
remove	1D@.207	ea	—	9.77	9.77
remove, then reinstall	7M@.771	ea	—	37.50	37.50

Electrical

	Craft@Hrs	Unit	Material	Labor	Total
Light fixture, mobile-home type.					
replace, average quality	7M@.590	ea	19.90	28.70	48.60
replace, high quality	7M@.590	ea	32.30	28.70	61.00
remove	1D@.426	ea	—	20.10	20.10
remove, then reinstall	7M@.633	ea	—	30.80	30.80

Finish Carpentry

	Craft@Hrs	Unit	Material	Labor	Total
Casing (door or window).					
replace, 2-1/2" prefinished plastic	7M@.038	lf	.41	1.85	2.26
remove	1D@.017	lf	—	.80	.80
Ceiling cove.					
replace, 3/4" prefinished plastic	7M@.038	lf	.47	1.85	2.32
remove	1D@.017	lf	—	.80	.80
Gimp molding.					
replace, 3/4" prefinished plastic	7M@.038	lf	.41	1.85	2.26
replace, 1" prefinished plastic	7M@.038	lf	.57	1.85	2.42
remove	1D@.017	lf	—	.80	.80

	Craft@Hrs	Unit	Material	Labor	Total
Outside corner trim.					
replace, 1" prefinished plastic	7M@.038	lf	.41	1.85	2.26
remove	1D@.017	lf	—	.80	.80
Shoe molding.					
replace, 3/4" prefinished plastic	7M@.038	lf	.41	1.85	2.26
remove	1D@.017	lf	—	.80	.80

Foundations, Setup and Site Prep

	Craft@Hrs	Unit	Material	Labor	Total
Anchor. Galvanized					
replace, 30" auger type	7M@.410	ea	17.00	19.90	36.90
replace, 48" auger type	7M@.410	ea	20.70	19.90	40.60
remove	1D@.138	ea	—	6.51	6.51
replace, slab type (includes concrete slab)	7M@.514	ea	28.00	25.00	53.00
remove	1D@.209	ea	—	9.86	9.86
Adjustable jack.					
replace, from 1'8" to 3'	7M@.307	ea	53.50	14.90	68.40
replace, from 3'1" to 5'	7M@.307	ea	66.90	14.90	81.80
replace, from 4'6" to 7'9"	7M@.307	ea	86.90	14.90	101.80
remove	1D@.116	ea	—	5.48	5.48
remove, then reinstall	7M@.462	ea	—	22.50	22.50
Raise, setup, block, and level.					
replace, single wide	7M@18.4	ea	—	894.00	894.00
replace, double wide	7M@24.6	ea	—	1,200.00	1,200.00
replace, triple wide	7M@34.9	ea	—	1,700.00	1,700.00
Relevel.					
replace, single wide	7M@5.14	ea	56.90	250.00	306.90
replace, double wide	7M@12.3	ea	113.00	598.00	711.00
replace, triple wide	7M@19.5	ea	172.00	948.00	1,120.00
Site prep.					
replace, standard conditions	7M@54.4	ea	—	2,640.00	2,640.00
Tie-down strap. Galvanized. Frame type includes adapter and is 6' long. Roof type is 37' long.					
replace, frame type	7M@.205	ea	15.40	9.96	25.36
remove	1D@.168	ea	—	7.93	7.93
replace, roof type	7M@.256	ea	36.60	12.40	49.00
remove	1D@.255	ea	—	12.00	12.00
Utility hookup.					
replace, sewer & water	7M@24.0	ea	—	1,170.00	1,170.00
replace, electrical	7M@25.4	ea	—	1,230.00	1,230.00

	Craft@Hrs	Unit	Material	Labor	Total
HVAC					
Baseboard heater, mobile-home type, electric.					
replace, 1,500 watt	7M@.205	lf	15.20	9.96	25.16
replace, 2,000 watt	7M@.205	lf	20.80	9.96	30.76
replace, 2,500 watt	7M@.205	lf	23.50	9.96	33.46
remove	1D@.116	lf	—	5.48	5.48
remove, then reinstall	7M@.307	lf	—	14.90	14.90
Furnace, mobile-home type, gas or electric.					
replace, 35,000 btu	7M@5.14	ea	806.00	250.00	1,056.00
replace, 55,000 btu	7M@5.63	ea	1,220.00	274.00	1,494.00
replace, 70,000 btu	7M@6.15	ea	1,420.00	299.00	1,719.00
replace, 80,000 btu	7M@6.15	ea	1,480.00	299.00	1,779.00
replace, 100,000 btu	7M@6.15	ea	1,710.00	299.00	2,009.00
replace, 125,000 btu	7M@6.15	ea	1,850.00	299.00	2,149.00
remove	1D@1.45	ea	—	68.40	68.40
remove, then reinstall	7M@8.72	ea	—	424.00	424.00
Moving & Towing					
Transport fees.					
replace, average, per section, per mile	7M@.079	mi	—	3.84	3.84
replace, average, pilot car, per mile	7M@.024	mi	—	1.17	1.17
Plumbing					
Water heater, mobile-home style, electric.					
replace, 10 gallon	7M@1.54	ea	162.00	74.80	236.80
replace, 20 gallon	7M@1.54	ea	182.00	74.80	256.80
replace, 30 gallon	7M@1.54	ea	230.00	74.80	304.80
replace, 40 gallon	7M@1.54	ea	253.00	74.80	327.80
replace, 50 gallon	7M@1.54	ea	325.00	74.80	399.80
remove	1D@.869	ea	—	41.00	41.00
remove, then reinstall	7M@1.95	ea	—	94.80	94.80
Water heater, mobile-home style, gas.					
replace, 20 gallon	7M@1.90	ea	182.00	92.30	274.30
replace, 30 gallon	7M@1.90	ea	221.00	92.30	313.30
replace, 40 gallon	7M@1.90	ea	259.00	92.30	351.30
replace, 50 gallon	7M@1.90	ea	318.00	92.30	410.30
remove	1D@.869	ea	—	41.00	41.00
remove, then reinstall	7M@1.95	ea	—	94.80	94.80

Roofing

	Craft@Hrs	Unit	Material	Labor	Total
Drip edge.					
replace, aluminum	7M@.031	lf	.41	1.51	1.92
remove	1D@.004	lf	—	.19	.19
replace, galvanized	7M@.031	lf	.41	1.51	1.92
remove	1D@.004	lf	—	.19	.19
Flashing.					
replace, aluminum, flexible	7M@.031	lf	1.71	1.51	3.22
replace, galvanized	7M@.031	lf	1.54	1.51	3.05
remove	1D@.004	lf	—	.19	.19
J-rail.					
replace, aluminum	7M@.031	lf	.57	1.51	2.08
remove	1D@.004	lf	—	.19	.19
Roof coating.					
replace, complete roof (all types)	7M@.006	sf	.27	.29	.56
replace, seams and edges only	7M@.009	lf	.27	.44	.71
Roof repair.					
replace, patch	7M@.771	ea	7.18	37.50	44.68
replace, section	7M@.102	lf	6.49	4.96	11.45
remove	1D@.023	lf	—	1.09	1.09

Rough Carpentry

	Craft@Hrs	Unit	Material	Labor	Total
Belt rail.					
replace, 1" x 2"	7M@.020	lf	.27	.97	1.24
replace, 1" x 4"	7M@.026	lf	.57	1.26	1.83
remove	1D@.011	lf	—	.52	.52
replace, 1" x 2", including dado cut	7M@.041	lf	.27	1.99	2.26
replace, 1" x 4", including dado cut	7M@.046	lf	.57	2.24	2.81
remove	1D@.017	lf	—	.80	.80
Exterior wall, mobile home, with belt rail.					
replace, 2" x 2"	7M@.307	lf	9.72	14.90	24.62
replace, 2" x 4"	7M@.361	lf	13.60	17.50	31.10
remove	1D@.047	lf	—	2.22	2.22
replace, 2" x 6"	7M@.410	lf	19.70	19.90	39.60
remove	1D@.047	lf	—	2.22	2.22
Interior wall, mobile home.					
replace, 2" x 2"	7M@.205	lf	8.73	9.96	18.69
replace, 2" x 4"	7M@.277	lf	11.90	13.50	25.40
replace, 2" x 6"	7M@.287	lf	17.70	13.90	31.60
remove	1D@.041	lf	—	1.94	1.94

	Craft@Hrs	Unit	Material	Labor	Total
Joist system, mobile home, per joist, 2" x 6".					
replace, 10'	7M@.216	ea	15.00	10.50	25.50
remove	1D@.139	ea	—	6.56	6.56
replace, 12'	7M@.259	ea	18.20	12.60	30.80
remove	1D@.145	ea	—	6.84	6.84
replace, 14'	7M@.302	ea	21.00	14.70	35.70
remove	1D@.151	ea	—	7.13	7.13
replace, 16'	7M@.346	ea	24.20	16.80	41.00
remove	1D@.157	ea	—	7.41	7.41
Joist system, mobile home, per joist, 2" x 8".					
replace, 10'	7M@.220	ea	17.00	10.70	27.70
remove	1D@.145	ea	—	6.84	6.84
replace, 12'	7M@.261	ea	19.90	12.70	32.60
remove	1D@.151	ea	—	7.13	7.13
replace, 14'	7M@.307	ea	23.50	14.90	38.40
remove	1D@.157	ea	—	7.41	7.41
replace, 16'	7M@.352	ea	27.00	17.10	44.10
remove	1D@.163	ea	—	7.69	7.69
Joist system, mobile home, per sf.					
replace, 2" x 6", 16" on center	7M@.021	sf	1.54	1.02	2.56
replace, 2" x 8", 16" on center	7M@.024	sf	1.84	1.17	3.01
remove	1D@.014	sf	—	.66	.66
replace, 2" x 6", 24" on center	7M@.018	sf	1.12	.87	1.99
replace, 2" x 8", 24" on center	7M@.017	sf	1.42	.83	2.25
remove	1D@.011	sf	—	.52	.52
Rafter, mobile home, 2" x 8".					
replace, 10'	7M@.307	ea	15.30	14.90	30.20
replace, 12'	7M@.327	ea	18.50	15.90	34.40
replace, 14'	7M@.349	ea	21.70	17.00	38.70
replace, 16'	7M@.391	ea	24.50	19.00	43.50
remove	1D@.127	ea	—	5.99	5.99
Truss, mobile home, bow roof.					
replace, 10'	7M@.361	ea	36.90	17.50	54.40
replace, 12'	7M@.391	ea	44.00	19.00	63.00
replace, 14'	7M@.422	ea	51.80	20.50	72.30
replace, 16'	7M@.462	ea	57.60	22.50	80.10
remove	1D@.151	ea	—	7.13	7.13
Truss, mobile home, gable roof.					
replace, 10'	7M@.410	ea	38.50	19.90	58.40
replace, 12'	7M@.431	ea	45.30	20.90	66.20
replace, 14'	7M@.450	ea	53.80	21.90	75.70
replace, 16'	7M@.462	ea	61.50	22.50	84.00
remove	1D@.151	ea	—	7.13	7.13

Siding

Horizontal lap siding, mobile home, all styles.

	Craft@Hrs	Unit	Material	Labor	Total
replace, aluminum	7M@.014	sf	.72	.68	1.40
remove	1D@.011	sf	—	.52	.52
replace, vinyl	7M@.014	sf	.57	.68	1.25
remove	1D@.011	sf	—	.52	.52

Vertical siding, mobile home style, all patterns.

	Craft@Hrs	Unit	Material	Labor	Total
replace, aluminum	7M@.014	sf	.84	.68	1.52
remove	1D@.011	sf	—	.52	.52

Shutters, mobile home style, per pair, 9" wide.

	Craft@Hrs	Unit	Material	Labor	Total
replace, 24" tall	7M@.205	ea	41.90	9.96	51.86
replace, 28" tall	7M@.205	ea	43.40	9.96	53.36
replace, 35" tall	7M@.205	ea	46.10	9.96	56.06
replace, 39" tall	7M@.205	ea	47.70	9.96	57.66
replace, 41" tall	7M@.205	ea	49.60	9.96	59.56
replace, 53" tall	7M@.205	ea	53.50	9.96	63.46
replace, 71" tall	7M@.205	ea	63.40	9.96	73.36
remove	1D@.081	ea	—	3.82	3.82
remove, then reinstall	7M@.307	ea	—	14.90	14.90

Shutters, mobile home style, per pair, 12" wide.

	Craft@Hrs	Unit	Material	Labor	Total
replace, 24" tall	7M@.256	ea	45.00	12.40	57.40
replace, 28" tall	7M@.256	ea	48.50	12.40	60.90
replace, 35" tall	7M@.256	ea	52.00	12.40	64.40
replace, 39" tall	7M@.256	ea	53.50	12.40	65.90
replace, 47" tall	7M@.256	ea	56.90	12.40	69.30
replace, 55" tall	7M@.256	ea	53.50	12.40	65.90
replace, 63" tall	7M@.256	ea	65.30	12.40	77.70
replace, 71" tall	7M@.256	ea	68.50	12.40	80.90
replace, 80" tall	7M@.256	ea	70.40	12.40	82.80
remove	1D@.081	ea	—	3.82	3.82
remove, then reinstall	7M@.361	ea	—	17.50	17.50

Steps

Redwood frame, steps, and side handrails (no landing).

	Craft@Hrs	Unit	Material	Labor	Total
replace, 24" high x 48" wide, 3 steps	7M@6.42	ea	146.00	312.00	458.00
replace, 28" high x 48" wide, 4 steps	7M@6.54	ea	170.00	318.00	488.00
replace, 32" high x 48" wide, 4 steps	7M@6.85	ea	197.00	333.00	530.00
remove	1D@.429	ea	—	20.20	20.20

Redwood frame, steps, and railing, 48" x 48" landing.

	Craft@Hrs	Unit	Material	Labor	Total
replace, 24" high x 48" wide, 3 steps	7M@7.52	ea	187.00	365.00	552.00
replace, 28" high x 48" wide, 4 steps	7M@7.71	ea	210.00	375.00	585.00
replace, 32" high x 48" wide, 4 steps	7M@8.13	ea	244.00	395.00	639.00
remove	1D@.522	ea	—	24.60	24.60

	Craft@Hrs	Unit	Material	Labor	Total
Steel frame, steps, and side handrails (no landing).					
replace, 24" high x 48" wide, 3 steps	7M@1.02	ea	199.00	49.60	248.60
replace, 28" high x 48" wide, 4 steps	7M@1.02	ea	235.00	49.60	284.60
replace, 32" high x 48" wide, 4 steps	7M@1.02	ea	253.00	49.60	302.60
remove	1D@.370	ea	—	17.50	17.50
Steel frame and railing, wood steps with 27" x 36" landing.					
replace, 24" high x 36" wide, 2 steps	7M@1.23	ea	269.00	59.80	328.80
replace, 28" high x 36" wide, 3 steps	7M@1.23	ea	307.00	59.80	366.80
replace, 32" high x 36" wide, 3 steps	7M@1.23	ea	319.00	59.80	378.80
remove	1D@.370	ea	—	17.50	17.50
Steel frame, steps, and railing, 27" x 72" landing.					
replace, 24" high x 72" wide, 3 steps	7M@1.25	ea	394.00	60.80	454.80
replace, 28" high x 72" wide, 4 steps	7M@1.25	ea	441.00	60.80	501.80
replace, 32" high x 72" wide, 4 steps	7M@1.25	ea	461.00	60.80	521.80
remove	1D@.370	ea	—	17.50	17.50
Steel frame and railing, wood steps with 4' x 5-1/2' landing.					
replace, 24" high x 48" wide, 2 steps	7M@2.00	ea	710.00	97.20	807.20
replace, 28" high x 48" wide, 3 steps	7M@2.00	ea	755.00	97.20	852.20
replace, 32" high x 48" wide, 3 steps	7M@2.00	ea	835.00	97.20	932.20
remove	1D@.522	ea	—	24.60	24.60
Steel frame and railing, wood steps with 9' x 5' landing.					
replace, 24" high x 36" wide, 2 steps	7M@2.05	ea	1,300.00	99.60	1,399.60
replace, 28" high x 36" wide, 3 steps	7M@2.05	ea	1,380.00	99.60	1,479.60
replace, 32" high x 36" wide, 3 steps	7M@2.05	ea	1,430.00	99.60	1,529.60
remove	1D@.522	ea	—	24.60	24.60

Skirting

	Craft@Hrs	Unit	Material	Labor	Total
Aluminum skirting, solid with vented panels.					
replace, 24" tall	7M@.056	lf	5.78	2.72	8.50
replace, 28" tall	7M@.056	lf	7.06	2.72	9.78
replace, 30" tall	7M@.056	lf	7.18	2.72	9.90
replace, 36" tall	7M@.056	lf	8.73	2.72	11.45
replace, 42" tall	7M@.056	lf	11.50	2.72	14.22
remove skirting	1D@.023	lf	—	1.09	1.09
replace, add for access doors	7M@.664	ea	73.60	32.30	105.90
remove access doors	1D@.116	ea	—	5.48	5.48
Vinyl skirting, simulated rock.					
replace, 30" tall	7M@.056	lf	13.60	2.72	16.32
replace, 36" tall	7M@.056	lf	17.20	2.72	19.92
replace, 48" tall	7M@.056	lf	23.60	2.72	26.32
remove skirting	1D@.023	lf	—	1.09	1.09
replace, add for access doors	7M@.664	ea	77.00	32.30	109.30
remove access doors	1D@.116	ea	—	5.48	5.48

	Craft@Hrs	Unit	Material	Labor	Total
Vinyl skirting, solid with vented panels.					
replace, 24" tall	7M@.056	lf	5.22	2.72	7.94
replace, 28" tall	7M@.056	lf	6.21	2.72	8.93
replace, 30" tall	7M@.056	lf	6.33	2.72	9.05
replace, 36" tall	7M@.056	lf	7.90	2.72	10.62
replace, 42" tall	7M@.056	lf	10.20	2.72	12.92
remove skirting	1D@.023	lf	—	1.09	1.09
replace, add for access doors	7M@.664	ea	66.20	32.30	98.50
remove access doors	1D@.116	ea	—	5.48	5.48
Wood skirting, rough-sawn plywood.					
replace, 24" tall	7M@.056	lf	3.65	2.72	6.37
replace, 28" tall	7M@.056	lf	4.36	2.72	7.08
replace, 30" tall	7M@.056	lf	4.50	2.72	7.22
replace, 36" tall	7M@.056	lf	5.49	2.72	8.21
replace, 42" tall	7M@.056	lf	7.18	2.72	9.90
remove skirting	1D@.023	lf	—	1.09	1.09
replace, add for access doors	7M@.664	ea	46.60	32.30	78.90
remove access doors	1D@.116	ea	—	5.48	5.48
Wood skirting, T1-11 or hardboard.					
replace, 24" tall	7M@.056	lf	3.38	2.72	6.10
replace, 28" tall	7M@.056	lf	4.23	2.72	6.95
replace, 30" tall	7M@.056	lf	4.23	2.72	6.95
replace, 36" tall	7M@.056	lf	5.22	2.72	7.94
replace, 42" tall	7M@.056	lf	6.77	2.72	9.49
remove skirting	1D@.023	lf	—	1.09	1.09
replace, add for access doors	7M@.664	ea	45.00	32.30	77.30
remove access doors	1D@.116	ea	—	5.48	5.48
Wall and Ceiling Panels					
Ceiling panel, drywall with textured finish.					
replace, 3/8", no battens	7M@.026	sf	.86	1.26	2.12
replace, 3/8" with battens	7M@.029	sf	1.00	1.41	2.41
remove	1D@.007	sf	—	.33	.33
Wall panel, drywall.					
replace, 1/2" with vinyl	7M@.028	sf	1.46	1.36	2.82
replace, 1/2" with wallpaper	7M@.028	sf	1.16	1.36	2.52
remove	1D@.007	sf	—	.33	.33
Ceiling and wall panel attachment.					
replace, batten	7M@.017	lf	.25	.83	1.08
replace, screw with rosette	7M@.005	ea	.34	.24	.58
replace, screw with wing clip	7M@.005	ea	.51	.24	.75
remove	1D@.002	ea	—	.09	.09

Windows

	Craft@Hrs	Unit	Material	Labor	Total
Awning window, mobile home.					
replace, 14" x 27" with 2 lites	7M@.801	ea	100.00	38.90	138.90
replace, 14" x 39" with 3 lites	7M@.801	ea	122.00	38.90	160.90
replace, 24" x 27" with 2 lites	7M@.801	ea	122.00	38.90	160.90
replace, 30" x 27" with 2 lites	7M@.801	ea	134.00	38.90	172.90
replace, 30" x 39" with 3 lites	7M@.801	ea	159.00	38.90	197.90
replace, 30" x 53" with 4 lites	7M@.801	ea	199.00	38.90	237.90
replace, 36" x 39" with 3 lites	7M@.801	ea	185.00	38.90	223.90
replace, 36" x 53" with 4 lites	7M@.801	ea	218.00	38.90	256.90
replace, 46" x 27" with 2 lites	7M@.801	ea	170.00	38.90	208.90
replace, 46" x 39" with 3 lites	7M@.801	ea	199.00	38.90	237.90
remove	1D@.348	ea	—	16.40	16.40
remove, then reinstall	7M@1.02	ea	—	49.60	49.60
Skylight, fixed, mobile home.					
replace, 14" x 22"	7M@1.79	ea	99.40	87.00	186.40
replace, 14" x 46"	7M@1.79	ea	174.00	87.00	261.00
replace, 22" x 22"	7M@1.79	ea	115.00	87.00	202.00
replace, 22" x 34"	7M@1.79	ea	404.00	87.00	491.00
remove	1D@.581	ea	—	27.40	27.40
remove, then reinstall	7M@2.25	ea	—	109.00	109.00
Skylight, ventilating, mobile home.					
replace, 14" x 22"	7M@2.25	ea	122.00	109.00	231.00
replace, 14" x 46"	7M@2.25	ea	198.00	109.00	307.00
replace, 22" x 22"	7M@2.25	ea	134.00	109.00	243.00
replace, 22" x 34"	7M@2.25	ea	465.00	109.00	574.00
remove	1D@.581	ea	—	27.40	27.40
remove, then reinstall	7M@2.56	ea	—	124.00	124.00

Time & Material Charts (selected items)

Manufactured Housing Materials

See Manufactured Housing material prices with the line items above.

Manufactured Housing Labor

Laborer	base wage	paid leave	true wage	taxes & ins.	total
MH repair specialist	$27.30	2.13	$29.43	19.17	$48.60
MH repair specialist's helper	$13.70	1.07	$14.77	12.83	$27.60

Paid leave is calculated based on two weeks paid vacation, one week sick leave, and seven paid holidays. Employer's matching portion of **FICA** is 7.65 percent. **FUTA** (Federal Unemployment) is .8 percent. **Worker's compensation** was calculated using a national average of 11.85 percent for a manufactured housing repair professional. **Unemployment insurance** was calculated using a national average of 8 percent. **Health insurance** was calculated based on a projected national average for 2020 of $1,280 per employee (and family when applicable) per month. Employer pays 80 percent for a per month cost of $1,024 per employee. **Retirement** is based on a 401(k) retirement program with employer matching of 50 percent. Employee contributions to the 401(k) plan are an average of 6 percent of the true wage. **Liability insurance** is based on a national average of 12.0 percent.

Masking & Moving

	Craft@Hrs	Unit	Material	Labor	Total
Minimum charge.					
for masking or moving work	3L@2.25	ea	22.40	93.20	115.60
Mask room.					
Small	3L@1.11	ea	4.40	46.00	50.40
Average size	3L@1.28	ea	4.98	53.00	57.98
Large	3L@1.52	ea	5.84	62.90	68.74
Very large	3L@1.85	ea	6.16	76.60	82.76
Per lf of wall	3L@.032	lf	.11	1.32	1.43
Mask window.					
Small	3L@.263	ea	2.20	10.90	13.10
Average size	3L@.308	ea	2.32	12.80	15.12
Large	3L@.371	ea	2.47	15.40	17.87
Very large	3L@.455	ea	2.60	18.80	21.40
Mask door or opening.					
Small	3L@.260	ea	2.43	10.80	13.23
Average size	3L@.303	ea	2.55	12.50	15.05
Large	3L@.364	ea	2.69	15.10	17.79
Very large	3L@.444	ea	2.80	18.40	21.20
Per lf of opening	3L@.030	lf	.07	1.24	1.31
Mask woodwork.					
Per lf	3L@.029	lf	.07	1.20	1.27
Mask electrical.					
Baseboard heater	3L@.222	ea	1.85	9.19	11.04
Light fixture	3L@.182	ea	1.42	7.53	8.95
Outlet or switch (remove cover)	3L@.114	ea	.34	4.72	5.06
Mask bathroom.					
Small	3L@1.16	ea	4.43	48.00	52.43
Average size	3L@1.35	ea	5.04	55.90	60.94
Large	3L@1.61	ea	5.93	66.70	72.63
Very large	3L@2.00	ea	6.31	82.80	89.11
Mask kitchen.					
Small	3L@1.20	ea	5.23	49.70	54.93
Average size	3L@1.41	ea	5.73	58.40	64.13
Large	3L@1.69	ea	6.25	70.00	76.25
Very large	3L@2.13	ea	6.53	88.20	94.73

	Craft@Hrs	Unit	Material	Labor	Total
Mask tile, marble, or stone.					
Per lf of edge	3L@.031	lf	.09	1.28	1.37
Move and cover room contents.					
Small	3L@1.25	ea	12.40	51.80	64.20
Average	3L@1.52	ea	13.90	62.90	76.80
Heavy or above average	3L@1.92	ea	15.80	79.50	95.30
Very heavy or above average	3L@2.70	ea	18.40	112.00	130.40

Time & Material Charts (selected items)

Masking & Moving Materials

See Masking & Moving material prices with the line items above.

Masking & Moving Labor

Laborer	base wage	paid leave	true wage	taxes & ins.	total
Laborer	$22.30	1.74	$24.04	17.36	$41.40

Paid leave is calculated based on two weeks paid vacation, one week sick leave, and seven paid holidays. Employer's matching portion of **FICA** is 7.65 percent. **FUTA** (Federal Unemployment) is .8 percent. **Worker's compensation** for the masking & moving trade was calculated using a national average of 14.05 percent. **Unemployment insurance** was calculated using a national average of 8 percent. **Health insurance** was calculated based on a projected national average for 2020 of $1,280 per employee (and family when applicable) per month. Employer pays 80 percent for a per month cost of $1,024 per employee. **Retirement** is based on a 401(k) retirement program with employer matching of 50 percent. Employee contributions to the 401(k) plan are an average of 6 percent of the true wage. **Liability insurance** is based on a national average of 12.0 percent.

Masonry

	Craft@Hrs	Unit	Material	Labor	Total
Minimum charge.					
for masonry work	4M@5.25	ea	227.00	327.00	554.00

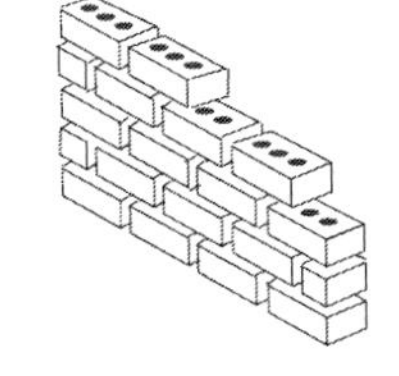

Brick walls. All brick walls are made from average to high quality face brick. Standard quality bricks will be up to 15% less and higher quality bricks will vary by as much as 45%. Brick sizes come in two general categories: non-modular and modular. Actual dimensions for non-modular standard bricks are 3-3/4" thick by 2-1/4" high by 8" long; non-modular oversize bricks are 3-3/4" thick by 2-3/4" high by 8" long; non-modular three-inch bricks are 3" thick by 2-3/4" high by 9-3/4" long. Three-inch bricks also come in lengths of 9-5/8" and 8-3/4" and widths vary from 2-5/8" to 3". They are normally used as veneer units. Modular bricks are designed to fit together with a minimum of cutting which saves labor and reduces waste. Nominal dimensions for modular bricks include the manufactured dimensions plus the thickness of the mortar joint for which the unit was designed (usually 1/2"). Actual brick sizes vary from manufacturer to manufacturer. The sizes in this section are typical.

	Craft@Hrs	Unit	Material	Labor	Total
Deduct for common brick.					
instead of average to high quality face brick	—	%	-16.0	—	—
4" wide brick wall. With 3/8" wide mortar joints. Includes 4% waste for brick and 25% waste for mortar.					
replace, standard non-modular brick	4M@.170	sf	8.64	10.60	19.24
replace, oversize non-modular brick	4M@.168	sf	7.19	10.40	17.59
replace, three-inch non-modular brick	4M@.167	sf	7.09	10.40	17.49
replace, standard brick	4M@.169	sf	6.12	10.50	16.62
replace, engineer brick	4M@.141	sf	6.36	8.77	15.13
replace, jumbo closure brick	4M@.120	sf	6.38	7.46	13.84
replace, double brick	4M@.101	sf	5.24	6.28	11.52
replace, Roman brick	4M@.160	sf	8.49	9.95	18.44
replace, Norman brick	4M@.130	sf	6.96	8.09	15.05

	Craft@Hrs	Unit	Material	Labor	Total
replace, Norwegian brick	4M@.111	sf	5.36	6.90	12.26
jumbo utility brick	4M@.102	sf	6.84	6.34	13.18
triple brick	4M@.096	sf	5.59	5.97	11.56
remove	1D@.087	sf	—	4.11	4.11

6" wide brick wall. With 3/8" wide mortar joints. Includes 4% waste for brick and 25% waste for mortar.

	Craft@Hrs	Unit	Material	Labor	Total
replace, Norwegian brick wall	4M@.124	sf	7.38	7.71	15.09
replace, Norman brick wall	4M@.148	sf	9.08	9.21	18.29
replace, jumbo brick wall	4M@.113	sf	8.26	7.03	15.29
remove	1D@.104	sf	—	4.91	4.91

8" wide brick wall. With 3/8" wide mortar joints. Includes 4% waste for brick and 25% waste for mortar.

	Craft@Hrs	Unit	Material	Labor	Total
replace, jumbo brick	4M@.127	sf	9.62	7.90	17.52
remove	1D@.132	sf	—	6.23	6.23

8" wide double wythe brick wall. With 3/8" wide mortar joints. Includes 4% waste for brick and 25% waste for mortar.

	Craft@Hrs	Unit	Material	Labor	Total
replace, standard non-modular brick	4M@.307	sf	17.90	19.10	37.00
replace, oversize non-modular brick	4M@.303	sf	14.80	18.80	33.60
replace, three-inch non-modular brick	4M@.300	sf	14.70	18.70	33.40
replace, standard brick	4M@.305	sf	12.90	19.00	31.90

	Craft@Hrs	Unit	Material	Labor	Total
replace, engineer brick	4M@.253	sf	13.20	15.70	28.90
replace, jumbo closure brick	4M@.216	sf	13.20	13.40	26.60
replace, double brick	4M@.181	sf	11.00	11.30	22.30
replace, Roman brick	4M@.287	sf	17.70	17.90	35.60
replace, Norman brick	4M@.234	sf	14.30	14.60	28.90
replace, Norwegian brick	4M@.199	sf	11.20	12.40	23.60
replace, jumbo utility brick	4M@.184	sf	14.10	11.40	25.50
replace, triple brick	4M@.173	sf	11.90	10.80	22.70
remove	1D@.132	sf	—	6.23	6.23

8" wide brick & block wall. 8" wide wall with 4" wide brick and 4" wide concrete block. With 3/8" wide mortar joints. Includes 4% waste for brick and block and 25% waste for mortar.

	Craft@Hrs	Unit	Material	Labor	Total
replace, standard non-modular brick	4M@.233	sf	12.50	14.50	27.00
replace, oversize non-modular brick	4M@.231	sf	11.00	14.40	25.40
replace, three-inch non-modular brick	4M@.230	sf	10.80	14.30	25.10
replace, standard brick	4M@.233	sf	9.91	14.50	24.41

	Craft@Hrs	Unit	Material	Labor	Total
replace, engineer brick	4M@.206	sf	10.20	12.80	23.00
replace, jumbo closure brick	4M@.188	sf	10.20	11.70	21.90
replace, double brick	4M@.170	sf	9.05	10.60	19.65
replace, Roman brick	4M@.224	sf	12.40	13.90	26.30
replace, Norman brick	4M@.197	sf	10.70	12.30	23.00
replace, Norwegian brick	4M@.180	sf	9.16	11.20	20.36
replace, jumbo utility brick	4M@.172	sf	10.60	10.70	21.30
replace, triple brick	4M@.166	sf	9.38	10.30	19.68
remove	1D@.132	sf	—	6.23	6.23

10" wide brick & block wall. 10" wide wall with 4" wide brick and 6" wide concrete block. With 3/8" wide mortar joints. Includes 4% waste for brick and block and 25% waste for mortar.

	Craft@Hrs	Unit	Material	Labor	Total
replace, standard non-modular brick	4M@.236	sf	13.00	14.70	27.70
replace, oversize non-modular brick	4M@.234	sf	11.40	14.60	26.00
replace, three-inch non-modular brick	4M@.233	sf	11.40	14.50	25.90

	Craft@Hrs	Unit	Material	Labor	Total
replace, standard brick	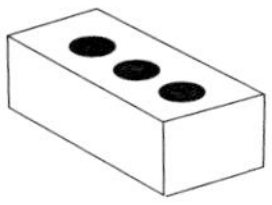 4M@.235	sf	10.50	14.60	25.10
replace, engineer brick	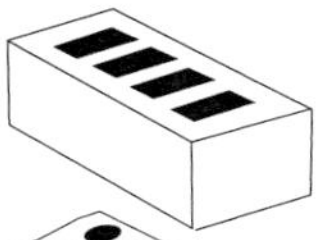 4M@.209	sf	10.70	13.00	23.70
replace, jumbo closure brick	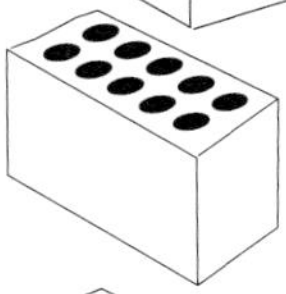 4M@.190	sf	10.70	11.80	22.50
replace, double brick	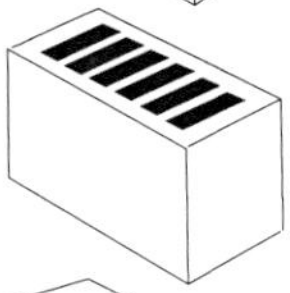 4M@.173	sf	9.57	10.80	20.37
replace, Roman brick	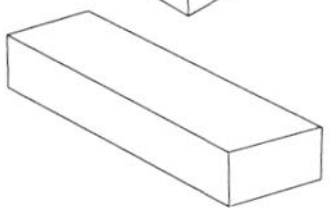 4M@.226	sf	12.90	14.10	27.00
replace, Norman brick	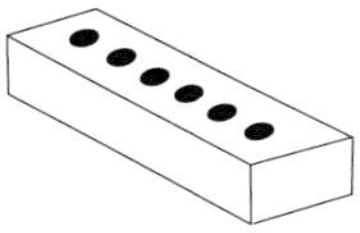 4M@.200	sf	11.20	12.40	23.60
replace, Norwegian brick	 4M@.182	sf	9.73	11.30	21.03
replace, jumbo utility brick	 4M@.175	sf	11.20	10.90	22.10
replace, triple brick	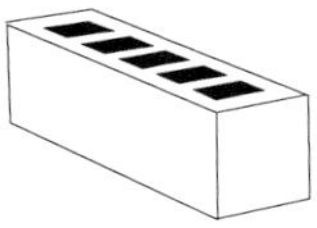 4M@.170	sf	9.93	10.60	20.53
remove	1D@.149	sf	—	7.03	7.03

12" wide brick & block wall. 12" wide wall with 4" wide brick and 8" wide concrete block. With 3/8" wide mortar joints. Includes 4% waste for brick and block and 25% waste for mortar.

	Craft@Hrs	Unit	Material	Labor	Total
replace, standard non-modular brick	4M@.240	sf	13.80	14.90	28.70
replace, oversize non-modular brick	4M@.239	sf	12.50	14.90	27.40
replace, three-inch non-modular brick	4M@.237	sf	12.50	14.70	27.20
replace, standard brick	4M@.240	sf	11.40	14.90	26.30
replace, engineer brick	4M@.214	sf	11.80	13.30	25.10
replace, jumbo closure brick	4M@.195	sf	11.90	12.10	24.00
replace, double brick	4M@.178	sf	10.50	11.10	21.60
replace, Roman brick	4M@.231	sf	13.70	14.40	28.10
replace, Norman brick	4M@.204	sf	12.40	12.70	25.10
replace, Norwegian brick	4M@.187	sf	10.70	11.60	22.30
replace, jumbo utility brick	4M@.180	sf	12.30	11.20	23.50
replace, triple brick	4M@.174	sf	10.80	10.80	21.60
remove	1D@.177	sf	—	8.35	8.35

12" wide triple wythe brick wall. With 3/8" wide mortar joints. Includes 4% waste for brick and 25% waste for mortar.

	Craft@Hrs	Unit	Material	Labor	Total
replace, standard non-modular brick	4M@.433	sf	26.50	26.90	53.40
replace, oversize non-modular brick	4M@.427	sf	21.80	26.60	48.40
replace, three-inch non-modular brick	4M@.424	sf	22.00	26.40	48.40
replace, standard brick	4M@.429	sf	19.00	26.70	45.70
replace, engineer brick	4M@.357	sf	19.60	22.20	41.80
replace, jumbo closure brick	4M@.305	sf	19.60	19.00	38.60
replace, double brick	4M@.255	sf	16.30	15.90	32.20
replace, Roman brick	4M@.407	sf	26.30	25.30	51.60
replace, Norman brick	4M@.330	sf	21.40	20.50	41.90
replace, Norwegian brick	4M@.282	sf	16.90	17.50	34.40
replace, jumbo utility brick	4M@.260	sf	21.10	16.20	37.30
replace, triple brick	4M@.244	sf	17.50	15.20	32.70
remove	1D@.177	sf	—	8.35	8.35

10" wide brick & block cavity wall. Made from 4" wide brick and 4" wide concrete block. With 2" dead air space (see page 239 for foam insulation in air space) and 3/8" wide mortar joints. Includes 4% waste for brick and block and 25% waste for mortar.

	Craft@Hrs	Unit	Material	Labor	Total
replace, standard non-modular brick	4M@.244	sf	12.60	15.20	27.80
replace, oversize non-modular brick	4M@.240	sf	11.10	14.90	26.00
replace, three-inch non-modular brick	4M@.240	sf	11.10	14.90	26.00
replace, standard brick	4M@.244	sf	10.10	15.20	25.30
replace, engineer brick	4M@.216	sf	10.30	13.40	23.70
replace, jumbo closure brick	4M@.196	sf	10.30	12.20	22.50
replace, double brick	4M@.178	sf	9.20	11.10	20.30
replace, Roman brick	4M@.234	sf	12.50	14.60	27.10
replace, Norman brick	4M@.206	sf	10.80	12.80	23.60
replace, Norwegian brick	4M@.188	sf	9.32	11.70	21.02
replace, jumbo utility brick	4M@.180	sf	10.70	11.20	21.90
replace, triple brick	4M@.174	sf	9.54	10.80	20.34
remove	1D@.132	sf	—	6.23	6.23

	Craft@Hrs	Unit	Material	Labor	Total
10" wide brick cavity wall. Made from 4" wide brick on both sides. With 2" dead air space (see cavity wall insulation below) and 3/8" wide mortar joints. Includes 4% waste for brick and 25% waste for mortar.					
replace, standard non-modular brick	4M@.321	sf	18.00	20.00	38.00
replace, oversize non-modular brick	4M@.317	sf	15.00	19.70	34.70
replace, three-inch non-modular brick	4M@.314	sf	14.80	19.50	34.30
replace, standard brick	4M@.319	sf	13.00	19.80	32.80
replace, engineer brick	4M@.266	sf	13.40	16.50	29.90
replace, jumbo closure brick	4M@.226	sf	13.40	14.10	27.50
replace, double brick	4M@.189	sf	11.20	11.80	23.00
replace, Roman brick	4M@.301	sf	17.80	18.70	36.50
replace, Norman brick	4M@.245	sf	14.60	15.20	29.80
replace, Norwegian brick	4M@.209	sf	11.40	13.00	24.40
replace, jumbo utility brick	4M@.193	sf	14.30	12.00	26.30
replace, triple brick	4M@.181	sf	12.00	11.30	23.30
remove	1D@.132	sf	—	6.23	6.23
Add for cavity wall insulation.					
2" polystyrene	4M@.001	sf	1.29	.06	1.35

	Craft@Hrs	Unit	Material	Labor	Total
Brick veneer. With 3/8" wide mortar joints. Includes 4% waste for brick and 25% waste for mortar. *common bond*					
replace, standard non-modular brick	4M@.144	sf	8.64	8.96	17.60
replace, oversize non-modular brick	4M@.143	sf	7.19	8.89	16.08
replace, three-inch non-modular brick	4M@.141	sf	7.09	8.77	15.86
replace, used brick	4M@.144	sf	12.50	8.96	21.46
replace, standard brick	4M@.144	sf	6.12	8.96	15.08
replace, engineer brick	4M@.119	sf	6.36	7.40	13.76
replace, jumbo closure brick	4M@.102	sf	6.38	6.34	12.72
replace, double brick	4M@.085	sf	5.24	5.29	10.53
replace, Roman brick	4M@.136	sf	8.49	8.46	16.95
replace, Norman brick	4M@.110	sf	6.96	6.84	13.80
replace, Norwegian brick	4M@.094	sf	5.36	5.85	11.21
replace, jumbo utility brick	4M@.087	sf	6.84	5.41	12.25
replace, triple brick	4M@.082	sf	5.59	5.10	10.69
remove	1D@.082	sf	—	3.87	3.87

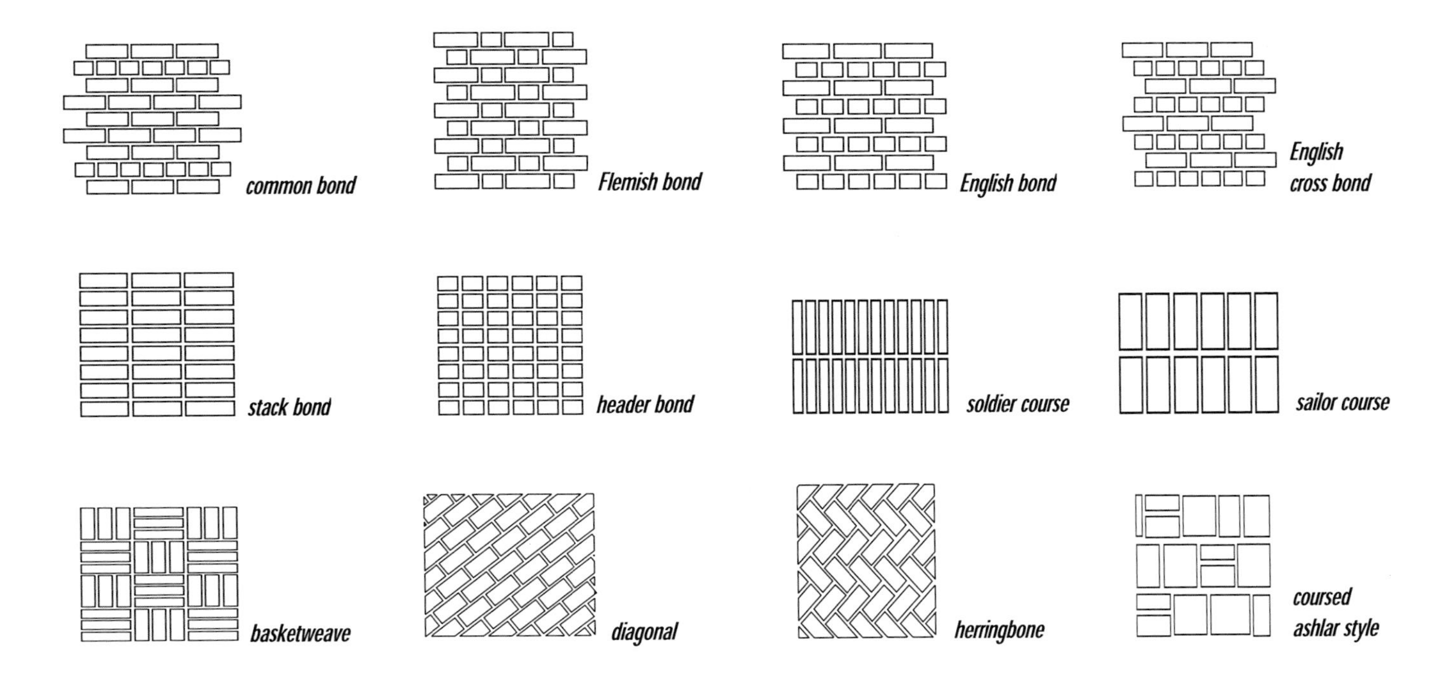

	Craft@Hrs	Unit	Material	Labor	Total
Add for other brick bonds. All brick prices above are for running bond. For other bonds add the percentage listed below.					
add for common bond (also called American bond)	—	%	—	16.0	—
add for Flemish bond	—	%	—	54.0	—
add for English bond	—	%	—	65.0	—
add for English cross bond (also called Dutch bond)	—	%	—	65.0	—
add for stack bond	—	%	—	8.0	—
add for false all header bond	—	%	—	116.0	—
add for all header bond	—	%	—	115.0	—
add for soldier course	—	%	—	15.0	—
add for sailor course	—	%	—	5.0	—
add for basketweave	—	%	—	122.0	—
add for herringbone weave	—	%	—	125.0	—
add for diagonal bond	—	%	—	90.0	—
add for coursed ashlar style brick bond with two sizes of bricks	—	%	—	70.0	—
add for curved brick walls	—	%	—	27.0	—
Add for openings in brick wall. Use only if sf of opening has been deducted from sf price. Includes angle iron lintel. Replace only.					
4" wide wall	4M@.260	lf	7.24	16.20	23.44
6" wide wall	4M@.263	lf	12.60	16.40	29.00
8" wide wall	4M@.265	lf	12.60	16.50	29.10
10" wide wall	4M@.270	lf	14.50	16.80	31.30
12" wide wall	4M@.272	lf	14.50	16.90	31.40

	Craft@Hrs	Unit	Material	Labor	Total
Add for raked joints.					
raked joints	4M@.006	sf	—	.37	.37

flat (jack arch) *elliptical* *semi-circular*

Brick arch. Per lf of opening. Replace only.

	Craft@Hrs	Unit	Material	Labor	Total
flat (jack arch)	4M@1.49	lf	22.80	92.70	115.50
elliptical	4M@1.85	lf	27.00	115.00	142.00
semi-circular	4M@1.96	lf	27.00	122.00	149.00
add to brace from below for repairs	4M@.556	lf	20.40	34.60	55.00
minimum	4M@10.0	ea	137.00	622.00	759.00

Concrete block walls. Unless otherwise noted all concrete block walls are made from 8" x 16" face block with 3/8" wide mortar joints. Cells are 36" on center and at each corner and include one length of #4 rebar in field cells and 2 lengths in corner cells for an average of 1.4 in each grouted cell. A horizontal bond beam is also calculated for every 8' of wall height and includes two lengths of #4 rebar. Ladder type horizontal wire reinforcing appears in every other course. Also includes 4% waste for block and 25% waste for mortar.

Concrete block wall.

	Craft@Hrs	Unit	Material	Labor	Total
replace, 4" wide	4M@.089	sf	3.28	5.54	8.82
replace, 6" wide	4M@.092	sf	3.83	5.72	9.55
replace, 8" wide	4M@.097	sf	4.82	6.03	10.85
replace, 10" wide	4M@.103	sf	5.95	6.41	12.36
replace, 12" wide	4M@.122	sf	7.56	7.59	15.15
remove, 4"	1D@.087	sf	—	4.11	4.11
remove, 6"	1D@.104	sf	—	4.91	4.91
remove, 8"	1D@.132	sf	—	6.23	6.23
remove, 10"	1D@.149	sf	—	7.03	7.03
remove, 12"	1D@.177	sf	—	8.35	8.35

Lightweight concrete block wall.

	Craft@Hrs	Unit	Material	Labor	Total
replace, 4" wide	4M@.083	sf	3.36	5.16	8.52
replace, 6" wide	4M@.086	sf	3.90	5.35	9.25
replace, 8" wide	4M@.091	sf	4.89	5.66	10.55
replace, 10" wide	4M@.097	sf	5.98	6.03	12.01
replace, 12" wide	4M@.114	sf	7.73	7.09	14.82
remove, 4"	1D@.087	sf	—	4.11	4.11
remove, 6"	1D@.104	sf	—	4.91	4.91
remove, 8"	1D@.132	sf	—	6.23	6.23
remove, 10"	1D@.149	sf	—	7.03	7.03
remove, 12"	1D@.177	sf	—	8.35	8.35

Slump block wall. 4" x 16" face.

	Craft@Hrs	Unit	Material	Labor	Total
replace, 4" wide	4M@.089	sf	5.51	5.54	11.05
replace, 6" wide	4M@.092	sf	6.73	5.72	12.45
replace, 8" wide	4M@.097	sf	8.34	6.03	14.37
replace, 10" wide	4M@.103	sf	9.91	6.41	16.32
replace, 12" wide	4M@.122	sf	12.50	7.59	20.09
remove, 4"	1D@.087	sf	—	4.11	4.11
remove, 6"	1D@.104	sf	—	4.91	4.91
remove, 8"	1D@.132	sf	—	6.23	6.23
remove, 10"	1D@.149	sf	—	7.03	7.03
remove, 12"	1D@.177	sf	—	8.35	8.35

Fluted block wall (fluted one side).

	Craft@Hrs	Unit	Material	Labor	Total
replace, 4" wide	4M@.089	sf	4.72	5.54	10.26
replace, 6" wide	4M@.092	sf	5.39	5.72	11.11
replace, 8" wide	4M@.097	sf	6.83	6.03	12.86
replace, 10" wide	4M@.103	sf	8.41	6.41	14.82
replace, 12" wide	4M@.122	sf	11.00	7.59	18.59
remove, 4"	1D@.087	sf	—	4.11	4.11
remove, 6"	1D@.104	sf	—	4.91	4.91
remove, 8"	1D@.132	sf	—	6.23	6.23
remove, 10"	1D@.149	sf	—	7.03	7.03
remove, 12"	1D@.177	sf	—	8.35	8.35

Fluted block wall (fluted two sides).

	Craft@Hrs	Unit	Material	Labor	Total
replace, 4" wide	4M@.089	sf	5.35	5.54	10.89
replace, 6" wide	4M@.092	sf	6.14	5.72	11.86
replace, 8" wide	4M@.097	sf	7.72	6.03	13.75
replace, 10" wide	4M@.103	sf	9.53	6.41	15.94
replace, 12" wide	4M@.122	sf	12.50	7.59	20.09
remove, 4"	1D@.087	sf	—	4.11	4.11
remove, 6"	1D@.104	sf	—	4.91	4.91
remove, 8"	1D@.132	sf	—	6.23	6.23
remove, 10"	1D@.149	sf	—	7.03	7.03
remove, 12"	1D@.177	sf	—	8.35	8.35

Glazed block wall (glazed one side).

	Craft@Hrs	Unit	Material	Labor	Total
replace, 4" wide	4M@.089	sf	18.60	5.54	24.14
replace, 6" wide	4M@.092	sf	19.30	5.72	25.02
replace, 8" wide	4M@.097	sf	20.20	6.03	26.23
replace, 10" wide	4M@.103	sf	21.40	6.41	27.81
replace, 12" wide	4M@.122	sf	22.40	7.59	29.99
remove, 4"	1D@.087	sf	—	4.11	4.11
remove, 6"	1D@.104	sf	—	4.91	4.91
remove, 8"	1D@.132	sf	—	6.23	6.23
remove, 10"	1D@.149	sf	—	7.03	7.03
remove, 12"	1D@.177	sf	—	8.35	8.35

	Craft@Hrs	Unit	Material	Labor	Total
Glazed block wall (glazed two sides).					
replace, 4" wide	4M@.089	sf	27.00	5.54	32.54
replace, 6" wide	4M@.092	sf	28.90	5.72	34.62
replace, 8" wide	4M@.097	sf	30.20	6.03	36.23
replace, 10" wide	4M@.103	sf	31.30	6.41	37.71
replace, 12" wide	4M@.122	sf	32.40	7.59	39.99
remove, 4"	1D@.087	sf	—	4.11	4.11
remove, 6"	1D@.104	sf	—	4.91	4.91
remove, 8"	1D@.132	sf	—	6.23	6.23
remove, 10"	1D@.149	sf	—	7.03	7.03
remove, 12"	1D@.177	sf	—	8.35	8.35
Split-face block wall.					
replace, 4" wide	4M@.089	sf	5.08	5.54	10.62
replace, 6" wide	4M@.092	sf	6.06	5.72	11.78
replace, 8" wide	4M@.097	sf	7.99	6.03	14.02
replace, 10" wide	4M@.103	sf	8.93	6.41	15.34
replace, 12" wide	4M@.122	sf	11.70	7.59	19.29
remove, 4"	1D@.087	sf	—	4.11	4.11
remove, 6"	1D@.104	sf	—	4.91	4.91
remove, 8"	1D@.132	sf	—	6.23	6.23
remove, 10"	1D@.149	sf	—	7.03	7.03
remove, 12"	1D@.177	sf	—	8.35	8.35
Split-rib block wall.					
replace, 4" wide	4M@.089	sf	4.80	5.54	10.34
replace, 6" wide	4M@.092	sf	5.26	5.72	10.98
replace, 8" wide	4M@.097	sf	7.91	6.03	13.94
replace, 10" wide	4M@.103	sf	9.02	6.41	15.43
replace, 12" wide	4M@.122	sf	10.20	7.59	17.79
remove, 4"	1D@.087	sf	—	4.11	4.11
remove, 6"	1D@.104	sf	—	4.91	4.91
remove, 8"	1D@.132	sf	—	6.23	6.23
remove, 10"	1D@.149	sf	—	7.03	7.03
remove, 12"	1D@.177	sf	—	8.35	8.35
Screen block. 12" x 12" face.					
4" wide screen block					
replace, pattern two sides	4M@.089	sf	9.37	5.54	14.91
replace, pattern four sides	4M@.089	sf	17.20	5.54	22.74
remove	1D@.087	sf	—	4.11	4.11

	Craft@Hrs	Unit	Material	Labor	Total
Other block colors.					
add for light ochre colored concrete block	—	%	11.0	—	—
add darker ochre colored concrete block	—	%	17.0	—	—
add for dark or bright colored concrete block	—	%	28.0	—	—
Interlocking block.					
add for interlocking concrete block	—	%	15.0	—	—
Block polystyrene insulation inserts. Additional cost to purchase block with polystyrene inserts already installed. Replace only.					
6" block	—	sf	1.47	—	1.47
8" block	—	sf	1.59	—	1.59
10" block	—	sf	1.78	—	1.78
12" block	—	sf	1.92	—	1.92
Block silicone treated perlite or vemiculite loose insulation. Includes 3% waste. Replace only.					
6" block	4M@.004	sf	.58	.25	.83
8" block	4M@.005	sf	.87	.31	1.18
10" block	4M@.005	sf	1.04	.31	1.35
12" block	4M@.005	sf	1.53	.31	1.84
Add for pilaster in block wall. 16" x 16" pilaster, single piece. For double piece 16" x 20" pilaster add **15%**. Replace only.					
per vertical lf	4M@.294	lf	21.10	18.30	39.40
Other block wall additions and deductions. Replace only.					
deduct for block walls used as backing	4M@.005	sf	—	.31	.31
deduct for wall with no horizontal wire reinforcement	4M@.001	sf	-.27	.06	.06
deduct for wall with no vertical reinforcement	4M@.006	sf	-.95	.37	.37
deduct for wall with vertical reinforcement every 48"	4M@.005	sf	-.27	.31	.31
add for wall with vertical reinforcement every 24"	4M@.005	sf	.32	.31	.63
deduct for wall with no bond beam reinforcement	4M@.004	sf	-.73	.25	.25
add for wall with bond beam reinforcement every 4'	4M@.006	sf	.45	.37	.82
add for high strength concrete block, 3,000 psi (add to materials cost only)	—	%	19.0	—	—
add for high strength concrete block, 5,000 psi (add to materials cost only)	—	%	21.0	—	—

block pilaster

	Craft@Hrs	Unit	Material	Labor	Total
Bond beam. Includes two lengths of #4 rebar. Replace only.					
6" wide	4M@.026	lf	2.21	1.62	3.83
8" wide	4M@.029	lf	2.58	1.80	4.38
10" wide	4M@.031	lf	3.01	1.93	4.94
12" wide	4M@.034	lf	3.50	2.11	5.61
Parging block walls.					
Parge block foundation wall	4P@.019	sf	.57	1.23	1.80
Grade beam cap on block wall. In some states, concrete block walls are capped with a grade beam. This not only adds strength but also levels the wall top. Because the mason does not take care to maintain a level wall, deduct **25%** from the labor cost for block walls capped with a grade beam. Up to 16" tall with four lengths of #4 rebar. Rebar from grouted cells is also bent into the grade beam.					
replace, 6" wide	4M@.106	lf	12.60	6.59	19.19
replace, 8" wide	4M@.110	lf	16.30	6.84	23.14
replace, 10" wide	4M@.115	lf	19.70	7.15	26.85
replace, 12" wide	4M@.120	lf	23.40	7.46	30.86
remove	1D@.363	lf	—	17.10	17.10
Masonry fence.					
Deduct for concrete block or brick wall installed as fence	—	%	-8.0	—	—
Clay backing tile wall. 12" x 12" face tile with 3/8" wide mortar joints. Includes 4% waste for tile and 25% waste for mortar.					
replace, 4" thick	4M@.112	sf	7.78	6.97	14.75
replace, 6" thick	4M@.127	sf	7.80	7.90	15.70
replace, 8" thick	4M@.145	sf	7.80	9.02	16.82
remove, 4"	1D@.087	sf	—	4.11	4.11
remove, 6"	1D@.104	sf	—	4.91	4.91
remove, 8"	1D@.132	sf	—	6.23	6.23
Structural tile wall. 8" x 16" face tile with 3/8" wide mortar joints. Includes 4% waste for tile and 25% waste for mortar.					
replace, 4" thick, glazed one side	4M@.204	sf	10.40	12.70	23.10
replace, 4" thick, glazed two sides	4M@.214	sf	15.90	13.30	29.20
replace, 6" thick, glazed one side	4M@.243	sf	13.70	15.10	28.80
replace, 6" thick, glazed two sides	4M@.253	sf	21.20	15.70	36.90
replace, 8" thick, glazed one side	4M@.263	sf	18.60	16.40	35.00
remove, 4"	1D@.087	sf	—	4.11	4.11
remove, 6"	1D@.104	sf	—	4.91	4.91
remove, 8"	1D@.132	sf	—	6.23	6.23

	Craft@Hrs	Unit	Material	Labor	Total
Gypsum partition tile wall. 12" x 30" face tile with 3/8" wide mortar joints. Includes 4% waste for tile and 25% waste for mortar.					
replace, 4" thick	4M@.057	sf	2.92	3.55	6.47
replace, 6" thick	4M@.083	sf	4.22	5.16	9.38
remove, 4"	1D@.087	sf	—	4.11	4.11
remove, 6"	1D@.132	sf	—	6.23	6.23

gypsum partition tile

Glass block wall. White mortar with 3/8" wide joints and ladder type wire reinforcing every other course. Includes 4% waste for glass blocks and 25% waste for mortar.

	Craft@Hrs	Unit	Material	Labor	Total
Thinline smooth-face glass block wall					
replace, 4" x 8"	4M@.333	sf	35.00	20.70	55.70
replace, 6" x 6"	4M@.310	sf	32.10	19.30	51.40
replace, 6" x 8"	4M@.267	sf	24.30	16.60	40.90
Smooth-face glass block wall					
replace, 4" x 8"	4M@.333	sf	43.60	20.70	64.30
replace, 6" x 6"	4M@.310	sf	38.60	19.30	57.90
replace, 6" x 8"	4M@.267	sf	35.00	16.60	51.60
replace, 8" x 8"	4M@.208	sf	27.20	12.90	40.10
replace, 12" x 12"	4M@.175	sf	31.70	10.90	42.60
remove	1D@.084	sf	—	3.96	3.96
Add for solar UV reflective glass block	—	%	80.0	—	—
Add for patterned face on glass block	—	%	4.0	—	—
Add for tinted glass block	—	%	12.0	—	—
Deduct for natural gray mortar in glass block wall	—	%	-2.0	—	—
Deduct for colored mortar in glass block wall	—	%	-1.0	—	—

Pavers. Pavers on sand base have sand embedded in joints with vibrating compactor. Mortar base pavers have grouted joints. Includes 4% waste for pavers and 25% waste for mortar when used.

	Craft@Hrs	Unit	Material	Labor	Total
Natural concrete pavers					
replace, sand base	4M@.147	sf	6.50	9.14	15.64
replace, mortar base	4M@.180	sf	6.50	11.20	17.70
Adobe pavers					
replace, sand base	4M@.147	sf	2.72	9.14	11.86
replace, mortar base	4M@.180	sf	2.72	11.20	13.92
Brick, standard grade paving					
replace, sand base	4M@.147	sf	3.59	9.14	12.73
replace, mortar base	4M@.180	sf	3.70	11.20	14.90
Brick, high grade paving					
replace, sand base	4M@.147	sf	4.76	9.14	13.90
replace, mortar base	4M@.180	sf	4.76	11.20	15.96
Brick, deluxe grade paving					
replace, sand base	4M@.147	sf	5.76	9.14	14.90
replace, mortar base	4M@.180	sf	5.88	11.20	17.08
remove, sand base	1D@.084	sf	—	3.96	3.96
remove, mortar base	1D@.120	sf	—	5.66	5.66
Paving made from full-size bricks					
replace, laid face up	4M@.185	sf	6.33	11.50	17.83
replace, laid edge up	4M@.208	sf	9.72	12.90	22.62
remove	1D@.090	sf	—	4.25	4.25

basketweave

	Craft@Hrs	Unit	Material	Labor	Total
Additional paver costs.					
add for curved edges	4M@.078	lf	2.18	4.85	7.03
add for steps installed over concrete	4M@.270	lf	1.03	16.80	17.83
add for steps installed over sand base	4M@.294	lf	.97	18.30	19.27
add for separate pattern at edges	4M@.076	lf	2.11	4.73	6.84
add for non-square with interlocking patterns	—	%	17.0	—	—
add for diagonal pattern	—	%	8.0	—	—
add for basketweave pattern	—	%	17.0	—	—
add for herringbone pattern	—	%	14.0	—	—

Stone walls. Stone walls are made from average to high quality stone. Prices are for stone quarried within 150 miles of the job site. Cast stone is made from a composite of crushed limestone and quartz sand. Granite is a good quality gray stone. Add **28%** for light red (pink), light purple, and light brown. Add **88%** for deep green, red, purple, blue, black, charcoal, and brown. Limestone is standard stock. Add **15%** for select stock and deduct **10%** for rustic stock. Marble is Grade A, average to high priced. Marble varies widely in price with little relationship to color and often even the quality of the stone. The more expensive grades are Italian. Sandstone is standard grade and varies from very hard rock to fairly soft. Use for brownstone work. Slate is standard grade, all colors.

Stone finishes are generally organized in this price book as natural, rough, and smooth. In practice there are many variations of each type of finish. Stone with natural finishes show the cleaving or sawing marks made in the quarry. Rough finishes are applied after the stone is quarried to achieve a specific rough look. Smooth finishes are achieved by polishing the stone. All molded work is finished smooth.

To estimate by the perch, multiply the cf price by 24.75. (A perch is 16-1/2' long, 1' high and 1-1/2' wide or 24-3/4 cubic feet.)

Rubble stone wall. Stone walls laid in a variety of rubble patterns, per cubic foot of stone. Includes an average of 1/3 cubic foot of mortar per cubic foot of wall.

	Craft@Hrs	Unit	Material	Labor	Total
replace, coral stone	4M@.518	cf	60.90	32.20	93.10
replace, fieldstone with no mortar	4M@.481	cf	44.60	29.90	74.50
replace, fieldstone	4M@.518	cf	72.90	32.20	105.10
replace, lava stone	4M@.518	cf	70.00	32.20	102.20
replace, river stone	4M@.518	cf	74.20	32.20	106.40
remove, no mortar	1D@.259	cf	—	12.20	12.20
remove, mortar	1D@.439	cf	—	20.70	20.70

rubble lay

Ashlar stone wall. Stone walls laid in a variety of ashlar patterns, per cubic foot of stone. Stone is 3-1/2" to 6" wide and various lengths and thickness. Includes mortar.

	Craft@Hrs	Unit	Material	Labor	Total
Limestone stone wall					
replace, natural finish	4M@.472	cf	58.70	29.40	88.10
replace, rough finish	4M@.472	cf	65.40	29.40	94.80
replace, smooth finish	4M@.472	cf	93.50	29.40	122.90
Marble stone wall					
replace, natural finish	4M@.472	cf	134.00	29.40	163.40
replace, rough finish	4M@.472	cf	145.00	29.40	174.40
replace, smooth finish	4M@.472	cf	180.00	29.40	209.40
Sandstone wall					
replace, natural finish	4M@.472	cf	56.50	29.40	85.90
replace, rough finish	4M@.472	cf	63.00	29.40	92.40
replace, smooth finish	4M@.472	cf	90.00	29.40	119.40
remove	1D@.439	cf	—	20.70	20.70

ashlar lay

semi-circular

elliptical arch

flat arch

	Craft@Hrs	Unit	Material	Labor	Total
Arch in stone wall. Replace only.					
flat arch in stone wall	4M@2.86	lf	137.00	178.00	315.00
elliptical arch in stone wall	4M@3.58	lf	163.00	223.00	386.00
semi-circular arch in stone wall	4M@3.70	lf	163.00	230.00	393.00
Add for curved stone wall	—	%	27.0	—	—
Stone rubble veneer.					
raplace, coral stone	4M@.243	sf	20.20	15.10	35.30
replace, field stone	4M@.243	sf	16.50	15.10	31.60
replace, flagstone	4M@.243	sf	10.40	15.10	25.50
replace, lava stone	4M@.243	sf	15.90	15.10	31.00
replace, river stone	4M@.243	sf	16.90	15.10	32.00
replace, sandstone	4M@.243	sf	23.40	15.10	38.50
remove	1D@.106	sf	—	5.00	5.00
Stone ashlar veneer.					
Flagstone veneer					
replace	4M@.236	sf	12.10	14.70	26.80
Limestone veneer					
replace, natural finish	4M@.236	sf	13.20	14.70	27.90
replace, rough finish	4M@.236	sf	14.60	14.70	29.30
replace, smooth finish	4M@.236	sf	20.70	14.70	35.40
Marble veneer					
replace, natural finish	4M@.236	sf	29.60	14.70	44.30
replace, rough finish	4M@.236	sf	32.30	14.70	47.00
replace, smooth finish	4M@.236	sf	39.70	14.70	54.40
Sandstone veneer					
replace, natural finish	4M@.236	sf	12.70	14.70	27.40
replace, rough finish	4M@.236	sf	13.90	14.70	28.60
replace, smooth finish	4M@.236	sf	19.80	14.70	34.50
Remove	1D@.106	sf	—	5.00	5.00

	Craft@Hrs	Unit	Material	Labor	Total

flat arch *elliptical arch* *semi-circular arch* *rustication*

Arch in stone veneer. Replace only.

	Craft@Hrs	Unit	Material	Labor	Total
flat arch	4M@1.75	lf	30.10	109.00	139.10
elliptical arch	4M@2.17	lf	36.00	135.00	171.00
semi-circular arch	4M@2.30	lf	36.00	143.00	179.00
add to support arch from below	4M@.556	lf	20.40	34.60	55.00
add for curved stone veneer	—	%	27.0	—	—
add for rusticated stone	—	%	55.0	—	—

Keystone. For use in stone or brick arches.

	Craft@Hrs	Unit	Material	Labor	Total
replace, concrete	4M@.606	ea	81.40	37.70	119.10
replace, cast stone	4M@.606	ea	111.00	37.70	148.70
replace, natural finish	4M@.606	ea	177.00	37.70	214.70
replace, with straight patterns	4M@.606	ea	228.00	37.70	265.70
replace, with complex straight patterns (Gothic)	4M@.606	ea	289.00	37.70	326.70
replace, with light hand carvings	4M@.606	ea	439.00	37.70	476.70
replace, with medium hand carvings	4M@.606	ea	564.00	37.70	601.70
replace, with heavy hand carvings	4M@.606	ea	791.00	37.70	828.70
remove	1D@.198	ea	—	9.35	9.35
remove for work, then reinstall	4M@1.02	ea	—	63.40	63.40

Quoin. In brick, stone, or stucco walls. Made from alternate courses of headers and stretchers. Price each includes one header and one stretcher (two stones). Cast stone quoins also include both sides. Quoins are plain, beveled, chamfered, rusticated, or rough tooled.

	Craft@Hrs	Unit	Material	Labor	Total
replace, cast stone	4M@.164	ea	178.00	10.20	188.20
replace, limestone	4M@.164	ea	202.00	10.20	212.20
replace, sandstone	4M@.164	ea	191.00	10.20	201.20
remove	1D@.196	ea	—	9.25	9.25
remove for work, then reinstall (per stone)	4M@.434	ea	—	27.00	27.00

stone quoin

Cultured stone veneer panels. 3/4" to 1-1/2" thick.

	Craft@Hrs	Unit	Material	Labor	Total
replace, smooth finish	4M@.263	sf	9.16	16.40	25.56
replace, rough finish	4M@.263	sf	12.40	16.40	28.80
replace, terrazzo style finish	4M@.263	sf	27.90	16.40	44.30
remove	1D@.074	sf	—	3.49	3.49

Ceramic veneer panels.

	Craft@Hrs	Unit	Material	Labor	Total
replace, precast	4M@.263	sf	22.40	16.40	38.80
remove	1D@.074	sf	—	3.49	3.49

	Craft@Hrs	Unit	Material	Labor	Total

Natural stone veneer panels. Granite, slate, and marble panels are 3/4" to 1-1/2" thick. Limestone and sandstone panels are 2" to 3" thick. See page 248 for more information about quality and finishes.

	Craft@Hrs	Unit	Material	Labor	Total
Granite veneer					
replace, natural finish	4M@.298	sf	41.30	18.50	59.80
replace, rough finish	4M@.298	sf	43.90	18.50	62.40
replace, smooth finish	4M@.298	sf	50.70	18.50	69.20
Limestone veneer					
replace, natural finish	4M@.298	sf	19.00	18.50	37.50
replace, rough finish	4M@.298	sf	28.50	18.50	47.00
replace, smooth finish	4M@.298	sf	37.50	18.50	56.00
Marble veneer					
replace, natural finish	4M@.298	sf	47.60	18.50	66.10
replace, rough finish	4M@.298	sf	61.80	18.50	80.30
replace, smooth finish	4M@.298	sf	77.60	18.50	96.10
Sandstone veneer					
replace, natural finish	4M@.298	sf	33.90	18.50	52.40
replace, rough finish	4M@.298	sf	37.20	18.50	55.70
replace, smooth finish	4M@.298	sf	41.50	18.50	60.00
Slate veneer					
replace, natural finish	4M@.298	sf	35.00	18.50	53.50
replace, rough finish	4M@.298	sf	38.00	18.50	56.50
replace, smooth finish	4M@.298	sf	42.60	18.50	61.10
Remove stone veneer panel	1D@.074	sf	—	3.49	3.49

Door architrave. Standard grade: Cast stone is 6" or less wide with simple straight patterns. High grade: Cast stone is 6" or less wide with more complex (Gothic) patterns. Deluxe grade: Same as high grade, but with arched top or flat top and decorative work around corners or center of header. Or, it may have straight patterns with cast stone up to 10" wide. Custom grade: Cast stone 6" or less wide with hand carvings on header. Or, it may be up to 10" wide with complex patterns (Gothic) and an arched top; or, it may have straight patterns with cast stone up to 12" wide. Custom deluxe grade: Cast stone 6" or less wide with heavy hand carvings on header (e.g. lion's face or human face) or lighter hand carvings throughout. Or, it may be up to 12" wide with complex straight patterns (Gothic). With arched top.

	Craft@Hrs	Unit	Material	Labor	Total
Cast stone architrave					
replace, standard grade	4M@.980	lf	35.30	61.00	96.30
replace, high grade	4M@.980	lf	38.60	61.00	99.60
replace, deluxe grade	4M@.980	lf	52.00	61.00	113.00
replace, custom grade	4M@.980	lf	63.90	61.00	124.90
replace, custom deluxe grade	4M@.980	lf	81.80	61.00	142.80
Limestone architrave					
replace, standard grade	4M@.980	lf	68.60	61.00	129.60
replace, high grade	4M@.980	lf	125.00	61.00	186.00
replace, deluxe grade	4M@.980	lf	180.00	61.00	241.00
replace, custom grade	4M@.980	lf	224.00	61.00	285.00
replace, custom deluxe grade	4M@.980	lf	275.00	61.00	336.00
Marble architrave					
replace, standard grade	4M@.980	lf	102.00	61.00	163.00
replace, high grade	4M@.980	lf	160.00	61.00	221.00
replace, deluxe grade	4M@.980	lf	215.00	61.00	276.00
replace, custom grade	4M@.980	lf	259.00	61.00	320.00
replace, custom deluxe grade	4M@.980	lf	306.00	61.00	367.00

	Craft@Hrs	Unit	Material	Labor	Total
Sandstone architrave					
replace, standard grade	4M@.980	lf	61.50	61.00	122.50
replace, high grade	4M@.980	lf	120.00	61.00	181.00
replace, deluxe grade	4M@.980	lf	173.00	61.00	234.00
replace, custom grade	4M@.980	lf	218.00	61.00	279.00
replace, custom deluxe grade	4M@.980	lf	265.00	61.00	326.00
Remove door architrave	1D@.131	lf	—	6.18	6.18

Window architrave. See Door architrave for more information on grades.

	Craft@Hrs	Unit	Material	Labor	Total
Cast stone architrave					
replace, standard grade	4M@1.00	lf	34.10	62.20	96.30
replace, high grade	4M@1.00	lf	37.70	62.20	99.90
replace, deluxe grade	4M@1.00	lf	50.90	62.20	113.10
replace, custom grade	4M@1.00	lf	62.10	62.20	124.30
replace, custom deluxe grade	4M@1.00	lf	79.60	62.20	141.80
Limestone architrave					
replace, standard grade	4M@1.00	lf	66.60	62.20	128.80
replace, high grade	4M@1.00	lf	123.00	62.20	185.20
replace, deluxe grade	4M@1.00	lf	175.00	62.20	237.20
replace, custom grade	4M@1.00	lf	218.00	62.20	280.20
replace, custom deluxe grade	4M@1.00	lf	264.00	62.20	326.20
Marble architrave					
replace, standard grade	4M@1.00	lf	99.50	62.20	161.70
replace, high grade	4M@1.00	lf	156.00	62.20	218.20
replace, deluxe grade	4M@1.00	lf	210.00	62.20	272.20
replace, custom grade	4M@1.00	lf	248.00	62.20	310.20
replace, custom deluxe grade	4M@1.00	lf	296.00	62.20	358.20
Sandstone architrave					
replace, standard grade	4M@1.00	lf	63.40	62.20	125.60
replace, high grade	4M@1.00	lf	120.00	62.20	182.20
replace, deluxe grade	4M@1.00	lf	172.00	62.20	234.20
replace, custom grade	4M@1.00	lf	214.00	62.20	276.20
replace, custom deluxe grade	4M@1.00	lf	261.00	62.20	323.20
Remove window architrave	1D@.131	lf	—	6.18	6.18

Additional stone architrave costs.

	Craft@Hrs	Unit	Material	Labor	Total
add for round or elliptical window	—	%	65.0	—	—
replace section cut to match	4M@3.34	ea	895.00	208.00	1,103.00

Cut trim or cornice stone. Per 4" of width. Use for friezes, architraves, cornices, string courses, band courses, and so on. Cut Stone: All horizontal patterns are shallow. Typical depth is 3/4" or less but may be as deep as 1". Complex horizontal patterns are deep Gothic style reliefs that require two passes through the stone planer. Vertical patterns are straight vertical cuts like dentil, meander, or fretwork. Light hand carvings are usually less than 3/4" deep and do not cover the entire face of the stone with detailed work. Medium hand carvings are heavier carvings like vermiculation, Acanthus, and so forth. Heavy hand carvings are heavy and deep carvings including a lion's face, a human face etc.

	Craft@Hrs	Unit	Material	Labor	Total
Cut stone trim					
replace, all horizontal patterns	4M@.370	lf	21.80	23.00	44.80
replace, complex horizontal patterns (Gothic)	4M@.370	lf	30.20	23.00	53.20
replace, with vertical patterns	4M@.370	lf	33.60	23.00	56.60
replace, with light hand carvings	4M@.370	lf	47.60	23.00	70.60
replace, with medium hand carvings	4M@.370	lf	58.90	23.00	81.90
replace, with heavy hand carvings	4M@.370	lf	77.10	23.00	100.10
Limestone trim					
replace, all horizontal patterns	4M@.370	lf	33.10	23.00	56.10
replace, complex horizontal patterns (Gothic)	4M@.370	lf	63.00	23.00	86.00
replace, with vertical patterns	4M@.370	lf	121.00	23.00	144.00
replace, with light hand carvings	4M@.370	lf	175.00	23.00	198.00
replace, with medium hand carvings	4M@.370	lf	219.00	23.00	242.00
replace, with heavy hand carvings	4M@.370	lf	266.00	23.00	289.00
Marble trim					
replace, all horizontal patterns	4M@.370	lf	63.20	23.00	86.20
replace, complex horizontal patterns (Gothic)	4M@.370	lf	93.20	23.00	116.20
replace, with vertical patterns	4M@.370	lf	150.00	23.00	173.00
replace, with light hand carvings	4M@.370	lf	204.00	23.00	227.00
replace, with medium hand carvings	4M@.370	lf	247.00	23.00	270.00
replace, with heavy hand carvings	4M@.370	lf	296.00	23.00	319.00
Sandstone trim					
replace, all horizontal patterns	4M@.370	lf	26.40	23.00	49.40
replace, complex horizontal patterns (Gothic)	4M@.370	lf	56.10	23.00	79.10
replace, with vertical patterns	4M@.370	lf	113.00	23.00	136.00
replace, with light hand carvings	4M@.370	lf	169.00	23.00	192.00
replace, with medium hand carvings	4M@.370	lf	211.00	23.00	234.00
replace, with heavy hand carvings	4M@.370	lf	260.00	23.00	283.00
Remove cut trim	1D@.104	lf	—	4.91	4.91
Replace section cut to match	4M@2.94	ea	735.00	183.00	918.00

horizontal patterns

	Craft@Hrs	Unit	Material	Labor	Total
Cut stone sill or stool. Up to 10" wide.					
replace, 1-1/2" bluestone	4M@.334	lf	22.20	20.80	43.00
replace, 4" cast stone	4M@.334	lf	15.50	20.80	36.30
replace, 1-1/2" granite	4M@.334	lf	24.10	20.80	44.90
replace, 4" granite	4M@.334	lf	43.50	20.80	64.30
replace, 1-1/2" limestone	4M@.334	lf	21.80	20.80	42.60
replace, 4" limestone	4M@.334	lf	39.00	20.80	59.80
replace, 1-1/2" marble	4M@.334	lf	25.80	20.80	46.60
replace, 1-1/2" sandstone	4M@.334	lf	15.90	20.80	36.70
replace, 4" sandstone	4M@.334	lf	31.20	20.80	52.00
replace, 1-1/2" slate	4M@.334	lf	25.90	20.80	46.70
remove	1D@.110	lf	—	5.19	5.19
Terra cotta wall cap.					
replace, 10" wide	4M@.185	lf	11.80	11.50	23.30
replace, 12" wide	4M@.185	lf	19.10	11.50	30.60
remove	1D@.063	lf	—	2.97	2.97
Aluminum wall cap.					
replace, all wall widths	4M@.181	lf	22.20	11.30	33.50
remove	1D@.019	lf	—	.90	.90
Wall coping stones. From 4" to 6" thick at center tapering or curved taper to edges. Widths over 8" include drip grooves on each side.					
Concrete wall coping stones					
replace, 10" wide	4M@.167	lf	24.10	10.40	34.50
replace, 12" wide	4M@.167	lf	25.10	10.40	35.50
replace, 14" wide	4M@.167	lf	32.10	10.40	42.50
Granite wall coping stones					
replace, 8" wide	4M@.185	lf	48.40	11.50	59.90
replace, 10" wide	4M@.185	lf	56.00	11.50	67.50
replace, 12" wide	4M@.185	lf	63.40	11.50	74.90
Limestone wall coping stones					
replace, 8" wide	4M@.185	lf	28.90	11.50	40.40
replace, 10" wide	4M@.185	lf	29.50	11.50	41.00
replace, 12" wide	4M@.185	lf	34.20	11.50	45.70
Marble coping stones					
replace, 8" wide	4M@.185	lf	41.00	11.50	52.50
replace, 10" wide	4M@.185	lf	50.80	11.50	62.30
replace, 12" wide	4M@.185	lf	63.40	11.50	74.90
Remove coping stones	1D@.104	lf	—	4.91	4.91

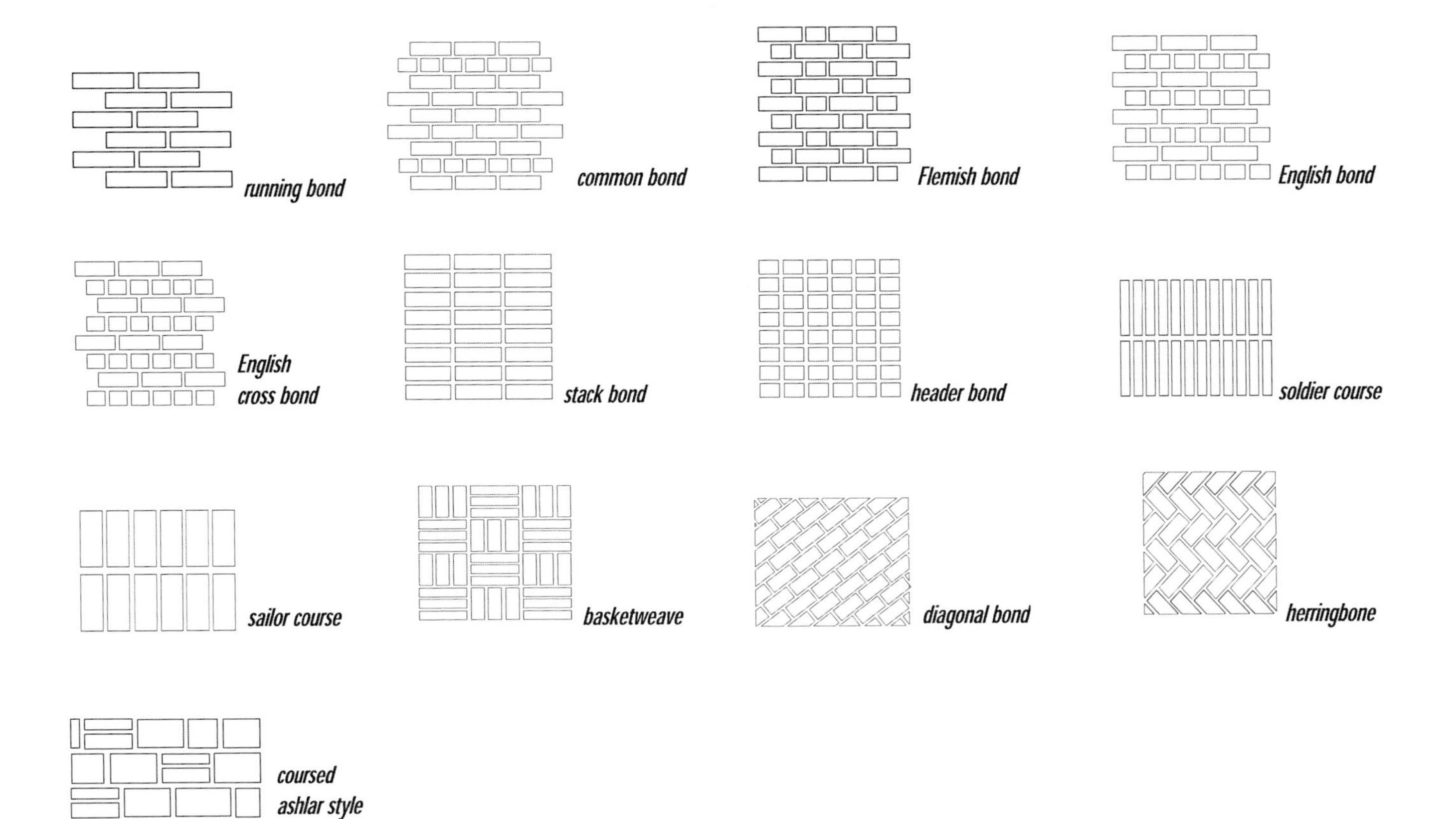

	Craft@Hrs	Unit	Material	Labor	Total
Repoint brick wall. Tuck pointing. To mask and grout deduct **12%**. Replace only.					
running bond	4M@.108	sf	.57	6.72	7.29
common bond (also called American bond)	4M@.111	sf	.57	6.90	7.47
Flemish bond	4M@.121	sf	.57	7.53	8.10
English bond	4M@.128	sf	.57	7.96	8.53
English cross bond (also called Dutch bond)	4M@.128	sf	.57	7.96	8.53
stack bond	4M@.082	sf	.57	5.10	5.67
all header bond	4M@.093	sf	.57	5.78	6.35
soldier course	4M@.106	sf	.57	6.59	7.16
sailor course	4M@.099	sf	.57	6.16	6.73
basketweave	4M@.164	sf	.57	10.20	10.77
herringbone weave	4M@.167	sf	.57	10.40	10.97
diagonal bond	4M@.144	sf	.57	8.96	9.53
coursed ashlar style bond with two sizes of bricks	4M@.120	sf	.57	7.46	8.03
add to repoint brick wall with hard mortar	—	%	12.0	—	—
add to repoint brick wall with very hard mortar	—	%	22.0	—	—
add to repoint brick wall with butter joint mortar	—	%	18.0	—	—
minimum	4M@5.00	ea	27.20	311.00	338.20

	Craft@Hrs	Unit	Material	Labor	Total
Repoint stone wall. Tuck pointing. Replace only.					
repoint rubble lay wall	4M@.278	sf	2.59	17.30	19.89
repoint ashlar wall	4M@.260	sf	2.00	16.20	18.20
minimum	4M@5.00	ea	27.20	311.00	338.20

ashlar lay

Salvage bricks. Existing bricks are removed and old mortar chipped away. An average of 75% of the bricks are salvaged. To replace the missing bricks, 25% of the salvaged bricks are sawn in half length wise and back filled with mortar to fit. Finish wall includes 50% full bricks and 50% backfilled half bricks. Normally not used for large walls, but on sections of damaged wall where bricks cannot be matched.

	Craft@Hrs	Unit	Material	Labor	Total
remove brick veneer, salvage bricks, and relay	4M@.877	sf	2.96	54.50	57.46
Brick wall repair. Replace only.					
replace single brick	4M@.430	sf	3.42	26.70	30.12
tooth in brick patch up to 4 sf	4M@.504	sf	63.20	31.30	94.50
tooth in brick patch over 4 sf	4M@.408	sf	15.90	25.40	41.30
minimum charge to repair brick wall	4M@11.5	ea	104.00	715.00	819.00
Stone wall repair. Replace only.					
replace single stone	4M@.451	ea	6.65	28.10	34.75
tooth in stone patch up to 4 sf	4M@.566	ea	123.00	35.20	158.20
tooth in stone patch over 4 sf	4M@.585	ea	30.40	36.40	66.80
remove segment of rubble wall and relay	5M@4.00	cf	7.34	273.00	280.34
remove segment of rubble veneer and relay	5M@1.92	sf	2.45	131.00	133.45
remove segment of ashlar wall and relay	5M@3.85	cf	5.34	263.00	268.34
remove segment of ashlar veneer and relay	5M@1.89	sf	1.79	129.00	130.79
minimum charge to repair stone wall	4M@12.7	ea	198.00	790.00	988.00

Stone carving. Work performed on site or in a shop within 150 miles of the site. Does not include clay or plaster model. Models for light carvings may run 15% less, and models for heavy carvings like faces may run 20% to 45% more. It is not unusual for models to cost as much as or more than the actual carvings. Replace only.

	Craft@Hrs	Unit	Material	Labor	Total
with light hand carvings	6M@.015	si	—	1.73	1.73
with medium hand carvings	6M@.022	si	—	2.53	2.53
with heavy hand carvings	6M@.031	si	—	3.57	3.57
add to match existing work	—	%	—	120.0	—
clay or plaster model for stone carvings	—	si	2.44	—	2.44
minimum charge for model for stone carvings	—	ea	251.00	—	251.00
Carved stone repair. Replace only.					
repair carved stone with epoxy	1M@.024	si	.37	1.71	2.08
repair carved stone with stucco	1M@.022	si	.03	1.56	1.59
minimum charge to repair carved stone	4M@4.58	ea	57.20	285.00	342.20

vermiculation

	Craft@Hrs	Unit	Material	Labor	Total
Stone repair. Replace only.					
epoxy repair and pin broken stone	1M@1.33	ea	14.80	94.60	109.40
reconstitute delaminating stone with epoxy & pins	1M@.013	si	.41	.92	1.33
reface damaged stone with grout mixed to match	1M@.149	si	.03	10.60	10.63
minimum charge for epoxy repair work	4M@3.17	ea	57.20	197.00	254.20
Clean masonry. Replace only.					
hand clean brick wall	3M@.042	sf	.08	2.12	2.20
hand clean stone wall	3M@.045	sf	.08	2.27	2.35
pressure spray brick wall	3M@.011	sf	.11	.55	.66
pressure spray stone wall	3M@.011	sf	.11	.55	.66
steam clean brick wall	3M@.015	sf	—	.76	.76
steam clean stone wall	3M@.016	sf	—	.81	.81
wet sandblast brick wall	3M@.046	sf	.53	2.32	2.85
wet sandblast stone wall	3M@.048	sf	.53	2.42	2.95
Cut opening in masonry wall. Cut masonry wall with concrete saw. Replace only.					
Cut opening in brick wall					
to 6" thick	5M@.179	lf	—	12.20	12.20
7" to 12" thick	5M@.200	lf	—	13.60	13.60
veneer	5M@.172	lf	—	11.70	11.70
Cut opening in stone wall					
to 6" thick	5M@.200	lf	—	13.60	13.60
to 7" to 12" thick	5M@.226	lf	—	15.40	15.40
stone veneer	5M@.192	lf	—	13.10	13.10
Shore masonry opening for repairs. Replace only.					
flat opening, per lf of header	5M@.385	lf	19.40	26.30	45.70
arched opening	5M@.504	lf	20.40	34.40	54.80
Concrete lintel. Replace only.					
4" wide wall	5M@.078	lf	15.70	5.32	21.02
6" wide wall	5M@.100	lf	21.10	6.82	27.92
8" wide wall	5M@.112	lf	22.80	7.64	30.44
10" wide wall	5M@.135	lf	39.00	9.21	48.21
12" wide wall	5M@.149	lf	51.70	10.20	61.90
add for key cast into lintel	—	ea	79.80	—	79.80

lintel with plain key

lintel with winged key

lintel with recessed key

	Craft@Hrs	Unit	Material	Labor	Total
Angle iron lintel. Replace only.					
4" wide wall	4M@.067	lf	6.30	4.17	10.47
6" wide wall	4M@.067	lf	12.60	4.17	16.77
8" wide wall	4M@.073	lf	13.40	4.54	17.94
10" wide wall	4M@.074	lf	15.00	4.60	19.60
12" wide wall	4M@.075	lf	21.80	4.67	26.47

Time & Material Charts (selected items)

Masonry Materials

	Craft@Hrs	Unit	Material	Labor	Total
Rebar					
#3 (3/8" .376 pound per lf), ($5.68 per 20' bar, 20 lf), 4% waste	—	lf	.29	—	.29
#4 (1/2" .668 pounds per lf), ($7.24 per 20' bar, 20 lf), 4% waste	—	lf	.37	—	.37
#5 (5/8" 1.043 pounds per lf), ($11.20 per 20' bar, 20 lf), 4% waste	—	lf	.58	—	.58
#6 (3/4" 1.502 pounds per lf), ($17.10 per 20' bar, 20 lf), 4% waste	—	lf	.91	—	.91
Mortar and grout supplies					
gypsum cement, ($27.00 per bag, 80 lb), 12% waste	—	lb	.37	—	.37
masonry cement, ($11.40 per bag, 70 lb), 12% waste	—	lb	.20	—	.20
white masonry cement, ($34.30 per bag, 70 lb), 12% waste	—	lb	.55	—	.55
hydrated lime, ($11.90 per bag, 50 lb), 12% waste	—	lb	.26	—	.26
double hydrated lime, ($16.00 per bag, 50 lb), 12% waste	—	lb	.34	—	.34
screened and washed sand, delivered, ($39.70 per ton, 1 ton), 0% waste	—	ton	39.70	—	39.70
Reinforcing wire strip					
4" ladder style, ($.25 lf, 1 lf), 2% waste	—	lf	.25	—	.25
12" ladder style, ($.31 lf, 1 lf), 2% waste	—	lf	.32	—	.32
4" truss style, ($.30 lf, 1 lf), 2% waste	—	lf	.31	—	.31
12" truss style, ($.37 lf, 1 lf), 2% waste	—	lf	.39	—	.39
add 30% for galvanized					

	Craft@Hrs	Unit	Material	Labor	Total
Masonry ties					
galvanized 22 gauge corrugated veneer wall tie, ($.30 ea, 1 ea), 2% waste	—	ea	.31	—	.31
galvanized 16 gauge corrugated veneer wall tie, ($.31 ea, 1 ea), 2% waste	—	ea	.32	—	.32
rectangular galvanized wall tie, ($.34 ea, 1 ea), 2% waste	—	ea	.35	—	.35
galvanized "Z" style cavity wall tie, ($.33 ea, 1 ea), 2% waste	—	ea	.34	—	.34
Grout					
for 6" wide bond beam, ($1.27 lf, 1 lf), 11% waste	—	lf	1.41	—	1.41
for 8" wide bond beam, ($1.61 lf, 1 lf), 11% waste	—	lf	1.78	—	1.78
for 10" wide bond beam, ($1.99 lf, 1 lf), 11% waste	—	lf	2.22	—	2.22
for 12" wide bond beam, ($2.44 lf, 1 lf), 11% waste	—	lf	2.70	—	2.70
for cells in 6" wide concrete block, ($1.55 lf, 1 lf), 11% waste	—	lf	1.72	—	1.72
for cells in 8" wide concrete block, ($2.27 lf, 1 lf), 11% waste	—	lf	2.52	—	2.52
for cells in 10" wide concrete block, ($2.87 lf, 1 lf), 11% waste	—	lf	3.18	—	3.18
for cells in 12" wide concrete block, ($3.65 lf, 1 lf), 11% waste	—	lf	4.07	—	4.07
3/8" brick mortar					
for 4" wide std non-modular or 3", ($13.00 cf, 25 sf), 25% waste	—	sf	.67	—	.67
for 4" wide oversize non-modular, ($13.00 cf, 26 sf), 25% waste	—	sf	.63	—	.63
for 4" wide engineer, ($13.00 cf, 27 sf), 25% waste	—	sf	.59	—	.59
for 4" wide jumbo closure, ($13.00 cf, 31 sf), 25% waste	—	sf	.53	—	.53
for 4" wide double, ($13.00 cf, 37 sf), 25% waste	—	sf	.44	—	.44
for 4" wide Roman, ($13.00 cf, 21 sf), 25% waste	—	sf	.76	—	.76
for 4" wide Norman, ($13.00 cf, 26 sf), 25% waste	—	sf	.63	—	.63
for 4" wide Norwegian, ($13.00 cf, 30 sf), 25% waste	—	sf	.54	—	.54
for 4" wide jumbo utility, ($13.00 cf, 36 sf), 25% waste	—	sf	.45	—	.45
for 4" wide triple, ($13.00 cf, 44 sf), 25% waste	—	sf	.35	—	.35

	Craft@Hrs	Unit	Material	Labor	Total
for 6" wide Norwegian, ($13.00 cf, 20 sf), 25% waste	—	sf	.81	—	.81
for 6" wide Norman, ($13.00 cf, 16.8 sf), 25% waste	—	sf	.98	—	.98
for 6" wide jumbo, ($13.00 cf, 24 sf), 25% waste	—	sf	.69	—	.69
for 8" wide jumbo, ($13.00 cf, 18 sf), 25% waste	—	sf	.92	—	.92
3/8" concrete block mortar (8" x 16" face)					
for 4" wide, ($13.00 cf, 40 sf), 25% waste	—	sf	.41	—	.41
for 6" wide, ($13.00 cf, 30 sf), 25% waste	—	sf	.54	—	.54
for 8" wide, ($13.00 cf, 25 sf), 25% waste	—	sf	.66	—	.66
for 10" wide, ($13.00 cf, 20.5 sf), 25% waste	—	sf	.80	—	.80
for 12" wide, ($13.00 cf, 16.6 sf), 25% waste	—	sf	.98	—	.98
3/8" concrete slump block mortar (4" x 16" face)					
for 4" wide, ($13.00 cf, 34 sf), 25% waste	—	sf	.47	—	.47
for 6" wide, ($13.00 cf, 24 sf), 25% waste	—	sf	.69	—	.69
for 8" wide, ($13.00 cf, 21 sf), 25% waste	—	sf	.76	—	.76
for 10" wide, ($13.00 cf, 16.5 sf), 25% waste	—	sf	.99	—	.99
for 12" wide, ($13.00 cf, 12.6 sf), 25% waste	—	sf	1.29	—	1.29
3/8" concrete screen block mortar (12" x 12" face)					
for 4" wide, ($13.00 cf, 35.6 sf), 25% waste	—	sf	.45	—	.45
3/8" backing tile mortar (12" x 12" face)					
for 4" thick, ($13.00 cf, 35.6 sf), 25% waste	—	sf	.45	—	.45
for 6" thick, ($13.00 cf, 26.5 sf), 25% waste	—	sf	.62	—	.62
for 8" thick, ($13.00 cf, 22.5 sf), 25% waste	—	sf	.73	—	.73
3/8" structural tile mortar (8" x 16" face)					
for 4" thick, ($13.00 cf, 40 sf), 25% waste	—	sf	.41	—	.41
for 6" thick, ($13.00 cf, 30 sf), 25% waste	—	sf	.54	—	.54
for 8" thick, ($13.00 cf, 25 sf), 25% waste	—	sf	.66	—	.66
3/8" gypsum partition tile mortar (12" x 30" face)					
for 4" thick, ($13.00 cf, 78 sf), 25% waste	—	sf	.22	—	.22
for 6" thick, ($13.00 cf, 66 sf), 25% waste	—	sf	.25	—	.25
3/8" white mortar for glass blocks					
for block with 4" x 8" face, ($16.50 cf, 26 sf), 25% waste	—	sf	.80	—	.80
for block with 6" x 6" face, ($16.50 cf, 29 sf), 25% waste	—	sf	.71	—	.71
for block with 6" x 8" face, ($16.50 cf, 36 sf), 25% waste	—	sf	.57	—	.57
for block with 8" x 8" face, ($16.50 cf, 44 sf), 25% waste	—	sf	.46	—	.46
for block with 12" x 12" face, ($16.50 cf, 35.6 sf), 25% waste	—	sf	.58	—	.58
deduct for natural gray mortar in glass block wall	—	%	-22.0	—	—
deduct for colored mortar in glass block wall	—	%	-8.0	—	—

	Craft@Hrs	Unit	Material	Labor	Total
Mortar for stone walls					
mortar for rubble stone wall, ($13.00 cf, 4.41 sf), 32% waste	—	cf	3.90	—	3.90
mortar for ashlar stone wall, ($13.00 cf, 5.5 sf), 25% waste	—	cf	2.97	—	2.97
mortar for rubble stone veneer, ($13.00 cf, 13.2 sf), 32% waste	—	sf	1.30	—	1.30
mortar for ashlar stone veneer, ($13.00 cf, 16.5 sf), 25% waste	—	sf	.99	—	.99
Brick					
deduct for common brick	—	%	-16.0	—	—
add **$852** per thousand for glazed brick					
4" wide standard non-modular, ($1,120.00 per thousand, 1,000 ea), 4% waste	—	ea	1.18	—	1.18
4" wide oversize non-modular, ($1,120.00 per thousand, 1,000 ea), 4% waste	—	ea	1.18	—	1.18
4" wide three-inch non-modular, ($1,110.00 per thousand, 1,000 ea), 4% waste	—	ea	1.17	—	1.17
4" wide standard, ($866.00 per thousand, 1,000 ea), 4% waste	—	ea	.91	—	.91
4" wide used, ($1,940.00 per thousand, 1,000 ea), 4% waste	—	ea	2.01	—	2.01
4" wide engineer, ($1,100.00 per thousand, 1,000 ea), 4% waste	—	ea	1.15	—	1.15
4" wide jumbo closure, ($1,350.00 per thousand, 1,000 ea), 4% waste	—	ea	1.41	—	1.41
4" wide double, ($1,430.00 per thousand, 1,000 ea), 4% waste	—	ea	1.50	—	1.50
4" wide Roman, ($1,560.00 per thousand, 1,000 ea), 4% waste	—	ea	1.62	—	1.62
4" wide Norman, ($1,500.00 per thousand, 1,000 ea), 4% waste	—	ea	1.55	—	1.55
4" wide Norwegian, ($1,350.00 per thousand, 1,000 ea), 4% waste	—	ea	1.41	—	1.41
4" wide jumbo utility, ($2,220.00 per thousand, 1,000 ea), 4% waste	—	ea	2.31	—	2.31
4" wide triple, ($2,340.00 per thousand, 1,000 ea), 4% waste	—	ea	2.44	—	2.44
6" wide Norwegian, ($2,140.00 per thousand, 1,000 ea), 4% waste	—	ea	2.23	—	2.23
6" wide Norman, ($2,590.00 per thousand, 1,000 ea), 4% waste	—	ea	2.69	—	2.69
6" wide jumbo, ($2,500.00 per thousand, 1,000 ea), 4% waste	—	ea	2.60	—	2.60
8" wide jumbo, ($3,400.00 per thousand, 1,000 ea), 4% waste	—	ea	3.55	—	3.55

	Craft@Hrs	Unit	Material	Labor	Total
Concrete block					
4" wide,					
($1,550.00 per thousand, 1,000 ea), 4% waste	—	ea	1.61	—	1.61
6" wide,					
($1,730.00 per thousand, 1,000 ea), 4% waste	—	ea	1.80	—	1.80
8" wide,					
($2,180.00 per thousand, 1,000 ea), 4% waste	—	ea	2.25	—	2.25
10" wide,					
($2,720.00 per thousand, 1,000 ea), 4% waste	—	ea	2.84	—	2.84
12" wide,					
($3,670.00 per thousand, 1,000 ea), 4% waste	—	ea	3.84	—	3.84
Lightweight concrete block					
4" wide,					
($1,750.00 per thousand, 1,000 ea), 4% waste	—	ea	1.84	—	1.84
6" wide,					
($1,940.00 per thousand, 1,000 ea), 4% waste	—	ea	2.01	—	2.01
8" wide,					
($2,440.00 per thousand, 1,000 ea), 4% waste	—	ea	2.53	—	2.53
10" wide,					
($3,010.00 per thousand, 1,000 ea), 4% waste	—	ea	3.13	—	3.13
12" wide,					
($4,120.00 per thousand, 1,000 ea), 4% waste	—	ea	4.29	—	4.29
Slump block					
4" wide,					
($1,650.00 per thousand, 1,000 ea), 4% waste	—	ea	1.72	—	1.72
6" wide,					
($1,970.00 per thousand, 1,000 ea), 4% waste	—	ea	2.04	—	2.04
8" wide,					
($2,460.00 per thousand, 1,000 ea), 4% waste	—	ea	2.56	—	2.56
10" wide,					
($2,890.00 per thousand, 1,000 ea), 4% waste	—	ea	3.00	—	3.00
12" wide,					
($3,720.00 per thousand, 1,000 ea), 4% waste	—	ea	3.87	—	3.87
Fluted block (fluted one side)					
4" wide,					
($2,830.00 per thousand, 1,000 ea), 4% waste	—	ea	2.94	—	2.94
6" wide,					
($3,160.00 per thousand, 1,000 ea), 4% waste	—	ea	3.29	—	3.29
8" wide,					
($3,970.00 per thousand, 1,000 ea), 4% waste	—	ea	4.13	—	4.13
10" wide,					
($4,990.00 per thousand, 1,000 ea), 4% waste	—	ea	5.17	—	5.17
12" wide,					
($6,680.00 per thousand, 1,000 ea), 4% waste	—	ea	6.95	—	6.95

	Craft@Hrs	Unit	Material	Labor	Total
Fluted block (fluted two sides)					
4" wide, ($3,390.00 per thousand, 1,000 ea), 4% waste	—	ea	3.54	—	3.54
6" wide, ($3,800.00 per thousand, 1,000 ea), 4% waste	—	ea	3.94	—	3.94
8" wide, ($4,790.00 per thousand, 1,000 ea), 4% waste	—	ea	4.99	—	4.99
10" wide, ($6,000.00 per thousand, 1,000 ea), 4% waste	—	ea	6.23	—	6.23
12" wide, ($8,100.00 per thousand, 1,000 ea), 4% waste	—	ea	8.43	—	8.43
Glazed block (glazed one side)					
4" wide, ($15,500.00 per thousand, 1,000 ea), 4% waste	—	ea	16.10	—	16.10
6" wide, ($15,700.00 per thousand, 1,000 ea), 4% waste	—	ea	16.30	—	16.30
8" wide, ($16,300.00 per thousand, 1,000 ea), 4% waste	—	ea	17.00	—	17.00
10" wide, ($16,900.00 per thousand, 1,000 ea), 4% waste	—	ea	17.50	—	17.50
12" wide, ($17,100.00 per thousand, 1,000 ea), 4% waste	—	ea	17.80	—	17.80
Glazed block (glazed two sides)					
4" wide, ($22,900.00 per thousand, 1,000 ea), 4% waste	—	ea	23.80	—	23.80
6" wide, ($24,400.00 per thousand, 1,000 ea), 4% waste	—	ea	25.30	—	25.30
8" wide, ($25,100.00 per thousand, 1,000 ea), 4% waste	—	ea	26.10	—	26.10
10" wide, ($25,900.00 per thousand, 1,000 ea), 4% waste	—	ea	26.90	—	26.90
12" wide, ($26,300.00 per thousand, 1,000 ea), 4% waste	—	ea	27.40	—	27.40
Split-face block					
4" wide, ($3,200.00 per thousand, 1,000 ea), 4% waste	—	ea	3.32	—	3.32
6" wide, ($3,780.00 per thousand, 1,000 ea), 4% waste	—	ea	3.92	—	3.92
8" wide, ($5,090.00 per thousand, 1,000 ea), 4% waste	—	ea	5.29	—	5.29
10" wide, ($5,370.00 per thousand, 1,000 ea), 4% waste	—	ea	5.58	—	5.58
12" wide, ($7,250.00 per thousand, 1,000 ea), 4% waste	—	ea	7.54	—	7.54

	Craft@Hrs	Unit	Material	Labor	Total
Split-rib block					
4" wide, ($2,900.00 per thousand, 1,000 ea), 4% waste	—	ea	3.01	—	3.01
6" wide, ($3,040.00 per thousand, 1,000 ea), 4% waste	—	ea	3.16	—	3.16
8" wide, ($4,940.00 per thousand, 1,000 ea), 4% waste	—	ea	5.15	—	5.15
10" wide, ($5,510.00 per thousand, 1,000 ea), 4% waste	—	ea	5.72	—	5.72
12" wide, ($6,100.00 per thousand, 1,000 ea), 4% waste	—	ea	6.35	—	6.35
Screen block					
4" wide, pattern two sides, ($8,410.00 per thousand, 1,000 ea), 4% waste	—	ea	8.74	—	8.74
4" wide, pattern four sides, ($16,000.00 per thousand, 1,000 ea), 4% waste	—	ea	16.50	—	16.50
Silicone treated perlite or vermiculite loose fill					
in 6" wide block, ($11.90 4 cf bag, 21 sf), 3% waste	—	sf	.57	—	.57
in 8" wide block, ($11.90 4 cf bag, 14.5 sf), 3% waste	—	sf	.84	—	.84
in 10" wide block, ($11.90 4 cf bag, 11.5 sf), 3% waste	—	sf	1.05	—	1.05
in 12" wide block, ($11.90 4 cf bag, 8 sf), 3% waste	—	sf	1.52	—	1.52
Glass block					
4" x 8" thinline smooth-face, ($32.60 sf, 1 sf), 4% waste	—	sf	33.90	—	33.90
6" x 6" thinline smooth-face, ($30.40 sf, 1 sf), 4% waste	—	sf	31.60	—	31.60
4" x 8" smooth-face, ($41.10 sf, 1 sf), 4% waste	—	sf	42.80	—	42.80
6" x 6" smooth-face, ($36.60 sf, 1 sf), 4% waste	—	sf	38.30	—	38.30
8" x 8" smooth-face, ($25.80 sf, 1 sf), 4% waste	—	sf	26.70	—	26.70
add for solar UV reflective	—	%	83.0	—	—
add for patterned face	—	%	11.0	—	—
add for tinted	—	%	55.0	—	—
Rubble stone					
coral stone rubble, ($54.00 cf, 1 cf), 4% waste	—	cf	56.30	—	56.30
fieldstone rubble, ($43.10 cf, 1 cf), 4% waste	—	cf	45.00	—	45.00
flagstone rubble, ($24.60 cf, 1 cf), 4% waste	—	cf	25.60	—	25.60
river stone rubble, ($44.00 cf, 1 cf), 4% waste	—	cf	45.80	—	45.80
lava stone rubble, ($41.30 cf, 1 cf), 4% waste	—	cf	43.00	—	43.00
sandstone rubble, ($63.20 cf, 1 cf), 4% waste	—	cf	65.70	—	65.70
Ashlar stone					
flagstone, ($25.90 cf, 1 cf), 4% waste	—	cf	26.90	—	26.90
limestone, natural finish, ($35.00 cf, 1 cf), 4% waste	—	cf	36.30	—	36.30
limestone, rough finish, ($39.10 cf, 1 cf), 4% waste	—	cf	40.70	—	40.70
limestone, smooth finish, ($56.80 cf, 1 cf), 4% waste	—	cf	59.10	—	59.10
marble, natural finish, ($83.00 cf, 1 cf), 4% waste	—	cf	86.30	—	86.30
marble, rough finish, ($89.50 cf, 1 cf), 4% waste	—	cf	93.10	—	93.10
marble, smooth finish, ($111.00 cf, 1 cf), 4% waste	—	cf	117.00	—	117.00

	Craft@Hrs	Unit	Material	Labor	Total
sandstone, natural finish, ($34.30 cf, 1 cf), 4% waste	—	cf	35.80	—	35.80
sandstone, rough finish, ($37.70 cf, 1 cf), 4% waste	—	cf	39.10	—	39.10
sandstone, smooth finish, ($54.40 cf, 1 cf), 4% waste	—	cf	56.70	—	56.70
Wall coping					
aluminum, all wall widths, ($21.20 lf, 1 lf), 4% waste	—	lf	22.10	—	22.10
10" wide concrete, ($22.70 lf, 1 lf), 2% waste	—	lf	23.20	—	23.20
14" wide concrete, ($30.20 lf, 1 lf), 2% waste	—	lf	30.90	—	30.90
8" wide granite, ($46.60 lf, 1 lf), 2% waste	—	lf	47.60	—	47.60
12" wide granite, ($61.20 lf, 1 lf), 2% waste	—	lf	62.60	—	62.60
8" wide limestone, ($27.50 lf, 1 lf), 2% waste	—	lf	28.10	—	28.10
12" wide limestone, ($32.70 lf, 1 lf), 2% waste	—	lf	33.40	—	33.40
8" wide marble, ($40.40 lf, 1 lf), 2% waste	—	lf	41.10	—	41.10
10" wide marble, ($62.20 lf, 1 lf), 2% waste	—	lf	63.50	—	63.50

Masonry Labor

Laborer	base wage	paid leave	true wage	taxes & ins.	total
Mason	$39.70	3.10	$42.80	28.30	$71.10
Mason's helper	$36.10	2.82	$38.92	26.28	$65.20
Hod carrier	$27.00	2.11	$29.11	21.29	$50.40
Plasterer	$37.30	2.91	$40.21	24.49	$64.70
Stone carver	$66.70	5.20	$71.90	43.10	$115.00
Demolition laborer	$25.90	2.02	$27.92	19.28	$47.20

Paid leave is calculated based on two weeks paid vacation, one week sick leave, and seven paid holidays. Employer's matching portion of **FICA** is 7.65 percent. **FUTA** (Federal Unemployment) is .8 percent. **Worker's compensation** was calculated using a national average of 19.41 percent for masonry workers and 13.28 percent for plastering workers. **Unemployment insurance** was calculated using a national average of 8 percent. **Health insurance** was calculated based on a projected national average for 2020 of $1,280 per employee (and family when applicable) per month. Employer pays 80 percent for a per month cost of $1,024 per employee. **Retirement** is based on a 401(k) retirement program with employer matching of 50 percent. Employee contributions to the 401(k) plan are an average of 6 percent of the true wage. **Liability insurance** is based on a national average of 12.0 percent.

Masonry Labor Productivity

	Craft@Hrs	Unit	Material	Labor	Total
Demolition of masonry					
remove 4" masonry wall	1D@.087	sf	—	4.11	4.11
remove 12" masonry wall	1D@.177	sf	—	8.35	8.35
remove brick veneer	1D@.082	sf	—	3.87	3.87
remove block wall pilaster	1D@.121	lf	—	5.71	5.71
remove block wall grade beam cap	1D@.363	lf	—	17.10	17.10
remove terra cotta wall cap	1D@.063	lf	—	2.97	2.97
remove coping stone	1D@.104	lf	—	4.91	4.91
remove glass block wall	1D@.084	sf	—	3.96	3.96
remove pavers on mortar base	1D@.120	sf	—	5.66	5.66
remove pavers on sand base	1D@.084	sf	—	3.96	3.96
remove stone rubble wall with no mortar	1D@.259	cf	—	12.20	12.20
remove stone rubble wall with mortar	1D@.439	cf	—	20.70	20.70
remove ashlar wall	1D@.439	cf	—	20.70	20.70
remove stone veneer	1D@.106	sf	—	5.00	5.00
remove keystone	1D@.198	ea	—	9.35	9.35

	Craft@Hrs	Unit	Material	Labor	Total
remove quoin	1D@.196	ea	—	9.25	9.25
remove veneer panels	1D@.074	sf	—	3.49	3.49
remove door or window architrave	1D@.131	lf	—	6.18	6.18
remove trim or cornice stones	1D@.104	ea	—	4.91	4.91
remove stone window sill	1D@.110	lf	—	5.19	5.19
remove precast concrete lintel	1D@.120	lf	—	5.66	5.66

Masonry crew

install masonry	mason	$71.10
install masonry	mason's helper	$65.20
install masonry	hod carrier / laborer	$50.40
install masonry	masonry crew	$62.20

Mason & helper

install masonry	mason	$71.10
install masonry	mason's helper	$65.20
install masonry	mason & helper	$68.20

	Craft@Hrs	Unit	Material	Labor	Total
Install brick wall					
4" wide standard non-modular	4M@.170	sf	—	10.60	10.60
4" wide oversize non-modular	4M@.168	sf	—	10.40	10.40
4" wide three-inch non-modular	4M@.167	sf	—	10.40	10.40
4" wide standard	4M@.169	sf	—	10.50	10.50
4" wide engineer	4M@.141	sf	—	8.77	8.77
4" wide jumbo closure	4M@.120	sf	—	7.46	7.46
4" wide double	4M@.101	sf	—	6.28	6.28
4" wide Roman	4M@.160	sf	—	9.95	9.95
4" wide Norman	4M@.130	sf	—	8.09	8.09
4" wide Norwegian	4M@.111	sf	—	6.90	6.90
4" wide jumbo utility	4M@.102	sf	—	6.34	6.34
4" wide triple	4M@.096	sf	—	5.97	5.97
6" wide Norwegian	4M@.124	sf	—	7.71	7.71
6" wide Norman	4M@.148	sf	—	9.21	9.21
6" wide jumbo	4M@.113	sf	—	7.03	7.03
8" wide jumbo	4M@.127	sf	—	7.90	7.90
Install 8" wide double wythe brick wall					
with standard non-modular brick	4M@.307	sf	—	19.10	19.10
with oversize non-modular brick	4M@.303	sf	—	18.80	18.80
with three-inch non-modular brick	4M@.300	sf	—	18.70	18.70
with standard brick	4M@.305	sf	—	19.00	19.00
with engineer brick	4M@.253	sf	—	15.70	15.70
with jumbo closure brick	4M@.216	sf	—	13.40	13.40
with double brick	4M@.181	sf	—	11.30	11.30
with Roman brick	4M@.287	sf	—	17.90	17.90
with Norman brick	4M@.234	sf	—	14.60	14.60
with Norwegian brick	4M@.199	sf	—	12.40	12.40
with jumbo utility	4M@.184	sf	—	11.40	11.40
with triple brick	4M@.173	sf	—	10.80	10.80

	Craft@Hrs	Unit	Material	Labor	Total
Install 8" wide wall with 4" wide brick and 4" wide concrete block					
with standard non-modular brick	4M@.233	sf	—	14.50	14.50
with standard brick	4M@.233	sf	—	14.50	14.50
with engineer brick	4M@.206	sf	—	12.80	12.80
with jumbo closure brick	4M@.188	sf	—	11.70	11.70
with double brick	4M@.170	sf	—	10.60	10.60
with Roman brick	4M@.224	sf	—	13.90	13.90
with Norman brick	4M@.197	sf	—	12.30	12.30
with Norwegian brick	4M@.180	sf	—	11.20	11.20
with jumbo utility brick	4M@.172	sf	—	10.70	10.70
with triple brick	4M@.166	sf	—	10.30	10.30
Install 10" wide wall with 4" wide brick and 6" wide concrete block					
with standard non-modular brick	4M@.236	sf	—	14.70	14.70
with standard brick	4M@.235	sf	—	14.60	14.60
with engineer brick	4M@.209	sf	—	13.00	13.00
with jumbo brick	4M@.190	sf	—	11.80	11.80
with double brick	4M@.173	sf	—	10.80	10.80
with Roman brick	4M@.226	sf	—	14.10	14.10
with Norman brick	4M@.200	sf	—	12.40	12.40
with Norwegian brick	4M@.182	sf	—	11.30	11.30
with jumbo utility brick	4M@.175	sf	—	10.90	10.90
with triple brick	4M@.170	sf	—	10.60	10.60
Install 12" wide wall with 4" wide brick and 8" wide concrete block					
with standard non-modular brick	4M@.240	sf	—	14.90	14.90
with standard brick	4M@.240	sf	—	14.90	14.90
with engineer brick	4M@.214	sf	—	13.30	13.30
with jumbo closure brick	4M@.195	sf	—	12.10	12.10
with double brick	4M@.178	sf	—	11.10	11.10
with Roman brick	4M@.231	sf	—	14.40	14.40
with Norman brick	4M@.204	sf	—	12.70	12.70
with Norwegian brick	4M@.187	sf	—	11.60	11.60
with jumbo utility brick	4M@.180	sf	—	11.20	11.20
with triple brick	4M@.174	sf	—	10.80	10.80
Install 12" wide triple wythe wall					
with standard non-modular brick	4M@.433	sf	—	26.90	26.90
with oversize non-modular brick	4M@.427	sf	—	26.60	26.60
with three-inch non-modular brick	4M@.424	sf	—	26.40	26.40
with standard brick	4M@.429	sf	—	26.70	26.70
with engineer brick	4M@.357	sf	—	22.20	22.20
with jumbo closure brick	4M@.305	sf	—	19.00	19.00
with double brick	4M@.255	sf	—	15.90	15.90
with Roman brick	4M@.407	sf	—	25.30	25.30
with Norman brick	4M@.330	sf	—	20.50	20.50
with Norwegian brick	4M@.282	sf	—	17.50	17.50
with jumbo utility brick	4M@.260	sf	—	16.20	16.20
with triple brick	4M@.244	sf	—	15.20	15.20

	Craft@Hrs	Unit	Material	Labor	Total
Install 10" wide cavity wall with 4" brick and 4" concrete block					
with standard non-modular brick	4M@.244	sf	—	15.20	15.20
with standard brick	4M@.243	sf	—	15.10	15.10
with engineer brick	4M@.216	sf	—	13.40	13.40
with jumbo closure brick	4M@.196	sf	—	12.20	12.20
with double brick	4M@.178	sf	—	11.10	11.10
with Roman brick	4M@.234	sf	—	14.60	14.60
with Norman brick	4M@.206	sf	—	12.80	12.80
with Norwegian brick	4M@.188	sf	—	11.70	11.70
with jumbo utility brick	4M@.180	sf	—	11.20	11.20
with triple brick	4M@.174	sf	—	10.80	10.80
Install 10" wide cavity wall with 4" brick on both sides					
with standard non-modular brick	4M@.321	sf	—	20.00	20.00
with oversize non-modular brick	4M@.317	sf	—	19.70	19.70
with three-inch non-modular brick	4M@.314	sf	—	19.50	19.50
with standard brick	4M@.319	sf	—	19.80	19.80
with engineer brick	4M@.266	sf	—	16.50	16.50
with jumbo closure brick	4M@.226	sf	—	14.10	14.10
with double brick	4M@.189	sf	—	11.80	11.80
with Roman brick	4M@.301	sf	—	18.70	18.70
with Norman brick	4M@.245	sf	—	15.20	15.20
with Norwegian brick	4M@.209	sf	—	13.00	13.00
with jumbo utility brick	4M@.193	sf	—	12.00	12.00
with triple brick	4M@.181	sf	—	11.30	11.30
Install brick veneer					
with standard brick	4M@.144	sf	—	8.96	8.96
with oversize non-modular brick	4M@.143	sf	—	8.89	8.89
with three-inch non-modular brick	4M@.141	sf	—	8.77	8.77
with standard brick	4M@.144	sf	—	8.96	8.96
with engineer brick	4M@.119	sf	—	7.40	7.40
with jumbo closure brick	4M@.102	sf	—	6.34	6.34
with double brick	4M@.085	sf	—	5.29	5.29
with Roman brick	4M@.136	sf	—	8.46	8.46
with Norman brick	4M@.110	sf	—	6.84	6.84
with Norwegian brick	4M@.094	sf	—	5.85	5.85
with jumbo utility brick	4M@.087	sf	—	5.41	5.41
with triple brick	4M@.082	sf	—	5.10	5.10
Install brick arch					
flat brick arch	4M@1.49	lf	—	92.70	92.70
elliptical brick arch	4M@1.85	lf	—	115.00	115.00
semi-circular brick arch	4M@1.96	lf	—	122.00	122.00
Install block wall with 8" x 16" face					
4" wide	4M@.089	sf	—	5.54	5.54
6" wide	4M@.092	sf	—	5.72	5.72
8" wide	4M@.097	sf	—	6.03	6.03
10" wide	4M@.103	sf	—	6.41	6.41
12" wide	4M@.122	sf	—	7.59	7.59

	Craft@Hrs	Unit	Material	Labor	Total
Install block wall with 4" x 16" face					
4" wide	4M@.102	sf	—	6.34	6.34
6" wide	4M@.106	sf	—	6.59	6.59
8" wide	4M@.112	sf	—	6.97	6.97
10" wide	4M@.119	sf	—	7.40	7.40
12" wide	4M@.140	sf	—	8.71	8.71
Install lightweight block wall with 8" x 16" face					
4" wide	4M@.083	sf	—	5.16	5.16
6" wide	4M@.086	sf	—	5.35	5.35
8" wide	4M@.091	sf	—	5.66	5.66
10" wide	4M@.097	sf	—	6.03	6.03
12" wide	4M@.114	sf	—	7.09	7.09
Form and pour grade-beam cap on block wall					
6" wide	4M@.106	lf	—	6.59	6.59
12" wide	4M@.120	lf	—	7.46	7.46
Install glass block					
4" x 8"	4M@.333	sf	—	20.70	20.70
6" x 8"	4M@.267	sf	—	16.60	16.60
8" x 8"	4M@.208	sf	—	12.90	12.90
12" x 12"	4M@.175	sf	—	10.90	10.90
Install pavers					
on sand base	4M@.147	sf	—	9.14	9.14
mortar base	4M@.180	sf	—	11.20	11.20
full-size bricks, laid face up	4M@.185	sf	—	11.50	11.50
full-size bricks, laid edge up	4M@.208	sf	—	12.90	12.90
add for paver steps installed over concrete	4M@.270	lf	—	16.80	16.80
add for paver steps installed over sand base	4M@.294	lf	—	18.30	18.30
add for separate pattern at edges of pavers	4M@.076	lf	—	4.73	4.73
Lay stone wall					
field stone rubble with no mortar	4M@.481	cf	—	29.90	29.90
rubble wall	4M@.518	cf	—	32.20	32.20
ashlar wall	4M@.472	cf	—	29.40	29.40
add to install flat arch	4M@2.86	lf	—	178.00	178.00
add to install elliptical arch	4M@3.58	lf	—	223.00	223.00
add to install semi-circular arch	4M@3.70	lf	—	230.00	230.00
Lay stone veneer					
rubble	4M@.243	sf	—	15.10	15.10
ashlar	4M@.236	sf	—	14.70	14.70
add to install flat arch	4M@1.75	lf	—	109.00	109.00
add to install elliptical arch	4M@2.17	lf	—	135.00	135.00
add to install semi-circular arch	4M@2.30	lf	—	143.00	143.00
Install veneer panels					
cultured stone	4M@.263	sf	—	16.40	16.40
natural stone	4M@.298	sf	—	18.50	18.50
Install stone architrave					
door	4M@.980	lf	—	61.00	61.00
window	4M@1.00	lf	—	62.20	62.20
Install trim or cornice stones					
install	4M@.370	lf	—	23.00	23.00

	Craft@Hrs	Unit	Material	Labor	Total
Install wall cap or coping					
terra cotta	4M@.185	lf	—	11.50	11.50
aluminum	4M@.181	lf	—	11.30	11.30
concrete coping	4M@.167	lf	—	10.40	10.40
stone coping	4M@.185	lf	—	11.50	11.50
Salvage brick veneer					
remove, salvage bricks, and relay	4M@.877	sf	—	54.50	54.50
Repoint brick wall					
running bond	4M@.108	sf	—	6.72	6.72
common bond (also called American bond)	4M@.111	sf	—	6.90	6.90
Flemish bond	4M@.121	sf	—	7.53	7.53
English bond	4M@.128	sf	—	7.96	7.96
English cross bond (also called Dutch bond)	4M@.128	sf	—	7.96	7.96
stack bond	4M@.082	sf	—	5.10	5.10
all header bond	4M@.093	sf	—	5.78	5.78
soldier course	4M@.106	sf	—	6.59	6.59
sailor course	4M@.099	sf	—	6.16	6.16
basketweave	4M@.164	sf	—	10.20	10.20
herringbone weave	4M@.167	sf	—	10.40	10.40
diagonal bond	4M@.144	sf	—	8.96	8.96
coursed ashlar style bond	4M@.120	sf	—	7.46	7.46
Repoint stone wall					
rubble	4M@.278	sf	—	17.30	17.30
ashlar	4M@.260	sf	—	16.20	16.20
Stone wall repair					
repair carved stone with epoxy	1M@.024	si	—	1.71	1.71
repair carved stone with stucco or grout	1M@.022	si	—	1.56	1.56
epoxy repair and pin broken stone	1M@1.33	ea	—	94.60	94.60
reconstitute delaminating stone with epoxy	1M@.013	si	—	.92	.92
reface stone with grout mixed to match	1M@.149	sf	—	10.60	10.60
remove segment of rubble wall and relay	5M@4.00	cf	—	273.00	273.00
remove segment of rubble veneer and relay	5M@1.92	sf	—	131.00	131.00
remove segment of ashlar wall and relay	5M@3.85	cf	—	263.00	263.00
remove segment of ashlar veneer and relay	5M@1.89	sf	—	129.00	129.00
Carve stone					
with light hand carvings	6M@.015	si	—	1.73	1.73
with medium hand carvings	6M@.022	si	—	2.53	2.53
with heavy hand carvings	6M@.031	si	—	3.57	3.57

Mold Remediation

Dealing with Mold

Mold remediation is an important consideration when dealing with almost any type of structural damage. The prices in this chapter are primarily based on the standards developed by the city of New York (recently updated) and by applying the EPA's guidelines for mold remediation in schools and commercial buildings to residential buildings.

Mitigation

Mold issues point out the importance of immediately removing any source of invading moisture. Mold can begin to grow immediately and can become a problem even when moisture is removed within 24 to 48 hours. Paying for after-hours mitigation work to remove the source of any moisture, dehumidifying, and aggressively drying items that sponsor mold is well worth the cost.

Testing

Not all molds are judged to be as harmful as others. Most of the harmful types of molds have been generally classified as "black molds." Black molds include aspergillus (more than 50 species), cladosporium, fusarium, stachybotrus chartarum, trichoderma, memnoniella, and penicillium. Molds may emit both spores and gas and many are still hazardous even when they are not alive. To get a clear picture of the types of molds present and the hazards they present, testing may be necessary.

Mold Specialists

All workers involved in mold remediation should be well trained and, when possible, certified. In many cases it is advantageous to hire an environmental consultant to help determine the best process and to oversee the work. The prices in this chapter assume that all work is done by qualified and certified staff supervised by an environmental consultant or someone on staff with similar credentials and abilities.

Containment

Severely mold-contaminated areas must be contained in ways that are similar to procedures used when dealing with asbestos and other types of hazardous materials. Contaminated areas must be sealed off from non-contaminated areas. Negative air pressure should be maintained in the contaminated area so air-borne mold spores and gas will not escape. Ventilating fans must filter any possible mold spores and gas from the air before it is ventilated to the exterior of the contaminated areas. Entry and exit from the contaminated area must be done through a decontamination area.

Personal Protection Equipment

All workers involved in mold remediation must wear personal protection equipment. Although the level and type of the mold involved may change some aspects of the personal protection equipment needed, in general, this chapter assumes workers wear a fit-tested half- or full-force respirator with a HEPA, organic/chemical cartridge. Although the New York standards suggest an N-95 rated mask in some circumstances, many specialists feel a better standard for their employees is to always require a respirator. If an N-95 mask is judged to be sufficient, care must be taken to ensure that the mask is actually N-95 rated. The N-95 mask looks very similar to other types of masks often worn when doing routine demolition. However, the N-95 mask is substantially more effective than standard masks.

Mold remediation specialists should also wear nitrile disposable gloves or, when working with debris that contains sharp edges, puncture-proof gloves. Workers should also wear level-B protective clothing that covers both the head and feet.

Levels of Remediation

The New York City Department of Health & Mental Hygiene, Bureau of Environmental & Occupational Disease Epidemiology has issued a document called Guidelines on Assessment and Remediation of Fungi in Indoor Environments (http://home2.nyc.gov/html/doh/html/epi/moldrpt1.shtml). We highly recommend reading this document and using it as a guideline. A key part of this document is the section titled Remediation. This section discusses the types of remediation recommended for different levels of mold growth. The prices in this chapter are based on this document's description of Level III mold growth and above, although some items also apply to Level II.

For More Information

A large body of information about mold remediation is now available on the Internet. Typing "Mold Remediation" in Google or a similar search engine will yield a wealth of results. We recommend viewing the New York City standards (http://home2.nyc.gov/html/doh/html/epi/moldrpt1.shtml) and visiting the EPA's web site (www.epa.gov/mold) as two excellent places to start.

	Craft@Hrs	Unit	Material	Labor	Total
Minimum charge.					
for mold remediation work when containment is required	9Z@6.00	ea	232.00	223.00	455.00
for mold remediation testing	9Z@1.25	ea	—	46.50	46.50
Testing.					
Anderson N-6 bioaerosol sampler	—	ea	136.00	—	136.00
Spore trap	—	ea	89.90	—	89.90
Prolab test kit (3 sampling methods)	—	ea	92.80	—	92.80
Swab/tape sampler	—	ea	98.10	—	98.10
Surface sampling test	—	sf	76.30	—	76.30
Air sampling test	—	ea	136.00	—	136.00
Environmental consultant.					
Environmental consultant, per hour	9Z@1.00	hr	—	37.20	37.20
Minimum charge for mold remediation testing	9Z@4.00	ea	—	149.00	149.00
Containment. Plastic cover is two layers of plastic attached with duct tape.					
Plastic cover attached to ceiling	9Z@.036	sf	1.74	1.34	3.08
Plastic cover attached to walls	9Z@.035	sf	1.64	1.30	2.94
Plastic cover and plywood over floor (2 layers)	9Z@.084	sf	3.23	3.12	6.35
Temporary containment walls with plastic cover	9Z@.142	sf	4.02	5.28	9.30
Airlock for containment area	9Z@7.41	ea	339.00	276.00	615.00

	Craft@Hrs	Unit	Material	Labor	Equip.	Total
Containment equipment.						
Prefabricated decontamination unit, rent per day	—	dy	—	—	126.00	126.00
Exhaust fan, HEPA filtered for containment area, rent per day	—	dy	—	—	169.00	169.00
Negative air machine with air scrubber, rent per day	—	dy	—	—	152.00	152.00
Dumpsters.						
Dumpster, with locking doors, 5 to 6 cy	—	ea	—	—	169.00	169.00
Dumpster, with locking doors, 10 to 12 cy	—	ea	—	—	270.00	270.00
Dumpster, with locking doors, 30 cy	—	ea	—	—	750.00	750.00

	Craft@Hrs	Unit	Material	Labor	Total

Tear out flooring. Removing mold-contaminated flooring includes cutting into strips or smaller pieces as needed, and bagging, or placing in a haz-mat drum for carpet tackless strip.

	Craft@Hrs	Unit	Material	Labor	Total
Remove mold-contaminated carpet	9Z@.006	sf	.10	.22	.32
Remove mold-contaminated carpet pad	9Z@.007	sf	.10	.26	.36
Remove mold-contaminated carpet tackless strip	9Z@.004	lf	.42	.15	.57
Remove mold-contaminated glue down carpet	9Z@.008	sf	.10	.30	.40
Remove mold-contaminated vinyl floor from underlayment	9Z@.008	sf	.10	.30	.40
Remove mold-contaminated vinyl floor from concrete	9Z@.009	sf	.10	.33	.43
Remove mold-contaminated underlayment	9Z@.008	sf	.26	.30	.56
Remove mold-contaminated wood floor	9Z@.012	sf	.26	.45	.71

Tear out wall finishes. Removing mold-contaminated wall finishes includes cutting into strips or breaking into smaller pieces as needed, and bagging, or placing in a haz-mat drum.

	Craft@Hrs	Unit	Material	Labor	Total
Remove mold-contaminated drywall	9Z@.006	sf	.26	.22	.48
Remove mold-contaminated plaster	9Z@.007	sf	.26	.26	.52
Remove mold-contaminated wood paneling	9Z@.004	sf	.26	.15	.41
Remove mold-contaminated trimwork	9Z@.004	lf	.10	.15	.25
Remove mold-contaminated concrete backer board	9Z@.008	sf	.26	.30	.56
Remove mold-contaminated wallpaper	9Z@.001	sf	—	.04	.04

Tear out door. Removing mold-contaminated doors includes cutting into smaller pieces, and bagging, or placing in a haz-mat drum.

	Craft@Hrs	Unit	Material	Labor	Total
Remove mold-contaminated hollow-core door	9Z@.067	ea	4.58	2.49	7.07
Remove mold-contaminated solid-core door	9Z@.091	ea	6.88	3.39	10.27

Tear out insulation. Removing mold-contaminated insulation includes placing in a bag for disposal.

	Craft@Hrs	Unit	Material	Labor	Total
Remove mold-contaminated batt insulation	9Z@.002	sf	.03	.07	.10
Remove mold-contaminated loose-fill insulation	9Z@.004	sf	.03	.15	.18

Tear out ceiling finishes. Removing mold-contaminated ceiling finishes includes breaking into smaller pieces and bagging or placing in a haz-mat drum.

	Craft@Hrs	Unit	Material	Labor	Total
Remove mold-contaminated acoustic ceiling tile	9Z@.006	sf	.10	.22	.32
Remove mold-contaminated ceiling furring strips	9Z@.009	sf	.10	.33	.43
Remove mold-contaminated wall furring strips	9Z@.008	sf	.10	.30	.40
Remove mold-contaminated suspended ceiling tile	9Z@.005	sf	.10	.19	.29

Tear out complete room. Includes stripping the mold-contaminated room to bare walls and sub-floor. All debris is cut or broken into smaller pieces then bagged or placed in haz-mat drums.

	Craft@Hrs	Unit	Material	Labor	Total
Strip mold-contaminated room	9Z@.166	sf	1.78	6.18	7.96
Strip mold-contaminated bathroom	9Z@.214	sf	2.32	7.96	10.28
Strip mold-contaminated kitchen	9Z@.187	sf	1.91	6.96	8.87
Strip mold-contaminated utility room	9Z@.173	sf	1.91	6.44	8.35
Strip mold-contaminated laundry room	9Z@.189	sf	1.91	7.03	8.94

Haz-mat drums & bags.

	Craft@Hrs	Unit	Material	Labor	Total
3 cf disposable fiber drum	—	ea	13.60	—	13.60
3 cf disposable bag	—	ea	1.60	—	1.60

	Craft@Hrs	Unit	Material	Labor	Total
Treat with antimicrobial spray. Includes treating the mold-contaminated surface, rinsing, wiping and drying.					
Treat floor with antimicrobial spray	9Z@.015	sf	.36	.56	.92
Treat walls with antimicrobial spray	9Z@.013	sf	.36	.48	.84
Treat ceiling with antimicrobial spray	9Z@.014	sf	.36	.52	.88
Treat trimwork with antimicrobial spray	9Z@.012	lf	.29	.45	.74
Treat door with antimicrobial spray	9Z@.295	ea	5.73	11.00	16.73
Treat suspended ceiling grid with antimicrobial spray	9Z@.012	sf	.27	.45	.72
Treat light fixture with antimicrobial spray	9Z@.185	ea	5.54	6.88	12.42
Treat switch/outlet & box with antimicrobial spray	9Z@.171	ea	3.32	6.36	9.68
Clean HVAC.					
Clean mold-contaminated ductwork, per diffuser or cold-air return	9Z@.655	ea	22.20	24.40	46.60
Clean mold-contaminated furnace	9Z@4.49	ea	18.00	167.00	185.00
Cleaning rule of thumb. A general rule of thumb for cleaning items not listed in this chapter.					
Add to cleaning prices to clean mold-contaminated items	—	%	—	28.0	—
Treat mold-contaminated framing with antimicrobial spray.					
Treat furring strips with antimicrobial spray	9Z@.002	sf	.13	.07	.20
Treat beams with antimicrobial spray	9Z@.006	lf	.13	.22	.35
Treat 2" x 4" framing with antimicrobial spray	9Z@.003	sf	.26	.11	.37
Treat 2" x 6" framing with antimicrobial spray	9Z@.003	sf	.32	.11	.43
Treat 2" x 8" framing with antimicrobial spray	9Z@.004	sf	.36	.15	.51
Treat 2" x 10" framing with antimicrobial spray	9Z@.004	sf	.42	.15	.57
Treat 2" x 12" framing with antimicrobial spray	9Z@.005	sf	.44	.19	.63
Treat mold-contaminated trusses with antimicrobial spray.					
Treat floor trusses with antimicrobial spray	9Z@.011	sf	.62	.41	1.03
Treat 4/12 trusses with antimicrobial spray	9Z@.049	sf	.76	1.82	2.58
Treat 6/12 trusses with antimicrobial spray	9Z@.066	sf	.90	2.46	3.36
Treat 8/12 trusses with antimicrobial spray	9Z@.084	sf	1.10	3.12	4.22
Treat 10/12 trusses with antimicrobial spray	9Z@.100	sf	1.50	3.72	5.22
Treat 12/12 trusses with antimicrobial spray	9Z@.118	sf	1.78	4.39	6.17
Treat 17/12 trusses with antimicrobial spray	9Z@.142	sf	1.97	5.28	7.25
Treat sheathing with antimicrobial spray	9Z@.002	sf	.27	.07	.34
Encapsulate with sealer. Cleaning mold-contaminated items and treating with an antimicriboial spray probably won't remove all mold spores or gas from the air. Since there is always some level of mold in the air, the goal is to reduce the mold to normal levels. In some cases it may be desirable to encapsulate cleaned areas to further reduce the likelihood of continuing problems from previously affected areas.					
Encapsulate cleaned floor with sealer	9Z@.008	sf	.42	.30	.72
Encapsulate cleaned wall with sealer	9Z@.007	sf	.42	.26	.68
Encapsulate cleaned ceiling with sealer	9Z@.007	sf	.42	.26	.68

	Craft@Hrs	Unit	Material	Labor	Total
Encapsulate framing.					
Encapsulate treated furring strips with sealer	9Z@.008	sf	.48	.30	.78
Encapsulate treated beams with sealer	9Z@.008	sf	.61	.30	.91
Encapsulate treated 2" x 4" framing with sealer	9Z@.006	sf	.72	.22	.94
Encapsulate treated 2" x 6" framing with sealer	9Z@.007	sf	.79	.26	1.05
Encapsulate treated 2" x 8" framing with sealer	9Z@.008	sf	.85	.30	1.15
Encapsulate treated 2" x 10" framing with sealer	9Z@.009	sf	.93	.33	1.26
Encapsulate treated 2" x 12" framing with sealer	9Z@.010	sf	1.03	.37	1.40
Encapsulate treated floor trusses with sealer	9Z@.016	sf	1.14	.60	1.74
Encapsulate treated 4/12 trusses with sealer	9Z@.019	sf	1.78	.71	2.49
Encapsulate treated 6/12 trusses with sealer	9Z@.037	sf	1.98	1.38	3.36
Encapsulate treated 8/12 trusses with sealer	9Z@.055	sf	2.16	2.05	4.21
Encapsulate treated 10/12 trusses with sealer	9Z@.074	sf	2.35	2.75	5.10
Encapsulate treated 12/12 trusses with sealer	9Z@.093	sf	2.53	3.46	5.99
Encapsulate treated 17/12 trusses with sealer	9Z@.112	sf	2.88	4.17	7.05
Encapsulate treated sheathing with sealer	9Z@.008	sf	.47	.30	.77
Encapsulation rule of thumb. A general rule of thumb for encapsulating items not listed in this chapter.					
Add to painting cost to encapsulate	—	%	—	34.0	—
Fog treatment.					
Treat area with antibacterial and antifungal fog	9Z@.001	cf	.01	.04	.05

Dehumidifier. Reducing the relative humidity in the room can be an important part of stopping the growth and spread of mold. Most experts want relative humidity at 60 percent or below and many shoot for around 40 percent.

	Craft@Hrs	Unit	Material	Labor	Equip.	Total
Dehumidifier unit, 5 gallon daily capacity, rental per day	—	dy	—	—	42.50	42.50
Dehumidifier unit, 10 gallon daily capacity, rental per day	—	dy	—	—	50.80	50.80
Dehumidifier unit, 19 gallon daily capacity, rental per day	—	dy	—	—	84.90	84.90
Dehumidifier unit, 24 gallon daily capacity, rental per day	—	dy	—	—	119.00	119.00
Dehumidifier unit, 28 gallon daily capacity, rental per day	—	dy	—	—	127.00	127.00
Drying fans.						
Drying fan, rental per day	—	dy	—	—	43.50	43.50
Drying fan, large, rental per day	—	dy	—	—	50.80	50.80
Drying fan, wall cavity, rental per day	—	dy	—	—	59.80	59.80

Time & Material Charts (selected items)
Mold Remediation Materials

See Mold Remediation material prices with the line items above.

Mold Remediation Labor

Laborer	base wage	paid leave	true wage	taxes & ins.	total
Mildew remediation specialist	$23.80	1.86	$25.66	20.24	$45.90
Mildew remediation assistant	$13.30	1.04	$14.34	14.16	$28.50

Paid leave is calculated based on two weeks paid vacation, one week sick leave, and seven paid holidays. Employer's matching portion of **FICA** is 7.65 percent. **FUTA** (Federal Unemployment) is .8 percent. **Worker's compensation** for the mold remediation trade was calculated using a national average of 23.10 percent. **Unemployment insurance** was calculated using a national average of 8 percent. **Health insurance** was calculated based on a projected national average for 2020 of $1,280 per employee (and family when applicable) per month. Employer pays 80 percent for a per month cost of $1,024 per employee. **Retirement** is based on a 401(k) retirement program with employer matching of 50 percent. Employee contributions to the 401(k) plan are an average of 6 percent of the true wage. **Liability insurance** is based on a national average of 12.0 percent.

Outbuildings

Metal storage shed. Prefabricated metal storage shed with baked enamel finish and gable roof. Includes assembly. Does not include concrete slab or footings. Add **16%** for gambrel roof. Add **30%** for vinyl covered metal.

	Craft@Hrs	Unit	Material	Labor	Total
8' x 6' metal storage shed					
replace	5C@6.51	ea	387.00	328.00	715.00
remove	1D@2.50	ea	—	118.00	118.00
8' x 10' metal storage shed					
replace	5C@7.01	ea	519.00	353.00	872.00
remove	1D@2.74	ea	—	129.00	129.00
9' x 10' metal storage shed					
replace	5C@7.51	ea	542.00	379.00	921.00
remove	1D@2.86	ea	—	135.00	135.00
10' x 12' metal storage shed					
replace	5C@8.01	ea	846.00	404.00	1,250.00
remove	1D@3.04	ea	—	143.00	143.00

Wood storage shed. Pre-built and delivered to site. Painted with trim, truss gambrel roof, panel siding, 1/2" roof sheathing, 3/4" tongue-and-groove floor sheathing with floor skids. Does not include footings or pier blocks. Deduct **12%** for gable roof. Deduct **15%** for gambrel roof on top of walls that are 5' tall or less.

	Craft@Hrs	Unit	Material	Labor	Total
8' x 6' wood storage shed					
replace	5C@7.01	ea	896.00	353.00	1,249.00
remove	1D@3.28	ea	—	155.00	155.00
8' x 10' wood storage shed					
replace	5C@7.51	ea	2,110.00	379.00	2,489.00
remove	1D@3.84	ea	—	181.00	181.00
9' x 10' wood storage shed					
replace	5C@8.01	ea	2,430.00	404.00	2,834.00
remove	1D@4.26	ea	—	201.00	201.00
10' x 12' wood storage shed					
replace	5C@8.50	ea	2,820.00	428.00	3,248.00
remove	1D@4.67	ea	—	220.00	220.00

Gazebo. High quality octagonal gazebos priced by diameter. Made from D grade or better kiln-dried western red cedar surfaced four sides. Wood cedar shingles. Prefabricated parts assembled on site. Standard grade: Cupola, simple fretwork and rails with cross-grain detailing. High grade: Cupola, fretwork and raised-panel rails. Deluxe grade: Cupola, Queen Anne fretwork and rails. Add **116%** to 15' diameter for 21' diameter triple roof gazebo, add **$255** for bench in 9' and 12' diameter gazebos. Add **$287** for bench in 15', add **$714** for bow roof (all sizes). Add **$415** for steps with balustrades. Deduct **25%** for no floor deck. Deduct **17%** if made from select #1 grade western red cedar.

	Craft@Hrs	Unit	Material	Labor	Total
9' gazebo					
replace, standard grade	1C@13.1	ea	9,450.00	886.00	10,336.00
replace, high grade	1C@15.0	ea	11,500.00	1,010.00	12,510.00
replace, deluxe grade	1C@17.5	ea	13,100.00	1,180.00	14,280.00
remove	1D@4.43	ea	—	209.00	209.00
12' gazebo					
replace, standard grade	1C@17.5	ea	13,500.00	1,180.00	14,680.00
replace, high grade	1C@17.5	ea	15,100.00	1,180.00	16,280.00
replace, deluxe grade	1C@17.5	ea	17,600.00	1,180.00	18,780.00
remove	1D@5.57	ea	—	263.00	263.00

single roof *double roof*

	Craft@Hrs	Unit	Material	Labor	Total
15' gazebo					
replace, standard grade	1C@21.0	ea	18,500.00	1,420.00	19,920.00
replace, high grade	1C@26.2	ea	22,300.00	1,770.00	24,070.00
replace, deluxe grade	1C@26.2	ea	24,800.00	1,770.00	26,570.00
remove	1D@7.13	ea	—	337.00	337.00

triple roof

Free-standing greenhouse. Residential style, aluminum frame. Automatic electric roof vent.

	Craft@Hrs	Unit	Material	Labor	Total
8' x 9' free-standing greenhouse					
replace	6C@15.0	ea	6,760.00	885.00	7,645.00
remove	1D@3.04	ea	—	143.00	143.00
8' x 11' free-standing greenhouse					
replace	6C@15.0	ea	7,380.00	885.00	8,265.00
remove	1D@3.44	ea	—	162.00	162.00
8' x 14' free-standing greenhouse					
replace	6C@17.5	ea	7,710.00	1,030.00	8,740.00
remove	1D@4.01	ea	—	189.00	189.00
8' x 17' free-standing greenhouse					
replace	6C@17.5	ea	9,200.00	1,030.00	10,230.00
remove	1D@4.33	ea	—	204.00	204.00

Lean-to greenhouse. Residential style, aluminum frame. Manually controlled vents.

	Craft@Hrs	Unit	Material	Labor	Total
4' x 8' lean-to greenhouse					
replace	6C@8.76	ea	2,870.00	517.00	3,387.00
remove	1D@2.86	ea	—	135.00	135.00
7' x 14' lean-to greenhouse					
replace	6C@13.1	ea	7,340.00	773.00	8,113.00
remove	1D@3.28	ea	—	155.00	155.00
8' x 16' lean-to greenhouse					
replace	6C@15.0	ea	8,980.00	885.00	9,865.00
remove	1D@3.49	ea	—	165.00	165.00

Screened swimming pool enclosure. Aluminum frame, vinyl screen.

	Craft@Hrs	Unit	Material	Labor	Total
replace	6C@.018	sf	8.85	1.06	9.91
remove	1D@.009	sf	—	.42	.42

Chickhee hut. Per square foot of floor. Typically found in the southern United States.

	Craft@Hrs	Unit	Material	Labor	Total
replace	5C@1.23	sf	35.30	62.00	97.30
remove	1D@.100	sf	—	4.72	4.72
Thatched roof for hut, per square foot of roof area					
replace	5C@.754	sf	6.92	38.00	44.92
remove	1D@.038	sf	—	1.79	1.79

Painting

Painting coats. Prices for one coat, two coats and three coats do **not** include: the primer coat, sealer coat or stain. To estimate painting, add the cost of the primer coat to the number of paint coats. For example, to estimate the cost to prime a wall then paint with two coats, add the prime price to the two coats price. To estimate the cost to stain an item then cover with two coats of varnish, add the stain cost to the two coats price (unless otherwise noted).

	Craft@Hrs	Unit	Material	Labor	Total
Minimum charge.					
for painting work	5F@3.00	ea	52.60	197.00	249.60
Acoustic ceilings.					
Paint acoustic ceiling texture					
prime	5F@.011	sf	.42	.72	1.14
1 coat	5F@.011	sf	.45	.72	1.17
2 coats	5F@.017	sf	.64	1.11	1.75
3 coats	5F@.023	sf	.84	1.51	2.35
Paint acoustical ceiling tile					
prime	5F@.011	sf	.42	.72	1.14
1 coat	5F@.011	sf	.45	.72	1.17
2 coats	5F@.017	sf	.64	1.11	1.75
3 coats	5F@.023	sf	.84	1.51	2.35
Awnings and carports.					
Paint aluminum carport or awning					
prime	5F@.006	sf	.30	.39	.69
1 coat	5F@.006	sf	.34	.39	.73
2 coats	5F@.010	sf	.45	.66	1.11
3 coats	5F@.013	sf	.61	.85	1.46
Drywall, plaster and stucco.					
Paint plaster or drywall					
prime	5F@.006	sf	.21	.39	.60
1 coat	5F@.006	sf	.22	.39	.61
2 coats	5F@.010	sf	.30	.66	.96
3 coats	5F@.013	sf	.39	.85	1.24
Paint stucco					
prime	5F@.014	sf	.34	.92	1.26
1 coat	5F@.015	sf	.36	.98	1.34
2 coats	5F@.023	sf	.49	1.51	2.00
3 coats	5F@.030	sf	.66	1.97	2.63

	Craft@Hrs	Unit	Material	Labor	Total

Columns. Deduct **26%** for pilasters, pilaster capitals, pilaster pedestals, and so on.

	Craft@Hrs	Unit	Material	Labor	Total
Paint column					
prime	5F@.017	lf	.51	1.11	1.62
1 coat	5F@.017	lf	.59	1.11	1.70
2 coats	5F@.027	lf	.81	1.77	2.58
3 coats	5F@.036	lf	1.08	2.36	3.44
Stain & varnish column					
stain	5F@.016	lf	.54	1.05	1.59
1 coat	5F@.016	lf	.61	1.05	1.66
2 coats	5F@.025	lf	.84	1.64	2.48
3 coats	5F@.033	lf	1.15	2.16	3.31
Paint capital, simple design					
prime	5F@.650	ea	28.90	42.60	71.50
1 coat	5F@.670	ea	31.90	43.90	75.80
2 coats	5F@1.03	ea	44.50	67.50	112.00
3 coats	5F@1.40	ea	59.10	91.70	150.80
Paint capital, complex design					
prime	5F@.839	ea	37.40	55.00	92.40
1 coat	5F@.863	ea	41.00	56.50	97.50
2 coats	5F@1.33	ea	57.00	87.10	144.10
3 coats	5F@1.80	ea	75.80	118.00	193.80
Stain & varnish capital, simple design					
Stain	5F@.670	ea	31.80	43.90	75.70
1 coat	5F@.691	ea	35.50	45.30	80.80
2 coats	5F@1.06	ea	49.30	69.40	118.70
3 coats	5F@1.44	ea	65.40	94.30	159.70
Stain & varnish capital, complex design					
stain	5F@.930	ea	40.90	60.90	101.80
1 coat	5F@.957	ea	45.30	62.70	108.00
2 coats	5F@1.47	ea	63.30	96.30	159.60
3 coats	5F@2.00	ea	84.00	131.00	215.00
Paint column pedestal					
prime	5F@.536	ea	23.90	35.10	59.00
1 coat	5F@.552	ea	26.30	36.20	62.50
2 coats	5F@.848	ea	36.80	55.50	92.30
3 coats	5F@1.15	ea	48.40	75.30	123.70
Stain & varnish column pedestal					
stain	5F@.590	ea	26.10	38.60	64.70
1 coat	5F@.608	ea	29.20	39.80	69.00
2 coats	5F@.935	ea	40.30	61.20	101.50
3 coats	5F@1.27	ea	53.40	83.20	136.60

Concrete.

	Craft@Hrs	Unit	Material	Labor	Total
Paint concrete floor					
prime	5F@.008	sf	.27	.52	.79
1 coat	5F@.008	sf	.30	.52	.82
2 coats	5F@.012	sf	.42	.79	1.21
3 coats	5F@.016	sf	.54	1.05	1.59

	Craft@Hrs	Unit	Material	Labor	Total
Paint concrete wall					
prime	5F@.008	sf	.27	.52	.79
1 coat	5F@.008	sf	.30	.52	.82
2 coats	5F@.013	sf	.42	.85	1.27
3 coats	5F@.017	sf	.54	1.11	1.65
Paint concrete step (per step)					
prime	5F@.023	ea	.83	1.51	2.34
1 coat	5F@.024	ea	.94	1.57	2.51
2 coats	5F@.037	ea	1.28	2.42	3.70
3 coats	5F@.050	ea	1.71	3.28	4.99

Doors. Unless otherwise noted, does not include jamb and casing. Includes painting both sides of door. Folding and by-passing doors are for both sides of each section. Add **20%** for half-round or elliptical top door.

	Craft@Hrs	Unit	Material	Labor	Total
Paint folding door					
prime	5F@.402	ea	16.50	26.30	42.80
1 coat	5F@.413	ea	18.30	27.10	45.40
2 coats	5F@.636	ea	25.30	41.70	67.00
3 coats	5F@.861	ea	33.80	56.40	90.20
Stain & varnish folding door					
stain	5F@.614	ea	24.50	40.20	64.70
1 coat	5F@.632	ea	27.00	41.40	68.40
2 coats	5F@.972	ea	37.60	63.70	101.30
3 coats	5F@1.32	ea	49.70	86.50	136.20
Paint half-louvered folding door					
prime	5F@.451	ea	18.30	29.50	47.80
1 coat	5F@.464	ea	19.90	30.40	50.30
2 coats	5F@.713	ea	28.40	46.70	75.10
3 coats	5F@.966	ea	37.80	63.30	101.10
Stain & varnish half-louvered folding door					
stain	5F@.688	ea	27.00	45.10	72.10
1 coat	5F@.709	ea	29.70	46.40	76.10
2 coats	5F@1.09	ea	41.50	71.40	112.90
3 coats	5F@1.48	ea	55.40	96.90	152.30
Paint full-louvered folding door					
prime	5F@.507	ea	19.90	33.20	53.10
1 coat	5F@.520	ea	22.70	34.10	56.80
2 coats	5F@.800	ea	31.40	52.40	83.80
3 coats	5F@1.09	ea	42.00	71.40	113.40
Stain & varnish full-louvered folding door					
stain	5F@.774	ea	29.70	50.70	80.40
1 coat	5F@.796	ea	33.20	52.10	85.30
2 coats	5F@1.22	ea	46.00	79.90	125.90
3 coats	5F@1.66	ea	60.80	109.00	169.80
Paint folding panel door					
prime	5F@.439	ea	18.10	28.80	46.90
1 coat	5F@.453	ea	19.70	29.70	49.40
2 coats	5F@.697	ea	27.60	45.70	73.30
3 coats	5F@.944	ea	37.20	61.80	99.00

	Craft@Hrs	Unit	Material	Labor	Total
Stain & varnish folding panel door					
stain	5F@.673	ea	26.30	44.10	70.40
1 coat	5F@.693	ea	29.40	45.40	74.80
2 coats	5F@1.07	ea	40.80	70.10	110.90
3 coats	5F@1.44	ea	54.20	94.30	148.50
Paint bypassing door					
prime	5F@.830	ea	21.00	54.40	75.40
1 coat	5F@.854	ea	23.20	55.90	79.10
2 coats	5F@1.32	ea	32.00	86.50	118.50
3 coats	5F@1.78	ea	42.80	117.00	159.80
Stain & varnish bypassing door					
stain	5F@1.27	ea	30.60	83.20	113.80
1 coat	5F@1.31	ea	34.00	85.80	119.80
2 coats	5F@2.01	ea	47.10	132.00	179.10
3 coats	5F@2.72	ea	62.90	178.00	240.90
Paint half-louvered bypassing door					
prime	5F@.933	ea	23.30	61.10	84.40
1 coat	5F@.960	ea	25.80	62.90	88.70
2 coats	5F@1.48	ea	35.80	96.90	132.70
3 coats	5F@2.00	ea	47.40	131.00	178.40
Stain & varnish half-louvered bypassing door					
stain	5F@1.43	ea	34.00	93.70	127.70
1 coat	5F@1.47	ea	37.90	96.30	134.20
2 coats	5F@2.26	ea	52.60	148.00	200.60
3 coats	5F@3.05	ea	69.80	200.00	269.80
Paint full-louvered bypassing door					
prime	5F@1.05	ea	25.80	68.80	94.60
1 coat	5F@1.08	ea	28.70	70.70	99.40
2 coats	5F@1.66	ea	39.90	109.00	148.90
3 coats	5F@2.24	ea	52.80	147.00	199.80
Stain & varnish full-louvered bypassing door					
stain	5F@1.60	ea	37.90	105.00	142.90
1 coat	5F@1.65	ea	42.10	108.00	150.10
2 coats	5F@2.53	ea	58.50	166.00	224.50
3 coats	5F@3.43	ea	77.70	225.00	302.70
Paint bypassing panel door					
prime	5F@.899	ea	23.20	58.90	82.10
1 coat	5F@.926	ea	25.70	60.70	86.40
2 coats	5F@1.42	ea	35.60	93.00	128.60
3 coats	5F@1.93	ea	47.10	126.00	173.10
Stain & varnish bypassing panel door					
stain	5F@1.38	ea	33.90	90.40	124.30
1 coat	5F@1.42	ea	37.80	93.00	130.80
2 coats	5F@2.18	ea	52.40	143.00	195.40
3 coats	5F@2.96	ea	69.50	194.00	263.50
Paint interior door					
prime	5F@.340	ea	12.50	22.30	34.80
1 coat	5F@.349	ea	13.80	22.90	36.70
2 coats	5F@.538	ea	19.00	35.20	54.20
3 coats	5F@.729	ea	26.00	47.70	73.70

	Craft@Hrs	Unit	Material	Labor	Total
Stain & varnish interior door					
stain	5F@.519	ea	18.60	34.00	52.60
1 coat	5F@.534	ea	21.10	35.00	56.10
2 coats	5F@.822	ea	29.20	53.80	83.00
3 coats	5F@1.11	ea	38.10	72.70	110.80
Paint half-louvered door					
prime	5F@.381	ea	13.80	25.00	38.80
1 coat	5F@.392	ea	15.40	25.70	41.10
2 coats	5F@.603	ea	21.50	39.50	61.00
3 coats	5F@.818	ea	28.40	53.60	82.00
Stain & varnish half-louvered door					
stain	5F@.583	ea	21.00	38.20	59.20
1 coat	5F@.601	ea	23.20	39.40	62.60
2 coats	5F@.924	ea	32.00	60.50	92.50
3 coats	5F@1.25	ea	42.80	81.90	124.70
Paint full-louvered door					
prime	5F@.428	ea	15.40	28.00	43.40
1 coat	5F@.439	ea	17.10	28.80	45.90
2 coats	5F@.677	ea	23.90	44.30	68.20
3 coats	5F@.917	ea	31.40	60.10	91.50
Stain & varnish full-louvered door					
stain	5F@.655	ea	23.30	42.90	66.20
1 coat	5F@.673	ea	25.80	44.10	69.90
2 coats	5F@1.04	ea	35.80	68.10	103.90
3 coats	5F@1.40	ea	47.40	91.70	139.10
Paint door jamb & casing					
prime	5F@.013	lf	.34	.85	1.19
1 coat	5F@.011	lf	.36	.72	1.08
2 coats	5F@.017	lf	.49	1.11	1.60
3 coats	5F@.023	lf	.66	1.51	2.17
Stain & varnish door jamb & casing					
stain	5F@.019	lf	.48	1.24	1.72
1 coat	5F@.017	lf	.53	1.11	1.64
2 coats	5F@.026	lf	.75	1.70	2.45
3 coats	5F@.035	lf	1.02	2.29	3.31
Paint French door					
prime	5F@.818	ea	21.50	53.60	75.10
1 coat	5F@.841	ea	23.90	55.10	79.00
2 coats	5F@1.30	ea	32.70	85.20	117.90
3 coats	5F@1.75	ea	43.50	115.00	158.50
Stain & varnish French door					
stain	5F@1.25	ea	31.10	81.90	113.00
1 coat	5F@1.29	ea	34.50	84.50	119.00
2 coats	5F@1.98	ea	48.00	130.00	178.00
3 coats	5F@2.68	ea	64.00	176.00	240.00
Paint full-lite door					
prime	5F@.587	ea	20.80	38.40	59.20
1 coat	5F@.605	ea	23.10	39.60	62.70
2 coats	5F@.930	ea	31.80	60.90	92.70
3 coats	5F@1.26	ea	42.30	82.50	124.80

	Craft@Hrs	Unit	Material	Labor	Total
Stain & varnish full-lite door					
stain	5F@.899	ea	30.50	58.90	89.40
1 coat	5F@.926	ea	33.90	60.70	94.60
2 coats	5F@1.42	ea	47.10	93.00	140.10
3 coats	5F@1.93	ea	62.70	126.00	188.70
Paint panel door					
prime	5F@.704	ea	22.70	46.10	68.80
1 coat	5F@.724	ea	25.10	47.40	72.50
2 coats	5F@1.11	ea	34.80	72.70	107.50
3 coats	5F@1.51	ea	46.60	98.90	145.50
Stain & varnish panel door					
stain	5F@1.08	ea	33.30	70.70	104.00
1 coat	5F@1.11	ea	37.40	72.70	110.10
2 coats	5F@1.70	ea	51.60	111.00	162.60
3 coats	5F@2.31	ea	68.90	151.00	219.90
Paint transom					
prime	5F@.370	ea	10.30	24.20	34.50
1 coat	5F@.381	ea	11.50	25.00	36.50
2 coats	5F@.585	ea	15.60	38.30	53.90
3 coats	5F@.794	ea	21.30	52.00	73.30
Stain & varnish transom					
stain	5F@.565	ea	15.10	37.00	52.10
1 coat	5F@.583	ea	16.90	38.20	55.10
2 coats	5F@.897	ea	23.30	58.80	82.10
3 coats	5F@1.21	ea	30.90	79.30	110.20
Paint storm door					
prime	5F@.379	ea	10.70	24.80	35.50
1 coat	5F@.390	ea	11.80	25.50	37.30
2 coats	5F@.601	ea	16.50	39.40	55.90
3 coats	5F@.814	ea	22.00	53.30	75.30
Stain & varnish storm door					
stain	5F@.581	ea	15.60	38.10	53.70
1 coat	5F@.599	ea	17.80	39.20	57.00
2 coats	5F@.919	ea	24.60	60.20	84.80
3 coats	5F@1.25	ea	32.10	81.90	114.00
Paint Dutch door					
prime	5F@.666	ea	22.50	43.60	66.10
1 coat	5F@.686	ea	25.00	44.90	69.90
2 coats	5F@1.06	ea	34.70	69.40	104.10
3 coats	5F@1.43	ea	46.20	93.70	139.90
Stain & varnish Dutch door					
stain	5F@1.02	ea	32.80	66.80	99.60
1 coat	5F@1.05	ea	36.80	68.80	105.60
2 coats	5F@1.61	ea	50.90	105.00	155.90
3 coats	5F@2.19	ea	67.90	143.00	210.90
Paint steel entry door					
prime	5F@.621	ea	19.80	40.70	60.50
1 coat	5F@.639	ea	22.40	41.90	64.30
2 coats	5F@.982	ea	31.00	64.30	95.30
3 coats	5F@1.33	ea	41.20	87.10	128.30

	Craft@Hrs	Unit	Material	Labor	Total
Paint wood entry door					
prime	5F@.742	ea	21.80	48.60	70.40
1 coat	5F@.763	ea	23.30	50.00	73.30
2 coats	5F@1.18	ea	31.70	77.30	109.00
3 coats	5F@1.59	ea	44.40	104.00	148.40
Stain & varnish wood entry door					
stain	5F@1.13	ea	31.10	74.00	105.10
1 coat	5F@1.17	ea	33.90	76.60	110.50
2 coats	5F@1.80	ea	43.50	118.00	161.50
3 coats	5F@2.43	ea	60.70	159.00	219.70
Paint entry door side lite					
prime	5F@.673	ea	17.10	44.10	61.20
1 coat	5F@.691	ea	18.90	45.30	64.20
2 coats	5F@1.06	ea	26.30	69.40	95.70
3 coats	5F@1.44	ea	35.00	94.30	129.30
Stain & varnish entry door side lite					
stain	5F@1.03	ea	25.00	67.50	92.50
1 coat	5F@1.06	ea	27.70	69.40	97.10
2 coats	5F@1.63	ea	38.50	107.00	145.50
3 coats	5F@2.20	ea	51.60	144.00	195.60
Paint entry door fan lite					
prime	5F@.635	ea	15.10	41.60	56.70
1 coat	5F@.655	ea	16.90	42.90	59.80
2 coats	5F@1.01	ea	23.40	66.20	89.60
3 coats	5F@1.36	ea	31.00	89.10	120.10
Stain & varnish entry door fan lite					
stain	5F@.971	ea	22.30	63.60	85.90
1 coat	5F@1.00	ea	24.80	65.50	90.30
2 coats	5F@1.54	ea	34.10	101.00	135.10
3 coats	5F@2.08	ea	45.30	136.00	181.30
Paint cafe doors					
prime	5F@.592	ea	16.90	38.80	55.70
1 coat	5F@.610	ea	18.50	40.00	58.50
2 coats	5F@.937	ea	25.80	61.40	87.20
3 coats	5F@1.27	ea	34.30	83.20	117.50
Stain & varnish cafe doors					
stain	5F@.906	ea	24.80	59.30	84.10
1 coat	5F@.933	ea	27.30	61.10	88.40
2 coats	5F@1.43	ea	38.00	93.70	131.70
3 coats	5F@1.94	ea	50.60	127.00	177.60
Paint wood sliding patio door					
prime	5F@.646	ea	7.26	42.30	49.56
1 coat	5F@.665	ea	8.08	43.60	51.68
2 coats	5F@1.02	ea	11.30	66.80	78.10
3 coats	5F@1.39	ea	14.90	91.00	105.90
Stain & varnish wood sliding patio door					
stain	5F@.988	ea	10.70	64.70	75.40
1 coat	5F@1.02	ea	11.80	66.80	78.60
2 coats	5F@1.56	ea	16.40	102.00	118.40
3 coats	5F@2.12	ea	21.80	139.00	160.80

	Craft@Hrs	Unit	Material	Labor	Total
Paint garage door					
prime	5F@.006	sf	.30	.39	.69
1 coat	5F@.006	sf	.34	.39	.73
2 coats	5F@.009	sf	.45	.59	1.04
3 coats	5F@.013	sf	.61	.85	1.46
Stain & varnish garage door					
stain	5F@.009	sf	.43	.59	1.02
1 coat	5F@.009	sf	.48	.59	1.07
2 coats	5F@.014	sf	.66	.92	1.58
3 coats	5F@.019	sf	.85	1.24	2.09

Medicine cabinet.

	Craft@Hrs	Unit	Material	Labor	Total
Paint medicine cabinet					
prime	5F@.404	ea	4.95	26.50	31.45
1 coat	5F@.417	ea	5.54	27.30	32.84
2 coats	5F@.639	ea	7.68	41.90	49.58
3 coats	5F@.868	ea	10.10	56.90	67.00
Stain & varnish medicine cabinet					
stain	5F@.619	ea	7.27	40.50	47.77
1 coat	5F@.637	ea	8.10	41.70	49.80
2 coats	5F@.978	ea	11.30	64.10	75.40
3 coats	5F@1.32	ea	15.00	86.50	101.50

Exterior light fixture post.

	Craft@Hrs	Unit	Material	Labor	Total
Paint exterior light-fixture post					
prime	5F@.079	ea	6.04	5.17	11.21
1 coat	5F@.081	ea	6.71	5.31	12.02
2 coats	5F@.125	ea	9.32	8.19	17.51
3 coats	5F@.170	ea	12.40	11.10	23.50

Fences. Wood fence prices are for one side only. Ornamental iron fence prices are for both sides.

	Craft@Hrs	Unit	Material	Labor	Total
Paint 4' high wood fence					
prime	5F@.051	lf	1.34	3.34	4.68
1 coat	5F@.053	lf	1.50	3.47	4.97
2 coats	5F@.081	lf	2.10	5.31	7.41
3 coats	5F@.109	lf	2.76	7.14	9.90
Seal or stain 4' high wood fence					
varnish	5F@.078	lf	1.96	5.11	7.07
1 coat	5F@.080	lf	2.21	5.24	7.45
2 coats	5F@.123	lf	3.06	8.06	11.12
3 coats	5F@.167	lf	4.06	10.90	14.96
Paint 6' high wood fence					
prime	5F@.077	lf	1.96	5.04	7.00
1 coat	5F@.079	lf	2.21	5.17	7.38
2 coats	5F@.122	lf	3.06	7.99	11.05
3 coats	5F@.165	lf	4.06	10.80	14.86

	Craft@Hrs	Unit	Material	Labor	Total
Seal or stain 6' high wood fence					
varnish	5F@.117	lf	2.92	7.66	10.58
1 coat	5F@.121	lf	3.22	7.93	11.15
2 coats	5F@.186	lf	4.50	12.20	16.70
3 coats	5F@.251	lf	5.97	16.40	22.37
Paint 8' high wood fence					
prime	5F@.102	lf	2.59	6.68	9.27
1 coat	5F@.105	lf	2.89	6.88	9.77
2 coats	5F@.162	lf	4.01	10.60	14.61
3 coats	5F@.219	lf	5.32	14.30	19.62
Seal or stain 8' high wood fence					
varnish	5F@.156	lf	3.81	10.20	14.01
1 coat	5F@.161	lf	4.23	10.50	14.73
2 coats	5F@.247	lf	5.90	16.20	22.10
3 coats	5F@.334	lf	7.81	21.90	29.71
Paint 3' high picket fence					
prime	5F@.041	lf	1.05	2.69	3.74
1 coat	5F@.042	lf	1.16	2.75	3.91
2 coats	5F@.064	lf	1.59	4.19	5.78
3 coats	5F@.087	lf	2.15	5.70	7.85
Paint 5' high picket fence					
prime	5F@.065	lf	1.69	4.26	5.95
1 coat	5F@.067	lf	1.85	4.39	6.24
2 coats	5F@.103	lf	2.58	6.75	9.33
3 coats	5F@.140	lf	3.43	9.17	12.60
Paint 48" high ornamental iron fence					
prime	5F@.077	lf	1.88	5.04	6.92
1 coat	5F@.079	lf	2.11	5.17	7.28
2 coats	5F@.121	lf	2.92	7.93	10.85
3 coats	5F@.164	lf	3.85	10.70	14.55
Paint 60" high ornamental iron fence					
prime	5F@.096	lf	2.51	6.29	8.80
1 coat	5F@.098	lf	2.80	6.42	9.22
2 coats	5F@.151	lf	3.89	9.89	13.78
3 coats	5F@.205	lf	5.18	13.40	18.58
Paint 72" high ornamental iron fence					
prime	5F@.115	lf	3.55	7.53	11.08
1 coat	5F@.118	lf	3.93	7.73	11.66
2 coats	5F@.182	lf	5.48	11.90	17.38
3 coats	5F@.247	lf	7.27	16.20	23.47

Finish carpentry.

	Craft@Hrs	Unit	Material	Labor	Total
Paint wood trim, simple design					
prime	5F@.011	lf	.21	.72	.93
1 coat	5F@.011	lf	.22	.72	.94
2 coats	5F@.017	lf	.30	1.11	1.41
3 coats	5F@.023	lf	.39	1.51	1.90

	Craft@Hrs	Unit	Material	Labor	Total
Stain & varnish wood trim, simple design					
stain	5F@.016	lf	.30	1.05	1.35
1 coat	5F@.017	lf	.22	1.11	1.33
2 coats	5F@.026	lf	.30	1.70	2.00
3 coats	5F@.035	lf	.39	2.29	2.68
Paint wood trim, ornate design					
prime	5F@.013	lf	.26	.85	1.11
1 coat	5F@.014	lf	.28	.92	1.20
2 coats	5F@.021	lf	.39	1.38	1.77
3 coats	5F@.028	lf	.51	1.83	2.34
Stain & varnish wood trim, ornate design					
stain	5F@.020	lf	.38	1.31	1.69
1 coat	5F@.021	lf	.42	1.38	1.80
2 coats	5F@.032	lf	.59	2.10	2.69
3 coats	5F@.044	lf	.77	2.88	3.65
Paint wood trim, very ornate design					
prime	5F@.017	lf	.30	1.11	1.41
1 coat	5F@.017	lf	.34	1.11	1.45
2 coats	5F@.026	lf	.45	1.70	2.15
3 coats	5F@.035	lf	.61	2.29	2.90
Stain & varnish wood trim, very ornate design					
stain	5F@.025	lf	.43	1.64	2.07
1 coat	5F@.026	lf	.48	1.70	2.18
2 coats	5F@.040	lf	.66	2.62	3.28
3 coats	5F@.054	lf	.85	3.54	4.39
Paint interior architrave, simple design					
prime	5F@.015	lf	.30	.98	1.28
1 coat	5F@.015	lf	.34	.98	1.32
2 coats	5F@.023	lf	.45	1.51	1.96
3 coats	5F@.032	lf	.61	2.10	2.71
Stain & varnish interior architrave, simple design					
stain	5F@.023	lf	.43	1.51	1.94
1 coat	5F@.023	lf	.48	1.51	1.99
2 coats	5F@.036	lf	.66	2.36	3.02
3 coats	5F@.048	lf	.85	3.14	3.99
Paint interior architrave, complex design					
prime	5F@.030	lf	.42	1.97	2.39
1 coat	5F@.030	lf	.45	1.97	2.42
2 coats	5F@.047	lf	.64	3.08	3.72
3 coats	5F@.063	lf	.84	4.13	4.97
Stain & varnish interior architrave, complex design					
stain	5F@.045	lf	.61	2.95	3.56
1 coat	5F@.047	lf	.69	3.08	3.77
2 coats	5F@.072	lf	1.00	4.72	5.72
3 coats	5F@.097	lf	1.27	6.35	7.62
Paint exterior architrave, simple design					
prime	5F@.033	lf	.57	2.16	2.73
1 coat	5F@.035	lf	.63	2.29	2.92
2 coats	5F@.053	lf	.85	3.47	4.32
3 coats	5F@.072	lf	1.16	4.72	5.88

	Craft@Hrs	Unit	Material	Labor	Total
Paint exterior architrave, complex design					
prime	5F@.067	lf	1.16	4.39	5.55
1 coat	5F@.069	lf	1.28	4.52	5.80
2 coats	5F@.106	lf	1.80	6.94	8.74
3 coats	5F@.143	lf	2.39	9.37	11.76
Paint exterior door surround, simple design					
prime	5F@.117	lf	2.32	7.66	9.98
1 coat	5F@.120	lf	2.57	7.86	10.43
2 coats	5F@.185	lf	3.56	12.10	15.66
3 coats	5F@.251	lf	4.71	16.40	21.11
Paint exterior door surround, complex design					
prime	5F@.287	lf	4.92	18.80	23.72
1 coat	5F@.296	lf	5.45	19.40	24.85
2 coats	5F@.455	lf	7.54	29.80	37.34
3 coats	5F@.617	lf	10.10	40.40	50.50
Paint exterior window surround, simple design					
prime	5F@.117	lf	2.32	7.66	9.98
1 coat	5F@.120	lf	2.57	7.86	10.43
2 coats	5F@.185	lf	3.56	12.10	15.66
3 coats	5F@.251	lf	4.71	16.40	21.11
Paint exterior window surround, complex design					
prime	5F@.287	lf	4.92	18.80	23.72
1 coat	5F@.296	lf	5.45	19.40	24.85
2 coats	5F@.455	lf	7.54	29.80	37.34
3 coats	5F@.617	lf	10.10	40.40	50.50
Paint closet rod, shelf, and brackets					
prime	5F@.127	lf	.36	8.32	8.68
1 coat	5F@.131	lf	.41	8.58	8.99
2 coats	5F@.201	lf	.54	13.20	13.74
3 coats	5F@.274	lf	.74	17.90	18.64
Paint closet organizer system					
prime	5F@.011	sf	.41	.72	1.13
1 coat	5F@.011	sf	.44	.72	1.16
2 coats	5F@.017	sf	.63	1.11	1.74
3 coats	5F@.023	sf	.83	1.51	2.34
Paint bookcase					
prime	5F@.014	sf	.90	.92	1.82
1 coat	5F@.015	sf	1.03	.98	2.01
2 coats	5F@.023	sf	1.39	1.51	2.90
3 coats	5F@.030	sf	1.85	1.97	3.82
Stain & varnish bookcase					
stain	5F@.022	sf	1.32	1.44	2.76
1 coat	5F@.023	sf	1.46	1.51	2.97
2 coats	5F@.035	sf	1.99	2.29	4.28
3 coats	5F@.047	sf	2.72	3.08	5.80
Paint fireplace mantel beam					
prime	5F@.018	lf	.45	1.18	1.63
1 coat	5F@.018	lf	.50	1.18	1.68
2 coats	5F@.028	lf	.71	1.83	2.54
3 coats	5F@.038	lf	.94	2.49	3.43

	Craft@Hrs	Unit	Material	Labor	Total
Stain & varnish mantel beam					
stain	5F@.027	lf	.69	1.77	2.46
1 coat	5F@.028	lf	.75	1.83	2.58
2 coats	5F@.043	lf	1.05	2.82	3.87
3 coats	5F@.059	lf	1.38	3.86	5.24
Paint fireplace mantel					
prime	5F@.025	lf	.61	1.64	2.25
1 coat	5F@.026	lf	.69	1.70	2.39
2 coats	5F@.040	lf	1.00	2.62	3.62
3 coats	5F@.054	lf	1.27	3.54	4.81
Stain & varnish mantel					
stain	5F@.038	lf	.90	2.49	3.39
1 coat	5F@.039	lf	1.03	2.55	3.58
2 coats	5F@.061	lf	1.39	4.00	5.39
3 coats	5F@.082	lf	1.85	5.37	7.22
Paint coffered ceiling, simple design					
prime	5F@.017	sf	.73	1.11	1.84
1 coat	5F@.018	sf	.80	1.18	1.98
2 coats	5F@.027	sf	1.14	1.77	2.91
3 coats	5F@.037	sf	1.46	2.42	3.88
Stain & varnish coffered ceiling, simple design					
stain	5F@.026	sf	1.06	1.70	2.76
1 coat	5F@.027	sf	1.18	1.77	2.95
2 coats	5F@.042	sf	1.61	2.75	4.36
3 coats	5F@.056	sf	2.17	3.67	5.84
Paint coffered ceiling, complex design					
prime	5F@.041	sf	1.28	2.69	3.97
1 coat	5F@.042	sf	1.44	2.75	4.19
2 coats	5F@.065	sf	1.97	4.26	6.23
3 coats	5F@.089	sf	2.65	5.83	8.48
Stain & varnish coffered ceiling, complex design					
stain	5F@.063	sf	1.88	4.13	6.01
1 coat	5F@.065	sf	2.11	4.26	6.37
2 coats	5F@.100	sf	2.92	6.55	9.47
3 coats	5F@.135	sf	3.85	8.84	12.69
Paint niche					
prime	5F@.538	ea	11.80	35.20	47.00
1 coat	5F@.554	ea	13.10	36.30	49.40
2 coats	5F@.852	ea	18.20	55.80	74.00
3 coats	5F@1.15	ea	24.50	75.30	99.80
Stain & varnish niche					
stain	5F@.823	ea	17.50	53.90	71.40
1 coat	5F@.845	ea	19.00	55.30	74.30
2 coats	5F@1.30	ea	26.70	85.20	111.90
3 coats	5F@1.76	ea	35.60	115.00	150.60
Paint gingerbread running trim					
prime	5F@.045	lf	1.83	2.95	4.78
1 coat	5F@.047	lf	1.99	3.08	5.07
2 coats	5F@.072	lf	2.84	4.72	7.56
3 coats	5F@.097	lf	3.78	6.35	10.13

	Craft@Hrs	Unit	Material	Labor	Total
Stain & varnish gingerbread running trim					
stain	5F@.069	lf	2.70	4.52	7.22
1 coat	5F@.071	lf	2.97	4.65	7.62
2 coats	5F@.110	lf	4.15	7.21	11.36
3 coats	5F@.148	lf	5.54	9.69	15.23
Paint gingerbread bracket					
prime	5F@.336	ea	2.76	22.00	24.76
1 coat	5F@.345	ea	3.09	22.60	25.69
2 coats	5F@.534	ea	4.30	35.00	39.30
3 coats	5F@.722	ea	5.69	47.30	52.99
Stain & varnish gingerbread bracket					
stain	5F@.516	ea	4.06	33.80	37.86
1 coat	5F@.529	ea	4.51	34.60	39.11
2 coats	5F@.814	ea	6.28	53.30	59.58
3 coats	5F@1.10	ea	8.30	72.10	80.40
Paint gingerbread corbel					
prime	5F@.330	ea	2.72	21.60	24.32
1 coat	5F@.341	ea	3.03	22.30	25.33
2 coats	5F@.522	ea	4.20	34.20	38.40
3 coats	5F@.709	ea	5.58	46.40	51.98
Stain & varnish gingerbread corbel					
stain	5F@.504	ea	4.00	33.00	37.00
1 coat	5F@.520	ea	4.45	34.10	38.55
2 coats	5F@.800	ea	6.19	52.40	58.59
3 coats	5F@1.08	ea	8.21	70.70	78.91
Paint gingerbread door or window header					
prime	5F@.054	lf	.90	3.54	4.44
1 coat	5F@.055	lf	1.03	3.60	4.63
2 coats	5F@.085	lf	1.39	5.57	6.96
3 coats	5F@.115	lf	1.85	7.53	9.38
Stain & varnish gingerbread door or window header					
stain	5F@.082	lf	1.32	5.37	6.69
1 coat	5F@.084	lf	1.46	5.50	6.96
2 coats	5F@.130	lf	1.99	8.52	10.51
3 coats	5F@.176	lf	2.72	11.50	14.22
Paint gingerbread spandrel					
prime	5F@.075	lf	.97	4.91	5.88
1 coat	5F@.077	lf	1.06	5.04	6.10
2 coats	5F@.119	lf	1.45	7.79	9.24
3 coats	5F@.161	lf	1.91	10.50	12.41
Stain & varnish gingerbread spandrel					
stain	5F@.115	lf	1.38	7.53	8.91
1 coat	5F@.118	lf	1.53	7.73	9.26
2 coats	5F@.182	lf	2.15	11.90	14.05
3 coats	5F@.247	lf	2.84	16.20	19.04
Paint gingerbread gable ornament					
prime	5F@.747	ea	19.60	48.90	68.50
1 coat	5F@.767	ea	22.10	50.20	72.30
2 coats	5F@1.18	ea	30.60	77.30	107.90
3 coats	5F@1.60	ea	40.60	105.00	145.60
add 51% to stain and varnish					

	Craft@Hrs	Unit	Material	Labor	Total
Paint gingerbread gable finial					
prime	5F@.253	ea	3.93	16.60	20.53
1 coat	5F@.260	ea	4.35	17.00	21.35
2 coats	5F@.401	ea	6.04	26.30	32.34
3 coats	5F@.543	ea	8.08	35.60	43.68
add 51% to stain and varnish					
Paint porch post					
prime	5F@.118	ea	6.28	7.73	14.01
1 coat	5F@.122	ea	6.97	7.99	14.96
2 coats	5F@.187	ea	9.66	12.20	21.86
3 coats	5F@.253	ea	12.80	16.60	29.40
Stain & varnish porch post					
stain	5F@.181	ea	9.19	11.90	21.09
1 coat	5F@.186	ea	10.20	12.20	22.40
2 coats	5F@.287	ea	14.30	18.80	33.10
3 coats	5F@.388	ea	18.80	25.40	44.20

Flooring. See Flooring for costs to prep wood floors for finish.

	Craft@Hrs	Unit	Material	Labor	Total
Seal stone floor					
seal	5F@.011	sf	.26	.72	.98
1 coat	5F@.011	sf	.28	.72	1.00
2 coats	5F@.017	sf	.39	1.11	1.50
3 coats	5F@.023	sf	.51	1.51	2.02
Paint wood floor					
prime	5F@.012	sf	.33	.79	1.12
1 coat	5F@.012	sf	.35	.79	1.14
2 coats	5F@.019	sf	.48	1.24	1.72
3 coats	5F@.026	sf	.64	1.70	2.34
Stain & varnish wood floor					
stain	5F@.018	sf	.45	1.18	1.63
1 coat	5F@.019	sf	.50	1.24	1.74
2 coats	5F@.029	sf	.71	1.90	2.61
3 coats	5F@.039	sf	.94	2.55	3.49

Paneling.

	Craft@Hrs	Unit	Material	Labor	Total
Paint wall paneling, simple pattern					
prime	5F@.012	sf	.48	.79	1.27
1 coat	5F@.012	sf	.53	.79	1.32
2 coats	5F@.019	sf	.75	1.24	1.99
3 coats	5F@.026	sf	1.02	1.70	2.72
Stain & varnish wall paneling, simple pattern					
stain	5F@.018	sf	.71	1.18	1.89
1 coat	5F@.019	sf	.77	1.24	2.01
2 coats	5F@.029	sf	1.08	1.90	2.98
3 coats	5F@.039	sf	1.44	2.55	3.99

	Craft@Hrs	Unit	Material	Labor	Total
Paint wall paneling, ornate pattern					
prime	5F@.017	sf	.64	1.11	1.75
1 coat	5F@.017	sf	.73	1.11	1.84
2 coats	5F@.027	sf	1.03	1.77	2.80
3 coats	5F@.036	sf	1.33	2.36	3.69
Stain & varnish wall paneling, ornate pattern					
stain	5F@.026	sf	.97	1.70	2.67
1 coat	5F@.027	sf	1.06	1.77	2.83
2 coats	5F@.041	sf	1.45	2.69	4.14
3 coats	5F@.056	sf	1.91	3.67	5.58
Paint wall paneling, very ornate pattern					
prime	5F@.028	sf	.77	1.83	2.60
1 coat	5F@.029	sf	.85	1.90	2.75
2 coats	5F@.044	sf	1.22	2.88	4.10
3 coats	5F@.060	sf	1.61	3.93	5.54
Stain & varnish wall paneling, very ornate pattern					
stain	5F@.043	sf	1.15	2.82	3.97
1 coat	5F@.044	sf	1.26	2.88	4.14
2 coats	5F@.068	sf	1.78	4.45	6.23
3 coats	5F@.093	sf	2.33	6.09	8.42

Masonry.

	Craft@Hrs	Unit	Material	Labor	Total
Paint concrete block (unpainted block)					
prime	5F@.014	sf	.48	.92	1.40
1 coat	5F@.015	sf	.53	.98	1.51
2 coats	5F@.023	sf	.75	1.51	2.26
3 coats	5F@.030	sf	1.02	1.97	2.99
deduct 26% for previously painted block or brick					
Paint brick (unpainted brick)					
prime	5F@.017	sf	.50	1.11	1.61
1 coat	5F@.017	sf	.57	1.11	1.68
2 coats	5F@.026	sf	.80	1.70	2.50
3 coats	5F@.035	sf	1.06	2.29	3.35
Paint stone wall					
prime	5F@.010	sf	.38	.66	1.04
1 coat	5F@.010	sf	.42	.66	1.08
2 coats	5F@.016	sf	.59	1.05	1.64
3 coats	5F@.021	sf	.77	1.38	2.15
Seal stone wall					
seal	5F@.008	sf	.50	.52	1.02
1 coat	5F@.009	sf	.57	.59	1.16
2 coats	5F@.013	sf	.80	.85	1.65
3 coats	5F@.018	sf	1.06	1.18	2.24

Roofing.

	Craft@Hrs	Unit	Material	Labor	Total
Paint metal roofing					
prime	5F@.017	sf	.53	1.11	1.64
1 coat	5F@.018	sf	.60	1.18	1.78
2 coats	5F@.027	sf	.83	1.77	2.60
3 coats	5F@.037	sf	1.14	2.42	3.56

	Craft@Hrs	Unit	Material	Labor	Total
Treat wood shingles or shakes with shingle oil					
treat	5F@.013	sf	.36	.85	1.21
1 coat	5F@.014	sf	.41	.92	1.33
2 coats	5F@.021	sf	.54	1.38	1.92
3 coats	5F@.028	sf	.74	1.83	2.57

Rough carpentry.

	Craft@Hrs	Unit	Material	Labor	Total
Seal wall framing, for odor control, smoke-stained framing					
seal	5F@.018	sf	.26	1.18	1.44
1 coat	5F@.018	sf	.28	1.18	1.46
2 coats	5F@.028	sf	.39	1.83	2.22
3 coats	5F@.038	sf	.51	2.49	3.00
Seal floor framing					
seal	5F@.022	sf	.34	1.44	1.78
1 coat	5F@.023	sf	.36	1.51	1.87
2 coats	5F@.035	sf	.49	2.29	2.78
3 coats	5F@.047	sf	.66	3.08	3.74
Seal roof framing					
seal	5F@.028	sf	.42	1.83	2.25
1 coat	5F@.029	sf	.45	1.90	2.35
2 coats	5F@.044	sf	.64	2.88	3.52
3 coats	5F@.059	sf	.84	3.86	4.70

Siding.

	Craft@Hrs	Unit	Material	Labor	Total
Paint exterior siding					
prime	5F@.016	sf	.38	1.05	1.43
1 coat	5F@.016	sf	.41	1.05	1.46
2 coats	5F@.025	sf	.57	1.64	2.21
3 coats	5F@.033	sf	.76	2.16	2.92
Stain exterior siding					
stain	5F@.015	sf	.61	.98	1.59
1 coat	5F@.015	sf	.69	.98	1.67
2 coats	5F@.023	sf	1.00	1.51	2.51
3 coats	5F@.032	sf	1.27	2.10	3.37
Paint window shutter (per shutter, per side)					
prime	5F@.379	ea	8.21	24.80	33.01
1 coat	5F@.390	ea	9.11	25.50	34.61
2 coats	5F@.601	ea	12.60	39.40	52.00
3 coats	5F@.814	ea	16.90	53.30	70.20
Stain window shutter (per shutter, per side)					
stain	5F@.581	ea	12.00	38.10	50.10
1 coat	5F@.596	ea	13.30	39.00	52.30
2 coats	5F@.917	ea	18.50	60.10	78.60
3 coats	5F@1.24	ea	24.80	81.20	106.00
Paint exterior fascia					
prime	5F@.011	lf	.11	.72	.83
1 coat	5F@.011	lf	.16	.72	.88
2 coats	5F@.017	lf	.26	1.11	1.37
3 coats	5F@.023	lf	.35	1.51	1.86

	Craft@Hrs	Unit	Material	Labor	Total
Stain exterior wood fascia					
stain	5F@.010	lf	.24	.66	.90
1 coat	5F@.010	lf	.26	.66	.92
2 coats	5F@.016	lf	.36	1.05	1.41
3 coats	5F@.021	lf	.48	1.38	1.86
Paint exterior soffit					
prime	5F@.011	sf	.38	.72	1.10
1 coat	5F@.012	sf	.42	.79	1.21
2 coats	5F@.018	sf	.59	1.18	1.77
3 coats	5F@.024	sf	.77	1.57	2.34
Stain exterior wood soffit					
stain	5F@.012	sf	.48	.79	1.27
1 coat	5F@.012	sf	.53	.79	1.32
2 coats	5F@.019	sf	.75	1.24	1.99
3 coats	5F@.026	sf	1.02	1.70	2.72
Paint rain gutter or downspout					
prime	5F@.018	lf	.33	1.18	1.51
1 coat	5F@.019	lf	.35	1.24	1.59
2 coats	5F@.029	lf	.48	1.90	2.38
3 coats	5F@.039	lf	.64	2.55	3.19

Stairs.

	Craft@Hrs	Unit	Material	Labor	Total
Paint stair balustrade					
prime	5F@.190	lf	4.46	12.40	16.86
1 coat	5F@.195	lf	4.97	12.80	17.77
2 coats	5F@.300	lf	6.95	19.70	26.65
3 coats	5F@.407	lf	9.22	26.70	35.92
Stain & varnish stair balustrade					
stain	5F@.290	lf	6.59	19.00	25.59
1 coat	5F@.298	lf	7.28	19.50	26.78
2 coats	5F@.459	lf	10.10	30.10	40.20
3 coats	5F@.621	lf	13.40	40.70	54.10
Paint stair riser					
prime	5F@.112	ea	3.77	7.34	11.11
1 coat	5F@.115	ea	4.18	7.53	11.71
2 coats	5F@.177	ea	5.79	11.60	17.39
3 coats	5F@.240	ea	7.72	15.70	23.42
Stain & varnish stair riser					
stain	5F@.171	ea	5.54	11.20	16.74
1 coat	5F@.176	ea	6.08	11.50	17.58
2 coats	5F@.271	ea	8.50	17.80	26.30
3 coats	5F@.368	ea	11.30	24.10	35.40
Paint stair tread					
prime	5F@.121	ea	4.18	7.93	12.11
1 coat	5F@.125	ea	4.63	8.19	12.82
2 coats	5F@.192	ea	6.43	12.60	19.03
3 coats	5F@.260	ea	8.53	17.00	25.53

	Craft@Hrs	Unit	Material	Labor	Total
Stain & varnish stair tread					
stain	5F@.185	ea	6.08	12.10	18.18
1 coat	5F@.191	ea	6.82	12.50	19.32
2 coats	5F@.294	ea	9.43	19.30	28.73
3 coats	5F@.399	ea	12.60	26.10	38.70
Paint stair bracket					
prime	5F@.092	ea	3.44	6.03	9.47
1 coat	5F@.094	ea	3.82	6.16	9.98
2 coats	5F@.145	ea	5.33	9.50	14.83
3 coats	5F@.196	ea	7.06	12.80	19.86
Stain & varnish stair bracket					
stain	5F@.140	ea	5.08	9.17	14.25
1 coat	5F@.144	ea	5.65	9.43	15.08
2 coats	5F@.222	ea	7.82	14.50	22.32
3 coats	5F@.300	ea	10.50	19.70	30.20

Window.

	Craft@Hrs	Unit	Material	Labor	Total
Paint small window (per side)					
prime	5F@.295	ea	11.90	19.30	31.20
1 coat	5F@.304	ea	13.20	19.90	33.10
2 coats	5F@.467	ea	18.30	30.60	48.90
3 coats	5F@.633	ea	24.60	41.50	66.10
Paint average size window					
prime	5F@.422	ea	18.20	27.60	45.80
1 coat	5F@.433	ea	19.80	28.40	48.20
2 coats	5F@.666	ea	27.70	43.60	71.30
3 coats	5F@.901	ea	37.40	59.00	96.40
Paint large window					
prime	5F@.534	ea	23.10	35.00	58.10
1 coat	5F@.549	ea	25.60	36.00	61.60
2 coats	5F@.845	ea	35.50	55.30	90.80
3 coats	5F@1.15	ea	47.10	75.30	122.40
Paint very large window					
prime	5F@.668	ea	32.70	43.80	76.50
1 coat	5F@.687	ea	36.70	45.00	81.70
2 coats	5F@1.06	ea	50.70	69.40	120.10
3 coats	5F@1.43	ea	67.10	93.70	160.80

Wallpaper.

	Craft@Hrs	Unit	Material	Labor	Total
Paint wallpaper					
prime	5F@.011	sf	.30	.72	1.02
1 coat	5F@.011	sf	.34	.72	1.06
2 coats	5F@.017	sf	.45	1.11	1.56
3 coats	5F@.023	sf	.61	1.51	2.12

Strip paint. Cost to strip paint that does not contain lead.

	Craft@Hrs	Unit	Material	Labor	Total
Strip paint or varnish from trim, simple design					
strip	5F@.081	lf	2.20	5.31	7.51
1 coat	5F@.083	lf	2.58	5.44	8.02
2 coats	5F@.126	lf	3.47	8.25	11.72
3 coats	5F@.172	lf	4.54	11.30	15.84

	Craft@Hrs	Unit	Material	Labor	Total
Strip paint or varnish from trim, complex design					
strip	5F@.153	lf	2.67	10.00	12.67
1 coat	5F@.158	lf	2.97	10.30	13.27
2 coats	5F@.291	lf	4.71	19.10	23.81
3 coats	5F@.513	lf	7.15	33.60	40.75
Strip paint or varnish from door (slab only)					
strip	5F@.529	ea	23.90	34.60	58.50
1 coat	5F@.545	ea	28.40	35.70	64.10
2 coats	5F@.845	ea	40.10	55.30	95.40
3 coats	5F@1.27	ea	59.60	83.20	142.80
Strip paint or varnish from door jamb and casing					
strip	5F@.092	lf	3.77	6.03	9.80
1 coat	5F@.095	lf	4.44	6.22	10.66
2 coats	5F@.146	lf	5.96	9.56	15.52
3 coats	5F@.198	lf	7.82	13.00	20.82

Time & Material Charts (selected items)
Painting Materials

	Craft@Hrs	Unit	Material	Labor	Total
Prime or seal					
primer or sealer, plaster or drywall ($34.70 gallon, 178 sf)	—	sf	.20	—	.20
primer or sealer, interior door ($34.70 gallon, 2.8 ea)	—	ea	12.40	—	12.40
primer or sealer, entry door ($34.70 gallon, 1.6 ea)	—	ea	21.70	—	21.70
primer or sealer, wood trim, simple design ($34.70 gallon, 178 lf)	—	lf	.20	—	.20
primer or sealer, exterior siding ($34.70 gallon, 94 sf)	—	sf	.38	—	.38
primer or sealer, stair balustrade ($34.70 gallon, 7.8 lf)	—	lf	4.46	—	4.46
Paint, one coat					
oil or latex paint, plaster or drywall ($44.40 gallon, 205 sf)	—	sf	.22	—	.22
oil or latex paint, interior door ($44.40 gallon, 3.2 ea)	—	ea	13.90	—	13.90
oil or latex paint, entry door ($44.40 gallon, 1.9 ea)	—	ea	23.30	—	23.30
oil or latex paint, wood trim, simple design ($44.40 gallon, 205 lf)	—	lf	.22	—	.22
oil or latex paint, exterior siding ($44.40 gallon, 108 sf)	—	sf	.42	—	.42
oil or latex paint, stair balustrade ($44.40 gallon, 8.9 lf)	—	lf	4.98	—	4.98

	Craft@Hrs	Unit	Material	Labor	Total
Paint, two coats					
oil or latex paint, plaster or drywall ($44.40 gallon, 151 sf)	—	sf	.29	—	.29
oil or latex paint, interior door ($44.40 gallon, 2.3 ea)	—	ea	19.30	—	19.30
oil or latex paint, entry door ($44.40 gallon, 1.4 ea)	—	ea	31.80	—	31.80
oil or latex paint, wood trim, simple design ($44.40 gallon, 151 lf)	—	lf	.29	—	.29
oil or latex paint, exterior siding ($44.40 gallon, 78 sf)	—	sf	.57	—	.57
oil or latex paint, stair balustrade ($44.40 gallon, 6.4 lf)	—	lf	6.93	—	6.93
Paint, three coats					
oil or latex paint, plaster or drywall ($44.40 gallon, 113 sf)	—	sf	.41	—	.41
oil or latex paint, interior door ($44.40 gallon, 1.7 ea)	—	ea	26.00	—	26.00
oil or latex paint, entry door ($44.40 gallon, 1.0 ea)	—	ea	44.40	—	44.40
oil or latex paint, wood trim, simple design ($44.40 gallon, 113 lf)	—	lf	.41	—	.41
oil or latex paint, exterior siding ($44.40 gallon, 58 sf)	—	sf	.76	—	.76
oil or latex paint, stair balustrade ($44.40 gallon, 4.8 lf)	—	lf	9.24	—	9.24
Stain					
stain, interior door ($46.80 gallon, 2.5 ea)	—	ea	18.80	—	18.80
stain, entry door ($46.80 gallon, 1.5 ea)	—	ea	31.30	—	31.30
stain, wood trim, simple design ($46.80 gallon, 159 lf)	—	lf	.29	—	.29
stain, exterior siding, one coat ($46.80 gallon, 77 sf)	—	sf	.61	—	.61
stain, stair balustrade ($46.80 gallon, 7.1 lf)	—	lf	6.59	—	6.59
Varnish, one coat					
varnish, clear, interior door ($60.70 gallon, 2.9 ea)	—	ea	20.90	—	20.90
varnish, clear, entry door ($60.70 gallon, 1.8 ea)	—	ea	33.80	—	33.80
varnish, clear, wood trim, simple design ($60.70 gallon, 284 lf)	—	lf	.22	—	.22
varnish, clear, stair balustrade ($60.70 gallon, 8.4 lf)	—	lf	7.22	—	7.22

	Craft@Hrs	Unit	Material	Labor	Total
Varnish, two coats					
varnish, clear, interior door ($60.70 gallon, 2.1 ea)	—	ea	29.00	—	29.00
varnish, clear, entry door ($60.70 gallon, 1.4 ea)	—	ea	43.30	—	43.30
varnish, clear, wood trim, simple design ($60.70 gallon, 208 lf)	—	lf	.29	—	.29
varnish, clear, stair balustrade ($60.70 gallon, 6.0 lf)	—	lf	10.10	—	10.10
Varnish, three coats					
varnish, clear, interior door ($60.70 gallon, 1.6 ea)	—	ea	37.90	—	37.90
varnish, clear, entry door ($60.70 gallon, 1.0 ea)	—	ea	60.70	—	60.70
varnish, clear, wood trim, simple design ($60.70 gallon, 156 lf)	—	lf	.39	—	.39
varnish, clear, stair balustrade ($60.70 gallon, 4.5 lf)	—	lf	13.50	—	13.50

Painting Labor

Laborer	base wage	paid leave	true wage	taxes & ins.	total
Painter	$37.50	2.93	$40.43	25.07	$65.50

Paid leave is calculated based on two weeks paid vacation, one week sick leave, and seven paid holidays. Employer's matching portion of **FICA** is 7.65 percent. **FUTA** (Federal Unemployment) is .8 percent. **Worker's compensation** for the painting trade was calculated using a national average of 14.50 percent. **Unemployment insurance** was calculated using a national average of 8 percent. **Health insurance** was calculated based on a projected national average for 2020 of $1,280 per employee (and family when applicable) per month. Employer pays 80 percent for a per month cost of $1,024 per employee. **Retirement** is based on a 401(k) retirement program with employer matching of 50 percent. Employee contributions to the 401(k) plan are an average of 6 percent of the true wage. **Liability insurance** is based on a national average of 12.0 percent.

Painting Labor Productivity	Craft@Hrs	Unit	Material	Labor	Total
Prime or seal					
acoustic ceiling texture	5F@.011	sf	—	.72	.72
plaster or drywall	5F@.006	sf	—	.39	.39
masonry, block	5F@.014	sf	—	.92	.92
concrete floor	5F@.008	sf	—	.52	.52
wood floor	5F@.012	sf	—	.79	.79
folding door	5F@.402	ea	—	26.30	26.30
bypassing door	5F@.830	ea	—	54.40	54.40
interior door	5F@.340	ea	—	22.30	22.30
door jamb & casing	5F@.013	lf	—	.85	.85
wood entry door	5F@.742	ea	—	48.60	48.60
wood sliding patio door	5F@.646	ea	—	42.30	42.30
garage door	5F@.006	sf	—	.39	.39
4' high wood fence (per side)	5F@.051	lf	—	3.34	3.34
48" high ornamental iron fence (both sides)	5F@.077	lf	—	5.04	5.04

	Craft@Hrs	Unit	Material	Labor	Total
wood trim, simple design	5F@.011	lf	—	.72	.72
wood trim, very ornate design	5F@.017	lf	—	1.11	1.11
seal smoke-stained frame walls	5F@.018	sf	—	1.18	1.18
seal smoke-stained joists	5F@.022	sf	—	1.44	1.44
seal smoke-stained rafters or trusses	5F@.028	sf	—	1.83	1.83
siding	5F@.016	sf	—	1.05	1.05
stucco	5F@.014	sf	—	.92	.92
stair balustrade	5F@.190	lf	—	12.40	12.40
small window	5F@.295	ea	—	19.30	19.30
very large window	5F@.668	ea	—	43.80	43.80
Paint, one coat					
acoustic ceiling texture	5F@.011	sf	—	.72	.72
plaster or drywall	5F@.006	sf	—	.39	.39
masonry, block	5F@.015	sf	—	.98	.98
concrete floor	5F@.008	sf	—	.52	.52
wood floor	5F@.012	sf	—	.79	.79
folding door	5F@.413	ea	—	27.10	27.10
bypassing door	5F@.854	ea	—	55.90	55.90
interior door	5F@.349	ea	—	22.90	22.90
door jamb & casing	5F@.011	lf	—	.72	.72
wood entry door	5F@.763	ea	—	50.00	50.00
wood sliding patio door	5F@.665	sf	—	43.60	43.60
garage door	5F@.006	sf	—	.39	.39
4' high wood fence (per side)	5F@.053	lf	—	3.47	3.47
48" high ornamental iron fence (both sides)	5F@.079	lf	—	5.17	5.17
wood trim, simple design	5F@.011	lf	—	.72	.72
wood trim, very ornate design	5F@.017	lf	—	1.11	1.11
siding	5F@.016	sf	—	1.05	1.05
stucco	5F@.015	sf	—	.98	.98
stair balustrade	5F@.195	lf	—	12.80	12.80
small window	5F@.304	ea	—	19.90	19.90
very large window	5F@.687	ea	—	45.00	45.00
Paint, two coats					
acoustic ceiling texture	5F@.017	sf	—	1.11	1.11
plaster or drywall	5F@.010	sf	—	.66	.66
masonry, block	5F@.023	sf	—	1.51	1.51
concrete floor	5F@.012	sf	—	.79	.79
wood floor	5F@.019	sf	—	1.24	1.24
folding door	5F@.636	ea	—	41.70	41.70
bypassing door	5F@1.32	ea	—	86.50	86.50
interior door	5F@.538	ea	—	35.20	35.20
door jamb & casing	5F@.017	lf	—	1.11	1.11
wood entry door	5F@1.18	ea	—	77.30	77.30
wood sliding patio door	5F@1.02	sf	—	66.80	66.80
garage door	5F@.009	sf	—	.59	.59
4' high wood fence (per side)	5F@.081	lf	—	5.31	5.31
48" high ornamental iron fence (both sides)	5F@.121	lf	—	7.93	7.93
wood trim, simple design	5F@.017	lf	—	1.11	1.11
wood trim, very ornate design	5F@.026	lf	—	1.70	1.70
siding	5F@.025	sf	—	1.64	1.64
stucco	5F@.023	sf	—	1.51	1.51

	Craft@Hrs	Unit	Material	Labor	Total
stair balustrade	5F@.300	lf	—	19.70	19.70
small window	5F@.467	ea	—	30.60	30.60
very large window	5F@1.06	ea	—	69.40	69.40
Paint, three coats					
acoustic ceiling texture	5F@.023	sf	—	1.51	1.51
plaster or drywall	5F@.013	sf	—	.85	.85
masonry, block	5F@.030	sf	—	1.97	1.97
concrete floor	5F@.016	sf	—	1.05	1.05
wood floor	5F@.026	sf	—	1.70	1.70
folding door	5F@.861	ea	—	56.40	56.40
bypassing door	5F@1.78	ea	—	117.00	117.00
interior door	5F@.729	ea	—	47.70	47.70
door jamb & casing	5F@.023	lf	—	1.51	1.51
wood entry door	5F@1.59	ea	—	104.00	104.00
wood sliding patio door	5F@1.39	sf	—	91.00	91.00
garage door	5F@.013	sf	—	.85	.85
4' high wood fence (per side)	5F@.109	lf	—	7.14	7.14
48" high ornamental iron fence (both sides)	5F@.164	lf	—	10.70	10.70
wood trim, simple design	5F@.023	lf	—	1.51	1.51
wood trim, very ornate design	5F@.035	lf	—	2.29	2.29
siding	5F@.033	sf	—	2.16	2.16
stucco	5F@.030	sf	—	1.97	1.97
stair balustrade	5F@.407	lf	—	26.70	26.70
small window	5F@.633	ea	—	41.50	41.50
very large window	5F@1.43	ea	—	93.70	93.70
Stain					
wood floor	5F@.018	sf	—	1.18	1.18
folding door	5F@.614	ea	—	40.20	40.20
bypassing door	5F@1.27	ea	—	83.20	83.20
interior door	5F@.519	ea	—	34.00	34.00
French door	5F@1.25	ea	—	81.90	81.90
panel door	5F@1.08	ea	—	70.70	70.70
door jamb & casing	5F@.019	lf	—	1.24	1.24
wood entry door	5F@1.13	ea	—	74.00	74.00
wood sliding patio door	5F@.988	ea	—	64.70	64.70
garage door	5F@.009	sf	—	.59	.59
4' high wood fence (per side)	5F@.078	lf	—	5.11	5.11
wood trim, simple design	5F@.016	lf	—	1.05	1.05
wood trim, very ornate design	5F@.025	lf	—	1.64	1.64
wood paneling, simple pattern	5F@.018	sf	—	1.18	1.18
wood paneling, very ornate pattern	5F@.043	sf	—	2.82	2.82
siding	5F@.015	sf	—	.98	.98
stair balustrade	5F@.290	lf	—	19.00	19.00
Varnish, one coat					
wood floor	5F@.019	sf	—	1.24	1.24
folding door	5F@.632	ea	—	41.40	41.40
bypassing door	5F@1.31	ea	—	85.80	85.80
interior door	5F@.534	ea	—	35.00	35.00

	Craft@Hrs	Unit	Material	Labor	Total
French door	5F@1.29	ea	—	84.50	84.50
panel door	5F@1.11	ea	—	72.70	72.70
door jamb & casing	5F@.017	lf	—	1.11	1.11
wood entry door	5F@1.17	ea	—	76.60	76.60
wood sliding patio door	5F@1.02	ea	—	66.80	66.80
garage door	5F@.009	sf	—	.59	.59
4' high wood fence (per side)	5F@.080	lf	—	5.24	5.24
wood trim, simple design	5F@.017	lf	—	1.11	1.11
wood trim, very ornate design	5F@.026	lf	—	1.70	1.70
wood paneling, simple pattern	5F@.019	sf	—	1.24	1.24
wood paneling, very ornate pattern	5F@.044	sf	—	2.88	2.88
siding	5F@.015	sf	—	.98	.98
stair balustrade	5F@.298	lf	—	19.50	19.50
Varnish, two coats					
wood floor	5F@.029	sf	—	1.90	1.90
folding door	5F@.972	ea	—	63.70	63.70
bypassing door	5F@2.01	ea	—	132.00	132.00
interior door	5F@.822	ea	—	53.80	53.80
French door	5F@1.98	ea	—	130.00	130.00
panel door	5F@1.70	ea	—	111.00	111.00
door jamb & casing	5F@.026	lf	—	1.70	1.70
wood entry door	5F@1.80	ea	—	118.00	118.00
wood sliding patio door	5F@1.56	ea	—	102.00	102.00
garage door	5F@.014	sf	—	.92	.92
4' high wood fence (per side)	5F@.123	lf	—	8.06	8.06
wood trim, simple design	5F@.026	lf	—	1.70	1.70
wood trim, very ornate design	5F@.040	lf	—	2.62	2.62
wood paneling, simple pattern	5F@.029	sf	—	1.90	1.90
wood paneling, very ornate pattern	5F@.068	sf	—	4.45	4.45
siding	5F@.023	sf	—	1.51	1.51
stair balustrade	5F@.459	lf	—	30.10	30.10
Varnish, three coats					
wood floor	5F@.039	sf	—	2.55	2.55
folding door	5F@1.32	ea	—	86.50	86.50
bypassing door	5F@2.72	ea	—	178.00	178.00
interior door	5F@1.11	ea	—	72.70	72.70
French door	5F@2.68	ea	—	176.00	176.00
panel door	5F@2.31	ea	—	151.00	151.00
door jamb & casing	5F@.035	lf	—	2.29	2.29
wood entry door	5F@2.43	ea	—	159.00	159.00
wood sliding patio door	5F@2.12	ea	—	139.00	139.00
garage door	5F@.019	sf	—	1.24	1.24
4' high wood fence (per side)	5F@.167	lf	—	10.90	10.90
wood trim, simple design	5F@.035	lf	—	2.29	2.29
wood trim, very ornate design	5F@.054	lf	—	3.54	3.54
wood paneling, simple pattern	5F@.039	sf	—	2.55	2.55
wood paneling, very ornate pattern	5F@.093	sf	—	6.09	6.09
siding	5F@.032	sf	—	2.10	2.10
stair balustrade	5F@.621	lf	—	40.70	40.70

Paneling

Minimum charge.

	Craft@Hrs	Unit	Material	Labor	Total
for paneling work	8P@3.25	ea	34.40	191.00	225.40
for frame-and-panel wall work	8P@4.00	ea	103.00	236.00	339.00

Hardboard paneling. Includes finish hardboard paneling, finish nails, construction adhesive as needed, and installation. Paneling is prefinished. Includes 5% waste.

	Craft@Hrs	Unit	Material	Labor	Total
replace, economy grade	8P@.033	sf	1.13	1.94	3.07
replace, standard grade	8P@.033	sf	1.38	1.94	3.32
replace, high grade	8P@.033	sf	2.11	1.94	4.05
replace, with simulated brick face	8P@.033	sf	1.45	1.94	3.39
replace, with simulated stone face	8P@.033	sf	1.53	1.94	3.47
remove	1D@.007	sf	—	.33	.33

Pegboard. Includes tempered pegboard, fasteners, spacers, construction adhesive as needed, and installation. Includes 5% waste.

	Craft@Hrs	Unit	Material	Labor	Total
replace	8P@.030	sf	1.09	1.77	2.86
remove	1D@.007	sf	—	.33	.33

Plywood paneling. Includes finish plywood, finish nails, construction adhesive as needed, and installation. Plywood is unfinished with top grade veneer. Includes 5% waste.

	Craft@Hrs	Unit	Material	Labor	Total
1/4" plywood paneling					
replace, ash	8P@.036	sf	4.79	2.12	6.91
replace, birch	8P@.036	sf	2.39	2.12	4.51
replace, cherry	8P@.036	sf	5.37	2.12	7.49
replace, hickory	8P@.036	sf	5.06	2.12	7.18
replace, knotty pine	8P@.036	sf	4.04	2.12	6.16
replace, mahogany	8P@.036	sf	4.39	2.12	6.51
replace, white maple	8P@.036	sf	3.56	2.12	5.68
replace, red or white oak	8P@.036	sf	2.62	2.12	4.74
replace, quartersawn red or white oak	8P@.036	sf	4.60	2.12	6.72
replace, walnut	8P@.036	sf	6.98	2.12	9.10
replace, rough-sawn cedar	8P@.036	sf	3.33	2.12	5.45
remove	1D@.008	sf	—	.38	.38

Add for wall molding on plywood paneling. Includes 6% waste. Add to plywood paneling prices above for moldings that form repeating squares on wall with radius corners. Add **11%** for slope-matched wall molding along stairways.

	Craft@Hrs	Unit	Material	Labor	Total
add for standard grade	—	%	44.0	—	—
add for high grade	—	%	53.0	—	—
add for deluxe grade	—	%	70.0	—	—
add for custom	—	%	84.0	—	—

Plywood paneling with grooves. Includes 5% waste. Includes prefinished plywood, finish nails, construction adhesive as needed, and installation.

	Craft@Hrs	Unit	Material	Labor	Total
1/4" plywood paneling with grooves					
replace, ash	8P@.036	sf	4.66	2.12	6.78
replace, birch	8P@.036	sf	2.34	2.12	4.46
replace, cherry	8P@.036	sf	5.23	2.12	7.35
replace, hickory	8P@.036	sf	4.92	2.12	7.04
replace, knotty pine	8P@.036	sf	3.97	2.12	6.09

	Craft@Hrs	Unit	Material	Labor	Total
replace, mahogany	8P@.036	sf	4.27	2.12	6.39
replace, white maple	8P@.036	sf	3.49	2.12	5.61
replace, red or white oak	8P@.036	sf	2.57	2.12	4.69
replace, quartersawn red or white oak	8P@.036	sf	4.44	2.12	6.56
replace, walnut	8P@.036	sf	6.81	2.12	8.93
replace, rough-sawn cedar	8P@.036	sf	3.24	2.12	5.36
remove	1D@.008	sf	—	.38	.38

Board-on-board or board-and-batten paneling. Includes boards, battens, finish nails, and installation. Board-and-batten and board-on-board paneling is installed vertically. Includes 4% waste. Prices for paneling are slightly higher than the same materials used in siding because there are more angles indoors.

	Craft@Hrs	Unit	Material	Labor	Total
replace, pine board-on-board	8P@.038	sf	5.81	2.24	8.05
replace, pine board-and-batten	8P@.042	sf	5.68	2.47	8.15
remove	1D@.011	sf	—	.52	.52

board-and-batten

board-on-board

Add for angled paneling installations.

	Craft@Hrs	Unit	Material	Labor	Total
add for chevron	—	%	—	10.0	—
add for herringbone	—	%	—	18.0	—

Tongue-&-groove paneling. Includes tongue-&-groove boards, finish nails, and installation. Boards are unfinished. Includes 5% waste.

	Craft@Hrs	Unit	Material	Labor	Total
replace, pine tongue-&-groove, simple pattern	8P@.047	sf	6.04	2.77	8.81
replace, pine tongue-&-groove, fancy pattern	8P@.047	sf	6.44	2.77	9.21
replace, pine tongue-&-groove, ornate pattern	8P@.047	sf	7.51	2.77	10.28
replace, cedar tongue-&-groove closet lining	8P@.047	sf	3.46	2.77	6.23
remove	1D@.011	sf	—	.52	.52

Add for wainscot installations.

	Craft@Hrs	Unit	Material	Labor	Total
add for paneling installed as wainscoting	—	%	—	14.0	—

Pine frame-and-panel wall. Includes framing lumber and furring strips as needed for backing, panels and/or finish plywood, moldings between panels, may include tongue-&-groove stiles and rails, crown, base, and other finish millwork such as casing dentil mold, cove, and so forth, framing nails, finish nails, glue, fasteners, and installation. Includes 5% waste.

	Craft@Hrs	Unit	Material	Labor	Total
replace, standard grade	8P@.725	sf	29.20	42.70	71.90
replace, high grade	8P@.870	sf	40.10	51.20	91.30
replace, deluxe grade	8P@1.05	sf	47.40	61.80	109.20
replace, custom grade	8P@1.25	sf	70.60	73.60	144.20
remove	1D@.019	sf	—	.90	.90

Oak frame-and-panel wall. Includes framing lumber and furring strips as needed for backing, panels and/or finish plywood, moldings between panels, may include tongue-&-groove stiles and rails, crown, base, and other finish millwork such as casing dentil mold, cove, and so forth, framing nails, finish nails, glue, fasteners, and installation. Includes 5% waste.

	Craft@Hrs	Unit	Material	Labor	Total
replace, standard grade	8P@.725	sf	39.30	42.70	82.00
replace, high grade	8P@.870	sf	54.00	51.20	105.20
replace, deluxe grade	8P@1.05	sf	63.50	61.80	125.30
replace, custom grade	8P@1.25	sf	94.60	73.60	168.20
remove	1D@.019	sf	—	.90	.90

Walnut frame-and-panel wall. Includes framing lumber and furring strips as needed for backing, panels and/or finish plywood, moldings between panels, may include tongue-&-groove stiles and rails, crown, base, and other finish millwork such as casing dentil mold, cove, and so forth, framing nails, finish nails, glue, fasteners, and installation. Includes 5% waste.

	Craft@Hrs	Unit	Material	Labor	Total
replace, standard grade	8P@.725	sf	65.00	42.70	107.70
replace, high grade	8P@.870	sf	88.90	51.20	140.10
replace, deluxe grade	8P@1.05	sf	105.00	61.80	166.80
replace, custom grade	8P@1.25	sf	155.00	73.60	228.60
remove	1D@.019	sf	—	.90	.90

Add for other wood species in frame-and-panel walls. For panel walls made from other woods add the percentages to the cost of oak panel walls.

	Craft@Hrs	Unit	Material	Labor	Total
add for quartersawn oak	—	%	12.0	—	—
add for cherry	—	%	17.0	—	—
add for teak	—	%	65.0	—	—
add for mahogany	—	%	10.0	—	—
add for hickory	—	%	14.0	—	—

Time & Material Charts (selected items)

Paneling Materials

	Craft@Hrs	Unit	Material	Labor	Total
Hardboard paneling					
economy grade ($34.60 sheet, 32 sf), 5% waste	—	sf	1.13	—	1.13
high grade ($59.80 sheet, 32 sf), 5% waste	—	sf	1.95	—	1.95
1/4" plywood paneling					
rough-sawn cedar ($101.00 sheet, 32 sf), 5% waste	—	sf	3.31	—	3.31
ash ($146.00 sheet, 32 sf), 5% waste	—	sf	4.80	—	4.80
birch ($72.60 sheet, 32 sf), 5% waste	—	sf	2.39	—	2.39
cherry ($161.00 sheet, 32 sf), 5% waste	—	sf	5.31	—	5.31
hickory ($153.00 sheet, 32 sf), 5% waste	—	sf	5.01	—	5.01
knotty pine ($124.00 sheet, 32 sf), 5% waste	—	sf	4.06	—	4.06
mahogany ($134.00 sheet, 32 sf), 5% waste	—	sf	4.36	—	4.36
white maple ($109.00 sheet, 32 sf), 5% waste	—	sf	3.55	—	3.55
red or white oak ($79.60 sheet, 32 sf), 5% waste	—	sf	2.61	—	2.61
quartersawn red or white oak ($139.00 sheet, 32 sf), 5% waste	—	sf	4.55	—	4.55
walnut ($215.00 sheet, 32 sf), 5% waste	—	sf	7.07	—	7.07
1/4" plywood paneling with grooves					
rough-sawn cedar ($98.60 sheet, 32 sf), 5% waste	—	sf	3.23	—	3.23
ash ($142.00 sheet, 32 sf), 5% waste	—	sf	4.66	—	4.66
birch ($70.80 sheet, 32 sf), 5% waste	—	sf	2.32	—	2.32
cherry ($157.00 sheet, 32 sf), 5% waste	—	sf	5.14	—	5.14
hickory ($151.00 sheet, 32 sf), 5% waste	—	sf	4.91	—	4.91
knotty pine ($120.00 sheet, 32 sf), 5% waste	—	sf	3.94	—	3.94
mahogany ($130.00 sheet, 32 sf), 5% waste	—	sf	4.25	—	4.25
white maple ($106.00 sheet, 32 sf), 5% waste	—	sf	3.47	—	3.47
red or white oak ($77.90 sheet, 32 sf), 5% waste	—	sf	2.57	—	2.57

	Craft@Hrs	Unit	Material	Labor	Total
quartersawn red or white oak ($135.00 sheet, 32 sf), 5% waste	—	sf	4.39	—	4.39
walnut ($210.00 sheet, 32 sf), 5% waste	—	sf	6.88	—	6.88
Pine board paneling					
board-on-board ($5.58 sf), 4% waste	—	sf	5.79	—	5.79
board-and-batten ($5.44 sf), 4% waste	—	sf	5.66	—	5.66
tongue-&-groove, simple ($5.77 sf), 4% waste	—	sf	5.99	—	5.99
tongue-&-groove, ornate ($7.23 sf), 4% waste	—	sf	7.52	—	7.52
Frame-and-panel wall					
pine, standard grade ($27.60 sf), 6% waste	—	sf	29.20	—	29.20
pine, custom grade ($66.20 sf), 6% waste	—	sf	70.20	—	70.20
oak, standard grade ($37.20 sf), 6% waste	—	sf	39.50	—	39.50
oak, custom grade ($89.10 sf), 6% waste	—	sf	94.50	—	94.50
walnut, standard grade ($61.20 sf), 6% waste	—	sf	64.80	—	64.80
walnut, custom grade ($147.00 sf), 6% waste	—	sf	154.00	—	154.00

Paneling Labor

Laborer	base wage	paid leave	true wage	taxes & ins.	total
Paneling installer	$38.00	2.96	$40.96	26.44	$67.40
Paneling installer's helper	$27.40	2.14	$29.54	20.86	$50.40
Demolition laborer	$25.90	2.02	$27.92	19.28	$47.20

Paid leave is calculated based on two weeks paid vacation, one week sick leave, and seven paid holidays. Employer's matching portion of **FICA** is 7.65 percent. **FUTA** (Federal Unemployment) is .8 percent. **Worker's compensation** for the paneling trade was calculated using a national average of 17.21 percent. **Unemployment insurance** was calculated using a national average of 8 percent. **Health insurance** was calculated based on a projected national average for 2020 of $1,280 per employee (and family when applicable) per month. Employer pays 80 percent for a per month cost of $1,024 per employee. **Retirement** is based on a 401(k) retirement program with employer matching of 50 percent. Employee contributions to the 401(k) plan are an average of 6 percent of the true wage. **Liability insurance** is based on a national average of 12.0 percent.

Paneling Labor Productivity

	Craft@Hrs	Unit	Material	Labor	Total
Demolition of paneling					
remove hardboard paneling	1D@.007	sf	—	.33	.33
remove pegboard	1D@.007	sf	—	.33	.33
remove plywood paneling	1D@.008	sf	—	.38	.38
remove board paneling	1D@.011	sf	—	.52	.52
remove tongue-&-groove paneling	1D@.011	sf	—	.52	.52
remove frame-and-panel wall	1D@.019	sf	—	.90	.90

Paneling installation crew

install paneling	paneling installer	$67.40
install paneling	installer's helper	$50.40
install paneling	paneling crew	$58.90

	Craft@Hrs	Unit	Material	Labor	Total
Install paneling					
hardboard	8P@.033	sf	—	1.94	1.94
pegboard	8P@.030	sf	—	1.77	1.77
plywood	8P@.036	sf	—	2.12	2.12
board-on-board	8P@.038	sf	—	2.24	2.24
board-and-batten	8P@.042	sf	—	2.47	2.47
tongue-&-groove	8P@.047	sf	—	2.77	2.77
Install frame-and-panel wall					
standard grade	8P@.725	sf	—	42.70	42.70
high grade	8P@.870	sf	—	51.20	51.20
deluxe grade	8P@1.05	sf	—	61.80	61.80
custom grade	8P@1.25	sf	—	73.60	73.60

Plaster & Stucco

Flexible molds. Cracks to decorative plaster that are longer than a hairline are often repaired using flexible molds. Flexible molds are made by constructing a wood box to fit around the area where the mold will be taken. A release agent is sprayed on the box and the original plaster to help the mold release when dried. Cold-cure silicone is usually used to make the mold, although many professionals use hot-melt mold-making compounds.

	Craft@Hrs	Unit	Material	Labor	Total
Minimum charge.					
for plaster wall or ceiling repair	6P@4.00	ea	74.50	246.00	320.50
Plaster repair.					
repair hair-line plaster crack	6P@.065	lf	.48	4.00	4.48
repair plaster crack less than 1" wide	6P@.081	lf	.76	4.98	5.74
patch small plaster hole less than 2" square	6P@.841	ea	2.25	51.70	53.95
patch plaster hole 2" to 10" square	6P@1.06	ea	10.10	65.20	75.30
patch plaster hole 10" to 20" square	6P@1.32	ea	17.40	81.20	98.60
patch plaster section	6P@.156	sf	3.69	9.59	13.28
repair plaster wall outside corner damage	6P@.289	lf	4.75	17.80	22.55
repair plaster wall inside corner damage	6P@.204	lf	3.92	12.50	16.42
Plaster pilaster or column repair.					
repair section of plaster pilaster or column	6P@.908	lf	32.80	55.80	88.60
Plaster molding or enrichment repair. Repair to enrichments, moldings, ceiling rose or medallion.					
repair crack less than 1/4"	6P@.071	lf	.48	4.37	4.85
take flexible mold of plaster molding	6P@.588	ea	63.80	36.20	100.00
create running mold for plaster molding	6P@4.00	ea	83.90	246.00	329.90
repair section	6P@1.28	ea	9.86	78.70	88.56
minimum charge	6P@3.50	ea	41.40	215.00	256.40

Acoustical plaster. Includes painted metal lath or gypsum lath, lath nails, acoustical plaster, corner bead, and installation. Installation includes placing of lath on frame walls, three coats over lath with corner bead on outside corners and a trowel finish. Includes expansion metal on large walls. ***Note:*** During and prior to the 1970s, acoustical plaster often contained asbestos. Acoustical plaster need not contain asbestos fibers and the contemporary mixes quoted here do not. However, acoustical plaster installations are now rare and many building owners replace acoustical plaster with other styles of like kind and quality.

	Craft@Hrs	Unit	Material	Labor	Total
Two coats of acoustical plaster on 3/8" gypsum lath					
replace, on wall	6P@.051	sf	1.97	3.14	5.11
replace, on ceiling	6P@.061	sf	1.97	3.75	5.72
remove	1D@.023	sf	—	1.09	1.09
Three coats of acoustical plaster on painted metal lath					
replace, on wall	6P@.057	sf	1.71	3.51	5.22
replace, on ceiling	6P@.068	sf	1.71	4.18	5.89
remove	1D@.024	sf	—	1.13	1.13

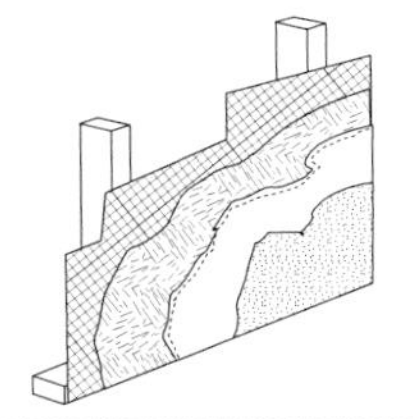

Gypsum plaster. Includes painted metal lath or gypsum lath, lath nails, gypsum plaster, corner bead, and installation. Installation includes placing of lath on frame walls, three coats over lath with corner bead on outside corners and a trowel finish. Includes expansion metal on large walls.

	Craft@Hrs	Unit	Material	Labor	Total
Two coats of gypsum plaster on 3/8" gypsum lath					
replace, on wall	6P@.052	sf	1.71	3.20	4.91
replace, on ceiling	6P@.062	sf	1.71	3.81	5.52
remove	1D@.023	sf	—	1.09	1.09
Three coats of gypsum plaster on painted metal lath					
replace, on wall	6P@.058	sf	1.46	3.57	5.03
replace, on ceiling	6P@.069	sf	1.46	4.24	5.70
remove	1D@.024	sf	—	1.13	1.13

Perlite or vermiculite plaster. Includes painted metal or gypsum lath, lath nails, plaster with perlite or vermiculite, corner bead, and installation. Installation includes placing of lath on frame walls, three coats over lath with corner bead on outside corners and a trowel finish. Includes expansion metal on large walls. Plaster with perlite and vermiculite is lighter than gypsum plaster and provides higher insulation value. Often used on firewalls and as a protective coat for structural steel beams and posts.

	Craft@Hrs	Unit	Material	Labor	Total
Two coats of perlite plaster on 3/8" gypsum lath					
replace, on wall	6P@.059	sf	1.80	3.63	5.43
replace, on ceiling	6P@.071	sf	1.80	4.37	6.17
remove	1D@.023	sf	—	1.09	1.09
Three coats of perlite plaster on painted metal lath					
replace, on wall	6P@.069	sf	2.93	4.24	7.17
replace, on ceiling	6P@.083	sf	2.93	5.10	8.03
remove	1D@.024	sf	—	1.13	1.13

Keene's cement plaster. Includes painted metal or gypsum lath, lath nails, Keene's plaster, corner bead, and installation. Installation includes placing of lath on frame walls, three coats over lath with corner bead on outside corners and a trowel finish. Includes expansion metal on large walls. Keene's plaster was developed in the late 19th century. It's a modified gypsum plaster with its chief benefit being superior hardness. It has often been used to make imitation marble and architectural casts. Keene's plaster is rarely used in contemporary construction.

	Craft@Hrs	Unit	Material	Labor	Total
Two coats of Keene's plaster on 3/8" gypsum lath					
replace, on wall	6P@.075	sf	2.13	4.61	6.74
replace, on ceiling	6P@.090	sf	2.13	5.54	7.67
remove	1D@.023	sf	—	1.09	1.09
Three coats of Keene's plaster on painted metal lath					
replace, on wall	6P@.082	sf	1.86	5.04	6.90
replace, on ceiling	6P@.093	sf	1.86	5.72	7.58
remove	1D@.024	sf	—	1.13	1.13

Thin-coat plaster. Includes gypsum lath, gypsum board nails or screws, plaster, tape, corner bead, and installation. Installation includes thin-coat over gypsum lath with corner bead on outside corners and a trowel finish. Currently the most common type of plaster installation in new homes (according to the Northwest Wall and Ceiling Bureau).

	Craft@Hrs	Unit	Material	Labor	Total
replace, on 3/8" thick gypsum lath	6P@.044	sf	.61	2.71	3.32
replace, on 1/2" thick gypsum lath	6P@.044	sf	.43	2.71	3.14
remove	1D@.021	sf	—	.99	.99

	Craft@Hrs	Unit	Material	Labor	Total
Additional plaster costs.					
add for curved wall	—	%	28.0	—	—
add for plaster installed on old-style wood lath	—	%	60.0	—	—

Plaster pilaster or column. Includes gypsum lath, plaster, tape, corner bead, and installation labor. Does not include framing of column. Plaster columns may also be prefabricated units designed to encase a structural post.

	Craft@Hrs	Unit	Material	Labor	Total
Square plaster pilaster					
replace pilaster, plain	6P@.189	lf	5.56	11.60	17.16
replace pilaster, rusticated	6P@.227	lf	5.57	14.00	19.57
remove	1D@.043	lf	—	2.03	2.03
Square plaster column					
replace column, plain	6P@.345	lf	11.10	21.20	32.30
replace column, rusticated	6P@.436	lf	11.10	26.80	37.90
remove	1D@.077	lf	—	3.63	3.63
Add for round pilaster or column	—	%	30.0	—	—

Cast-in-place plaster molding. Includes plaster and installation labor. For straight-pattern moldings cast in place on walls and/or ceilings using a running mold. Inside corners require a great deal of handwork by a skilled specialist who may be hard to find in some regions of North America. Outside corners are less complex. Does not include scaffolding or knife charges.

	Craft@Hrs	Unit	Material	Labor	Total
to 3" wide	6P@1.56	lf	9.96	95.90	105.86
over 3" to 6" wide	6P@1.69	lf	19.80	104.00	123.80
over 6" to 14" wide	6P@1.89	lf	37.00	116.00	153.00
add for inside corner	6P@3.23	ea	10.10	199.00	209.10
add for outside corner	6P@1.54	ea	2.69	94.70	97.39

Production plaster molding. Includes prefabricated plaster molding, fasteners, adhesive, caulk, and installation labor. Installation is per manufacturer's specs, usually with a combination of adhesive and fasteners. Touch-up material is provided to hide fasteners. Prefabricated plaster moldings are available from specialized suppliers in a wide variety of patterns. Be careful not to confuse plaster moldings with those made of imitation polyurethane or similar materials.

	Craft@Hrs	Unit	Material	Labor	Total
to 3" wide	6P@.263	lf	41.40	16.20	57.60
over 3" to 6" wide	6P@.270	lf	61.80	16.60	78.40
over 6" to 14" wide	6P@.287	lf	99.40	17.70	117.10
add for inside corner	6P@.333	ea	2.20	20.50	22.70
add for outside corner	6P@.333	ea	2.49	20.50	22.99

Plaster enrichment. Includes prefabricated plaster enrichment, fasteners, adhesive, and installation labor. Installation is per manufacturer's specs, usually with a combination of adhesive and fasteners. Touch-up material is provided to hide fasteners. Enrichments are placed on walls or ceilings to further accent friezes, cornices, ceiling medallions and so forth. Be careful not to confuse plaster enrichments with those made of imitation polyurethane or similar materials.

	Craft@Hrs	Unit	Material	Labor	Total
straight patterns	6P@.263	lf	37.90	16.20	54.10
ornate patterns	6P@.270	lf	57.00	16.60	73.60
very ornate patterns	6P@.287	lf	82.40	17.70	100.10

Plaster frieze. Includes prefabricated plaster frieze, fasteners, adhesive, and installation labor. Installation is per manufacturer's specs, usually with a combination of adhesive and fasteners. Touch-up material is provided to hide fasteners. Plaster friezes are usually flat with ornate patterns, and can be up to 14" wide. Be careful not to confuse plaster moldings with imitation polyurethane or similar moldings.

	Craft@Hrs	Unit	Material	Labor	Total
straight patterns	6P@.263	lf	63.00	16.20	79.20
ornate patterns	6P@.270	lf	95.10	16.60	111.70
very ornate patterns	6P@.287	lf	138.00	17.70	155.70

	Craft@Hrs	Unit	Material	Labor	Total

Plaster architrave. Includes prefabricated plaster architrave, fasteners, adhesive, and installation labor. Most complex plaster architraves are production units built by a specialty supplier.

	Craft@Hrs	Unit	Material	Labor	Total
plain	6P@.154	lf	4.45	9.47	13.92
rusticated	6P@.185	lf	4.75	11.40	16.15
straight patterns	6P@.477	lf	5.46	29.30	34.76
ornate patterns	6P@.699	lf	5.59	43.00	48.59

Plaster ceiling medallion. Includes prefabricated plaster ceiling medallion, fasteners, adhesive, and installation labor. Installation is per manufacturer's specs, usually with a combination of adhesive and fasteners. Touch-up material is provided to hide fasteners. Plaster ceiling medallions are usually round or oval in shape, designed for the center of the ceiling. Be careful not to confuse plaster medallions with imitation polyurethane or similar medallions.

	Craft@Hrs	Unit	Material	Labor	Total
simple pattern	6P@.908	ea	221.00	55.80	276.80
ornate pattern	6P@.952	ea	421.00	58.50	479.50
very ornate pattern	6P@1.00	ea	772.00	61.50	833.50

Plaster ceiling rose. Includes prefabricated plaster ceiling rose, fasteners, adhesive, and installation labor. Installation is per manufacturer's specs, usually with a combination of adhesive and fasteners. Touch-up material is provided to hide fasteners. Plaster ceiling roses are very ornate ceiling medallions common to Victorian homes.

	Craft@Hrs	Unit	Material	Labor	Total
small	6P@.908	ea	376.00	55.80	431.80
average size	6P@.952	ea	575.00	58.50	633.50
large	6P@1.00	ea	1,090.00	61.50	1,151.50

Stucco. Includes felt or building paper, metal lath, and control joint and corner metal.

	Craft@Hrs	Unit	Material	Labor	Total
Three coat Portland cement stucco					
replace, sand float finish	6P@.096	sf	1.50	5.90	7.40
replace, sand float finish, on masonry	6P@.075	sf	.57	4.61	5.18
replace, trowel float finish	6P@.142	sf	1.51	8.73	10.24
replace, trowel float finish, on masonry	6P@.121	sf	.60	7.44	8.04
remove	1D@.027	sf	—	1.27	1.27

Additional stucco costs.

	Craft@Hrs	Unit	Material	Labor	Total
add for application on soffits	—	%	26.0	—	—
add for colors	—	%	10.0	—	—
add for white cement	—	%	19.0	—	—
add for pebble dash (rough cast) finish	—	%	32.0	—	—
add for vermiculated stucco finish	—	%	88.0	—	—
patch single "stone" section in stucco rustication	6P@.769	ea	7.79	47.30	55.09
add for stucco quoin	6P@.624	ea	12.10	38.40	50.50
add for stucco key	6P@1.08	ea	18.50	66.40	84.90
add for stucco raised trim	6P@.149	lf	4.73	9.16	13.89
add for stucco raised curved trim	6P@.251	lf	5.47	15.40	20.87
add for stucco color change	6P@.054	ea	.32	3.32	3.64

Synthetic stucco. Includes cement board or polystyrene foam insulation board, adhesive primer, color coat stucco, and corner mesh.

	Craft@Hrs	Unit	Material	Labor	Total
replace, over 1/2" cement board	6P@.074	sf	1.43	4.55	5.98
replace, over 1" insulating foam board	6P@.074	sf	1.15	4.55	5.70
replace, over 2" insulating foam board	6P@.074	sf	1.24	4.55	5.79
remove	1D@.027	sf	—	1.27	1.27

	Craft@Hrs	Unit	Material	Labor	Total
Additional synthetic stucco costs.					
add for synthetic stucco quoin	6P@.935	ea	6.83	57.50	64.33
add for synthetic stucco key	6P@1.61	ea	10.50	99.00	109.50
add for synthetic stucco raised trim	6P@.223	lf	2.66	13.70	16.36
add for synthetic stucco raised curved trim	6P@.373	lf	3.08	22.90	25.98
add for synthetic stucco color change	6P@.081	lf	.15	4.98	5.13
Minimum charge.					
for stucco repair	6P@5.25	ea	86.30	323.00	409.30
Stucco repair. For repair of standard or synthetic stucco.					
repair hair-line stucco crack	6P@.066	lf	.49	4.06	4.55
repair stucco crack less than 1" wide	6P@.083	lf	.78	5.10	5.88
patch small stucco hole less than 2" square	6P@.855	ea	2.31	52.60	54.91
patch stucco hole 2" to 10" square	6P@1.09	ea	10.30	67.00	77.30
patch stucco hole 10" to 20" square	6P@1.33	ea	17.90	81.80	99.70
patch stucco section	6P@.159	sf	3.79	9.78	13.57
repair stucco wall outside corner damage	6P@.296	lf	4.85	18.20	23.05
repair stucco wall inside corner damage	6P@.208	lf	4.00	12.80	16.80

Stucco pilaster or column. Includes stucco and installation labor. Use as additional costs for stucco columns when estimating standard or synthetic stucco. Does not include framing of column. Stucco columns may also be prefabricated units.

	Craft@Hrs	Unit	Material	Labor	Total
Square stucco pilaster					
replace pilaster, plain	6P@.192	lf	5.75	11.80	17.55
replace pilaster, rusticated	6P@.230	lf	5.79	14.10	19.89
remove	1D@.048	lf	—	2.27	2.27
Square stucco column					
replace column, plain	6P@.349	lf	11.60	21.50	33.10
replace column, rusticated	6P@.441	lf	11.60	27.10	38.70
remove	1D@.083	lf	—	3.92	3.92
Additional stucco pilaster or column costs.					
add for round stucco pilaster or column	—	%	30.0	—	—
repair section of stucco pilaster or column	6P@.925	ea	32.80	56.90	89.70
minimum charge	6P@1.50	ea	58.00	92.30	150.30

Stucco architrave. Includes foam, stucco, fasteners, adhesive, caulk, and installation labor. Many stucco architraves are built from shaped foam covered with synthetic stucco. Some complex stucco architraves are production units built by a specialty supplier.

	Craft@Hrs	Unit	Material	Labor	Total
plain	6P@.156	lf	4.61	9.59	14.20
rusticated	6P@.188	lf	4.67	11.60	16.27
straight patterns	6P@.482	lf	4.96	29.60	34.56
ornate patterns	6P@.708	lf	5.64	43.50	49.14
take flexible mold of existing architrave	6P@.595	ea	63.80	36.60	100.40
create mold for architrave	6P@4.00	ea	83.90	246.00	329.90
repair section	6P@1.35	ea	10.30	83.00	93.30
minimum charge	6P@3.75	ea	41.40	231.00	272.40

	Craft@Hrs	Unit	Material	Labor	Total
Time & Material Charts (selected items)					
Plaster and Stucco Materials					
Gauging plaster, ($35.70 100 lb. bag, 1 ea), 21% waste	—	ea	43.20	—	43.20
Gypsum plaster, ($31.20 80 lb. bag, 1 ea), 21% waste	—	ea	38.00	—	38.00
Thin-coat plaster, ($19.50 50 lb. bag, 1 ea), 21% waste	—	ea	23.50	—	23.50
Keene's cement, ($40.70 100 lb. bag, 1 ea), 21% waste	—	ea	49.40	—	49.40
Perlite or vermiculite plaster, ($27.10 100 lb. bag, 1 ea), 21% waste	—	ea	32.80	—	32.80
3/8" perforated gypsum lath, ($27.00 sheet, 32 sf), 4% waste	—	sf	.87	—	.87
1/2" perforated gypsum lath, ($28.70 sheet, 32 sf), 4% waste	—	sf	.93	—	.93
Painted metal lath, ($.43 sf, 1 sf), 4% waste	—	sf	.46	—	.46
Painted stucco mesh, ($.68 sf, 1 sf), 4% waste	—	sf	.70	—	.70

Plaster & Stucco Labor

Laborer	base wage	paid leave	true wage	taxes & ins.	total
Plasterer	$37.30	2.91	$40.21	24.49	$64.70
Plasterer's helper	$33.20	2.59	$35.79	22.51	$58.30
Demolition laborer	$25.90	2.02	$27.92	19.28	$47.20

Paid leave is calculated based on two weeks paid vacation, one week sick leave, and seven paid holidays. Employer's matching portion of FICA is 7.65 percent. **FUTA** (Federal Unemployment) is .8 percent. **Worker's compensation** for the plaster and stucco trade was calculated using a national average of 13.28 percent. **Unemployment insurance** was calculated using a national average of 8 percent. **Health insurance** was calculated based on a projected national average for 2020 of $1,280 per employee (and family when applicable) per month. Employer pays 80 percent for a per month cost of $1,024 per employee. **Retirement** is based on a 401(k) retirement program with employer matching of 50 percent. Employee contributions to the 401(k) plan are an average of 6 percent of the true wage. **Liability insurance** is based on a national average of 12.0 percent.

	Craft@Hrs	Unit	Material	Labor	Total
Plaster & Stucco Labor Productivity					
Demolition of plaster and stucco					
remove plaster on gypsum lath	1D@.023	sf	—	1.09	1.09
remove plaster on metal lath	1D@.024	sf	—	1.13	1.13
remove thin-coat plaster on gypsum	1D@.021	sf	—	.99	.99
remove plaster pilaster	1D@.043	lf	—	2.03	2.03
remove plaster column	1D@.077	lf	—	3.63	3.63

	Craft@Hrs	Unit	Material	Labor	Total
remove stucco	1D@.027	sf	—	1.27	1.27
remove synthetic stucco on cement board	1D@.027	sf	—	1.27	1.27
remove stucco pilaster	1D@.048	lf	—	2.27	2.27
remove stucco column	1D@.083	lf	—	3.92	3.92

Plaster installation crew

install plaster	plasterer	$64.70
install plaster	plasterer's helper	$58.30
install plaster	plastering crew	$61.50

	Craft@Hrs	Unit	Material	Labor	Total
Install acoustical plaster					
two coat on 3/8" gypsum lath on wall	6P@.051	sf	—	3.14	3.14
two coat on 3/8" gypsum lath on ceiling	6P@.061	sf	—	3.75	3.75
three coat on painted metal lath on wall	6P@.057	sf	—	3.51	3.51
three coat on painted metal lath on ceiling	6P@.068	sf	—	4.18	4.18
Install gypsum plaster					
two coat on 3/8" gypsum lath on wall	6P@.052	sf	—	3.20	3.20
two coat on 3/8" gypsum lath on ceiling	6P@.062	sf	—	3.81	3.81
three coat on painted metal lath on wall	6P@.058	sf	—	3.57	3.57
three coat on painted metal lath on ceiling	6P@.069	sf	—	4.24	4.24
Install perlite or vermiculite plaster					
two coat on 3/8" gypsum lath on wall	6P@.059	sf	—	3.63	3.63
two coat on 3/8" gypsum lath on ceiling	6P@.071	sf	—	4.37	4.37
three coat on painted metal lath on wall	6P@.069	sf	—	4.24	4.24
three coat on painted metal lath on ceiling	6P@.083	sf	—	5.10	5.10
Install Keene's cement plaster					
two coat on 3/8" gypsum lath on wall	6P@.075	sf	—	4.61	4.61
two coat on 3/8" gypsum lath on ceiling	6P@.090	sf	—	5.54	5.54
three coat on painted metal lath on wall	6P@.082	sf	—	5.04	5.04
three coat on painted metal lath on ceiling	6P@.093	sf	—	5.72	5.72
Install thin-coat plaster					
on 3/8" thick gypsum lath	6P@.044	sf	—	2.71	2.71
on 1/2" thick gypsum lath	6P@.044	sf	—	2.71	2.71
Install three-coat Portland cement stucco					
sand float finish	6P@.096	sf	—	5.90	5.90
sand float finish, on masonry	6P@.075	sf	—	4.61	4.61
trowel float finish	6P@.142	sf	—	8.73	8.73
trowel float finish, on masonry	6P@.121	sf	—	7.44	7.44
Install synthetic stucco					
all applications	6P@.074	sf	—	4.55	4.55

Plumbing

	Craft@Hrs	Unit	Material	Labor	Total
Minimum charge.					
for plumbing work	7P@2.75	ea	62.60	213.00	275.60
Black steel supply pipe. Field threaded with fitting or coupling every 8.5 lf and hanger every 10 lf.					
1/2" pipe	7P@.127	lf	3.71	9.83	13.54
3/4" pipe	7P@.136	lf	3.95	10.50	14.45
1" pipe	7P@.158	lf	5.27	12.20	17.47
2" pipe	7P@.231	lf	10.20	17.90	28.10
Replace section of black steel supply pipe. Includes cut-out in drywall, replacement pipe section and couplings. Does not include wall repair (see Drywall).					
all sizes	7P@1.02	ea	43.70	78.90	122.60
Brass water supply pipe. Field threaded with fitting or coupling every 8.5 lf and hanger every 10 lf.					
1/2" pipe	7P@.172	lf	10.90	13.30	24.20
3/4"pipe	7P@.180	lf	14.50	13.90	28.40
1" pipe	7P@.192	lf	19.40	14.90	34.30
2" pipe	7P@.210	lf	48.00	16.30	64.30
Replace section of brass water supply pipe. Includes cut-out in drywall, replacement pipe section and couplings. Does not include wall repair (see Drywall).					
all sizes	7P@1.39	ea	90.10	108.00	198.10
Copper water supply pipe. With sweated fitting or coupling every 8.5 lf and hanger every 10 lf.					
Type K copper pipe					
1/2" pipe	7P@.104	lf	3.78	8.05	11.83
3/4"pipe	7P@.109	lf	3.87	8.44	12.31
1" pipe	7P@.122	lf	7.28	9.44	16.72
2" pipe	7P@.202	lf	16.90	15.60	32.50
Type L copper pipe					
1/2" pipe	7P@.104	lf	3.30	8.05	11.35
3/4" pipe	7P@.109	lf	3.56	8.44	12.00
1" pipe	7P@.122	lf	6.27	9.44	15.71
2" pipe	7P@.202	lf	15.60	15.60	31.20
Type M copper pipe					
1/2" pipe	7P@.104	lf	2.73	8.05	10.78
3/4" pipe	7P@.109	lf	3.56	8.44	12.00
1" pipe	7P@.122	lf	4.94	9.44	14.38
Replace section of copper water supply pipe. Includes cut-out in drywall, replacement pipe section and couplings. Does not include wall repair (see Drywall).					
all types and sizes	7P@.827	ea	33.10	64.00	97.10
Galvanized steel water supply pipe. Field threaded with fitting or coupling every 8.5 lf and hanger every 10 lf.					
1/2" pipe	7P@.129	lf	4.46	9.98	14.44
3/4" pipe	7P@.136	lf	4.83	10.50	15.33
1" pipe	7P@.155	lf	6.96	12.00	18.96
2" pipe	7P@.231	lf	12.70	17.90	30.60

	Craft@Hrs	Unit	Material	Labor	Total
Replace section of galvanized steel water supply pipe. Includes cut-out in drywall, replacement pipe section and couplings. Does not include wall repair (see Drywall).					
all sizes	7P@1.04	ea	53.30	80.50	133.80
CPVC water supply pipe, schedule 40. With fitting or coupling every 8.5 lf and hanger every 3.5 lf.					
1/2" pipe	7P@.153	lf	5.76	11.80	17.56
3/4" pipe	7P@.162	lf	7.02	12.50	19.52
1" pipe	7P@.189	lf	8.58	14.60	23.18
PVC cold water supply pipe, schedule 40. With fitting or coupling every 8.5 lf and hanger every 3.5 lf.					
1/2" pipe	7P@.153	lf	3.35	11.80	15.15
3/4" pipe	7P@.162	lf	3.46	12.50	15.96
1" pipe	7P@.189	lf	4.02	14.60	18.62
Replace section of PVC or CPVC water supply pipe. Includes cut-out in drywall, replacement pipe section and couplings. Does not include wall repair (see Drywall).					
all sizes	7P@1.23	ea	26.20	95.20	121.40
Minimum charge.					
for water supply pipe work	7P@1.82	ea	93.80	141.00	234.80
Cast-iron DWV pipe. Lead and oakum joints with fitting every 9.5 lf.					
2" pipe	7P@.236	lf	9.46	18.30	27.76
3" pipe	7P@.248	lf	12.90	19.20	32.10
4" pipe	7P@.270	lf	17.20	20.90	38.10
5" pipe	7P@.304	lf	22.70	23.50	46.20
6" pipe	7P@.317	lf	27.60	24.50	52.10
Replace section of cast-iron DWV pipe. Includes cut-out in drywall, replacement pipe section and couplings. Does not include wall repair (see Drywall).					
all sizes	7P@1.67	ea	166.00	129.00	295.00
Repack cast-iron pipe joint with oakum and caulk.					
all sizes	7P@.181	ea	20.00	14.00	34.00
Cast-iron no-hub DWV pipe. With fitting or coupling every 9.5 lf.					
2" pipe	7P@.210	lf	11.20	16.30	27.50
3" pipe	7P@.223	lf	15.20	17.30	32.50
4" pipe	7P@.242	lf	18.70	18.70	37.40
5" pipe	7P@.272	lf	27.10	21.10	48.20
6" pipe	7P@.283	lf	31.60	21.90	53.50
Replace section of cast-iron no-hub DWV pipe. Includes cut-out in drywall, replacement pipe section and couplings. Does not include wall repair (see Drywall).					
all sizes	7P@1.49	ea	128.00	115.00	243.00
Repair pin-hole leak in cast-iron no-hub DWV pipe with epoxy.					
all sizes	7P@.544	ea	14.00	42.10	56.10

	Craft@Hrs	Unit	Material	Labor	Total
PVC DWV pipe. With fitting or coupling every 8.5 lf and hanger every 10 lf.					
1-1/4" pipe	7P@.198	lf	4.44	15.30	19.74
1-1/2" pipe	7P@.229	lf	4.94	17.70	22.64
2" pipe	7P@.251	lf	5.26	19.40	24.66
3" pipe	7P@.281	lf	7.69	21.70	29.39
4" pipe	7P@.310	lf	9.86	24.00	33.86
5" pipe	7P@.347	lf	13.40	26.90	40.30
6" pipe	7P@.379	lf	19.30	29.30	48.60
Replace section of PVC DWV pipe. Includes cut-out in drywall, replacement pipe section and couplings. Does not include wall repair (see Drywall).					
all sizes	7P@1.45	ea	63.30	112.00	175.30
Minimum charge for DWV line work.					
all types	7P@1.54	ea	100.00	119.00	219.00
Exterior lines. Includes excavation up to 8' deep, backfill, laying pipe, pipe, and fittings. Does not include mobilization or shoring.					
Sewer line					
4" pipe	7P@1.64	lf	40.90	127.00	167.90
6" pipe	7P@1.69	lf	57.80	131.00	188.80
Water supply line					
1" pipe	7P@1.64	lf	27.60	127.00	154.60
1-1/2" pipe	7P@1.69	lf	32.70	131.00	163.70
Minimum charge for exterior line work	7P@10.0	ea	162.00	774.00	936.00
Interior water supply shut-off valve with pressure valve.					
for 1-1/2" line	7P@1.10	ea	636.00	85.10	721.10
for 2" line	7P@1.10	ea	819.00	85.10	904.10
Repair underground line. Includes excavation to specified depth, backfill, laying pipe, pipe, and fittings. Does not include mobilization or shoring.					
4' deep or less	7P@1.67	lf	85.90	129.00	214.90
4' to 8' deep	7P@2.16	lf	85.90	167.00	252.90
8' to 12' deep	7P@3.04	lf	85.90	235.00	320.90
12' to 16' deep	7P@4.35	lf	85.90	337.00	422.90
Break-out existing concrete slab and install pipe. Saw concrete slab and break-out, hand excavate, lay pipe, hand backfill, and patch concrete. Includes sand bed for pipe as necessary. Does not include pipe, couplings or fittings.					
all sizes and types of pipe	7P@2.23	lf	27.70	173.00	200.70
minimum charge	7P@12.5	ea	172.00	968.00	1,140.00
Complete house supply lines, waste lines and finish fixtures. Includes all rough sewer, drain, vent, and water supply lines for entire house with installation. Installation includes tie-in to sewer line, tie-in to city water meter, pressure regulator when needed, main shut off valve and successful completion of all pressure tests. Includes finish fixtures installed with all lines connected and ready to pass final inspection. Grades refer to quality of plumbing fixtures. For residential structures only.					
economy grade (PVC DWV, PVC & CPVC supply)	7P@.081	sf	3.53	6.27	9.80
standard grade (PVC DWV, Cu supply)	7P@.085	sf	5.68	6.58	12.26
high grade (PVC DWV, Cu supply)	7P@.086	sf	6.15	6.66	12.81
deluxe grade (cast-iron DWV, Cu supply)	7P@.091	sf	7.08	7.04	14.12
custom deluxe grade (cast-iron DWV, Cu supply)	9P@.108	sf	7.80	6.38	14.18

	Craft@Hrs	Unit	Material	Labor	Total

Complete house rough plumbing (no fixtures). Includes all rough sewer, drain, vent, and water supply lines for entire house with installation. Installation includes tie-in to sewer line, tie-in to city water meter, pressure regulator when needed, main shut off valve and successful completion of all pressure tests. Lines are capped at locations and not connected to fixtures. Does not include fixtures.

	Craft@Hrs	Unit	Material	Labor	Total
economy grade (PVC DWV, PVC & CPVC supply)	7P@.038	sf	.93	2.94	3.87
standard grade (PVC DWV, Cu supply)	7P@.042	sf	1.37	3.25	4.62
high grade (cast-iron DWV, Cu supply)	7P@.045	sf	1.98	3.48	5.46

Complete house plumbing fixtures. Includes finish fixtures installed with all lines connected and ready to pass final inspection. Grades refer to quality of plumbing fixtures. For residential structures only. Does not include rough plumbing.

	Craft@Hrs	Unit	Material	Labor	Total
economy grade	7P@.043	sf	2.61	3.33	5.94
standard grade	7P@.044	sf	4.30	3.41	7.71
high grade	7P@.044	sf	4.76	3.41	8.17
deluxe grade	7P@.045	sf	5.10	3.48	8.58
custom deluxe grade	9P@.054	sf	5.78	3.19	8.97

Rough plumbing, bathroom. Does not include fixtures.

	Craft@Hrs	Unit	Material	Labor	Total
Two fixture bathroom					
(typically 1 toilet and 1 sink)					
plastic supply & plastic DWV	7P@19.6	ea	531.00	1,520.00	2,051.00
copper supply & plastic DWV	7P@20.4	ea	643.00	1,580.00	2,223.00
copper supply & cast-iron DWV	7P@21.3	ea	734.00	1,650.00	2,384.00
Three fixture bathroom					
(typically 1 toilet, 1 sink and a shower or bathtub)					
plastic supply & plastic DWV	7P@25.1	ea	690.00	1,940.00	2,630.00
copper supply & plastic DWV	7P@26.3	ea	830.00	2,040.00	2,870.00
copper supply & cast-iron DWV	7P@28.7	ea	946.00	2,220.00	3,166.00
Four fixture bathroom					
(typically 1 toilet, 1 sink, 1 shower and a bathtub)					
plastic supply & plastic DWV	7P@33.4	ea	852.00	2,590.00	3,442.00
copper supply & plastic DWV	7P@35.8	ea	1,030.00	2,770.00	3,800.00
copper supply & cast-iron DWV	7P@38.5	ea	1,190.00	2,980.00	4,170.00
Five fixture bathroom					
(typically 1 toilet, 2 sinks, 1 shower and a bathtub)					
plastic supply & plastic DWV	7P@41.8	ea	1,030.00	3,240.00	4,270.00
copper supply & plastic DWV	7P@43.5	ea	1,240.00	3,370.00	4,610.00
copper supply & cast-iron DWV	7P@45.4	ea	1,400.00	3,510.00	4,910.00

Rough plumbing by room. Includes all rough sewer, drain, vent, and water supply lines with installation. Lines are capped at locations and not connected to fixtures. Does not include fixtures.

	Craft@Hrs	Unit	Material	Labor	Total
Laundry room plumbing					
(includes supply valves and recessed wall box)					
plastic supply & plastic DWV	7P@6.49	ea	227.00	502.00	729.00
copper supply & plastic DWV	7P@6.81	ea	274.00	527.00	801.00
copper supply & cast-iron DWV	7P@7.19	ea	321.00	557.00	878.00
Laundry room with sink plumbing					
(includes above plus plumbing for 1 to 2 hole laundry sink)					
plastic supply & plastic DWV	7P@12.7	ea	538.00	983.00	1,521.00
copper supply & plastic DWV	7P@13.0	ea	649.00	1,010.00	1,659.00
copper supply & cast-iron DWV	7P@13.3	ea	741.00	1,030.00	1,771.00

	Craft@Hrs	Unit	Material	Labor	Total
Kitchen plumbing					
(includes supply valves, recessed wall box and plumbing for sink)					
plastic supply & plastic DWV	7P@8.20	ea	389.00	635.00	1,024.00
copper supply & plastic DWV	7P@8.54	ea	439.00	661.00	1,100.00
copper supply & cast-iron DWV	7P@8.78	ea	486.00	680.00	1,166.00
Kitchen with ice maker plumbing					
(includes ice maker with line from sink to refrigerator)					
plastic supply & plastic DWV	7P@9.81	ea	446.00	759.00	1,205.00
copper supply & plastic DWV	7P@10.1	ea	494.00	782.00	1,276.00
copper supply & cast-iron DWV	7P@10.4	ea	538.00	805.00	1,343.00

Rough plumbing by fixture.

	Craft@Hrs	Unit	Material	Labor	Total
Wet bar plumbing					
plastic supply & plastic DWV	7P@7.41	ea	197.00	574.00	771.00
copper supply & plastic DWV	7P@7.69	ea	227.00	595.00	822.00
copper supply & cast-iron DWV	7P@7.94	ea	274.00	615.00	889.00
Bathroom sink plumbing					
plastic supply & plastic DWV	7P@7.41	ea	197.00	574.00	771.00
copper supply & plastic DWV	7P@7.69	ea	235.00	595.00	830.00
copper supply & cast-iron DWV	7P@7.94	ea	281.00	615.00	896.00
Kitchen sink plumbing					
plastic supply & plastic DWV	7P@8.20	ea	338.00	635.00	973.00
copper supply & plastic DWV	7P@8.54	ea	383.00	661.00	1,044.00
copper supply & cast-iron DWV	7P@8.78	ea	430.00	680.00	1,110.00
Laundry room sink plumbing					
plastic supply & plastic DWV	7P@6.49	ea	187.00	502.00	689.00
copper supply & plastic DWV	7P@6.81	ea	227.00	527.00	754.00
copper supply & cast-iron DWV	7P@7.19	ea	274.00	557.00	831.00
Bathtub or shower plumbing					
plastic supply & plastic DWV	7P@6.90	ea	281.00	534.00	815.00
copper supply & plastic DWV	7P@7.26	ea	346.00	562.00	908.00
copper supply & cast-iron DWV	7P@7.58	ea	407.00	587.00	994.00
Toilet plumbing					
plastic supply & plastic DWV	7P@5.85	ea	243.00	453.00	696.00
copper supply & plastic DWV	7P@6.21	ea	297.00	481.00	778.00
copper supply & cast-iron DWV	7P@6.45	ea	352.00	499.00	851.00
Bidet plumbing					
plastic supply & plastic DWV	7P@6.49	ea	274.00	502.00	776.00
copper supply & plastic DWV	7P@6.90	ea	338.00	534.00	872.00
copper supply & cast-iron DWV	7P@7.13	ea	401.00	552.00	953.00
Clothes washer plumbing					
plastic supply & plastic DWV	7P@7.26	ea	197.00	562.00	759.00
copper supply & plastic DWV	7P@7.52	ea	235.00	582.00	817.00
copper supply & cast-iron DWV	7P@7.75	ea	281.00	600.00	881.00
Dishwasher plumbing					
plastic supply & plastic DWV	7P@1.39	ea	34.60	108.00	142.60
copper supply & plastic DWV	7P@1.49	ea	43.90	115.00	158.90
copper supply & cast-iron DWV	7P@1.59	ea	53.10	123.00	176.10

	Craft@Hrs	Unit	Material	Labor	Total
Exterior hose bibb (freeze-proof)					
plastic supply	7P@3.55	ea	120.00	275.00	395.00
copper supply	7P@3.51	ea	162.00	272.00	434.00
Evaporative cooler plumbing					
copper supply line	7P@2.68	ea	88.90	207.00	295.90
Refrigerator with ice maker and/or water dispenser plumbing					
plastic supply line	7P@2.10	ea	21.70	163.00	184.70
copper supply line	7P@2.21	ea	56.40	171.00	227.40
Floor drain					
in-floor French drain	7P@3.23	ea	170.00	250.00	420.00
in-floor drain connected to sewer	7P@5.27	ea	414.00	408.00	822.00

Faucet. Some faucet quality "rules of thumb." Economy: Light gauge metal, chrome plated. May have some plastic components. Little or no pattern. Standard: Heavier gauge metal, chrome or brass plated with little or no pattern or wood handles made of ash or oak. High: Brass, chrome over brass, or nickel over brass with minimal detail or plated hardware with ornate detail or European style curved plastic or with porcelain components. Deluxe: Brass, chrome over brass, or nickel over brass with ornate detail.

Bathroom sink faucet. Includes bathroom sink faucet, feeder tubes, and installation.

	Craft@Hrs	Unit	Material	Labor	Total
replace, economy grade	7P@.557	ea	158.00	43.10	201.10
replace, standard grade	7P@.557	ea	281.00	43.10	324.10
replace, high grade	7P@.557	ea	531.00	43.10	574.10
replace, deluxe grade	7P@.557	ea	916.00	43.10	959.10
remove	1D@.356	ea	—	16.80	16.80
remove for work, then reinstall	7P@.833	ea	—	64.50	64.50

Bidet faucet. Includes bidet faucet, feeder tubes, and installation.

	Craft@Hrs	Unit	Material	Labor	Total
replace, standard grade	7P@.595	ea	461.00	46.10	507.10
replace, high grade	7P@.595	ea	734.00	46.10	780.10
replace, deluxe grade	7P@.595	ea	1,220.00	46.10	1,266.10
remove	1D@.356	ea	—	16.80	16.80
remove for work, then reinstall	7P@.869	ea	—	67.30	67.30

Kitchen sink faucet. Includes kitchen sink faucet, feeder tubes, and installation.

	Craft@Hrs	Unit	Material	Labor	Total
replace, economy grade	7P@.561	ea	170.00	43.40	213.40
replace, standard grade	7P@.561	ea	310.00	43.40	353.40
replace, high grade	7P@.561	ea	596.00	43.40	639.40
replace, deluxe grade	7P@.561	ea	1,030.00	43.40	1,073.40
remove	1D@.356	ea	—	16.80	16.80
remove for work, then reinstall	7P@.869	ea	—	67.30	67.30

Laundry sink faucet. Includes laundry sink faucet, feeder tubes, and installation.

	Craft@Hrs	Unit	Material	Labor	Total
replace, economy grade	7P@.546	ea	93.80	42.30	136.10
replace, standard grade	7P@.546	ea	170.00	42.30	212.30
replace, high grade	7P@.546	ea	235.00	42.30	277.30
replace, deluxe grade	7P@.546	ea	414.00	42.30	456.30
remove	1D@.356	ea	—	16.80	16.80
remove for work, then reinstall	7P@.820	ea	—	63.50	63.50

	Craft@Hrs	Unit	Material	Labor	Total
Shower faucet. Includes shower faucet and installation.					
replace, economy grade	7P@.576	ea	211.00	44.60	255.60
replace, standard grade	7P@.576	ea	376.00	44.60	420.60
replace, high grade	7P@.576	ea	643.00	44.60	687.60
replace, deluxe grade	7P@.576	ea	957.00	44.60	1,001.60
remove	1D@.356	ea	—	16.80	16.80
remove for work, then reinstall	7P@.884	ea	—	68.40	68.40
Bathtub faucet. Includes bathtub faucet and installation.					
replace, economy grade	7P@.567	ea	235.00	43.90	278.90
replace, standard grade	7P@.567	ea	422.00	43.90	465.90
replace, high grade	7P@.567	ea	723.00	43.90	766.90
replace, deluxe grade	7P@.567	ea	1,060.00	43.90	1,103.90
remove	1D@.356	ea	—	16.80	16.80
remove for work, then reinstall	7P@.878	ea	—	68.00	68.00
Bathtub with shower faucet. Includes bathtub with shower faucet and installation.					
replace, economy grade	7P@.606	ea	257.00	46.90	303.90
replace, standard grade	7P@.606	ea	461.00	46.90	507.90
replace, high grade	7P@.606	ea	792.00	46.90	838.90
replace, deluxe grade	7P@.606	ea	1,180.00	46.90	1,226.90
remove	1D@.401	ea	—	18.90	18.90
remove for work, then reinstall	7P@.901	ea	—	69.70	69.70
Wet bar sink faucet. Includes wet bar sink faucet, feeder tubes, and installation.					
replace, economy grade	7P@.542	ea	140.00	42.00	182.00
replace, standard grade	7P@.542	ea	250.00	42.00	292.00
replace, high grade	7P@.542	ea	476.00	42.00	518.00
replace, deluxe grade	7P@.542	ea	814.00	42.00	856.00
remove	1D@.356	ea	—	16.80	16.80
remove for work, then reinstall	7P@.801	ea	—	62.00	62.00
Refurbish and repack faucet.					
all types	7P@.454	ea	31.00	35.10	66.10
Bathroom sink. With visible plumbing, including water feeds, P-trap, escutcheons, to wall or floor. Does not include faucet.					
Wall-hung porcelain-enamel cast iron bathroom sink					
replace, oval	7P@.878	ea	679.00	68.00	747.00
replace, round	7P@.878	ea	673.00	68.00	741.00
replace, hexagonal	7P@.878	ea	728.00	68.00	796.00
Wall-hung porcelain-enamel steel bathroom sink					
replace, oval	7P@.878	ea	572.00	68.00	640.00
replace, round	7P@.878	ea	564.00	68.00	632.00
replace, hexagonal	7P@.878	ea	611.00	68.00	679.00
Wall-hung vitreous china bathroom sink					
replace, oval	7P@.878	ea	549.00	68.00	617.00
replace, oval with pattern	7P@.878	ea	723.00	68.00	791.00
replace, round	7P@.878	ea	538.00	68.00	606.00

	Craft@Hrs	Unit	Material	Labor	Total
replace, round with pattern	7P@.878	ea	703.00	68.00	771.00
replace, hexagonal	7P@.878	ea	564.00	68.00	632.00
remove	1D@.367	ea	—	17.30	17.30
Remove wall-hung bathroom sink for work, then reinstall, all types	7P@1.09	ea	—	84.40	84.40
Self-rimming vitreous china bathroom sink					
replace, oval	7P@.910	ea	494.00	70.40	564.40
replace, oval with pattern	7P@.910	ea	649.00	70.40	719.40
replace, round	7P@.910	ea	486.00	70.40	556.40
replace, round with pattern	7P@.910	ea	634.00	70.40	704.40
replace, hexagonal	7P@.910	ea	509.00	70.40	579.40
Self-rimming stainless steel bathroom sink					
replace, oval	7P@.910	ea	531.00	70.40	601.40
replace, oval, scalloped	7P@.910	ea	611.00	70.40	681.40
Self-rimming polished brass bathroom skin					
replace, oval	7P@.910	ea	584.00	70.40	654.40
replace, oval with hammered finish	7P@.910	ea	663.00	70.40	733.40
replace, oval with scalloped finish	7P@.910	ea	734.00	70.40	804.40
remove	1D@.367	ea	—	17.30	17.30
Remove self-rimming bathroom sink for work, then reinstall, all types	7P@1.14	ea	—	88.20	88.20
Under-counter vitreous china bathroom sink					
replace, oval	7P@.953	ea	517.00	73.80	590.80
replace, oval with pattern	7P@.953	ea	679.00	73.80	752.80
replace, round	7P@.953	ea	509.00	73.80	582.80
replace, round sink with pattern	7P@.953	ea	663.00	73.80	736.80
replace, hexagonal	7P@.953	ea	531.00	73.80	604.80
remove	1D@.367	ea	—	17.30	17.30
Remove under-counter sink for work, then reinstall, all types	7P@1.18	ea	—	91.30	91.30

Kitchen sink. With visible plumbing, including water feeds, P-trap, escutcheons, to wall or floor. Does not include faucet.

	Craft@Hrs	Unit	Material	Labor	Total
Porcelain-enamel cast-iron kitchen sink					
replace, 24" by 21" single-bowl	7P@.925	ea	538.00	71.60	609.60
replace, 30" by 21" single-bowl	7P@.925	ea	663.00	71.60	734.60
replace, 32" by 21" double-bowl	7P@.925	ea	792.00	71.60	863.60
replace, 42" by 21" triple-bowl	7P@.925	ea	1,700.00	71.60	1,771.60
Porcelain-enamel steel kitchen sink					
replace, 24" by 21" single-bowl	7P@.925	ea	274.00	71.60	345.60
replace, 30" by 21" single-bowl	7P@.925	ea	321.00	71.60	392.60
replace, 32" by 21" double-bowl	7P@.925	ea	330.00	71.60	401.60
replace, 42" by 21" triple-bowl	7P@.925	ea	1,480.00	71.60	1,551.60
Stainless steel kitchen sink					
replace, 19" by 18" single-bowl	7P@.925	ea	407.00	71.60	478.60
replace, 25" by 22" single-bowl	7P@.925	ea	439.00	71.60	510.60
replace, 33" by 22" double-bowl	7P@.925	ea	584.00	71.60	655.60
replace, 43" by 22" double-bowl	7P@.925	ea	626.00	71.60	697.60
replace, 43" by 22" triple-bowl	7P@.925	ea	845.00	71.60	916.60
remove	1D@.370	ea	—	17.50	17.50
Remove kitchen sink for work, then reinstall all types	7P@1.11	ea	—	85.90	85.90

	Craft@Hrs	Unit	Material	Labor	Total
Laundry sink. With visible plumbing, including water feeds, P-trap, escutcheons, to wall or floor. Does not include faucet.					
Stainless steel laundry sink in countertop					
replace, 22" x 17" single	7P@.901	ea	577.00	69.70	646.70
replace, 19" x 22" single	7P@.901	ea	620.00	69.70	689.70
replace, 33" x 22" double	7P@.901	ea	663.00	69.70	732.70
Porcelain-enamel cast-iron laundry sink on black iron frame					
replace, 24" x 20" single	7P@.833	ea	814.00	64.50	878.50
replace, 24" x 23" single	7P@.833	ea	883.00	64.50	947.50
Plastic laundry sink with plastic legs					
replace, 18" x 23" single	7P@.833	ea	217.00	64.50	281.50
replace, 20" x 24" single	7P@.833	ea	281.00	64.50	345.50
replace, 36" x 23" double	7P@.833	ea	338.00	64.50	402.50
replace, 40" x 24" double	7P@.833	ea	461.00	64.50	525.50
remove	1D@.370	ea	—	17.50	17.50
Remove laundry sink for work, then reinstall					
all types	7P@1.12	ea	—	86.70	86.70
Wet bar sink. With visible plumbing, including water feeds, P-trap, escutcheons, to wall or floor. Does not include faucet.					
Self-rimming wet bar sink					
replace, vitreous china	7P@.910	ea	889.00	70.40	959.40
replace, vitreous china with pattern	7P@.910	ea	1,060.00	70.40	1,130.40
replace, porcelain-enamel cast iron	7P@.910	ea	698.00	70.40	768.40
replace, porcelain-enamel steel	7P@.910	ea	389.00	70.40	459.40
replace, stainless steel	7P@.910	ea	304.00	70.40	374.40
replace, scalloped stainless steel	7P@.910	ea	439.00	70.40	509.40
replace, polished brass	7P@.910	ea	602.00	70.40	672.40
replace, hammered finish brass	7P@.910	ea	690.00	70.40	760.40
replace, scalloped polished brass	7P@.910	ea	774.00	70.40	844.40
remove	1D@.356	ea	—	16.80	16.80
remove for work, then reinstall	7P@1.14	ea	—	88.20	88.20
Under-counter wet bar sink					
replace, vitreous china	7P@.953	ea	549.00	73.80	622.80
replace, vitreous china with pattern	7P@.953	ea	711.00	73.80	784.80
replace, porcelain-enamel cast iron	7P@.953	ea	679.00	73.80	752.80
replace, porcelain-enamel steel	7P@.953	ea	383.00	73.80	456.80
replace, stainless steel	7P@.953	ea	297.00	73.80	370.80
replace, scalloped stainless steel	7P@.953	ea	430.00	73.80	503.80
replace, polished brass	7P@.953	ea	584.00	73.80	657.80
replace, hammered finish brass	7P@.953	ea	673.00	73.80	746.80
replace, scalloped polished brass	7P@.953	ea	752.00	73.80	825.80
remove	1D@.356	ea	—	16.80	16.80
remove for work, then reinstall	7P@1.18	ea	—	91.30	91.30
Additional sink costs.					
add for brass supply and waste lines	—	ea	206.00	—	206.00
add for almond colored porcelain enamel	—	%	33.0	—	—
add for colored porcelain enamel	—	%	65.0	—	—

	Craft@Hrs	Unit	Material	Labor	Total
Sink finish plumbing. With visible plumbing, including water feeds, P-trap, escutcheons, to wall or floor. Does not include faucet.					
replace, chrome	7P@.475	ea	187.00	36.80	223.80
replace, brass	7P@.475	ea	304.00	36.80	340.80
replace, chrome-plated brass	7P@.475	ea	389.00	36.80	425.80
Toilet finish plumbing. Includes cold water supply, shut-off valve, wax ring, and toilet bolts.					
replace, chrome	7P@.345	ea	125.00	26.70	151.70
replace, brass	7P@.345	ea	204.00	26.70	230.70
Bidet finish plumbing. Includes water supply lines with shut-off valves.					
replace, standard grade	7P@1.47	ea	634.00	114.00	748.00
replace, deluxe grade	7P@1.47	ea	929.00	114.00	1,043.00
remove	1D@.360	ea	—	17.00	17.00
remove for work, then reinstall	7P@2.38	ea	3.06	184.00	187.06
Toilet. Includes toilet, wax bowl ring, fastener bolts with caps, and installation. Economy grade toilets are typically white with round bowls. Standard grade toilets are typically white with elongated bowls. High and deluxe grades are typically low-tank toilets, in a variety of colors, with elongated bowls. Deluxe grade may also be "jet" toilets which combine low-water consumption with a direct-fed jet, siphon-assisted blow out action.					
replace, economy grade	7P@1.33	ea	243.00	103.00	346.00
replace, standard grade	7P@1.33	ea	297.00	103.00	400.00
replace, high grade	7P@1.33	ea	430.00	103.00	533.00
replace, deluxe grade	7P@1.33	ea	741.00	103.00	844.00
remove	1D@.346	ea	—	16.30	16.30
remove for work, then reinstall	7P@2.16	ea	3.06	167.00	170.06
Toilet seat.					
replace, plain round	7P@.223	ea	38.90	17.30	56.20
replace, hardwood	7P@.223	ea	38.90	17.30	56.20
replace, with pattern	7P@.223	ea	62.60	17.30	79.90
replace, elongated	7P@.223	ea	86.00	17.30	103.30
replace, padded	7P@.223	ea	70.30	17.30	87.60
remove	1D@.139	ea	—	6.56	6.56
remove for work, then reinstall	7P@.323	ea	—	25.00	25.00
Bathtub. Includes standard chrome drain and overflow assembly. Does not include faucet.					
48" long by 44" deep bathtub					
replace, cast-iron corner	7P@2.93	ea	2,810.00	227.00	3,037.00
replace, acrylic fiberglass corner	7P@1.79	ea	1,160.00	139.00	1,299.00
60" long by 32" deep bathtub					
replace, acrylic fiberglass	7P@1.79	ea	525.00	139.00	664.00
replace, porcelain-enamel cast iron	7P@2.93	ea	1,290.00	227.00	1,517.00
replace, porcelain-enamel steel	7P@2.93	ea	517.00	227.00	744.00
72" long by 36" deep bathtub					
replace, acrylic fiberglass	7P@1.79	ea	1,080.00	139.00	1,219.00
replace, porcelain-enamel cast iron	7P@2.93	ea	2,670.00	227.00	2,897.00
replace, porcelain-enamel steel	7P@2.93	ea	814.00	227.00	1,041.00
Remove bathtub					
cast-iron or steel	1D@1.61	ea	—	76.00	76.00
fiberglass	1D@1.02	ea	—	48.10	48.10

	Craft@Hrs	Unit	Material	Labor	Total
Remove for work, then reinstall					
fiberglass	7P@3.45	ea	—	267.00	267.00
cast-iron or steel	7P@5.27	ea	—	408.00	408.00

Bathtub with whirlpool jets. Includes standard chrome drain and overflow assembly and jets and electrical pump hookup. Does not include faucet or electrical rough wiring.

	Craft@Hrs	Unit	Material	Labor	Total
60" long by 32" deep whirlpool bathtub					
replace, acrylic fiberglass	7P@3.45	ea	2,740.00	267.00	3,007.00
replace, porcelain-enamel cast-iron	7P@4.00	ea	6,560.00	310.00	6,870.00
72" long by 36" deep whirlpool bathtub					
replace, acrylic fiberglass	7P@3.45	ea	5,720.00	267.00	5,987.00
replace, porcelain-enamel cast-iron	7P@4.00	ea	13,600.00	310.00	13,910.00
Fiberglass whirlpool bathtub					
replace, 60" long by 30" deep	7P@3.45	ea	3,760.00	267.00	4,027.00
replace, 66" long by 48" deep	7P@3.45	ea	5,010.00	267.00	5,277.00
replace, 72" long by 36" deep	7P@3.45	ea	4,760.00	267.00	5,027.00
replace, 72" long by 42" deep	7P@3.45	ea	6,020.00	267.00	6,287.00
replace, 83" long by 65" deep	7P@3.45	ea	10,300.00	267.00	10,567.00
remove, cast-iron or steel	1D@2.13	ea	—	101.00	101.00
remove, fiberglass	1D@1.45	ea	—	68.40	68.40
Remove whirlpool bathtub for work, then reinstall					
fiberglass	7P@6.66	ea	—	515.00	515.00
cast-iron or steel for work	7P@9.10	ea	—	704.00	704.00

Bathtub and shower combination. Includes manufactured integrated bathtub and shower unit, assembly of unit, attachment hardware and fasteners, plumbers putty, P-trap with connectors, and installation. Installation includes placement of tub, placement of drain assembly, and connection to waste line with P-trap. Does not include wall framing, tub/shower faucet, or installation of the tub/shower faucet. High and deluxe quality usually indicates thicker fiberglass with glossier finishes and more ornate soap and wash cloth compartments.

	Craft@Hrs	Unit	Material	Labor	Total
Acrylic fiberglass bathtub and shower					
replace, standard grade	7P@3.23	ea	728.00	250.00	978.00
replace, high grade	7P@3.23	ea	1,160.00	250.00	1,410.00
replace, deluxe grade	7P@3.23	ea	1,450.00	250.00	1,700.00
remove	1D@1.69	ea	—	79.80	79.80
remove for work, then reinstall	7P@5.57	ea	—	431.00	431.00

Shower stall. Includes manufactured shower stall, assembly of shower stall, attachment hardware and fasteners, plumbers putty, P-trap with connectors, and installation. Installation includes placement of stall with concrete beneath pan, placement of drain assembly, and connection to waste line with P-trap. Does not include wall framing, shower faucet, or installation of the shower faucet.

	Craft@Hrs	Unit	Material	Labor	Total
32" wide by 32" deep shower stall					
replace, acrylic fiberglass	7P@2.70	ea	602.00	209.00	811.00
replace, acrylic fiberglass with terrazzo base	7P@2.70	ea	723.00	209.00	932.00
replace, metal	7P@2.70	ea	281.00	209.00	490.00
48" wide by 35" deep shower stall					
replace, acrylic fiberglass	7P@2.70	ea	690.00	209.00	899.00
60" wide by 35" deep shower stall					
replace, acrylic fiberglass	7P@2.70	ea	759.00	209.00	968.00
remove	1D@1.22	ea	—	57.60	57.60
Remove shower stall for work, then reinstall	7P@4.35	ea	—	337.00	337.00

	Craft@Hrs	Unit	Material	Labor	Total

Shower pan. Includes pan, waterproof membrane when needed, drain assembly, and installation. Does not include framing. Does not include installation of tile or tile mortar bed.

	Craft@Hrs	Unit	Material	Labor	Total
typical size	7P@1.02	ea	86.00	78.90	164.90
large size	7P@1.02	ea	132.00	78.90	210.90

Shower base. Includes prefabricated shower base, concrete mix when needed, plumbers putty, caulk, attachment hardware and fasteners, P-trap with connectors, and installation. Installation includes placement of shower base, placement of drain assembly, and connection to waste line with P-trap. Some lower-grade shower bases require concrete to support the base. Does not include framing.

	Craft@Hrs	Unit	Material	Labor	Total
fiberglass	7P@1.05	ea	197.00	81.30	278.30
terrazzo style	7P@1.05	ea	703.00	81.30	784.30
Corner entry shower base					
fiberglass	7P@1.05	ea	235.00	81.30	316.30
terrazzo style	7P@1.05	ea	774.00	81.30	855.30

Bathtub surround. Includes bathtub surround, construction adhesive and/or attachment hardware, caulk, and installation. Surround includes coverage of the three walls surrounding bathtub: two widths and one length. Does not include wall framing, doors, or shower curtain rod.

	Craft@Hrs	Unit	Material	Labor	Total
Cultured stone bathtub surround					
replace, all types	7P@2.00	ea	938.00	155.00	1,093.00
Fiberglass bathtub surround					
replace, standard grade	7P@1.54	ea	549.00	119.00	668.00
replace, high grade	7P@1.54	ea	860.00	119.00	979.00
replace, deluxe grade	7P@1.54	ea	1,230.00	119.00	1,349.00
remove	1D@.893	ea	—	42.10	42.10
Remove for work, then reinstall bathtub surround					
all types	7P@2.63	ea	—	204.00	204.00

Glass shower. Includes wall panels, glass panels, integrated door with hardware, shower base, concrete mix (when needed), and installation. Installation includes placement of shower base, placement of drain assembly, and connection to waste line with P-trap. Fiberglass panels are placed on two walls, glass on one wall, and a glass door on one wall. Corner entry showers have two glass walls and a corner entry door. Some lower-grade shower bases require concrete to support the base. Does not include shower faucet or installation of shower faucet.

	Craft@Hrs	Unit	Material	Labor	Total
Glass shower with cultured slabs on walls					
replace, standard grade	7P@2.33	ea	1,230.00	180.00	1,410.00
replace, high grade	7P@2.33	ea	1,370.00	180.00	1,550.00
Corner entry glass shower with cultured slabs on walls					
replace, standard grade	7P@2.70	ea	1,600.00	209.00	1,809.00
replace, high grade	7P@2.70	ea	1,780.00	209.00	1,989.00
Glass shower with fiberglass panels on walls					
replace, standard grade	7P@2.08	ea	1,100.00	161.00	1,261.00
replace, high grade	7P@2.08	ea	1,260.00	161.00	1,421.00
Corner entry glass shower with fiberglass panels on walls					
replace, standard grade	7P@2.44	ea	1,450.00	189.00	1,639.00
replace, high grade	7P@2.44	ea	1,650.00	189.00	1,839.00
remove	1D@1.22	ea	—	57.60	57.60
Remove glass shower for work, then reinstall					
all types	7P@3.13	ea	—	242.00	242.00

	Craft@Hrs	Unit	Material	Labor	Total

Sliding glass bathtub door. Includes sliding bathtub doors, track, hardware, caulk, and installation. Installation includes assembly of unit per manufacturer's specs and caulking.

	Craft@Hrs	Unit	Material	Labor	Total
Sliding glass bathtub door with mill finish trim					
replace	7P@1.37	ea	352.00	106.00	458.00
replace, with fancy glass	7P@1.37	ea	446.00	106.00	552.00
Sliding glass bathtub door with gold finish trim					
replace	7P@1.37	ea	430.00	106.00	536.00
replace, with fancy glass	7P@1.37	ea	549.00	106.00	655.00
Sliding glass bathtub door with brass trim					
replace	7P@1.37	ea	814.00	106.00	920.00
replace, with etched glass	7P@1.37	ea	1,030.00	106.00	1,136.00
remove	1D@.516	ea	—	24.40	24.40
remove for work, then reinstall	7P@2.28	ea	—	176.00	176.00

Folding plastic bathtub door. Includes folding bathtub door, track, hardware, caulk, and installation. Installation includes assembly of unit per manufacturer's specs and caulking.

	Craft@Hrs	Unit	Material	Labor	Total
replace, typical	7P@1.25	ea	346.00	96.80	442.80
remove	1D@.502	ea	—	23.70	23.70
remove for work, then reinstall	7P@1.69	ea	—	131.00	131.00

Shower door. Includes door, hardware, caulk, and installation. Installation includes assembly of unit per manufacturer's specs and caulking.

	Craft@Hrs	Unit	Material	Labor	Total
replace, solid glass	7P@1.20	ea	414.00	92.90	506.90
replace, mill finish trim with fancy glass	7P@1.20	ea	470.00	92.90	562.90
replace, gold finish trim	7P@1.20	ea	461.00	92.90	553.90
replace, gold finish trim with fancy glass	7P@1.20	ea	525.00	92.90	617.90
replace, brass finish trim	7P@1.20	ea	783.00	92.90	875.90
replace, brass trim with etched glass	7P@1.20	ea	883.00	92.90	975.90
remove	1D@.484	ea	—	22.80	22.80
remove door for work, then reinstall	7P@1.64	ea	—	127.00	127.00

Gas water heater. Includes gas water heater and installation. Installation includes placement on site and connection to existing gas lines, water lines, and vent pipes. Does not include drain pan.

	Craft@Hrs	Unit	Material	Labor	Total
replace, 10 gallon	7P@3.34	ea	501.00	259.00	760.00
replace, 20 gallon	7P@3.34	ea	538.00	259.00	797.00
replace, 30 gallon	7P@3.34	ea	611.00	259.00	870.00
replace, 40 gallon	7P@3.34	ea	634.00	259.00	893.00
replace, 50 gallon	7P@3.34	ea	734.00	259.00	993.00
replace, 75 gallon	7P@3.34	ea	1,300.00	259.00	1,559.00
remove	1D@1.11	ea	—	52.40	52.40
remove for work, then reinstall	7P@5.01	ea	—	388.00	388.00

Electric water heater. Includes electrical water heater and installation. Installation includes placement on site and connection to existing electrical, water lines, and vent pipes. Does not include drain pan.

	Craft@Hrs	Unit	Material	Labor	Total
replace, 10 gallon	7P@3.13	ea	389.00	242.00	631.00
replace, 20 gallon	7P@3.13	ea	494.00	242.00	736.00
replace, 30 gallon	7P@3.13	ea	564.00	242.00	806.00
replace, 40 gallon	7P@3.13	ea	602.00	242.00	844.00
replace, 50 gallon	7P@3.13	ea	723.00	242.00	965.00
replace, 75 gallon	7P@3.13	ea	1,320.00	242.00	1,562.00
remove	1D@1.06	ea	—	50.00	50.00
replace element	7P@1.28	ea	150.00	99.10	249.10
remove for work, then reinstall	7P@4.75	ea	—	368.00	368.00

Water softener. Includes water softener, water softener salt, and installation. Installation includes placement on site and connection to existing water lines. Typically includes one to two bags of water softener salt.

	Craft@Hrs	Unit	Material	Labor	Total
Automatic two tank					
replace, up to 30 grains/gallon	7P@3.34	ea	883.00	259.00	1,142.00
replace, up to 100 grains/gallon	7P@3.34	ea	1,870.00	259.00	2,129.00
remove	1D@.408	ea	—	19.30	19.30
remove for work, then reinstall	7P@5.27	ea	—	408.00	408.00

Septic system with leach field. Includes mobilization, excavation, backfill, gravel, crushed stone, pipe, couplings, connectors, building paper, tank, and distribution box.

	Craft@Hrs	Unit	Material	Labor	Total
Septic system and leach field with concrete tank					
1,000 gallon with 1,000 sf leach field	7P@77.7	ea	4,070.00	6,010.00	10,080.00
1,000 gallon with 2,000 sf leach field	7P@112	ea	5,840.00	8,670.00	14,510.00
1,250 gallon with 1,000 sf leach field	7P@79.7	ea	4,390.00	6,170.00	10,560.00
1,250 gallon with 2,000 sf leach field	7P@114	ea	6,200.00	8,820.00	15,020.00
1,500 gallon with 1,000 sf leach field	7P@80.7	ea	4,700.00	6,250.00	10,950.00
1,500 gallon with 2,000 sf leach field	7P@115	ea	6,430.00	8,900.00	15,330.00
Septic system and leach field with fiberglass tank					
1,000 gallon with 1,000 sf leach field	7P@77.5	ea	3,890.00	6,000.00	9,890.00
1,000 gallon with 2,000 sf leach field	7P@112	ea	5,720.00	8,670.00	14,390.00
1,250 gallon with 1,000 sf leach field	7P@79.2	ea	4,220.00	6,130.00	10,350.00
1,250 gallon with 2,000 sf leach field	7P@114	ea	6,020.00	8,820.00	14,840.00
1,500 gallon with 1,000 sf leach field	7P@80.3	ea	4,540.00	6,220.00	10,760.00
1,500 gallon with 2,000 sf leach field	7P@115	ea	6,260.00	8,900.00	15,160.00
Septic system and leach field with polyethylene tank					
1,000 gallon with 1,000 sf leach field	7P@77.5	ea	3,580.00	6,000.00	9,580.00
1,000 gallon with 2,000 sf leach field	7P@112	ea	5,380.00	8,670.00	14,050.00
1,250 gallon with 1,000 sf leach field	7P@79.2	ea	3,890.00	6,130.00	10,020.00
1,250 gallon with 2,000 sf leach field	7P@114	ea	5,720.00	8,820.00	14,540.00
1,500 gallon with 1,000 sf leach field	7P@80.3	ea	4,220.00	6,220.00	10,440.00
1,500 gallon with 2,000 sf leach field	7P@115	ea	5,960.00	8,900.00	14,860.00

	Craft@Hrs	Unit	Material	Labor	Total
Submersible water pump. Includes wiring and placement for well up to 100' deep.					
replace, 1/2 hp	7P@5.52	ea	860.00	427.00	1,287.00
replace, 3/4 hp	7P@5.52	ea	938.00	427.00	1,365.00
replace, 1 hp	7P@5.52	ea	957.00	427.00	1,384.00
replace, 1-1/2 hp	7P@5.52	ea	1,200.00	427.00	1,627.00
replace, 2 hp	7P@5.52	ea	1,300.00	427.00	1,727.00
replace, 3 hp	7P@5.52	ea	1,700.00	427.00	2,127.00
replace, 5 hp	7P@5.52	ea	1,970.00	427.00	2,397.00
remove	1D@1.75	ea	—	82.60	82.60
remove for work then reinstall	7P@7.69	ea	—	595.00	595.00
Sump pump. Basement style installation, includes up to 45 lf of 1-1/2" PVC pipe and wiring hookup. Does not include rough electrical.					
Plastic pipe sump pump					
replace, 1/4 hp	7P@1.39	ea	321.00	108.00	429.00
replace, 1/3 hp	7P@1.39	ea	383.00	108.00	491.00
replace, 1/2 hp	7P@1.39	ea	509.00	108.00	617.00
Cast-iron pipe sump pump					
replace, 1/4 hp	7P@1.39	ea	389.00	108.00	497.00
replace, 1/3 hp	7P@1.39	ea	439.00	108.00	547.00
replace, 1/2 hp	7P@1.39	ea	572.00	108.00	680.00
remove	1D@.561	ea	—	26.50	26.50
remove for work, then reinstall	7P@2.42	ea	—	187.00	187.00
Antique style sink faucet. Antique style faucets are usually brass, chrome over brass, or nickel over brass. Often used in homes that are not historical.					
replace, standard grade	7P@.557	ea	281.00	43.10	324.10
replace, high grade	7P@.557	ea	476.00	43.10	519.10
replace, deluxe grade	7P@.557	ea	643.00	43.10	686.10
replace, custom grade	7P@.557	ea	946.00	43.10	989.10
replace, custom deluxe grade	7P@.557	ea	1,340.00	43.10	1,383.10
remove	1D@.356	ea	—	16.80	16.80
remove for work, then reinstall	7P@.833	ea	—	64.50	64.50
Antique style bathtub faucet. Antique style are usually brass, chrome over brass, or nickel over brass with ornate detail. Often used in homes that are not historical.					
replace, economy grade	7P@.893	ea	358.00	69.10	427.10
replace, standard grade	7P@.893	ea	620.00	69.10	689.10
replace, high grade	7P@.893	ea	830.00	69.10	899.10
replace, deluxe grade	7P@.893	ea	1,230.00	69.10	1,299.10
replace, custom grade	7P@.893	ea	1,700.00	69.10	1,769.10
remove	1D@.360	ea	—	17.00	17.00
remove for work, then reinstall	7P@1.14	ea	—	88.20	88.20
Antique style bathtub faucet with hand-held shower. Antique styles are usually brass, chrome over brass, or nickel over brass with ornate detail. Often used in homes that are not historical.					
replace, standard grade	7P@.989	ea	564.00	76.50	640.50
replace, high grade	7P@.989	ea	962.00	76.50	1,038.50
replace, deluxe grade	7P@.989	ea	1,290.00	76.50	1,366.50
replace, custom grade	7P@.989	ea	1,870.00	76.50	1,946.50
remove	1D@.363	ea	—	17.10	17.10
remove for work, then reinstall	7P@1.19	ea	—	92.10	92.10

	Craft@Hrs	Unit	Material	Labor	Total
Antique style bathtub faucet with shower conversion. With exposed shower supply pipe and circular shower curtain rod.					
replace, standard grade	7P@2.16	ea	723.00	167.00	890.00
replace, high grade	7P@2.16	ea	1,050.00	167.00	1,217.00
replace, deluxe grade	7P@2.16	ea	1,270.00	167.00	1,437.00
replace, custom grade	7P@2.16	ea	1,780.00	167.00	1,947.00
remove	1D@.505	ea	—	23.80	23.80
remove for work, then reinstall	7P@3.13	ea	—	242.00	242.00
Antique style tub supply lines. Exposed lines are used on antique style bathtubs.					
Claw-foot bathtub supply lines					
replace, chrome-plated brass	7P@1.12	ea	291.00	86.70	377.70
replace, brass	7P@1.12	ea	250.00	86.70	336.70
remove	1D@.429	ea	—	20.20	20.20
Antique style bathtub drain.					
Claw-foot bathtub drain					
replace, chrome-plated brass	7P@1.15	ea	267.00	89.00	356.00
replace, brass	7P@1.15	ea	211.00	89.00	300.00
remove	1D@.436	ea	—	20.60	20.60
Antique style bathtub free-standing supply lines.					
Free-standing water feeds for claw-foot or slipper bathtub					
replace, chrome-plated brass	7P@1.22	ea	414.00	94.40	508.40
replace, brass	7P@1.22	ea	486.00	94.40	580.40
remove	1D@.484	ea	—	22.80	22.80
Pedestal sink. Antique or contemporary style.					
replace, economy grade	7P@1.22	ea	617.00	94.40	711.40
replace, standard grade	7P@1.22	ea	723.00	94.40	817.40
replace, high grade	7P@1.22	ea	968.00	94.40	1,062.40
replace, deluxe grade	7P@1.22	ea	1,660.00	94.40	1,754.40
replace, custom grade	7P@1.22	ea	2,440.00	94.40	2,534.40
replace, custom deluxe grade	7P@1.22	ea	3,110.00	94.40	3,204.40
remove	1D@.484	ea	—	22.80	22.80
remove for work, then reinstall	7P@2.13	ea	—	165.00	165.00

	Craft@Hrs	Unit	Material	Labor	Total
Antique style pillbox toilet. Round tank with beaded rim.					
replace	7P@2.23	ea	3,100.00	173.00	3,273.00
remove	1D@.429	ea	—	20.20	20.20
remove for work, then reinstall	7P@3.45	ea	—	267.00	267.00
Antique style low-tank toilet. Identifiable by pipe (usually brass) running from tank to bowl.					
Low-tank toilet					
replace, porcelain tank	7P@2.23	ea	1,650.00	173.00	1,823.00
replace, oak tank	7P@2.22	ea	1,390.00	172.00	1,562.00
remove	1D@.429	ea	—	20.20	20.20
remove for work, then reinstall	7P@3.45	ea	3.06	267.00	270.06
Antique style high-tank toilet. Tank mounted high on wall. Includes brass pipe and connectors, pull chain, hardwood toilet seat, brass tank support brackets. Wood tanks include plastic liner.					
High-tank toilet					
replace, porcelain tank	7P@2.63	ea	2,040.00	204.00	2,244.00
replace, oak tank	7P@2.63	ea	1,780.00	204.00	1,984.00
remove	1D@.602	ea	—	28.40	28.40
remove for work, then reinstall	7P@3.70	ea	3.06	286.00	289.06
Antique bathtub. Does not include supply lines or faucet. Includes **$134** for crating and **$321** for shipping.					
Antique claw-foot or ball-foot cast iron (as is condition)					
replace, 60"	7P@5.27	ea	1,200.00	408.00	1,608.00
replace, 66"	7P@5.57	ea	1,300.00	431.00	1,731.00
Refinished antique bathtub. Interior and lip refinished exterior surfaces painted. Does not include visible supply lines or faucet. Includes **$134** for crating and **$321** for shipping.					
Antique claw-foot or ball-foot cast iron					
replace, 48"	7P@4.75	ea	3,580.00	368.00	3,948.00
replace, 52"	7P@5.01	ea	3,460.00	388.00	3,848.00
replace, 60"	7P@5.27	ea	3,100.00	408.00	3,508.00
replace, 66"	7P@5.57	ea	3,210.00	431.00	3,641.00

	Craft@Hrs	Unit	Material	Labor	Total
Reproduction antique bathtub. Claw foot or ball foot. New tubs built in the antique style. Does not include supply lines or faucet. Includes **$134** for crating and **$321** for shipping. (**$223** shipping for fiberglass tubs.)					
Cast-iron reproduction bathtub					
replace, 60"	7P@5.27	ea	2,500.00	408.00	2,908.00
replace, 66"	7P@5.57	ea	3,100.00	431.00	3,531.00
remove	1D@1.82	ea	—	85.90	85.90
Fiberglass reproduction bathtub					
replace, 60"	7P@3.13	ea	3,480.00	242.00	3,722.00
replace, 66"	7P@3.34	ea	4,700.00	259.00	4,959.00
remove	1D@1.45	ea	—	68.40	68.40
remove for work, then reinstall	7P@5.57	ea	—	431.00	431.00

slipper bathtub

	Craft@Hrs	Unit	Material	Labor	Total
Reproduction slipper bathtub. Claw foot or ball foot. Does not include supply lines or faucet. Includes **$176** for crating and **$321** for shipping. (**$215** shipping for fiberglass tubs.)					
replace, 60" cast-iron	7P@5.27	ea	4,390.00	408.00	4,798.00
replace, 66" fiberglass	7P@5.57	ea	8,060.00	431.00	8,491.00
remove, cast-iron	1D@1.82	ea	—	85.90	85.90
remove, fiberglass	1D@1.45	ea	—	68.40	68.40
Additional antique bathtub costs.					
add for solid brass legs	—	ea	140.00	—	140.00
add for brass-plated legs	—	ea	78.30	—	78.30
add for oak trim around lip of antique bathtub	—	ea	2,430.00	—	2,430.00

Time & Material Charts (selected items)

Plumbing Materials

	Craft@Hrs	Unit	Material	Labor	Total
Black steel pipe					
1/2" pipe	—	lf	1.92	—	1.92
3/4" pipe	—	lf	2.27	—	2.27
1" pipe	—	lf	3.21	—	3.21
2" pipe	—	lf	10.20	—	10.20
Brass pipe					
1/2" pipe	—	lf	6.56	—	6.56
3/4" pipe	—	lf	8.75	—	8.75
1" pipe	—	lf	14.30	—	14.30
2" pipe	—	lf	22.40	—	22.40
Type K copper pipe					
1/2" pipe	—	lf	3.73	—	3.73
3/4" pipe	—	lf	4.95	—	4.95
1" pipe	—	lf	6.43	—	6.43
2" pipe	—	lf	15.60	—	15.60
Type L copper pipe					
1/2" pipe	—	lf	2.18	—	2.18
3/4" pipe	—	lf	3.61	—	3.61
1" pipe	—	lf	4.95	—	4.95
2" pipe	—	lf	12.10	—	12.10

	Craft@Hrs	Unit	Material	Labor	Total
Type M copper pipe					
1/2" pipe	—	lf	1.68	—	1.68
3/4" pipe	—	lf	2.77	—	2.77
1" pipe	—	lf	3.76	—	3.76
Galvanized steel pipe					
1/2" pipe	—	lf	2.33	—	2.33
3/4" pipe	—	lf	2.76	—	2.76
1" pipe	—	lf	3.89	—	3.89
2" pipe	—	lf	12.50	—	12.50
CPVC pipe					
1/2" pipe	—	lf	1.34	—	1.34
3/4" pipe	—	lf	1.66	—	1.66
1" pipe	—	lf	1.99	—	1.99
PVC pipe					
1/2" pipe	—	lf	.83	—	.83
3/4" pipe	—	lf	.91	—	.91
1" pipe	—	lf	.99	—	.99
1-1/2" pipe	—	lf	1.34	—	1.34
2" pipe	—	lf	1.64	—	1.64
3" pipe	—	lf	2.35	—	2.35
4" pipe	—	lf	3.05	—	3.05
5" pipe	—	lf	4.43	—	4.43
6" pipe	—	lf	5.27	—	5.27
Cast-iron pipe					
2" pipe	—	lf	8.59	—	8.59
3" pipe	—	lf	12.00	—	12.00
4" pipe	—	lf	16.60	—	16.60
5" pipe	—	lf	47.60	—	47.60
6" pipe	—	lf	61.10	—	61.10
No-hub cast-iron pipe					
2" pipe	—	lf	12.00	—	12.00
3" pipe	—	lf	15.40	—	15.40
4" pipe	—	lf	19.80	—	19.80
5" pipe	—	lf	28.70	—	28.70

Plumbing Labor

Laborer	base wage	paid leave	true wage	taxes & ins.	total
Plumber	$46.10	3.60	$49.70	27.70	$77.40
Plumber's helper	$34.20	2.67	$36.87	22.23	$59.10
Demolition laborer	$25.90	2.02	$27.92	19.28	$47.20

Paid leave is calculated based on two weeks paid vacation, one week sick leave, and seven paid holidays. Employer's matching portion of **FICA** is 7.65 percent. **FUTA** (Federal Unemployment) is .8 percent. **Worker's compensation** for the plumbing trade was calculated using a national average of 11.24 percent. **Unemployment insurance** was calculated using a national average of 8 percent. **Health insurance** was calculated based on a projected national average for 2020 of $1,280 per employee (and family when applicable) per month. Employer pays 80 percent for a per month cost of $1,024 per employee. **Retirement** is based on a 401(k) retirement program with employer matching of 50 percent. Employee contributions to the 401(k) plan are an average of 6 percent of the true wage. **Liability insurance** is based on a national average of 12.0 percent.

Plumbing Labor Productivity

	Craft@Hrs	Unit	Material	Labor	Total
Demolition of plumbing					
remove faucet	1D@.356	ea	—	16.80	16.80
remove bathroom sink	1D@.367	ea	—	17.30	17.30
remove kitchen or laundry sink	1D@.370	ea	—	17.50	17.50
remove wet bar sink	1D@.356	ea	—	16.80	16.80
remove toilet	1D@.346	ea	—	16.30	16.30
remove fiberglass bathtub	1D@1.02	ea	—	48.10	48.10
remove cast-iron or steel bathtub	1D@1.61	ea	—	76.00	76.00
remove fiberglass bathtub with whirlpool jets	1D@1.45	ea	—	68.40	68.40
remove cast-iron or steel bathtub with whirlpool jets	1D@2.13	ea	—	101.00	101.00
remove bathtub and shower combination	1D@1.69	ea	—	79.80	79.80
remove shower stall	1D@1.22	ea	—	57.60	57.60
remove bathtub surround	1D@.893	ea	—	42.10	42.10
remove glass shower	1D@1.22	ea	—	57.60	57.60
remove sliding glass bathtub door	1D@.516	ea	—	24.40	24.40
remove shower door	1D@.484	ea	—	22.80	22.80
remove gas water heater	1D@1.11	ea	—	52.40	52.40
remove electric water heater	1D@1.06	ea	—	50.00	50.00
remove water softener	1D@.408	ea	—	19.30	19.30
remove submersible pump	1D@1.75	ea	—	82.60	82.60
remove sump pump	1D@.561	ea	—	26.50	26.50
remove antique style sink faucet	1D@.356	ea	—	16.80	16.80
remove antique style bathtub faucet	1D@.360	ea	—	17.00	17.00
remove claw-foot bathtub supply lines	1D@.429	ea	—	20.20	20.20
remove claw-foot bathtub drain	1D@.436	ea	—	20.60	20.60
remove free-standing water feeds	1D@.484	ea	—	22.80	22.80
remove pedestal sink	1D@.484	ea	—	22.80	22.80
remove antique style pillbox toilet	1D@.429	ea	—	20.20	20.20
remove antique style low-tank toilet	1D@.429	ea	—	20.20	20.20
remove antique style high-tank toilet	1D@.602	ea	—	28.40	28.40
remove cast-iron claw-foot bathtub	1D@1.82	ea	—	85.90	85.90
remove fiberglass claw-foot bathtub	1D@1.45	ea	—	68.40	68.40
Install field threaded pipe					
1/2"	7P@.127	lf	—	9.83	9.83
3/4"	7P@.136	lf	—	10.50	10.50
1"	7P@.158	lf	—	12.20	12.20
2"	7P@.231	lf	—	17.90	17.90
Install copper pipe					
1/2"	7P@.104	lf	—	8.05	8.05
3/4"	7P@.109	lf	—	8.44	8.44
1"	7P@.122	lf	—	9.44	9.44
2"	7P@.202	lf	—	15.60	15.60

	Craft@Hrs	Unit	Material	Labor	Total
Install plastic pipe					
1/2"	7P@.153	lf	—	11.80	11.80
3/4"	7P@.162	lf	—	12.50	12.50
1"	7P@.189	lf	—	14.60	14.60
2"	7P@.251	lf	—	19.40	19.40
3"	7P@.281	lf	—	21.70	21.70
4"	7P@.310	lf	—	24.00	24.00
5"	7P@.347	lf	—	26.90	26.90
6"	7P@.379	lf	—	29.30	29.30
Install cast-iron pipe					
2"	7P@.236	lf	—	18.30	18.30
3"	7P@.248	lf	—	19.20	19.20
4"	7P@.270	lf	—	20.90	20.90
5"	7P@.304	lf	—	23.50	23.50
6"	7P@.317	lf	—	24.50	24.50
Install no-hub cast-iron pipe					
2"	7P@.210	lf	—	16.30	16.30
3"	7P@.223	lf	—	17.30	17.30
4"	7P@.242	lf	—	18.70	18.70
6"	7P@.283	lf	—	21.90	21.90
Install faucet					
bathroom sink	7P@.557	ea	—	43.10	43.10
kitchen sink	7P@.561	ea	—	43.40	43.40
laundry sink	7P@.546	ea	—	42.30	42.30
shower	7P@.576	ea	—	44.60	44.60
tub	7P@.567	ea	—	43.90	43.90
tub with shower	7P@.606	ea	—	46.90	46.90
Install bathroom sink					
wall-hung	7P@.878	ea	—	68.00	68.00
self-rimming	7P@.910	ea	—	70.40	70.40
under-counter	7P@.953	ea	—	73.80	73.80
Install kitchen sink	7P@.925	ea	—	71.60	71.60
Install laundry sink					
in countertop	7P@.901	ea	—	69.70	69.70
on legs or frame	7P@.833	ea	—	64.50	64.50
Install wet bar sink					
self-rimming	7P@.910	ea	—	70.40	70.40
under-counter	7P@.953	ea	—	73.80	73.80
Install fixtures					
toilet	7P@1.33	ea	—	103.00	103.00
cast-iron or steel bathtub	7P@2.93	ea	—	227.00	227.00
fiberglass bathtub	7P@1.79	ea	—	139.00	139.00
cast-iron or steel bathtub with whirlpool jets	7P@4.00	ea	—	310.00	310.00
fiberglass bathtub with whirlpool jets	7P@3.45	ea	—	267.00	267.00
fiberglass bathtub & shower combination	7P@3.23	ea	—	250.00	250.00
shower stall	7P@2.70	ea	—	209.00	209.00

	Craft@Hrs	Unit	Material	Labor	Total
Install shower or bathtub door					
sliding bathtub door	7P@1.37	ea	—	106.00	106.00
shower door	7P@1.20	ea	—	92.90	92.90
Install water heater					
gas	7P@3.34	ea	—	259.00	259.00
electric	7P@3.13	ea	—	242.00	242.00
Install water softener	7P@3.34	ea	—	259.00	259.00
Install septic system					
with concrete tank					
1,000 gallon with 1,000 sf leach field	7P@77.7	ea	3,660.00	6,010.00	9,670.00
1,000 gallon with 2,000 sf leach field	7P@112	ea	5,260.00	8,670.00	13,930.00
with fiberglass tank					
1,000 gallon with 1,000 sf leach field	7P@77.5	ea	3,500.00	6,000.00	9,500.00
1,000 gallon with 2,000 sf leach field	7P@112	ea	5,140.00	8,670.00	13,810.00
with polyethylene tank					
1,000 gallon with 1,000 sf leach field	7P@77.5	ea	3,220.00	6,000.00	9,220.00
1,000 gallon with 2,000 sf leach field	7P@112	ea	4,840.00	8,670.00	13,510.00
Install water pump					
submersible	7P@5.52	ea	—	427.00	427.00
automatic sump	7P@1.39	ea	—	108.00	108.00
Install claw-foot bathtub faucet					
faucet	7P@.893	ea	—	69.10	69.10
faucet with shower conversion	7P@2.16	ea	—	167.00	167.00
Install pedestal sink	7P@1.22	ea	—	94.40	94.40
Install antique style toilet					
pillbox or low tank	7P@2.23	ea	—	173.00	173.00
high-tank	7P@2.63	ea	—	204.00	204.00
Install antique style bathtub					
52" cast-iron	7P@5.01	ea	—	388.00	388.00
60" cast-iron	7P@5.27	ea	—	408.00	408.00
66" cast-iron	7P@5.57	ea	—	431.00	431.00
60" reproduction fiberglass	7P@3.13	ea	—	242.00	242.00
66" reproduction fiberglass	7P@3.34	ea	—	259.00	259.00

Retaining Walls

Retaining walls. All walls include excavation and backfill. Measure the sf of surface area. Includes sand and/or gravel base and 2' to 3' of walls buried beneath exposed surface.

	Craft@Hrs	Unit	Material	Labor	Equip.	Total
Minimum charge.						
for retaining wall work	3R@3.50	ea	250.00	233.00	181.00	664.00
Concrete retaining walls. See Masonry for brick veneer, wall caps and wall coping. Includes 4" perforated pipe drain line at inside base that is set in gravel.						
replace, typical	3R@.227	sf	11.50	15.10	4.76	31.36
remove	1D@.345	sf	—	16.30	—	16.30
Pile retaining walls with wood lagging.						
replace, wood	3R@.095	sf	8.94	6.34	2.80	18.08
replace, steel	3R@.104	sf	9.31	6.94	2.80	19.05
remove, wood	1D@.185	sf	—	8.73	—	8.73
remove, steel	1D@.204	sf	—	9.63	—	9.63
Anchored tieback, for all retaining wall types.						
Add for anchored tiebacks						
replace	3R@2.44	ea	176.00	163.00	55.00	394.00
remove	1D@1.81	ea	—	85.40	—	85.40
Railroad tie retaining walls.						
replace, with tie tiebacks	3R@.244	sf	9.14	16.30	2.69	28.13
replace, no tie tiebacks	3R@.167	sf	8.27	11.10	2.40	21.77
remove, with tie tiebacks	1D@.199	sf	—	9.39	—	9.39
remove, no tie tiebacks	1D@.188	sf	—	8.87	—	8.87
Cedar tie retaining walls.						
replace, with tie tiebacks	3R@.251	sf	11.50	16.70	2.50	30.70
replace, no tie tiebacks	3R@.169	sf	11.10	11.30	2.32	24.72
remove, with tie tiebacks	1D@.188	sf	—	8.87	—	8.87
remove, no tie tiebacks	1D@.178	sf	—	8.40	—	8.40
Stone. Includes 4" perforated pipe drain line at inside base set in gravel.						
Random stone retaining walls						
replace, dry set	1M@.385	sf	26.30	27.40	2.80	56.50
replace, mortar set	1M@.333	sf	27.40	23.70	2.80	53.90
Cut stone retaining wall						
replace, dry set	1M@.327	sf	33.10	23.20	2.80	59.10
replace, mortar set	1M@.323	sf	34.20	23.00	2.80	60.00
Remove stone retaining wall	1D@.384	sf	—	18.10	—	18.10

	Craft@Hrs	Unit	Material	Labor	Equip.	Total
Interlocking masonry block retaining walls. Knock-out cap is same as solid cap except 3" to 4" back the cap is cut out to about half its thickness to allow soil and vegetation to overlap the top of the wall.						
replace, typical	3R@.313	sf	15.90	20.90	2.80	39.60
remove	1D@.333	sf	—	15.70	—	15.70
add for solid cap, 4" x 16" x 10"	3R@.048	lf	7.83	3.20	—	11.03
add for knock-out cap, 4" x 16" x 10" per lf of wall	3R@.048	lf	9.03	3.20	—	12.23
Add for fabric reinforcement. One common brand name is called GeoGrid. Fabric ties retaining walls into hillside.						
4-1/2' deep every 2' of wall height	3R@.007	sf	5.89	.47	—	6.36
4-1/2' deep every 3' of wall height	3R@.005	sf	3.91	.33	—	4.24
6' deep every 2' of wall height	3R@.007	sf	7.11	.47	—	7.58
6' deep every 3' of wall height	3R@.005	sf	4.74	.33	—	5.07
Shotcrete slope stabilization.						
Slope stabilization with wire mesh (per inch deep)	3R@.104	sf	2.69	6.94	.44	10.07

Time & Material Charts (selected items)
Retaining Walls Materials

See Retaining Walls material prices above.

Retaining Walls Labor

Laborer	base wage	paid leave	true wage	taxes & ins.	total
Retaining wall installer	$37.00	2.89	$39.89	26.81	$66.70
Mason	$39.70	3.10	$42.80	28.30	$71.10
Equipment operator	$52.10	4.06	$56.16	32.74	$88.90
Demolition laborer	$25.90	2.02	$27.92	19.28	$47.20

Paid leave is calculated based on two weeks paid vacation, one week sick leave, and seven paid holidays. Employer's matching portion of **FICA** is 7.65 percent. **FUTA** (Federal Unemployment) is .8 percent. **Worker's compensation** for the retaining walls trade was calculated using a national average of 19.41 percent for the retaining wall installer; 19.41 percent for the mason, and 15.32 percent for the equipment operator. **Unemployment insurance** was calculated using a national average of 8 percent. **Health insurance** was calculated based on a projected national average for 2020 of $1,280 per employee (and family when applicable) per month. Employer pays 80 percent for a per month cost of $1,024 per employee. **Retirement** is based on a 401(k) retirement program with employer matching of 50 percent. Employee contributions to the 401(k) plan are an average of 6 percent of the true wage. **Liability insurance** is based on a national average of 12.0 percent.

Retaining Walls Labor Productivity	Craft@Hrs	Unit	Material	Labor	Total
Install retaining wall					
concrete	3R@.227	sf	—	15.10	15.10
railroad tie with tie tiebacks	3R@.244	sf	—	16.30	16.30
railroad tie, no tie tiebacks	3R@.167	sf	—	11.10	11.10
cedar tie with tie tiebacks	3R@.251	sf	—	16.70	16.70
cedar tie, no tie tiebacks	3R@.169	sf	—	11.30	11.30
interlocking masonry block	3R@.313	sf	—	20.90	20.90

	Craft@Hrs	Unit	Material	Labor	Total

Roofing

Roofing. Costs are based on a typical 2,550 sf house with a 6 in 12 slope roof or less. Roof height is less than 16' so ropes and other safety equipment are not required. Includes costs for ridges and rakes. Typical roof has an intersecting or L-shaped roof and two dormers. Also includes boots and flashing.

Minimum charge.

	Craft@Hrs	Unit	Material	Labor	Total
for roofing work	6R@5.50	ea	129.00	401.00	530.00

Add for steep and complex roofs. Cut-up roofs contain four to six intersecting roof lines. Very cut-up roofs contain seven or more.

	Craft@Hrs	Unit	Material	Labor	Total
add for steep roofs 6-12 to 8-12 slope (all roof types)	—	%	—	35.0	—
add for steep roofs greater than 8-12 slope (all roof types)	—	%	—	50.0	—
add for cut-up roof	—	%	—	30.0	—
add for very cut-up roof	—	%	—	50.0	—

Aluminum shingles. Includes natural finish aluminum shingles, pipe flashing, roofing felt, valley metal, drip edge, ridge cap, rust-resistant nails, and installation. Installation on roof 16' tall or less.

	Craft@Hrs	Unit	Material	Labor	Total
replace, .02" thick	6R@1.06	sq	437.00	77.30	514.30
replace, .03" thick	6R@1.06	sq	465.00	77.30	542.30
remove, all thicknesses	1D@1.04	sq	—	49.10	49.10
add for colored aluminum shingles (all colors)	—	%	20.0	—	—

Galvanized steel shingles. Includes galvanized steel shingles, pipe flashing, roofing felt, valley metal, drip edge, ridge cap, rust-resistant nails, and installation. Installation on roof 16' tall or less.

	Craft@Hrs	Unit	Material	Labor	Total
replace, 26 gauge	6R@1.11	sq	325.00	80.90	405.90
replace, 24 gauge	6R@1.11	sq	363.00	80.90	443.90
replace, 22 gauge	6R@1.11	sq	376.00	80.90	456.90
remove, all gauges	1D@1.04	sq	—	49.10	49.10
add for shingles with baked enamel factory finish	—	%	21.0	—	—
add for metal shingles with 1" polystyrene insulation	—	%	10.0	—	—

Minimum charge.

	Craft@Hrs	Unit	Material	Labor	Total
for metal shingle roof repair	6R@2.50	ea	81.50	182.00	263.50

Asphalt shingles. Includes asphalt shingles, pipe flashing, roofing felt, valley metal, drip edge, ridge cap, rust-resistant nails, and installation. Installation on roof 16' tall or less. Prices are for standard three-tab shingles. (20-year shinges are not available in all areas.) Add **14%** for 4 to 6 tab. Add **16%** for thatch style edge. Add **15%** for shingles with cut-outs. Add **10%** for textured shingles.

	Craft@Hrs	Unit	Material	Labor	Total
replace, 20 year (210 to 230 lb)	6R@1.24	sq	136.00	90.40	226.40
replace, 25 year (220 to 240 lb)	6R@1.24	sq	171.00	90.40	261.40
replace, 30 year (240 to 260 lb)	6R@1.27	sq	199.00	92.60	291.60
replace, 40 year (260 to 300 lb)	6R@1.27	sq	232.00	92.60	324.60
remove	1D@1.02	sq	—	48.10	48.10

	Craft@Hrs	Unit	Material	Labor	Total
Laminated asphalt shingles.					
replace, standard grade (250 to 300 lb)	6R@2.06	sq	197.00	150.00	347.00
replace, standard grade (300 to 390 lb)	6R@2.06	sq	227.00	150.00	377.00
replace, high grade (260 to 300 lb)	6R@2.45	sq	280.00	179.00	459.00
replace, high grade (300 to 400 lb)	6R@2.45	sq	320.00	179.00	499.00
remove, all grades	1D@1.02	sq	—	48.10	48.10
T-lock, diamond and hexagonal asphalt shingles. (Not available in all areas.)					
replace, T-lock (210 to 230 lb)	6R@1.27	sq	115.00	92.60	207.60
replace, diamond (210 to 230 lb)	6R@1.27	sq	115.00	92.60	207.60
replace, hexagonal (210 to 230 lb)	6R@1.27	sq	120.00	92.60	212.60
remove, all styles	1D@1.02	sq	—	48.10	48.10
Minimum charge.					
for asphalt shingle roof repair					
all types and grades	6R@2.50	ea	58.20	182.00	240.20
Roll roofing. Includes roll roofing, galvanized roofing nails, seaming tar, and installation. Overlap according to manufacturer's specifications —usually a few inches. Installation on roof 16' tall or less. Add **82%** for double selvage lay.					
replace, 90 lb	6R@.645	sq	103.00	47.00	150.00
replace, 110 lb	6R@.645	sq	123.00	47.00	170.00
replace, 140 lb	6R@.645	sq	144.00	47.00	191.00
remove, all weights	1D@.735	sq	—	34.70	34.70
Slate roofing. Includes roofing slates, galvanized nails, boots and flashing, roofing felt, valley metal, drip edge, ridge cap, and installation. Installation on roof 16' tall or less. Does not include furring strips.					
replace, Vermont unfading green, clear	6R@4.00	sq	1,210.00	292.00	1,502.00
replace, Vermont unfading purple, clear	6R@4.00	sq	1,240.00	292.00	1,532.00
replace, Vermont unfading variegated purple, clear	6R@4.00	sq	1,390.00	292.00	1,682.00
replace, Vermont unfading dark gray or black, clear	6R@4.00	sq	1,280.00	292.00	1,572.00
replace, Vermont unfading red, clear	6R@4.00	sq	3,080.00	292.00	3,372.00
replace, Pennsylvania black weathering, clear	6R@4.00	sq	1,180.00	292.00	1,472.00
replace, Vermont weathering green, clear	6R@4.00	sq	1,140.00	292.00	1,432.00
remove, all types	1D@3.70	sq	—	175.00	175.00
deduct for slate with ribbons of foreign color	—	%	-15.0	—	—
minimum charge for slate roof repair	6R@4.00	ea	448.00	292.00	740.00
replace single slate	6R@.385	ea	32.50	28.10	60.60
Wood shingles. Includes wood shingles, boots, roofing felt, valley metal, drip edge, ridge cap, rust-resistant nails, and installation. Installation on roof 16' tall or less.					
Blue label cedar shingles					
replace, 16" (Royals)	6R@2.86	sq	460.00	208.00	668.00
replace, 18" (Perfections)	6R@2.55	sq	489.00	186.00	675.00

thatch serrated Dutch weave pyramid

	Craft@Hrs	Unit	Material	Labor	Total
Red label cedar shingles					
replace, 16" (Royals)	6R@2.86	sq	446.00	208.00	654.00
replace, 18" (Perfections)	6R@2.55	sq	471.00	186.00	657.00
remove, all types	1D@1.20	sq	—	56.60	56.60
minimum charge for wood shingle repair	6R@3.00	ea	113.00	219.00	332.00
replace single wood shingle	6R@.207	ea	24.20	15.10	39.30
deduct for black label shingles	—	%	-6.0	—	—
add for 16" wood shingles with 4" exposure	—	%	20.0	—	—
deduct for 16" shingles with 5-1/2" exposure	—	%	-17.0	—	—
deduct for 16" shingles with 6" exposure	—	%	-20.0	—	—
add for shingles treated with fire retardant	—	%	18.0	—	—
add for shingles treated with mildew retardant	—	%	16.0	—	—
add for shingle staggered installation	—	%	16.0	—	—
add for shingle thatch installation	—	%	26.0	—	—
add for shingle serrated installation	—	%	24.0	—	—
add for shingle Dutch weave installation	—	%	38.0	—	—
add for shingle pyramid installation	—	%	45.0	—	—

Wood shakes. Includes shakes, boots, roofing felt, valley metal, drip edge, ridge cap, rust-resistant nails, and installation. Installation on roof 16' tall or less.

	Craft@Hrs	Unit	Material	Labor	Total
Handsplit and resawn shakes					
replace, 18" medium	6R@3.34	sq	300.00	243.00	543.00
replace, 18" heavy	6R@3.34	sq	330.00	243.00	573.00
replace, 24" medium	6R@2.86	sq	403.00	208.00	611.00
replace, 24" heavy	6R@2.86	sq	446.00	208.00	654.00
Taper split shakes					
replace, 24"	6R@2.86	sq	317.00	208.00	525.00
Straight split shakes					
replace, 18"	6R@3.34	sq	273.00	243.00	516.00
replace, 24"	6R@2.86	sq	286.00	208.00	494.00
remove, all types and sizes	1D@1.20	sq	—	56.60	56.60
add for 8" shake exposure	—	%	20.0	—	—
deduct for 12" shake exposure	—	%	-20.0	—	—
add for staggered shake installation	—	%	16.0	—	—
minimum charge for shake roof repair	6R@3.00	ea	113.00	219.00	332.00
replace single shake	6R@.208	ea	32.50	15.20	47.70

	Craft@Hrs	Unit	Material	Labor	Total

Fiber and cement shingles. Includes fiber and cement shingles, galvanized nails, boots, roofing felt, valley metal, drip edge, ridge cap, and installation. Installation on roof 16' tall or less. Does not include furring strips. Older shingles of this type may contain asbestos. This is a hazardous material which requires special techniques and equipment to remove.

	Craft@Hrs	Unit	Material	Labor	Total
replace, weathering shake type	6R@2.38	sq	149.00	174.00	323.00
replace, Spanish tile type	6R@2.50	sq	184.00	182.00	366.00
remove, all non-asbestos types	1D@1.11	sq	—	52.40	52.40
minimum charge for fiber and cement shingle roof repair	6R@2.50	ea	68.00	182.00	250.00

Granular coated metal tile. Includes granular coated metal tile, boots, roofing felt, valley metal, drip edge, ridge cap, rust-resistant nails, and installation. Installation on roof 16' tall or less.

	Craft@Hrs	Unit	Material	Labor	Total
replace, simple pattern	6R@4.27	sq	494.00	311.00	805.00
replace, complex pattern	6R@4.27	sq	598.00	311.00	909.00
remove, all types	1D@1.04	sq	—	49.10	49.10

Flat clay tile shingle roof. Includes flat clay roofing tile, galvanized nails, boots, roofing felt, valley metal, drip edge, ridge cap, and installation. Installation on roof 16' tall or less. Does not include furring strips. Old style clay tile shingles.

	Craft@Hrs	Unit	Material	Labor	Total
replace, terra cotta red	6R@3.23	sq	167.00	235.00	402.00
replace, glazed red	6R@3.23	sq	354.00	235.00	589.00
remove, all types	1D@3.45	sq	—	163.00	163.00

Mission tile roof. Includes 30 lb felt underlayment, doubled felt on rakes and eaves, bird stop, booster tiles, ridge and rake tiles.

	Craft@Hrs	Unit	Material	Labor	Total
replace, terra cotta red	6R@5.56	sq	337.00	405.00	742.00
replace, peach	6R@5.56	sq	383.00	405.00	788.00
replace, white	6R@5.56	sq	448.00	405.00	853.00
replace, color blends	6R@5.56	sq	420.00	405.00	825.00
replace, glazed white	6R@5.56	sq	918.00	405.00	1,323.00
replace, glazed gray	6R@5.56	sq	926.00	405.00	1,331.00
replace, glazed burgundy	6R@5.56	sq	930.00	405.00	1,335.00
replace, glazed terra cotta red	6R@5.56	sq	927.00	405.00	1,332.00
replace, glazed teal	6R@5.56	sq	1,150.00	405.00	1,555.00
replace, glazed blue	6R@5.56	sq	1,200.00	405.00	1,605.00
remove, all types and colors	1D@3.45	sq	—	163.00	163.00
add for vertical furring strips under all cap tiles	6R@3.26	sq	177.00	238.00	415.00
add for tiles attached with wire	6R@2.22	sq	30.00	162.00	192.00
add for tiles attached with stainless steel nails	—	sq	41.30	—	41.30
add for tiles attached with brass or copper nails	—	sq	55.40	—	55.40
add for tiles attached to braided wire runners	6R@2.18	sq	53.60	159.00	212.60
add for tiles attached with hurricane clips or locks	6R@3.03	sq	36.50	221.00	257.50
add for adhesive between tiles	6R@.154	sq	28.70	11.20	39.90

Spanish tile roof. Includes 30 lb felt underlayment, doubled felt on rakes and eaves, bird stop, ridge and rake tiles.

	Craft@Hrs	Unit	Material	Labor	Total
replace, terra cotta red	6R@3.34	sq	212.00	243.00	455.00
replace, peach	6R@3.34	sq	260.00	243.00	503.00
replace, white	6R@3.34	sq	324.00	243.00	567.00
replace, color blends	6R@3.34	sq	296.00	243.00	539.00
replace, glazed white	6R@3.34	sq	831.00	243.00	1,074.00
replace, glazed gray	6R@3.34	sq	837.00	243.00	1,080.00

	Craft@Hrs	Unit	Material	Labor	Total
replace, glazed burgundy	6R@3.34	sq	1,050.00	243.00	1,293.00
replace, glazed terra cotta red	6R@3.34	sq	841.00	243.00	1,084.00
replace, glazed teal	6R@3.34	sq	1,060.00	243.00	1,303.00
replace, glazed blue	6R@3.34	sq	1,090.00	243.00	1,333.00
remove, all types and colors	1D@3.45	sq	—	163.00	163.00
add for tiles attached with wire	6R@1.75	sq	25.10	128.00	153.10
add for tiles attached with stainless-steel nails	—	sq	34.90	—	34.90
add for tiles attached with brass or copper nails	—	sq	46.60	—	46.60
add for tiles attached to braided wire runners	6R@1.47	sq	44.20	107.00	151.20
add for tiles attached with hurricane clips or locks	6R@1.82	sq	27.80	133.00	160.80
add for tiles attached using mortar set method	6R@2.63	sq	82.20	192.00	274.20
add for adhesive between tiles	6R@.137	sq	28.70	9.99	38.69

Other clay tile grades. Deduct from the cost of Spanish or mission tiles.

	Craft@Hrs	Unit	Material	Labor	Total
Deduct					
for ASTM grade 2 moderate weathering tiles	—	%	-5.0	—	—
for ASTM grade 3 negligible weathering tiles	—	%	-11.0	—	—
for low profile Spanish or Mission tiles	—	%	-12.0	—	—

Furring strips. For roofing tiles.

	Craft@Hrs	Unit	Material	Labor	Total
replace, vertically laid only	6R@.606	sq	35.40	44.20	79.60
remove	1D@.478	sq	—	22.60	22.60
replace, vertically and horizontally laid	6R@.877	sq	58.30	63.90	122.20
remove	1D@.840	sq	—	39.60	39.60

Corrugated concrete tile. Includes corrugated concrete tile, edge cap as needed, boots, roofing felt, valley metal, drip edge, ridge cap, fasteners, and installation. Installation on roof 16' tall or less. Does not include furring strips.

	Craft@Hrs	Unit	Material	Labor	Total
replace, natural gray	6R@3.13	sq	179.00	228.00	407.00
replace, black	6R@3.13	sq	205.00	228.00	433.00
replace, browns and terra cotta reds	6R@3.13	sq	218.00	228.00	446.00
replace, bright reds	6R@3.13	sq	231.00	228.00	459.00
replace, greens	6R@3.13	sq	253.00	228.00	481.00
replace, blues	6R@3.13	sq	267.00	228.00	495.00
remove, all colors	1D@3.85	sq	—	182.00	182.00

Flat concrete tile. Includes flat concrete tile, boots, roofing felt, valley metal, drip edge, ridge cap, fasteners, and installation. Installation on roof 16' tall or less. Does not include furring strips.

	Craft@Hrs	Unit	Material	Labor	Total
replace, natural gray	6R@3.03	sq	165.00	221.00	386.00
replace, black	6R@3.03	sq	186.00	221.00	407.00
replace, browns and terra cotta reds	6R@3.03	sq	204.00	221.00	425.00
replace, bright reds	6R@3.03	sq	213.00	221.00	434.00
replace, greens	6R@3.03	sq	231.00	221.00	452.00
replace, blues	6R@3.03	sq	240.00	221.00	461.00
remove, all colors	1D@3.85	sq	—	182.00	182.00
add for glazed tiles	—	%	2.0	—	—
add for painted tiles	—	%	5.0	—	—

	Craft@Hrs	Unit	Material	Labor	Total

Aluminum sheet corrugated. Includes corrugated aluminum panels, rust-resistant nails, purlins with corrugated pattern, end caps when needed, boots, roofing felt, valley metal, drip edge, ridge cap, fasteners, and installation. Weights are per SF. Includes 4% waste.

	Craft@Hrs	Unit	Material	Labor	Total
Corrugated aluminum roofing					
Natural finish corrugated aluminum roofing					
replace, 0.016" thick	6R@.024	sf	1.56	1.75	3.31
replace, 0.019" thick	6R@.024	sf	1.67	1.75	3.42
Colored finish corrugated aluminum roofing					
replace, 0.016" thick	6R@.024	sf	1.94	1.75	3.69
replace, 0.019" thick	6R@.024	sf	2.13	1.75	3.88
Remove, all types	1D@.010	sf	—	.47	.47

Aluminum sheet ribbed. Includes aluminum ribbed panels, rust-resistant nails, purlins with ribbed pattern as needed, end caps when needed, boots, roofing felt, valley metal, drip edge, ridge cap, fasteners, and installation. Weights are per SF. Includes 4% waste. Installation on roof 16' tall or less.

	Craft@Hrs	Unit	Material	Labor	Total
Ribbed aluminum roofing					
Natural finish ribbed aluminum roofing					
replace, 0.016" thick	6R@.026	sf	1.62	1.90	3.52
replace, 0.019" thick	6R@.026	sf	2.21	1.90	4.11
replace, 0.032" thick	6R@.026	sf	3.18	1.90	5.08
replace, 0.04" thick	6R@.026	sf	4.14	1.90	6.04
replace, 0.05" thick	6R@.026	sf	4.89	1.90	6.79
Colored finish ribbed aluminum roofing					
replace, 0.016" thick	6R@.026	sf	1.94	1.90	3.84
replace, 0.019" thick	6R@.026	sf	2.17	1.90	4.07
replace, 0.032" thick	6R@.026	sf	3.25	1.90	5.15
replace, 0.04" thick	6R@.026	sf	5.52	1.90	7.42
replace, 0.05" thick	6R@.026	sf	6.33	1.90	8.23
Remove, all types	1D@.010	sf	—	.47	.47

Fiberglass corrugated. Includes fiberglass panels, rust-resistant nails, purlins with corrugated pattern, end caps when needed, boots, roofing felt, valley metal, drip edge, ridge cap, fasteners, and installation. Weights are per SF. Includes 4% waste.

	Craft@Hrs	Unit	Material	Labor	Total
Corrugated fiberglass roofing					
replace, 8 ounce	6R@.024	sf	4.11	1.75	5.86
replace, 12 ounce	6R@.024	sf	5.50	1.75	7.25
remove, all types	1D@.009	sf	—	.42	.42

Galvanized steel corrugated roofing. Includes corrugated galvanized steel panels, rust-resistant nails, purlins with corrugated pattern, end caps when needed, boots, roofing felt, valley metal, drip edge, ridge cap, fasteners, and installation. Weights are per SF. Includes 4% waste.

	Craft@Hrs	Unit	Material	Labor	Total
replace, 30 gauge	6R@.025	sf	1.55	1.82	3.37
replace, 28 gauge	6R@.025	sf	1.61	1.82	3.43
replace, 26 gauge	6R@.025	sf	1.67	1.82	3.49
replace, 24 gauge	6R@.025	sf	1.91	1.82	3.73
remove, all gauges	1D@.010	sf	—	.47	.47

	Craft@Hrs	Unit	Material	Labor	Total
Galvanized steel ribbed roofing.					
Natural finish galvanized steel ribbed roofing					
replace, 30 gauge	6R@.026	sf	1.61	1.90	3.51
replace, 28 gauge	6R@.026	sf	1.65	1.90	3.55
replace, 26 gauge	6R@.026	sf	1.77	1.90	3.67
replace, 24 gauge	6R@.026	sf	2.01	1.90	3.91
remove, all gauges	1D@.010	sf	—	.47	.47
Colored finish galvanized steel ribbed roofing					
replace, 30 gauge	6R@.026	sf	2.21	1.90	4.11
replace, 28 gauge	6R@.026	sf	2.33	1.90	4.23
replace, 26 gauge	6R@.026	sf	2.50	1.90	4.40
replace, 24 gauge	6R@.026	sf	2.77	1.90	4.67
remove, all gauges	1D@.010	sf	—	.47	.47

Copper. Includes cold-rolled copper sheets, boots, roofing felt, valley metal, drip edge, ridge cap, fasteners, anchoring clips, solder, welding rods, protective coatings, separators, sealants as recommended by copper sheet manufacturer, and installation. Installation on roof 16' tall or less.

	Craft@Hrs	Unit	Material	Labor	Total
Standing seam copper roofing, includes copper batten caps and copper batten end caps when needed					
replace, 16 ounce	6R@6.67	sq	1,030.00	486.00	1,516.00
replace, 18 ounce	6R@7.14	sq	1,140.00	521.00	1,661.00
replace, 20 ounce	6R@7.69	sq	1,310.00	561.00	1,871.00
Batten seam copper roofing					
replace, 16 ounce	6R@7.69	sq	1,060.00	561.00	1,621.00
replace, 18 ounce	6R@8.33	sq	1,170.00	607.00	1,777.00
replace, 20 ounce	6R@9.09	sq	1,330.00	663.00	1,993.00
Flat seam copper roofing					
replace, 16 ounce	6R@7.14	sq	945.00	521.00	1,466.00
replace, 18 ounce	6R@7.69	sq	1,060.00	561.00	1,621.00
replace, 20 ounce	6R@8.33	sq	1,180.00	607.00	1,787.00
Remove, all types	1D@1.03	sq	—	48.60	48.60

standing seam

batten seam

flat seam

Lead roofing. Includes cold-rolled lead sheets, boots, roofing felt, valley metal, drip edge, ridge cap, fasteners, anchoring clips, solder, welding rods, protective coatings, separators, sealants as recommended by lead sheet manufacturer, and installation. Installation on roof 16' tall or less.

	Craft@Hrs	Unit	Material	Labor	Total
Batten seam					
replace, 3 lb	6R@7.14	sq	1,150.00	521.00	1,671.00
Flat seam					
replace, 3 lb	6R@7.69	sq	1,050.00	561.00	1,611.00
Remove, all types	1D@1.03	sq	—	48.60	48.60

Stainless steel roofing. Includes cold-rolled stainless steel sheets, boots, roofing felt, valley metal, drip edge, ridge cap, fasteners, anchoring clips, solder, protective coatings, sealants as recommended by manufacturer, and installation. Installation on roof 16' tall or less.

	Craft@Hrs	Unit	Material	Labor	Total
Standing seam stainless steel roofing					
replace, 28 gauge	6R@7.14	sq	756.00	521.00	1,277.00
replace, 26 gauge	6R@6.67	sq	843.00	486.00	1,329.00

	Craft@Hrs	Unit	Material	Labor	Total
Batten seam stainless steel roofing					
replace, 28 gauge	6R@7.69	sq	777.00	561.00	1,338.00
replace, 26 gauge	6R@7.14	sq	866.00	521.00	1,387.00
Flat seam stainless steel roofing					
replace, 28 gauge	6R@7.14	sq	719.00	521.00	1,240.00
replace, 26 gauge	6R@6.67	sq	803.00	486.00	1,289.00
Remove all types of stainless steel roofing	1D@1.03	sq	—	48.60	48.60

Stainless steel roofing coated finish.

	Craft@Hrs	Unit	Material	Labor	Total
add for terne coated stainless steel	—	%	13.0	—	—
add for lead coated stainless steel	—	%	14.0	—	—

Terne roofing. Includes cold-rolled terne sheets, boots, roofing felt, valley metal, drip edge, ridge cap, fasteners, anchoring clips, solder, protective coatings, sealants as recommended by manufacturer, and installation. Installation on roof 16' tall or less. Terne is short for Terneplate. Terneplate is sheet iron or steel that is plated with an alloy that is one part tin and three to four parts lead.

	Craft@Hrs	Unit	Material	Labor	Total
Standing seam					
replace, 40 lb	6R@8.33	sq	1,670.00	607.00	2,277.00
Batten seam					
replace, 40 lb	6R@7.69	sq	1,740.00	561.00	2,301.00
Flat seam					
replace, 40 lb	6R@9.09	sq	1,620.00	663.00	2,283.00
Remove, all types	1D@1.03	sq	—	48.60	48.60

Seamed metal roof repair. For repairs to standing seam, batten seam, or flat seam metal roofs.

	Craft@Hrs	Unit	Material	Labor	Total
solder patch into metal roof (less than 1' square)	4R@1.00	ea	73.10	78.90	152.00
minimum charge	4R@4.00	ea	178.00	316.00	494.00

Built-up roofing. Includes base sheet, ply sheets, hot-mop tar, and installation. Does not include wrap up parapet wall, or finish coat. Unless indicated, does not include ballast.

	Craft@Hrs	Unit	Material	Labor	Equip.	Total
replace, 3 ply	6R@3.34	sq	91.30	243.00	17.20	351.50
replace, 4 ply	6R@4.00	sq	128.00	292.00	19.10	439.10
replace, 5 ply	6R@4.76	sq	190.00	347.00	21.40	558.40
remove, 3 ply	1D@2.44	sq	—	115.00	—	115.00
remove, 4 ply	1D@3.03	sq	—	143.00	—	143.00
remove, 5 ply	1D@3.85	sq	—	182.00	—	182.00
minimum charge						
for hot asphalt work	6R@5.00	ea	290.00	365.00	—	655.00
add for gravel surface						
embedded in flood coat	6R@.400	sq	23.00	29.20	2.64	54.84
add to remove built-up roof with ballast	1D@.847	sq	—	40.00	—	40.00
add for flood coat	6R@.714	sq	28.60	52.10	7.08	87.78
add to paint with aluminum UV coating	6R@.834	sq	27.80	60.80	—	88.60
add to paint with aluminum UV coating						
with fiber	6R@.834	sq	34.20	60.80	—	95.00
add for mineral faced cap sheet	—	sq	28.60	—	—	28.60

	Craft@Hrs	Unit	Material	Labor	Equip.	Total
remove gravel, flood coat, then replace gravel	6R@2.48	sq	61.30	181.00	28.70	271.00
add per lf of parapet wall, includes cant strip and roofing wrap up wall	6R@.226	lf	16.20	16.50	1.26	33.96

SCPE elastomeric roofing. Includes SCPE elastomeric roofing rolls and fasteners. Installation is according to manufacturer's specs using fasteners they provide. Chlorosulfonated polyethylene-hypalon roofing is abbreviated SCPE. Ballasted roofs contain 1/2 ton of stone per square.

	Craft@Hrs	Unit	Material	Labor	Equip.	Total
45 mil						
replace, loose laid and ballasted with stone	6R@1.00	sq	279.00	72.90	5.10	357.00
remove	1D@2.04	sq	—	96.30	—	96.30
replace, attached at seams with batten strips	6R@.834	sq	291.00	60.80	8.12	359.92
remove	1D@.980	sq	—	46.30	—	46.30
replace, fully attached to deck	6R@1.04	sq	308.00	75.80	15.40	399.20
remove	1D@1.56	sq	—	73.60	—	73.60

EPDM elastomeric roofing. Includes EPDM roofing rolls and fasteners. Installation is according to manufacturer's specs using fasteners they provide. Ethylene propylene diene monomer roofing is abbreviated EPDM. Ballasted roofs contain 1/2-ton of stone per square.

	Craft@Hrs	Unit	Material	Labor	Equip.	Total
45 mil						
replace, loose laid and ballasted with stone	6R@2.45	sq	156.00	179.00	5.10	340.10
remove	1D@2.04	sq	—	96.30	—	96.30
replace, attached to deck at seams	6R@2.00	sq	134.00	146.00	8.12	288.12
remove	1D@.980	sq	—	46.30	—	46.30
replace, fully attached to deck	6R@2.86	sq	186.00	208.00	15.40	409.40
remove	1D@1.56	sq	—	73.60	—	73.60
55 mil						
replace, loose laid and ballasted with stone	6R@2.51	sq	179.00	183.00	5.10	367.10
remove	1D@2.04	sq	—	96.30	—	96.30
replace, attached to deck at seams	6R@2.04	sq	159.00	149.00	8.12	316.12
remove	1D@.980	sq	—	46.30	—	46.30
replace, fully attached to deck	6R@2.95	sq	214.00	215.00	15.40	444.40
remove	1D@1.56	sq	—	73.60	—	73.60
60 mil						
replace, loose laid and ballasted with stone	6R@2.55	sq	192.00	186.00	5.10	383.10
remove	1D@2.04	sq	—	96.30	—	96.30
replace, attached to deck at seams	6R@2.08	sq	164.00	152.00	8.12	324.12
remove	1D@.980	sq	—	46.30	—	46.30
replace, fully attached to deck	6R@3.03	sq	217.00	221.00	15.40	453.40
remove	1D@1.56	sq	—	73.60	—	73.60

	Craft@Hrs	Unit	Material	Labor	Equip.	Total

PVC roofing. Includes PVC roofing rolls and fasteners. Installation is according to manufacturer's specs using fasteners they provide. Polyvinyl chloride roofing is abbreviated PVC. Ballasted roofs contain 1/2 ton of stone per square.

	Craft@Hrs	Unit	Material	Labor	Equip.	Total
45 mil						
replace, loose laid and ballasted with stone	6R@.587	sq	186.00	42.80	5.10	233.90
remove	1D@2.04	sq	—	96.30	—	96.30
replace, attached to deck at seams	6R@.909	sq	235.00	66.30	8.12	309.42
remove	1D@.980	sq	—	46.30	—	46.30
replace, fully attached to deck	6R@1.33	sq	254.00	97.00	15.40	366.40
remove	1D@1.56	sq	—	73.60	—	73.60
add for 45 mil reinforced PVC	—	%	22.0	—	—	—
48 mil						
replace, loose laid and ballasted with stone	6R@.624	sq	205.00	45.50	5.10	255.60
remove	1D@2.04	sq	—	96.30	—	96.30
replace, attached to deck at seams	6R@1.00	sq	241.00	72.90	8.12	322.02
remove	1D@.980	sq	—	46.30	—	46.30
replace, fully attached to deck	6R@1.35	sq	267.00	98.40	15.40	380.80
remove single ply roof fully adhered, no ballast	1D@1.56	sq	—	73.60	—	73.60
60 mil						
replace, loose laid and ballasted with stone	6R@.666	sq	279.00	48.60	5.10	332.70
remove single ply roof with ballast	1D@2.04	sq	—	96.30	—	96.30
replace, attached to deck at seams	6R@1.02	sq	295.00	74.40	8.12	377.52
remove single ply roof partially adhered, no ballast	1D@.980	sq	—	46.30	—	46.30
replace, fully attached to deck	6R@1.39	sq	338.00	101.00	15.40	454.40
remove single ply roof fully adhered, no ballast	1D@1.56	sq	—	73.60	—	73.60

Modified bitumen roofing. Includes modified bitumen rolls, hot tar as needed, propane as needed, fasteners as needed, and installation. Installation is according to manufacturer's specs using fasteners they provide. Modified bitumen rolls come in two styles: hot-mop or torch-down. Hot mop styles are adhered with a coat of hot tar. Torch-down styles are melted in place with a torch along the seams. Unless indicated, does not include ballast or additional hot-mop coats.

	Craft@Hrs	Unit	Material	Labor	Equip.	Total
120 mil						
replace, loose laid and ballasted with stone	6R@.990	sq	143.00	72.20	12.60	227.80
remove	1D@2.04	sq	—	96.30	—	96.30
replace, attached to deck at seams	6R@1.15	sq	134.00	83.80	12.60	230.40
remove	1D@.980	sq	—	46.30	—	46.30
replace, fully attached to deck with torch	6R@1.49	sq	132.00	109.00	15.30	256.30
replace, fully attached to deck with hot asphalt	6R@1.49	sq	148.00	109.00	19.70	276.70
remove	1D@1.56	sq	—	73.60	—	73.60

	Craft@Hrs	Unit	Material	Labor	Equip.	Total
150 mil						
replace, loose laid and ballasted with stone	6R@1.00	sq	151.00	72.90	12.60	236.50
remove	1D@2.04	sq	—	96.30	—	96.30
replace, attached to deck at seams	6R@1.18	sq	143.00	86.00	12.60	241.60
remove	1D@.980	sq	—	46.30	—	46.30
replace, fully attached to deck with torch	6R@1.52	sq	139.00	111.00	15.30	265.30
replace, fully attached to deck with hot asphalt	6R@1.56	sq	153.00	114.00	20.30	287.30
remove	1D@1.56	sq	—	73.60	—	73.60
160 mil						
replace, loose laid and ballasted with stone	6R@1.02	sq	153.00	74.40	12.60	240.00
remove	1D@2.04	sq	—	96.30	—	96.30
replace, attached to deck at seams	6R@1.20	sq	148.00	87.50	12.60	248.10
remove	1D@.980	sq	—	46.30	—	46.30
replace, fully attached to deck with torch	6R@1.54	sq	146.00	112.00	15.30	273.30
replace, fully attached to deck with hot asphalt	6R@1.59	sq	165.00	116.00	20.30	301.30
remove	1D@1.56	sq	—	73.60	—	73.60

	Craft@Hrs	Unit	Material	Labor	Total
Roofing felt.					
replace, 15 pound	6R@.085	sq	7.30	6.20	13.50
replace, 30 pound	6R@.085	sq	13.20	6.20	19.40
remove	6R@.089	sq	—	6.49	6.49
Ice and water shield					
replace	6R@.008	sf	.86	.58	1.44
Drip edge					
replace	6R@.006	lf	.81	.44	1.25
remove	6R@.003	lf	—	.22	.22
Gravel stop					
replace	6R@.006	lf	1.38	.44	1.82
remove	6R@.003	lf	—	.22	.22
Step flashing					
replace	6R@.040	lf	2.26	2.92	5.18
Chimney saw-kerf flashing					
replace	6R@.158	lf	11.20	11.50	22.70
minimum charge for replace	6R@1.00	ea	67.40	72.90	140.30
Valley metal					
replace, galvanized or aluminum	6R@.017	lf	3.00	1.24	4.24
replace, copper valley metal	6R@.017	lf	12.10	1.24	13.34
remove	6R@.006	ea	—	.44	.44

	Craft@Hrs	Unit	Material	Labor	Total
Pipe jack					
replace	6R@.175	ea	14.80	12.80	27.60
remove	6R@.205	ea	—	14.90	14.90
Roof vent					
replace, turbine	6R@.295	ea	55.90	21.50	77.40
replace, turtle	6R@.253	ea	27.00	18.40	45.40
replace, gable	6R@.253	ea	34.10	18.40	52.50
remove	6R@.253	ea	—	18.40	18.40
remove for work, then reinstall	6R@.490	ea	—	35.70	35.70
Furnace vent cap					
replace	6R@.205	ea	19.50	14.90	34.40
remove	6R@.257	ea	—	18.70	18.70
remove for work, then reinstall	6R@.355	ea	—	25.90	25.90
Repair leak with mastic					
replace, roof leak	6R@.452	ea	5.37	33.00	38.37
replace, flashing leak	6R@.436	ea	5.37	31.80	37.17

Time & Material Charts (selected items)

Roofing Materials

See Roofing material prices above.

Roofing Labor

Laborer	base wage	paid leave	true wage	taxes & ins.	total
Roofer / slater	$41.00	3.20	$44.20	34.70	$78.90
Roofer's helper	$34.20	2.67	$36.87	30.03	$66.90
Demolition laborer	$25.90	2.02	$27.92	19.28	$47.20

Paid leave is calculated based on two weeks paid vacation, one week sick leave, and seven paid holidays. Employer's matching portion of **FICA** is 7.65 percent. **FUTA** (Federal Unemployment) is .8 percent. **Worker's compensation** for the roofing trade was calculated using a national average of 34.11 percent. **Unemployment insurance** was calculated using a national average of 8 percent. **Health insurance** was calculated based on a projected national average for 2020 of $1,280 per employee (and family when applicable) per month. Employer pays 80 percent for a per month cost of $1,024 per employee. **Retirement** is based on a 401(k) retirement program with employer matching of 50 percent. Employee contributions to the 401(k) plan are an average of 6 percent of the true wage. **Liability insurance** is based on a national average of 12.0 percent.

Roofing Labor Productivity

	Craft@Hrs	Unit	Material	Labor	Total
Demolition of roofing					
remove metal shingles	1D@1.04	sq	—	49.10	49.10
remove asphalt shingles	1D@1.02	sq	—	48.10	48.10
remove roll roofing	1D@.735	sq	—	34.70	34.70
remove slate shingles	1D@3.70	sq	—	175.00	175.00
remove wood shingles or shakes	1D@1.20	sq	—	56.60	56.60
remove fiber and cement shingles	1D@1.11	sq	—	52.40	52.40
remove mission or Spanish tile shingles	1D@3.45	sq	—	163.00	163.00

	Craft@Hrs	Unit	Material	Labor	Total
remove vertically laid furring strips	1D@.478	sq	—	22.60	22.60
remove vertical & horizontal laid furring strips	1D@.840	sq	—	39.60	39.60
remove concrete shingles	1D@3.85	sq	—	182.00	182.00
remove preformed metal roofing	1D@.010	sf	—	.47	.47
remove preformed fiberglass roofing	1D@.009	sf	—	.42	.42
remove seamed metal roofing	1D@1.03	sq	—	48.60	48.60
remove built-up 3 ply roofing	1D@2.44	sq	—	115.00	115.00
remove built-up 4 ply roofing	1D@3.03	sq	—	143.00	143.00
remove built-up 5 ply roofing	1D@3.85	sq	—	182.00	182.00
add to remove built-up roof with ballast	1D@.847	sq	—	40.00	40.00
remove single ply roof partially adhered	1D@.980	sq	—	46.30	46.30
remove single ply roof fully adhered	1D@1.56	sq	—	73.60	73.60
remove single ply roof with ballast	1D@2.04	sq	—	96.30	96.30

Roofing Crew

install roofing	roofer	$78.90
install roofing	roofer's helper	$66.90
install roofing	roofing crew	$72.90

	Craft@Hrs	Unit	Material	Labor	Total
Install shingles					
aluminum shingles	6R@1.06	sq	—	77.30	77.30
galvanized steel shingles	6R@1.11	sq	—	80.90	80.90
20 year to 25 year asphalt	6R@1.24	sq	—	90.40	90.40
30 year to 40 year asphalt	6R@1.27	sq	—	92.60	92.60
standard grade laminated asphalt	6R@2.06	sq	—	150.00	150.00
high grade laminated asphalt	6R@2.45	sq	—	179.00	179.00
T-lock asphalt	6R@1.27	sq	—	92.60	92.60
hexagonal asphalt	6R@1.27	sq	—	92.60	92.60
diamond asphalt	6R@1.27	sq	—	92.60	92.60
Install roll roofing					
all weights	6R@.645	sq	—	47.00	47.00
Install slate					
typical	6R@4.00	sq	—	292.00	292.00
replace single slate shingle	6R@.385	ea	—	28.10	28.10
Install wood shingles					
16"	6R@2.86	sq	—	208.00	208.00
18"	6R@2.55	sq	—	186.00	186.00
replace single wood shingle	6R@.208	ea	—	15.20	15.20
Install wood shakes					
18"	6R@3.34	sq	—	243.00	243.00
24"	6R@2.86	sq	—	208.00	208.00
replace single shake	6R@.208	ea	—	15.20	15.20
Install fiber and cement shingles					
weathering shake style	6R@2.38	sq	—	174.00	174.00
Spanish tile style	6R@2.50	sq	—	182.00	182.00

	Craft@Hrs	Unit	Material	Labor	Total
Install clay tile					
flat	6R@3.23	sq	—	235.00	235.00
mission tile	6R@5.56	sq	—	405.00	405.00
Spanish tile	6R@3.34	sq	—	243.00	243.00
Install tile furring strips					
vertically laid only	6R@.606	sq	—	44.20	44.20
vertically and horizontally laid	6R@.877	sq	—	63.90	63.90
Install concrete tile					
corrugated	6R@3.13	sq	—	228.00	228.00
flat	6R@3.03	sq	—	221.00	221.00
Install pre-formed roofing					
corrugated aluminum or fiberglass	6R@.024	sf	—	1.75	1.75
corrugated galvanized steel	6R@.025	sf	—	1.82	1.82
ribbed (all types)	6R@.026	sf	—	1.90	1.90
Install standing seam metal roofing					
16 ounce copper	6R@6.67	sq	—	486.00	486.00
20 ounce copper	6R@7.69	sq	—	561.00	561.00
28 gauge stainless steel	6R@7.14	sq	—	521.00	521.00
40 lb terne	6R@8.33	sq	—	607.00	607.00
Install batten seam metal roofing					
16 ounce copper	6R@7.69	sq	—	561.00	561.00
20 ounce copper	6R@9.09	sq	—	663.00	663.00
28 gauge stainless steel	6R@7.69	sq	—	561.00	561.00
26 gauge stainless steel	6R@7.14	sq	—	521.00	521.00
3 lb lead	6R@7.14	sq	—	521.00	521.00
40 lb terne	6R@7.69	sq	—	561.00	561.00
Install flat seam metal roofing					
16 ounce copper	6R@7.14	sq	—	521.00	521.00
20 ounce copper	6R@8.33	sq	—	607.00	607.00
28 gauge stainless steel	6R@7.14	sq	—	521.00	521.00
26 gauge stainless steel	6R@6.67	sq	—	486.00	486.00
3 lb lead	6R@7.69	sq	—	561.00	561.00
40 lb terne	6R@9.09	sq	—	663.00	663.00
Seamed metal roofing repair					
solder patch (less than 1' square)	4R@1.00	ea	—	78.90	78.90
Install built-up roof					
3 ply	6R@3.34	sq	—	243.00	243.00
4 ply	6R@4.00	sq	—	292.00	292.00
5 ply	6R@4.76	sq	—	347.00	347.00

Rough Carpentry

Minimum charge.

	Craft@Hrs	Unit	Material	Labor	Total
minimum for rough carpentry work	6C@3.00	ea	68.50	177.00	245.50

Lumber prices. Some lumber prices, such as oak timbers, vary by as much as 60% depending on the region. These prices will be 15% to 30% lower in the Northeast, South, and most Great Lake states and provinces. In many areas of the Western, Midwest and Mountain states and provinces, oak timbers are not available unless specially shipped for a particular job. For the most accurate results, compare your local prices with the prices found in the material charts for this section starting on page 379 and create a customized materials factor.

Interior partition wall, per linear foot. Includes 5% waste. Based on an average wall with 4 openings and 5 corners per 56 lf of wall. A net of 2.2 studs is added for each corner and opening. All walls have two top plates and a single bottom plate.

	Craft@Hrs	Unit	Material	Labor	Total
2" x 4" interior partition wall, per lf					
replace, 8' tall 16" on center	6C@.241	lf	11.40	14.20	25.60
replace, 8' tall 24" on center	6C@.205	lf	9.53	12.10	21.63
remove, 8' tall 16" or 24" on center	1D@.045	lf	—	2.12	2.12
replace, 10' tall 16" on center	6C@.244	lf	13.60	14.40	28.00
replace, 10' tall 24" on center	6C@.207	lf	11.30	12.20	23.50
remove, 10' tall 16" or 24" on center	1D@.057	lf	—	2.69	2.69
2" x 6" interior partition wall, per lf					
replace, 8' tall 16" on center	6C@.244	lf	16.20	14.40	30.60
replace, 8' tall 24" on center	6C@.207	lf	13.90	12.20	26.10
remove, 8' tall 16" or 24" on center	1D@.048	lf	—	2.27	2.27
replace, 10' tall 16" on center	6C@.247	lf	19.60	14.60	34.20
replace, 10' tall 24" on center	6C@.209	lf	16.00	12.30	28.30
remove, 10' tall 16" or 24" on center	1D@.060	lf	—	2.83	2.83
2" x 8" interior partition wall, per lf					
replace, 8' tall 16" on center	6C@.247	lf	21.70	14.60	36.30
replace, 8' tall 24" on center	6C@.209	lf	18.00	12.30	30.30
remove, 8' tall 16" or 24" on center	1D@.049	lf	—	2.31	2.31
replace, 10' tall 16" on center	6C@.250	lf	25.60	14.80	40.40
replace, 10' tall 24" on center	6C@.211	lf	21.30	12.40	33.70
remove, 10' tall 16" or 24" on center	1D@.061	lf	—	2.88	2.88

Interior bearing walls, per linear foot. Includes studs, top plates, bottom plate, headers, and installation labor. Based on an average wall with 4 openings and 5 corners per 56 LF of wall. A net of 2.2 studs is added for each corner and opening. (Four openings are two 3' wide, one 4' wide, and one 5' wide for a total of 15 LF.) Headers in openings are made from sandwiched 2" x 10" boards with 1/2" CDX plywood stiffeners between. All walls have two top plates and a single bottom plate.

	Craft@Hrs	Unit	Material	Labor	Total
2" x 4" interior bearing wall, per lf					
replace, 8' tall 12" on center	6C@.307	lf	16.40	18.10	34.50
replace, 8' tall 16" on center	6C@.261	lf	14.80	15.40	30.20
remove, 8' tall 12" or 16" on center	1D@.045	lf	—	2.12	2.12
replace, 10' tall 12" on center	6C@.312	lf	19.10	18.40	37.50
replace, 10' tall 16" on center	6C@.265	lf	16.70	15.60	32.30
remove, 10' tall 12" or 16" on center	1D@.057	lf	—	2.69	2.69

	Craft@Hrs	Unit	Material	Labor	Total
2" x 6" interior bearing wall, per lf					
replace, 8' tall 12" on center	6C@.310	lf	28.10	18.30	46.40
replace, 8' tall 16" on center	6C@.263	lf	25.60	15.50	41.10
remove, 8' tall 12" or 16" on center	1D@.048	lf	—	2.27	2.27
replace, 10' tall 12" on center	6C@.315	lf	31.70	18.60	50.30
replace, 10' tall 16" on center	6C@.267	lf	28.40	15.80	44.20
remove, 10' tall 12" or 16" on center	1D@.060	lf	—	2.83	2.83
2" x 8" interior bearing wall, per lf					
replace, 8' tall 12" on center	6C@.312	lf	36.90	18.40	55.30
replace, 8' tall 16" on center	6C@.266	lf	33.50	15.70	49.20
remove, 8' tall 12" or 16" on center	1D@.049	lf	—	2.31	2.31
replace, 10' tall 12" on center	6C@.317	lf	41.40	18.70	60.10
replace, 10' tall 16" on center	6C@.270	lf	37.20	15.90	53.10
remove, 10' tall 12" or 16" on center	1D@.061	lf	—	2.88	2.88

Exterior walls, per linear foot. Includes 5% waste. Includes studs, top plates, bottom plate, headers, metal let-in bracing, and installation labor. Based on an average wall with 4 openings and 5 corners per 56 LF of wall. A net of 2.2 studs is added for each corner and opening. (Four openings are two 3' wide, one 4' wide, and one 5' wide for a total of 15 LF.) Headers in openings are made from sandwiched 2" x 10" boards with 1/2" CDX plywood stiffeners between boards. Includes 4 metal let-in braces for every 56 LF of wall. All walls have two top plates and a single bottom plate. Does not include exterior sheathing.

	Craft@Hrs	Unit	Material	Labor	Total
2" x 4" exterior wall, per lf					
replace, 8' tall 12" on center	6C@.312	lf	16.80	18.40	35.20
replace, 8' tall 16" on center	6C@.267	lf	15.20	15.80	31.00
replace, 8' tall 24" on center	6C@.216	lf	13.10	12.70	25.80
remove, 8' tall 12", 16" or 24" on center	1D@.045	lf	—	2.12	2.12
replace, 10' tall 12" on center	6C@.317	lf	19.60	18.70	38.30
replace, 10' tall 16" on center	6C@.270	lf	17.10	15.90	33.00
replace, 10' tall 24" on center	6C@.217	lf	14.90	12.80	27.70
remove, 10' tall 12", 16" or 24" on center	1D@.057	lf	—	2.69	2.69
2" x 6" exterior wall, per lf					
replace, 8' tall 12" on center	6C@.315	lf	28.40	18.60	47.00
replace, 8' tall 16" on center	6C@.270	lf	25.90	15.90	41.80
replace, 8' tall 24" on center	6C@.218	lf	23.20	12.90	36.10
remove, 8' tall 12", 16" or 24" on center	1D@.045	lf	—	2.12	2.12
replace, 10' tall 12" on center	6C@.320	lf	32.20	18.90	51.10
replace, 10' tall 16" on center	6C@.272	lf	28.80	16.00	44.80
replace, 10' tall 24" on center	6C@.221	lf	25.60	13.00	38.60
remove, 10' tall 12", 16" or 24" on center	1D@.057	lf	—	2.69	2.69
2" x 8" exterior wall, per lf					
replace, 8' tall 12" on center	6C@.320	lf	37.20	18.90	56.10
replace, 8' tall 16" on center	6C@.272	lf	33.80	16.00	49.80
replace, 8' tall 24" on center	6C@.220	lf	30.20	13.00	43.20
remove, 8' tall 12", 16" or 24" on center	1D@.045	lf	—	2.12	2.12
replace, 10' tall 12" on center	6C@.323	lf	41.90	19.10	61.00
replace, 10' tall 16" on center	6C@.275	lf	37.60	16.20	53.80
replace, 10' tall 24" on center	6C@.223	lf	33.50	13.20	46.70
remove, 10' tall 12", 16" or 24" on center	1D@.057	lf	—	2.69	2.69

	Craft@Hrs	Unit	Material	Labor	Total

Interior partition walls, per square foot. Includes 5% waste. Includes studs, top plates, bottom plate, and installation labor. Based on an average wall with 4 openings and 5 corners per 56 LF of wall. A net of 2.2 studs is added for each corner and opening. All walls have two top plates and a single bottom plate.

	Craft@Hrs	Unit	Material	Labor	Total
2" x 4" interior partition walls, per sf					
replace, 16" on center	6C@.030	sf	1.42	1.77	3.19
replace, 24" on center	6C@.026	sf	1.20	1.53	2.73
remove, 16" or 24" on center	1D@.006	sf	—	.28	.28
2" x 6" interior partition walls, per sf					
replace, 16" on center	6C@.030	sf	2.06	1.77	3.83
replace, 24" on center	6C@.026	sf	1.71	1.53	3.24
remove, 16" or 24" on center	1D@.006	sf	—	.28	.28
2" x 8" interior partition walls, per sf					
replace, 16" on center	6C@.031	sf	5.19	1.83	7.02
replace, 24" on center	6C@.026	sf	4.66	1.53	6.19
remove, 16" or 24" on center	1D@.006	sf	—	.28	.28

Sloping interior partition wall, per square foot. Includes 5% waste. Based on an average wall with one 3' wide opening, two 4' 6" wide openings, and 2 corners per 24 lf of wall. A net of 2.4 studs is added per corner or opening. All walls have two top plates and a single bottom plate.

	Craft@Hrs	Unit	Material	Labor	Total
2" x 4" sloping interior partition wall, per sf					
replace, 16" on center	6C@.041	sf	1.13	2.42	3.55
replace, 24" on center	6C@.035	sf	.97	2.07	3.04
remove, 16" or 24" on center	1D@.006	sf	—	.28	.28
2" x 6" sloping interior partition wall, per sf					
replace, 16" on center	6C@.042	sf	1.60	2.48	4.08
replace, 24" on center	6C@.036	sf	1.37	2.12	3.49
remove, 16" or 24" on center	1D@.006	sf	—	.28	.28
2" x 8" sloping interior partition wall, per sf					
replace, 16" on center	6C@.042	sf	2.12	2.48	4.60
replace, 24" on center	6C@.036	sf	1.76	2.12	3.88
remove, 16" or 24" on center	1D@.006	sf	—	.28	.28

Sloping interior bearing wall, per square foot. Includes 5% waste. Based on an average wall with one 3' wide opening, two 4' 6" wide openings, and 2 corners per 24 lf of wall. A net of 2.4 studs is added per corner or opening. Headers in openings are made from sandwiched 2" x 10" boards with 1/2" CDX plywood stiffeners between boards. All walls have two top plates and a single bottom plate.

	Craft@Hrs	Unit	Material	Labor	Total
2" x 4" sloping interior bearing wall, per sf					
replace, 12" on center	6C@.059	sf	1.62	3.48	5.10
replace, 16" on center	6C@.050	sf	1.47	2.95	4.42
remove, 12" or 16" on center	1D@.006	sf	—	.28	.28
2" x 6" sloping interior bearing wall, per sf					
replace, 12" on center	6C@.059	sf	2.24	3.48	5.72
replace, 16" on center	6C@.050	sf	1.98	2.95	4.93
remove, 12" or 16" on center	1D@.006	sf	—	.28	.28
2" x 8" sloping interior bearing wall, per sf					
replace, 12" on center	6C@.060	sf	2.73	3.54	6.27
replace, 16" on center	6C@.051	sf	2.44	3.01	5.45
remove, 12" or 16" on center	1D@.006	sf	—	.28	.28

	Craft@Hrs	Unit	Material	Labor	Total

Interior bearing wall, per square foot. Includes 5% waste. Based on an average wall with 4 openings and 5 corners per 56 lf. A net of 2.2 studs is added for each corner and opening. Openings are two 3' wide, one 4' wide, and one 5' wide for a total of 15 lf. Headers are made from sandwiched 2" x 10" boards with 1/2" CDX plywood stiffeners. All walls have two top plates and a single bottom plate.

	Craft@Hrs	Unit	Material	Labor	Total
2" x 4" interior bearing walls, per sf					
replace, 12" on center	6C@.038	sf	2.08	2.24	4.32
replace, 16" on center	6C@.033	sf	1.82	1.95	3.77
remove, 12" or 16" on center	1D@.006	sf	—	.28	.28
2" x 6" interior bearing walls, per sf					
replace, 12" on center	6C@.039	sf	3.53	2.30	5.83
replace, 16" on center	6C@.033	sf	3.16	1.95	5.11
remove, 12" or 16" on center	1D@.006	sf	—	.28	.28
2" x 8" interior bearing walls, per sf					
replace, 12" on center	6C@.039	sf	4.61	2.30	6.91
replace, 16" on center	6C@.033	sf	4.17	1.95	6.12
remove, 12" or 16" on center	1D@.006	sf	—	.28	.28

Exterior wall, per square foot. Includes 5% waste. Based on an average wall with 4 openings and 5 corners per 56 lf of wall. A net of 2.2 studs is added for each corner and opening. Openings are two 3' wide, one 4' wide, and one 5' wide for a total of 15 lf. Headers in openings are made from sandwiched 2" x 10" boards with 1/2" CDX plywood stiffeners between boards. Includes 4 metal let-in braces for every 56 lf of wall. All walls have two top plates and a single bottom plate. Does not include exterior sheathing.

	Craft@Hrs	Unit	Material	Labor	Total
2" x 4" exterior walls, per sf					
replace, 12" on center	6C@.040	sf	2.14	2.36	4.50
replace, 16" on center	6C@.034	sf	1.89	2.01	3.90
replace, 24" on center	6C@.027	sf	1.62	1.59	3.21
remove, 12", 16" or 24" on center	1D@.006	sf	—	.28	.28
2" x 6" exterior walls, per sf					
replace, 12" on center	6C@.040	sf	3.59	2.36	5.95
replace, 16" on center	6C@.034	sf	3.22	2.01	5.23
replace, 24" on center	6C@.027	sf	2.88	1.59	4.47
remove, 12", 16" or 24" on center	1D@.006	sf	—	.28	.28
2" x 8" exterior walls, per sf					
replace, 12" on center	6C@.040	sf	4.66	2.36	7.02
replace, 16" on center	6C@.034	sf	4.23	2.01	6.24
replace, 24" on center	6C@.028	sf	3.78	1.65	5.43
remove, 12", 16" or 24" on center	1D@.006	sf	—	.28	.28

Sloping exterior wall, per square foot. Includes 5% waste. Based on an average wall with one 3' wide opening, two 4' 6" wide openings, and 2 corners per 24 lf. A net of 2.4 studs is added per corner or opening. Headers in openings are made from sandwiched 2" x 10" boards with 1/2" CDX plywood stiffeners between boards. Includes 3 metal let-in braces for every 24 lf of wall. All walls have two top plates and a single bottom plate.

	Craft@Hrs	Unit	Material	Labor	Total
2" x 4" sloping exterior walls, per sf					
replace, 12" on center	6C@.054	sf	1.91	3.19	5.10
replace, 16" on center	6C@.046	sf	1.71	2.71	4.42
replace, 24" on center	6C@.038	sf	1.53	2.24	3.77
remove, 12", 16" or 24" on center	1D@.006	sf	—	.28	.28
2" x 6" sloping exterior walls, per sf					
replace, 12" on center	6C@.054	sf	2.56	3.19	5.75
replace, 16" on center	6C@.046	sf	2.28	2.71	4.99
replace, 24" on center	6C@.038	sf	2.00	2.24	4.24
remove, 12", 16" or 24" on center	1D@.006	sf	—	.28	.28

	Craft@Hrs	Unit	Material	Labor	Total
2" x 8" sloping exterior walls, per sf					
replace, 12" on center	6C@.055	sf	3.13	3.25	6.38
replace, 16" on center	6C@.047	sf	2.77	2.77	5.54
replace, 24" on center	6C@.038	sf	2.43	2.24	4.67
remove, 12", 16" or 24" on center	1D@.006	sf	—	.28	.28

Additional wall framing costs.

	Craft@Hrs	Unit	Material	Labor	Total
Replace only					
add for beam pocket in exterior or bearing wall	6C@1.11	ea	56.70	65.50	122.20
add to frame round- or elliptical-top door or window, per lf of opening	6C@.587	lf	4.61	34.60	39.21
add to frame round window, per lf window diameter	6C@.638	lf	5.32	37.60	42.92
add to frame bay window, per lf of wall	6C@.582	lf	5.32	34.30	39.62
add to frame bow window, per lf of wall	6C@.648	lf	5.32	38.20	43.52

California Earthquake Code. Items that include "CEC" in the description are priced according to California earthquake code requirements. Although these Universal Building Code standards are also required in other areas, they are most commonly associated with efforts initiated in California to improve the construction and engineering of structures in quake zones.

Additional costs for CEC bracing. Additional cost to nail, brace, and shear panel to California Earthquake Code standards. Per sf of wall.

	Craft@Hrs	Unit	Material	Labor	Total
Replace only					
brace & shear panel interior bearing wall	6C@.011	sf	.37	.65	1.02
brace & shear panel sloping interior wall	6C@.014	sf	.41	.83	1.24
brace & shear panel exterior wall	6C@.011	sf	.45	.65	1.10
brace & shear panel sloping exterior wall	6C@.015	sf	.50	.89	1.39

Wall framing, per board foot. Includes 7% waste.

	Craft@Hrs	Unit	Material	Labor	Total
replace, interior wall	6C@.034	bf	1.53	2.01	3.54
remove, interior wall	1D@.007	bf	—	.33	.33
replace, sloping interior wall	6C@.047	bf	1.60	2.77	4.37
remove, sloping interior wall	1D@.007	bf	—	.33	.33
replace, exterior wall	6C@.037	bf	1.88	2.18	4.06
remove, exterior wall	1D@.007	bf	—	.33	.33
replace, sloping exterior wall	6C@.051	bf	2.00	3.01	5.01
remove, sloping exterior wall	1D@.007	bf	—	.33	.33

Top plate. Includes 5% waste. Includes top plate and installation labor. Does not include labor to remove existing top plate or bracing.

	Craft@Hrs	Unit	Material	Labor	Total
Replace top plate					
2" x 4"	6C@.017	lf	1.03	1.00	2.03
2" x 6"	6C@.018	lf	1.42	1.06	2.48
2" x 8"	6C@.018	lf	2.08	1.06	3.14
Remove, all sizes	1D@.013	lf	—	.61	.61

Foundation sill plate. Includes foam sealer strip, hole drilling for anchor bolt, and fastening.

	Craft@Hrs	Unit	Material	Labor	Total
Replace 2" x 4" foundation sill plate					
treated wood	6C@.067	lf	1.62	3.95	5.57
con common redwood	6C@.065	lf	1.92	3.84	5.76
Replace 2" x 6" foundation sill plate					
treated wood	6C@.067	lf	2.44	3.95	6.39
con common redwood	6C@.065	lf	2.45	3.84	6.29

	Craft@Hrs	Unit	Material	Labor	Total
Replace 2" x 8" foundation sill plate					
treated-wood	6C@.068	lf	3.25	4.01	7.26
con common redwood	6C@.066	lf	3.40	3.89	7.29
Remove, all sizes and wood species	1D@.045	lf	—	2.12	2.12

Wall header beam. Includes 5% waste. 2 x and 4 x sizes made from doubled 2 x dimensional lumber with 1/2" CDX plywood stiffeners. 6 x sizes made from tripled 2 x dimensional lumber with two 1/2" CDX plywood stiffeners. 8 x sizes made from quadrupled 2 x dimensional lumber with three 1/2" CDX plywood stiffeners.

	Craft@Hrs	Unit	Material	Labor	Total
Header beam for 2" x 4" wall, replace					
4" x 6"	6C@.040	lf	6.98	2.36	9.34
4" x 8"	6C@.042	lf	7.92	2.48	10.40
4" x 10"	6C@.043	lf	11.10	2.54	13.64
4" x 12"	6C@.045	lf	13.50	2.66	16.16
Header beam for 2" x 6" wall, replace					
6" x 6"	6C@.061	lf	12.50	3.60	16.10
6" x 8"	6C@.065	lf	14.00	3.84	17.84
6" x 10"	6C@.069	lf	19.20	4.07	23.27
6" x 12"	6C@.074	lf	23.70	4.37	28.07
Header beam for 2" x 8" wall, replace					
8" x 6"	6C@.073	lf	18.00	4.31	22.31
8" x 8"	6C@.078	lf	20.20	4.60	24.80
8" x 10"	6C@.085	lf	22.40	5.02	27.42
8" x 12"	6C@.093	lf	33.80	5.49	39.29
Remove, all sizes	1D@.040	lf	—	1.89	1.89

Posts. Includes post and installation labor. Also includes 4% waste.

	Craft@Hrs	Unit	Material	Labor	Total
4" x 4" posts, replace					
pine	6C@.042	lf	3.96	2.48	6.44
con common redwood	6C@.042	lf	4.85	2.48	7.33
treated pine	6C@.042	lf	5.19	2.48	7.67
4" x 6" posts, replace					
pine	6C@.043	lf	4.62	2.54	7.16
con common redwood	6C@.043	lf	5.21	2.54	7.75
treated pine	6C@.043	lf	6.00	2.54	8.54
4" x 8" posts, replace					
pine	6C@.043	lf	7.39	2.54	9.93
con common redwood	6C@.043	lf	7.47	2.54	10.01
treated pine	6C@.043	lf	9.59	2.54	12.13
6" x 6" posts, replace					
pine	6C@.044	lf	9.72	2.60	12.32
con common redwood	6C@.044	lf	8.38	2.60	10.98
treated pine	6C@.044	lf	12.60	2.60	15.20
6" x 8" posts, replace					
pine	6C@.045	lf	12.90	2.66	15.56
con common redwood	6C@.045	lf	11.20	2.66	13.86
treated pine	6C@.045	lf	16.60	2.66	19.26
6" x 10" posts, replace					
pine	6C@.047	lf	16.00	2.77	18.77
con common redwood	6C@.047	lf	14.00	2.77	16.77
treated pine	6C@.047	lf	21.30	2.77	24.07

	Craft@Hrs	Unit	Material	Labor	Total
8" x 8" posts, replace					
pine	6C@.048	lf	17.40	2.83	20.23
con common redwood	6C@.048	lf	14.90	2.83	17.73
treated pine	6C@.048	lf	22.70	2.83	25.53
Remove, all sizes and wood species	1D@.059	lf	—	2.78	2.78

Precast concrete pier. Includes precast concrete pier and installation labor. Installation includes hand excavation as needed.

	Craft@Hrs	Unit	Material	Labor	Total
replace	6C@.261	ea	27.60	15.40	43.00
remove	1D@.313	ea	—	14.80	14.80

Adjustable steel jackpost. Includes adjustable jack post, fasteners, and installation labor.

	Craft@Hrs	Unit	Material	Labor	Total
Replace					
51" to 90", 13,000 pound load maximum	6C@.357	ea	76.40	21.10	97.50
20" to 36", 16,000 pound load maximum	6C@.243	ea	50.00	14.30	64.30
48" to 100", 16,000 pound load maximum	6C@.386	ea	86.40	22.80	109.20
37" to 60", 17,500 pound load maximum	6C@.272	ea	63.10	16.00	79.10
56" to 96" to 25,000 pound load maximum	6C@.352	ea	98.20	20.80	119.00
Remove, all sizes	1D@.313	ea	—	14.80	14.80

Lally column. Includes 3-1/2" diameter concrete-filled lally column, fasteners, and installation labor.

	Craft@Hrs	Unit	Material	Labor	Total
Replace					
6' to 8' high	6C@.526	ea	120.00	31.00	151.00
8' to 10' high	6C@.556	ea	135.00	32.80	167.80
10' to 12' high	6C@.587	ea	164.00	34.60	198.60
Remove, all sizes	1D@.356	ea	—	16.80	16.80

Glue-laminated beam, per board foot. Includes glue-laminated beam, brackets and fasteners, and installation labor. Also includes costs for equipment — usually a fork lift — to place the beam. Includes 3% waste.

	Craft@Hrs	Unit	Material	Labor	Equip.	Total
replace	6C@.060	bf	6.54	3.54	.58	10.66
remove	1D@.025	bf	—	1.18	—	1.18

Glue-laminated beam, per linear foot. Includes 3% waste.

	Craft@Hrs	Unit	Material	Labor	Equip.	Total
3-1/8" wide glue-laminated beam, replace						
3-1/8" x 7-1/2"	6C@.145	lf	12.70	8.56	1.07	22.33
3-1/8" x 9"	6C@.146	lf	15.30	8.61	1.07	24.98
3-1/8" x 10-1/2"	6C@.172	lf	17.80	10.10	1.07	28.97
3-1/8" x 12"	6C@.189	lf	20.40	11.20	1.07	32.67
3-1/8" x 13-1/2"	6C@.211	lf	23.00	12.40	1.07	36.47
3-1/8" x 15"	6C@.212	lf	25.80	12.50	1.07	39.37
3-1/8" x 16-1/2"	6C@.214	lf	28.00	12.60	1.07	41.67
3-1/8" x 18"	6C@.216	lf	30.40	12.70	1.07	44.17

	Craft@Hrs	Unit	Material	Labor	Equip.	Total
3-1/2" wide glue-laminated beam, replace						
3-1/2" x 9"	6C@.146	lf	17.00	8.61	1.07	26.68
3-1/2" x 12"	6C@.174	lf	22.90	10.30	1.07	34.27
3-1/2" x 15"	6C@.194	lf	28.40	11.40	1.07	40.87
3-1/2" x 19-1/2"	6C@.221	lf	37.10	13.00	1.07	51.17
3-1/2" x 21"	6C@.223	lf	39.70	13.20	1.07	53.97
5-1/8" wide glue-laminated beam, replace						
5-1/8" x 7-1/2"	6C@.173	lf	21.20	10.20	1.07	32.47
5-1/8" x 9"	6C@.175	lf	24.90	10.30	1.07	36.27
5-1/8" x 10-1/2"	6C@.194	lf	29.00	11.40	1.07	41.47
5-1/8" x 12"	6C@.218	lf	34.00	12.90	1.07	47.97
5-1/8" x 13-1/2"	6C@.221	lf	37.60	13.00	1.07	51.67
5-1/8" x 15"	6C@.224	lf	41.90	13.20	1.07	56.17
5-1/8" x 16-1/2"	6C@.227	lf	46.10	13.40	1.07	60.57
5-1/8" x 18"	6C@.231	lf	50.10	13.60	1.07	64.77
5-1/8" x 19-1/2"	6C@.234	lf	50.70	13.80	1.07	65.57
5-1/8" x 21"	6C@.238	lf	58.20	14.00	1.07	73.27
5-1/8" x 22-1/2"	6C@.242	lf	63.00	14.30	1.07	78.37
5-1/8" x 24"	6C@.245	lf	66.80	14.50	1.07	82.37
6-3/4" wide glue-laminated beam, replace						
6-3/4" x 18"	6C@.244	lf	66.00	14.40	1.07	81.47
6-3/4" x 19-1/2"	6C@.250	lf	71.70	14.80	1.07	87.57
6-3/4" x 24"	6C@.267	lf	87.80	15.80	1.07	104.67
Remove, all widths and sizes	1D@.222	lf	—	10.50	—	10.50

Micro-laminated beam. Includes micro-laminated beam, brackets, fasteners, and installation labor. Micro-laminated beams are often doubled as beams. Includes 4% waste.

	Craft@Hrs	Unit	Material	Labor	Total
1-3/4" wide micro-laminated beam, replace					
1-3/4" x 7-1/4"	6C@.048	lf	7.47	2.83	10.30
1-3/4" x 9-1/2"	6C@.050	lf	9.79	2.95	12.74
1-3/4" x 11-7/8"	6C@.053	lf	12.30	3.13	15.43
1-3/4" x 14"	6C@.056	lf	14.40	3.30	17.70
1-3/4" x 16"	6C@.059	lf	16.20	3.48	19.68
1-3/4" x 18"	6C@.063	lf	18.40	3.72	22.12
Remove, all sizes	1D@.106	lf	—	5.00	5.00

Pine beam, select structural grade. Includes beam, brackets, fasteners, and installation labor. Also includes costs for equipment — usually a fork lift — to place the beam. Deduct **12%** for #1 and better grade pine. Rough sawn. Installed as posts, beams, joists, and purlins. Pine species vary by region. Western coastal areas: Douglas fir, hemlock fir. Mountain states and provinces: spruce, pine, Douglas fir. South: southern yellow pine. Midwest: spruce, pine, Douglas fir, southern yellow pine. Southwest: southern yellow pine. Northeast: spruce, pine, Douglas fir, and hemlock fir.

	Craft@Hrs	Unit	Material	Labor	Equip.	Total
4" wide beam, replace						
4" x 6"	6C@.078	lf	5.40	4.60	1.07	11.07
4" x 8"	6C@.092	lf	7.21	5.43	1.07	13.71
4" x 10"	6C@.114	lf	9.02	6.73	1.07	16.82
4" x 12"	6C@.149	lf	10.90	8.79	1.07	20.76
4" x 14"	6C@.178	lf	12.60	10.50	1.07	24.17
4" x 16"	6C@.180	lf	14.40	10.60	1.07	26.07

	Craft@Hrs	Unit	Material	Labor	Equip.	Total
6" wide, replace						
6" x 6"	6C@.102	lf	8.11	6.02	1.07	15.20
6" x 8"	6C@.115	lf	10.90	6.79	1.07	18.76
6" x 10"	6C@.132	lf	13.60	7.79	1.07	22.46
6" x 12"	6C@.154	lf	16.00	9.09	1.07	26.16
6" x 14"	6C@.169	lf	18.90	9.97	1.07	29.94
6" x 16"	6C@.189	lf	21.70	11.20	1.07	33.97
8" wide beam, replace						
8" x 8"	6C@.139	lf	14.40	8.20	1.07	23.67
8" x 10"	6C@.156	lf	17.90	9.20	1.07	28.17
8" x 12"	6C@.182	lf	21.70	10.70	1.07	33.47
8" x 14"	6C@.193	lf	25.00	11.40	1.07	37.47
8" x 16"	6C@.199	lf	28.80	11.70	1.07	41.57
10" wide beam, replace						
10" x 10"	6C@.176	lf	22.70	10.40	1.07	34.17
10" x 12"	6C@.196	lf	26.80	11.60	1.07	39.47
10" x 14"	6C@.203	lf	31.50	12.00	1.07	44.57
10" x 16"	6C@.210	lf	36.20	12.40	1.07	49.67
12" wide beam, replace						
12" x 12"	6C@.204	lf	32.30	12.00	1.07	45.37
12" x 14"	6C@.213	lf	37.80	12.60	1.07	51.47
12" x 16"	6C@.222	lf	43.00	13.10	1.07	57.17
Remove, all widths and sizes	1D@.222	lf	—	10.50	—	10.50

Oak beam. Includes beam, brackets, fasteners, and installation labor. Also includes costs for equipment — usually a fork lift — to place the beam. Installed as posts, beams, joists, and purlins. Includes 4% waste.

	Craft@Hrs	Unit	Material	Labor	Equip.	Total
4" wide oak beam, replace						
4" x 6"	6C@.078	lf	13.20	4.60	1.07	18.87
4" x 8"	6C@.092	lf	17.50	5.43	1.07	24.00
4" x 10"	6C@.114	lf	22.30	6.73	1.07	30.10
4" x 12"	6C@.149	lf	26.40	8.79	1.07	36.26
4" x 14"	6C@.178	lf	30.70	10.50	1.07	42.27
4" x 16"	6C@.180	lf	35.40	10.60	1.07	47.07
6" wide oak beam, replace						
6" x 6"	6C@.102	lf	20.00	6.02	1.07	27.09
6" x 8"	6C@.115	lf	27.30	6.79	1.07	35.16
6" x 10"	6C@.132	lf	33.30	7.79	1.07	42.16
6" x 12"	6C@.154	lf	39.60	9.09	1.07	49.76
6" x 14"	6C@.169	lf	46.40	9.97	1.07	57.44
6" x 16"	6C@.189	lf	53.00	11.20	1.07	65.27
8" wide oak beam, replace						
8" x 8"	6C@.139	lf	35.40	8.20	1.07	44.67
8" x 10"	6C@.156	lf	44.10	9.20	1.07	54.37
8" x 12"	6C@.182	lf	53.00	10.70	1.07	64.77
8" x 14"	6C@.193	lf	61.70	11.40	1.07	74.17
8" x 16"	6C@.199	lf	70.60	11.70	1.07	83.37
10" wide oak beam, replace						
10" x 10"	6C@.176	lf	55.00	10.40	1.07	66.47
10" x 12"	6C@.196	lf	66.10	11.60	1.07	78.77
10" x 14"	6C@.203	lf	77.20	12.00	1.07	90.27
10" x 16"	6C@.210	lf	88.00	12.40	1.07	101.47

	Craft@Hrs	Unit	Material	Labor	Equip.	Total
12" wide oak beam, replace						
12" x 12"	6C@.204	lf	79.50	12.00	1.07	92.57
12" x 14"	6C@.213	lf	92.70	12.60	1.07	106.37
12" x 16"	6C@.222	lf	107.00	13.10	1.07	121.17
Remove, all widths and sizes	1D@.222	lf	—	10.50	—	10.50

Additional beam costs.	Craft@Hrs	Unit	Material	Labor	Total
for hand-hewn pine beam	—	si	.03	—	.03
for curved pine post & beam framing member	4B@.909	lf	—	66.40	66.40
for hand-hewn oak beam	—	si	.04	—	.04
for curved oak post & beam framing member	4B@1.20	lf	—	87.60	87.60

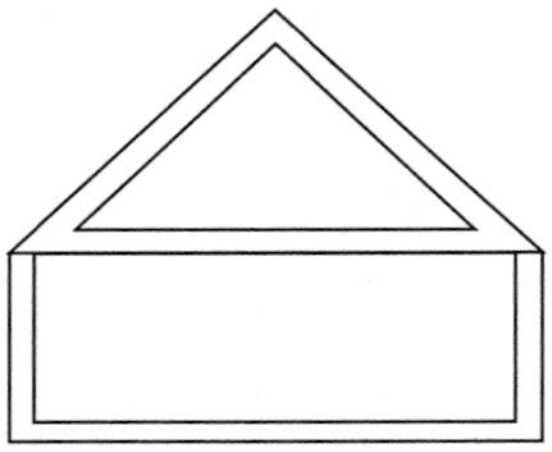

simple truss

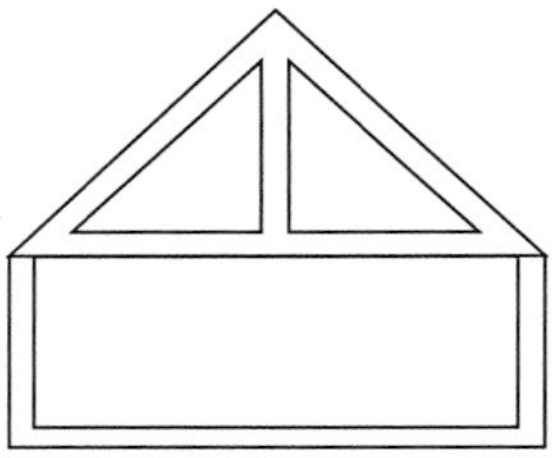

king post truss

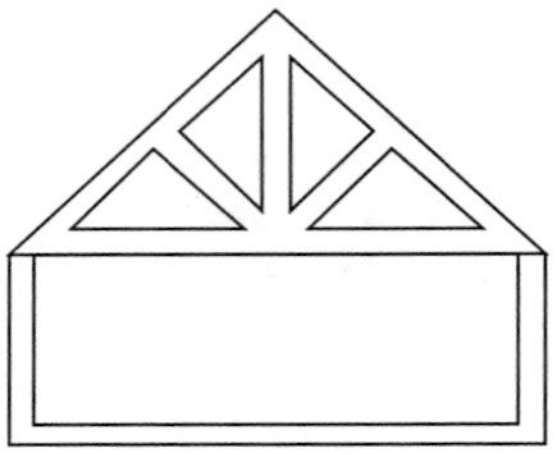

king post & struts truss

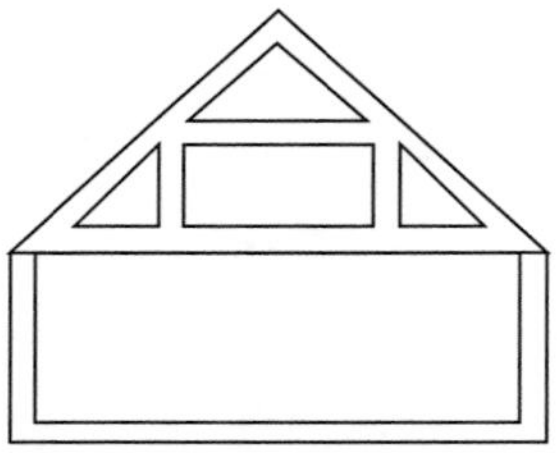

queen post truss

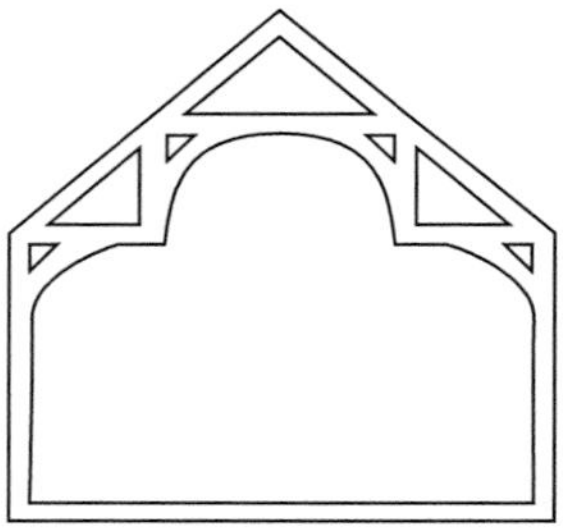

hammer beam truss

scissors truss

Pine post & beam bent with truss. Includes beams, posts, and installation labor. All connections between beams and posts are made from cut joints. Joints are held together with dowels that are often hand-cut from the beam material. Includes glue as needed. Does not include connecting members or their joinery. This price should not be used for similar systems made of beams that are joined with butt and miter joints often held together by steel brackets. With pine posts and beams, per SF of bent.

	Craft@Hrs	Unit	Material	Labor	Equip.	Total
Replace single-story bent with						
simple truss	4B@.527	sf	4.61	38.50	.67	43.78
king post truss	4B@.569	sf	5.11	41.50	.67	47.28
king post & struts truss	4B@.618	sf	5.75	45.10	.67	51.52
queen post truss	4B@.609	sf	5.37	44.50	.67	50.54
hammer beam truss	4B@1.05	sf	5.67	76.70	.67	83.04
scissors truss	4B@.989	sf	6.00	72.20	.67	78.87
Remove, single-story bent with truss, all types	1D@.071	sf	—	3.35	—	3.35

	Craft@Hrs	Unit	Material	Labor	Equip.	Total
Replace two-story bent with						
simple truss	4B@.571	sf	4.24	41.70	.67	46.61
king post truss	4B@.620	sf	4.58	45.30	.67	50.55
king post & struts truss	4B@.679	sf	4.98	49.60	.67	55.25
queen post truss	4B@.671	sf	4.73	49.00	.67	54.40
hammer beam truss	4B@1.25	sf	4.95	91.30	.67	96.92
scissors truss	4B@1.04	sf	5.09	75.90	.67	81.66
Remove, two-story bent with truss, all types	1D@.071	sf	—	3.35	—	3.35
Add for overhang drop	4B@20.0	ea	258.00	1,460.00	.67	1,718.67

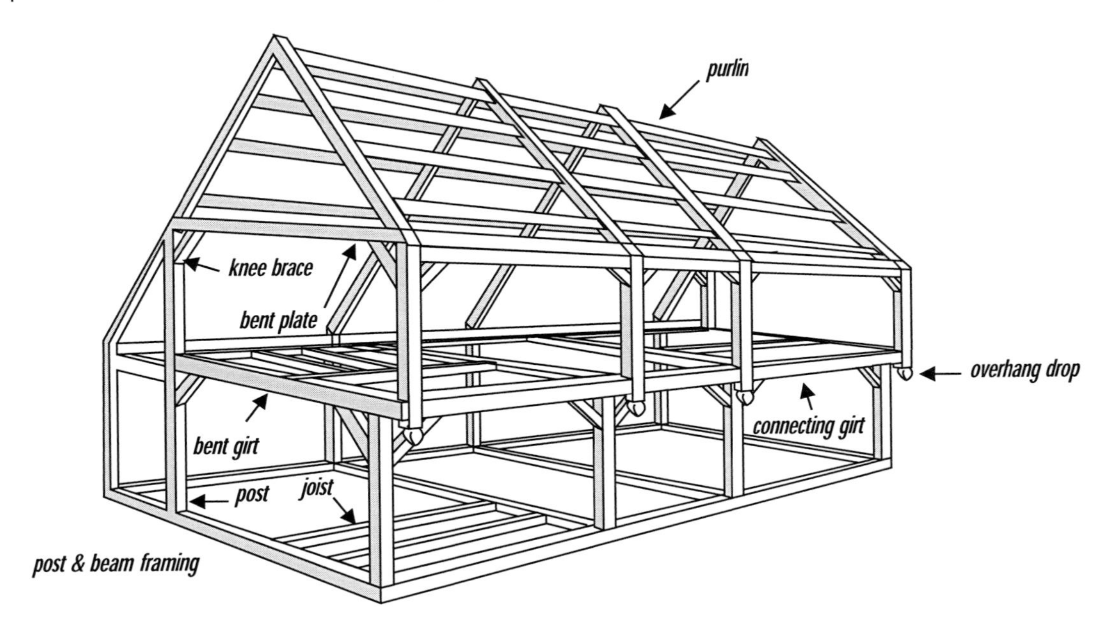

Pine connecting members between bents. Includes floor joists, roof purlins, diagonal braces, and beams. Per sf of area beneath roof. Measure only one floor.

	Craft@Hrs	Unit	Material	Labor	Equip.	Total
between single-story bents & trusses	4B@1.11	sf	8.03	81.00	—	89.03
remove	1D@.214	sf	—	10.10	—	10.10
between two-story bents & trusses	4B@1.85	sf	12.00	135.00	—	147.00
remove	1D@.286	sf	—	13.50	—	13.50

Oak post & beam bent with truss. Includes beams, posts, and installation labor. All connections between beams are made from cut joints. Joints are held together with pegs that are often hand-cut from the beam material. Includes glue as needed. Does not include connecting members or their joinery. This price should not be used for similar systems made of posts and beams that are joined with butt and miter joints often held together by steel brackets. With oak posts and beams, per SF of bent.

	Craft@Hrs	Unit	Material	Labor	Equip.	Total
Replace single-story bent with						
simple truss	4B@.527	sf	11.20	38.50	.67	50.37
king post truss	4B@.569	sf	12.50	41.50	.67	54.67
king post & struts truss	4B@.618	sf	14.00	45.10	.67	59.77
queen post truss	4B@.609	sf	13.00	44.50	.67	58.17
hammer beam truss	4B@1.05	sf	13.90	76.70	.67	91.27
scissors truss	4B@.989	sf	14.50	72.20	.67	87.37

	Craft@Hrs	Unit	Material	Labor	Equip.	Total
Remove, single-story bent with truss, all types	1D@.071	sf	—	3.35	—	3.35
Replace two-story bent with						
simple truss	4B@.571	sf	10.40	41.70	.67	52.77
king post truss	4B@.620	sf	11.20	45.30	.67	57.17
king post & struts truss	4B@.679	sf	12.00	49.60	.67	62.27
queen post truss	4B@.671	sf	11.60	49.00	.67	61.27
hammer beam truss	4B@1.25	sf	11.90	91.30	.67	103.87
scissors truss	4B@1.04	sf	12.40	75.90	.67	88.97
Remove, two-story bent with truss, all types	1D@.071	sf	—	3.35	—	3.35
Add for overhang drop in two-story bent	4B@20.0	ea	562.00	1,460.00	.67	2,022.67

Oak connecting members between bents. Includes floor joists, roof purlins, diagonal braces, and beams. Per sf of area beneath roof. Measure only one floor.

	Craft@Hrs	Unit	Material	Labor	Equip.	Total
between single-story bents & trusses	4B@1.11	sf	19.60	81.00	—	100.60
remove	1D@.214	sf	—	10.10	—	10.10
between two-story bents & trusses	4B@1.85	sf	29.10	135.00	—	164.10
remove	1D@.286	sf	—	13.50	—	13.50
Minimum charge for post & beam framing work	4B@20.0	ea	525.00	1,460.00	885.00	2,870.00

sill corner half-lap

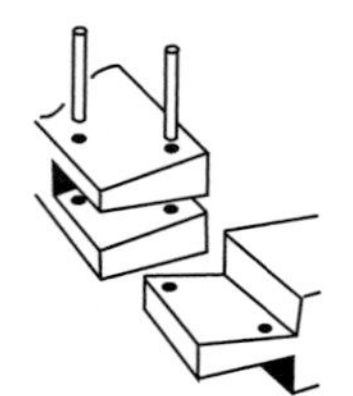

sill corner dovetail

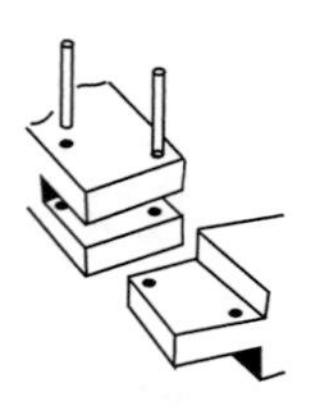

sill corner tongue & fork

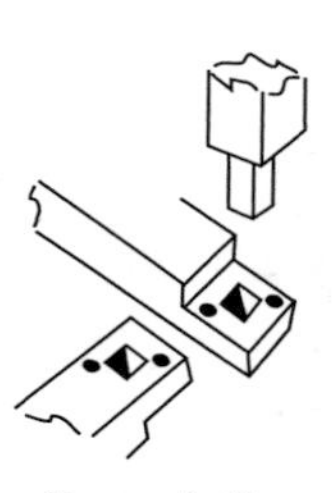

sill corner half-lap with through tenon

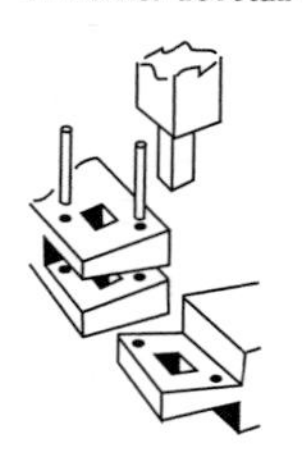

sill corner dovetail with through tenon

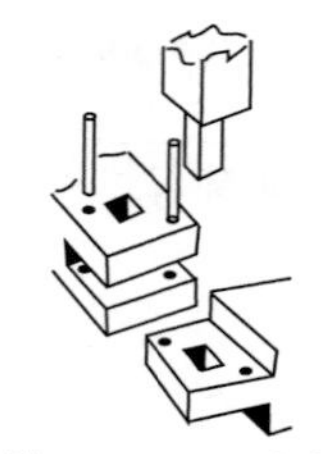

sill corner tongue & fork with through tenon

Post & beam joints. A labor-only cost to cut the described joint. Does not include the beams, bracing of existing systems, or equipment costs to place replacement posts or beams. Includes cutting, drilling, and joining with pegs.

	Craft@Hrs	Unit	Material	Labor	Total
Sill corner joints, replace					
half-lap joint	2B@1.10	ea	—	93.70	93.70
half-lap joint with through tenon	2B@2.77	ea	—	236.00	236.00
dovetail joint	2B@1.99	ea	—	170.00	170.00
dovetail joint with through tenon	2B@3.86	ea	—	329.00	329.00
tongue and fork joint	2B@1.79	ea	—	153.00	153.00
tongue and fork joint with through tenon	2B@3.44	ea	—	293.00	293.00

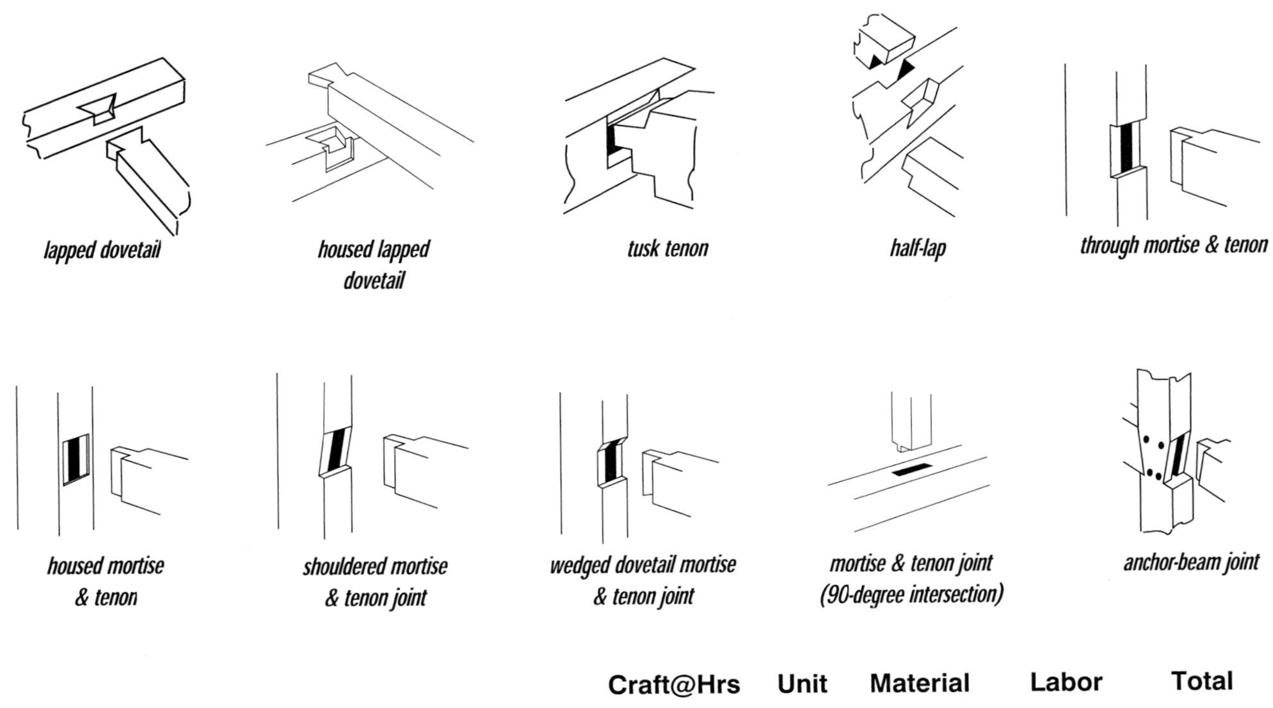

	Craft@Hrs	Unit	Material	Labor	Total
Joints for connecting two timbers at 90-degree angles, replace					
lapped dovetail joint	2B@2.05	ea	—	175.00	175.00
housed lapped dovetail joint	2B@2.55	ea	—	217.00	217.00
tusk tenon joint	2B@2.77	ea	—	236.00	236.00
half-lap joint	2B@1.08	ea	—	92.00	92.00
through mortise & tenon joint with shoulders	2B@2.26	ea	—	193.00	193.00
housed mortise & tenon joint	2B@2.37	ea	—	202.00	202.00
shouldered mortise & tenon joint	2B@2.34	ea	—	199.00	199.00
wedged dovetail mortise & tenon joint	2B@2.37	ea	—	202.00	202.00
mortise & tenon joint (90-degree intersection)	2B@1.25	ea	—	107.00	107.00
anchor-beam joint	2B@2.93	ea	—	250.00	250.00

framed overhang

through half-lap

mortise & tenon knee brace

collar tie lapped half dovetail

	Craft@Hrs	Unit	Material	Labor	Total
Other post & beam joints, replace only					
framed overhang joint	2B@10.0	ea	—	852.00	852.00
through half-lap joint	2B@1.20	ea	—	102.00	102.00
mortise & tenon knee brace joint	2B@1.56	ea	—	133.00	133.00
collar tie lapped half dovetail joint	2B@1.96	ea	—	167.00	167.00

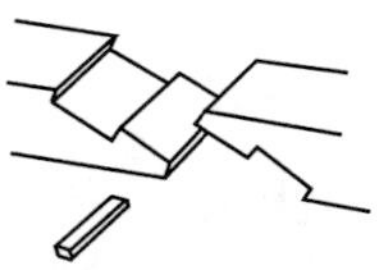

stopped splayed scarf 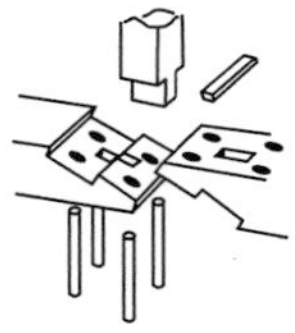 *stopped splayed scarf with through tenon* 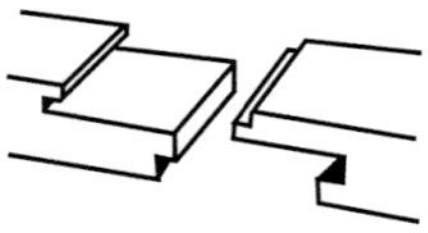 *bladed scarf* 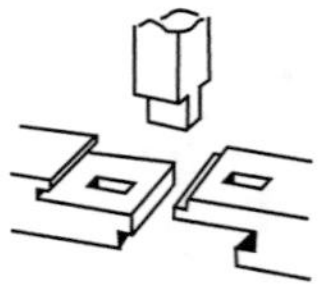 *bladed scarf with through tenon* 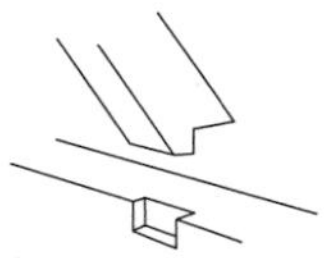 *rafter foot housed bird's mouth*

	Craft@Hrs	Unit	Material	Labor	Total
Joints for splicing together two level beams, replace only					
stopped splayed scarf joint	2B@3.22	ea	—	274.00	274.00
stopped splayed scarf joint with through tenon	2B@5.00	ea	—	426.00	426.00
bladed scarf joint	2B@2.45	ea	—	209.00	209.00
bladed scarf joint with through tenon	2B@4.35	ea	—	371.00	371.00

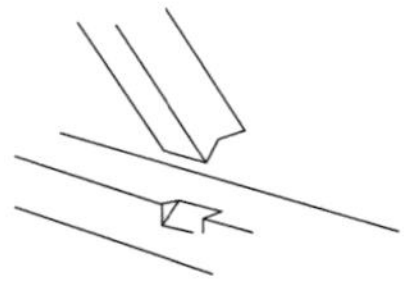

rafter foot beveled shoulder bird's mouth 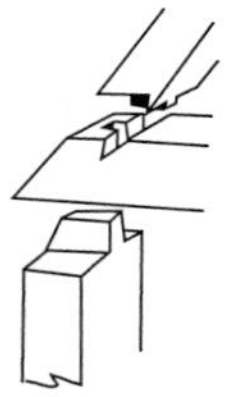 *rafter foot bird's mouth, purlin, and post* 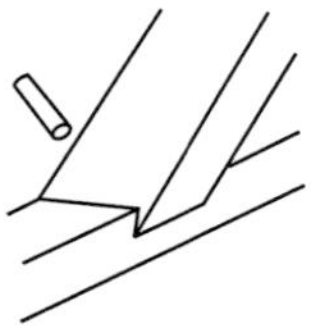 *rafter foot bird's mouth* 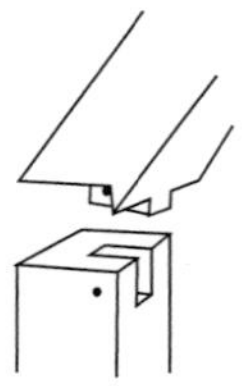 *rafter foot bird's mouth with tenon*  *rafter peak tongue & fork*

	Craft@Hrs	Unit	Material	Labor	Total
Rafter joints, replace only					
rafter foot housed bird's mouth joint	2B@1.39	ea	—	118.00	118.00
rafter foot beveled shoulder bird's mouth joint	2B@1.30	ea	—	111.00	111.00
rafter foot bird's mouth, purlin, and post joint	2B@3.13	ea	—	267.00	267.00
rafter foot bird's mouth joint	2B@1.00	ea	—	85.20	85.20
rafter foot bird's mouth joint with tenon	2B@2.86	ea	—	244.00	244.00
rafter peak tongue & fork joint	2B@1.79	ea	—	153.00	153.00

Exterior deck joist system. Includes 5% waste. Complete joist system. Does not include decking, railing, or stairs. Includes footings, excavation for footings, column stirrups and connectors, posts, and beams on outside edge of deck, connection with joist hangers against house, and joists.

	Craft@Hrs	Unit	Material	Labor	Total
Replace exterior deck joist system					
2" x 6" joists	6C@.067	sf	3.14	3.95	7.09
2" x 8" joists	6C@.071	sf	3.51	4.19	7.70
2" x 10" joists	6C@.077	sf	4.42	4.54	8.96
2" x 12" joists	6C@.083	sf	4.85	4.90	9.75
Remove, all sizes	1D@.013	sf	—	.61	.61

Exterior decking. Includes 2" x 4" decking lumber, rust-resistant screws or nails, and installation labor. Deduct **8%** for 5/4 treated wood decking (decking that is 1-1/4" thick instead of 1-1/2" thick). Includes 5% waste.

	Craft@Hrs	Unit	Material	Labor	Total
2" x 4" redwood exterior decking, replace					
construction common	6C@.040	sf	6.15	2.36	8.51
construction heart	6C@.040	sf	7.10	2.36	9.46
2" x 6" redwood exterior deck, replace					
construction common	6C@.037	sf	4.96	2.18	7.14
construction heart	6C@.037	sf	5.73	2.18	7.91
2" x 8" redwood exterior deck, replace					
construction common	6C@.033	sf	5.29	1.95	7.24
construction heart	6C@.033	sf	6.13	1.95	8.08
Treated exterior wood decking, replace					
2" x 4"	6C@.040	sf	5.29	2.36	7.65
2" x 6"	6C@.037	sf	5.08	2.18	7.26
2" x 8"	6C@.033	sf	5.03	1.95	6.98
Remove, all sizes and types	1D@.026	sf	—	1.23	1.23

Exterior deck railing. Includes 2" x 6" top rail, often routed with a bull-nose, 2" x 2" balusters placed approximately 4" on center, fasteners, and installation labor. Does not include posts. Also includes 5% waste.

	Craft@Hrs	Unit	Material	Labor	Total
Replace exterior deck railing					
redwood construction common	6C@.275	lf	11.10	16.20	27.30
redwood construction heart	6C@.275	lf	12.70	16.20	28.90
treated wood	6C@.275	lf	10.80	16.20	27.00
Remove, all types	1D@.034	lf	—	1.60	1.60

Exterior deck stairs with railing, per step. Includes up to three stringers made from 2" x 12" lumber. Each step has up to three 2" x 4" treads. Balusters are 2" x 2" placed approximately 4" on center. Balusters attach to side stringers on the bottom and a top rail at the top. Top rail is 2" x 6" often routed with a bullnose edge. Does not include foundation system for bottom of stairs. Includes 5% waste.

	Craft@Hrs	Unit	Material	Labor	Total
2' 6" wide deck stairs, replace					
construction common redwood	6C@.540	ea	46.40	31.90	78.30
construction heart redwood	6C@.540	ea	49.90	31.90	81.80
treated wood	6C@.540	ea	46.00	31.90	77.90
3' wide exterior deck stairs, replace					
construction common redwood	6C@.587	ea	49.60	34.60	84.20
construction heart redwood	6C@.587	ea	53.00	34.60	87.60
treated wood	6C@.587	ea	48.60	34.60	83.20
Remove, all widths and types	1D@.061	ea	—	2.88	2.88

exterior deck stairs

Exterior deck landing. Includes 2" x 10" joist system with 2" x 6" board decking, beams, posts, stirrups, footings, excavation for footings, all fasteners, and installation labor. Includes 5% waste.

	Craft@Hrs	Unit	Material	Labor	Total
Replace exterior deck landing					
construction common decking	6C@.074	sf	16.10	4.37	20.47
construction heart decking	6C@.074	sf	18.00	4.37	22.37
treated wood	6C@.074	sf	15.90	4.37	20.27
Remove, all types	1D@.015	sf	—	.71	.71

	Craft@Hrs	Unit	Material	Labor	Total

Furring strips. Includes 1" x 2" furring strips, construction adhesive, case hardened nails or powder-actuated fasteners, and installation labor. Furring strips placed as fireblocking at top and bottom of walls are included. Fireblocking furring strips inside the wall for taller walls are also included. Also includes 5% waste.

	Craft@Hrs	Unit	Material	Labor	Total
Replace 1" x 2" furring strips applied to wood					
12" on center	6C@.015	sf	.27	.89	1.16
16" on center	6C@.013	sf	.22	.77	.99
24" on center	6C@.011	sf	.16	.65	.81
Remove 1" x 2" furring strips applied to wood	1D@.011	sf	—	.52	.52
Replace 1" x 2" furring strips applied to masonry or concrete					
12" on center	6C@.020	sf	.27	1.18	1.45
16" on center	6C@.016	sf	.22	.94	1.16
24" on center	6C@.014	sf	.16	.83	.99
Remove 1" x 2" furring strips applied to masonry or concrete	1D@.012	sf	—	.57	.57
Replace 2" x 2" furring strips applied to wood					
12" on center	6C@.015	sf	.68	.89	1.57
16" on center	6C@.013	sf	.50	.77	1.27
24" on center	6C@.011	sf	.40	.65	1.05
Remove 2" x 2" furring strips applied to wood	1D@.011	sf	—	.52	.52
Replace 2" x 2" furring strips applied to masonry or concrete					
12" on center	6C@.020	sf	.68	1.18	1.86
16" on center	6C@.016	sf	.50	.94	1.44
24" on center	6C@.014	sf	.40	.83	1.23
Remove 2" x 2" furring strips applied to masonry or concrete	1D@.012	sf	—	.57	.57

Joist system, per square foot. Includes 5% waste. All joist systems include joists, rim joists, blocking, and steel X braces at center span. (Except 2" x 4" joists which include solid blocking at center of span.)

	Craft@Hrs	Unit	Material	Labor	Total
2" x 4" joist system, replace					
12" on center	6C@.018	sf	1.24	1.06	2.30
16" on center	6C@.014	sf	1.03	.83	1.86
24" on center	6C@.011	sf	.72	.65	1.37
2" x 6" joist system, replace					
12" on center	6C@.025	sf	1.81	1.48	3.29
16" on center	6C@.019	sf	1.45	1.12	2.57
24" on center	6C@.013	sf	1.07	.77	1.84
2" x 8" joist system, replace					
12" on center	6C@.029	sf	2.24	1.71	3.95
16" on center	6C@.024	sf	1.75	1.42	3.17
24" on center	6C@.018	sf	1.25	1.06	2.31
2" x 10" joist system, replace					
12" on center	6C@.036	sf	3.35	2.12	5.47
replace, 16" on center	6C@.029	sf	2.63	1.71	4.34
replace, 24" on center	6C@.022	sf	1.81	1.30	3.11
2" x 12" joist system, replace					
12" on center	6C@.045	sf	3.80	2.66	6.46
16" on center	6C@.033	sf	2.99	1.95	4.94
24" on center	6C@.029	sf	2.08	1.71	3.79
Remove joist system, all sizes	1D@.014	sf	—	.66	.66

	Craft@Hrs	Unit	Material	Labor	Total
Joist system, per board foot.					
Replace, all sizes	6C@.021	bf	1.78	1.24	3.02
Remove, all sizes	1D@.006	bf	—	.28	.28

Laminated lumber I joist. Includes 3% waste. The lf price includes I joist, rim joists, blocking, and steel X braces at center span.

	Craft@Hrs	Unit	Material	Labor	Total
1-3/4" flange laminated lumber I joist					
Replace					
9-1/2" deep	6C@.034	lf	3.38	2.01	5.39
11-7/8" deep	6C@.036	lf	4.09	2.12	6.21
14" deep	6C@.037	lf	5.01	2.18	7.19
2-5/16" flange laminated lumber I joist					
Replace					
11-7/8" deep	6C@.036	lf	5.80	2.12	7.92
14" deep	6C@.038	lf	7.10	2.24	9.34
16" deep	6C@.039	lf	8.38	2.30	10.68
Remove I joist, all flange sizes and depths	1D@.014	sf	—	.66	.66

Floor truss. The lf price includes cat walks, diagonal braces, fasteners and rim materials.

	Craft@Hrs	Unit	Material	Labor	Total
Replace					
for typical load	6C@.091	lf	5.94	5.37	11.31
for heavy load	6C@.094	lf	6.40	5.55	11.95
Remove, all types	1D@.014	sf	—	.66	.66

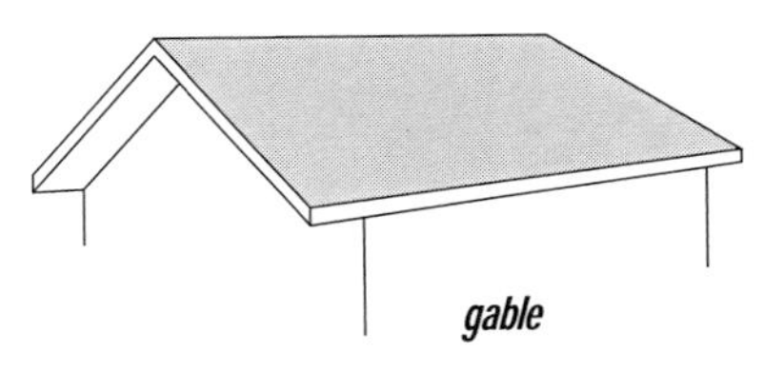

Hand-framed gable roof. Includes 5% waste. Measure the sf of area beneath roof. (The sf of bottom floor plus sf under overhangs.) Do not measure the sf of roof. Includes rafters, collar ties, a ridge board that is 2" wider than the rafter (except for 2" x 12" rafters, which have a ridge made from a 1-1/2" x 11-7/8" micro-laminated beam), rough fascia board, blocking, and wall fasteners. If roof has a hip or a Dutch hip, see page 371. There is no additional charge for a Dutch gable.

	Craft@Hrs	Unit	Material	Labor	Total
2" x 4" rafters for gable roof					
4/12 slope, replace	6C@.034	sf	.87	2.01	2.88
remove	1D@.016	sf	—	.76	.76
6/12 slope, replace	6C@.036	sf	.92	2.12	3.04
remove	1D@.017	sf	—	.80	.80
8/12 slope, replace	6C@.040	sf	1.01	2.36	3.37
remove	1D@.019	sf	—	.90	.90
10/12 slope, replace	6C@.044	sf	1.05	2.60	3.65
remove	1D@.024	sf	—	1.13	1.13
12/12 slope, replace	6C@.049	sf	1.11	2.89	4.00
remove	1D@.032	sf	—	1.51	1.51
14/12 slope, replace	6C@.054	sf	1.17	3.19	4.36
remove	1D@.038	sf	—	1.79	1.79
16/12 slope, replace	6C@.060	sf	1.25	3.54	4.79
remove	1D@.050	sf	—	2.36	2.36
18/12 slope, replace	6C@.067	sf	1.35	3.95	5.30
remove	1D@.077	sf	—	3.63	3.63

	Craft@Hrs	Unit	Material	Labor	Total
2" x 6" rafters for gable roof					
4/12 slope, replace	6C@.035	sf	1.25	2.07	3.32
remove	1D@.016	sf	—	.76	.76
6/12 slope, replace	6C@.038	sf	1.30	2.24	3.54
remove	1D@.017	sf	—	.80	.80
8/12 slope, replace	6C@.042	sf	1.39	2.48	3.87
remove	1D@.019	sf	—	.90	.90
10/12 slope, replace	6C@.046	sf	1.47	2.71	4.18
remove	1D@.024	sf	—	1.13	1.13
12/12 slope, replace	6C@.051	sf	1.57	3.01	4.58
remove	1D@.032	sf	—	1.51	1.51
14/12 slope, replace	6C@.056	sf	1.66	3.30	4.96
remove	1D@.038	sf	—	1.79	1.79
16/12 slope, replace	6C@.062	sf	1.79	3.66	5.45
remove	1D@.050	sf	—	2.36	2.36
18/12 slope, replace	6C@.070	sf	1.91	4.13	6.04
remove	1D@.077	sf	—	3.63	3.63
2" x 8" rafters for gable roof					
4/12 slope, replace	6C@.036	sf	1.79	2.12	3.91
remove	1D@.016	sf	—	.76	.76
6/12 slope, replace	6C@.039	sf	1.88	2.30	4.18
remove	1D@.017	sf	—	.80	.80
8/12 slope, replace	6C@.043	sf	1.98	2.54	4.52
remove	1D@.019	sf	—	.90	.90
10/12 slope, replace	6C@.047	sf	2.12	2.77	4.89
remove	1D@.024	sf	—	1.13	1.13
12/12 slope, replace	6C@.052	sf	2.28	3.07	5.35
remove	1D@.032	sf	—	1.51	1.51
14/12 slope, replace	6C@.058	sf	2.42	3.42	5.84
remove	1D@.038	sf	—	1.79	1.79
16/12 slope, replace	6C@.064	sf	2.59	3.78	6.37
remove	1D@.050	sf	—	2.36	2.36
18/12 slop, replace	6C@.072	sf	2.73	4.25	6.98
remove	1D@.077	sf	—	3.63	3.63
2" x 10" rafters for gable roof					
4/12 slope, replace	6C@.037	sf	2.34	2.18	4.52
remove	1D@.016	sf	—	.76	.76
6/12 slope, replace	6C@.040	sf	2.43	2.36	4.79
remove	1D@.017	sf	—	.80	.80
8/12 slope, replace	6C@.044	sf	2.59	2.60	5.19
remove	1D@.019	sf	—	.90	.90
10/12 slope, replace	6C@.049	sf	2.73	2.89	5.62
remove	1D@.024	sf	—	1.13	1.13
12/12 slope, replace	6C@.054	sf	2.95	3.19	6.14
remove	1D@.032	sf	—	1.51	1.51
14/12 slope, replace	6C@.059	sf	3.16	3.48	6.64
remove	1D@.038	sf	—	1.79	1.79
16/12 slope, replace	6C@.066	sf	3.39	3.89	7.28
remove	1D@.050	sf	—	2.36	2.36
18/12 slope, replace	6C@.075	sf	3.66	4.43	8.09
remove	1D@.077	sf	—	3.63	3.63

	Craft@Hrs	Unit	Material	Labor	Total
2" x 12" rafters for gable roof					
4/12 slope, replace	6C@.038	sf	3.16	2.24	5.40
remove	1D@.016	sf	—	.76	.76
6/12 slope, replace	6C@.042	sf	3.30	2.48	5.78
remove	1D@.017	sf	—	.80	.80
8/12 slope, replace	6C@.046	sf	3.49	2.71	6.20
remove	1D@.019	sf	—	.90	.90
10/12 slope, replace	6C@.050	sf	3.68	2.95	6.63
remove	1D@.024	sf	—	1.13	1.13
12/12 slope, replace	6C@.056	sf	3.90	3.30	7.20
remove	1D@.032	sf	—	1.51	1.51
14/12 slope, replace	6C@.061	sf	4.13	3.60	7.73
remove	1D@.038	sf	—	1.79	1.79
16/12 slope, replace	6C@.068	sf	4.39	4.01	8.40
remove	1D@.050	sf	—	2.36	2.36
18/12 slope, replace	6C@.077	sf	4.68	4.54	9.22
remove	1D@.077	sf	—	3.63	3.63

Additional costs for hand-framed roofs.

	Craft@Hrs	Unit	Material	Labor	Total
for each lf of valley	6C@.786	lf	18.80	46.40	65.20
for hip roof, per lf of width	6C@.212	lf	2.95	12.50	15.45
for Dutch-hip, per lf of width	6C@.229	lf	3.26	13.50	16.76

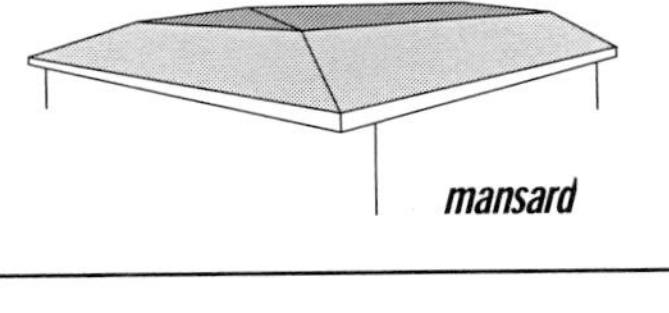

Hand-framed mansard roof, per cubic foot of roof. Includes 7% waste. Roof area equals roof system width times height times length. Includes rafters, collar ties, ridge boards, hip rafters, rough fascia board, blocking, and wall fasteners.

	Craft@Hrs	Unit	Material	Labor	Total
Replace mansard roof					
2" x 4" rafters	6C@.005	cf	.12	.30	.42
2" x 6" rafters	6C@.005	cf	.21	.30	.51
2" x 8" rafters	6C@.005	cf	.28	.30	.58
2" x 10" rafters	6C@.005	cf	.36	.30	.66
2" x 12" rafters	6C@.006	cf	.41	.35	.76
Remove, all sizes	1D@.002	cf	—	.09	.09
Add for each lf of valley	6C@1.08	lf	22.70	63.70	86.40

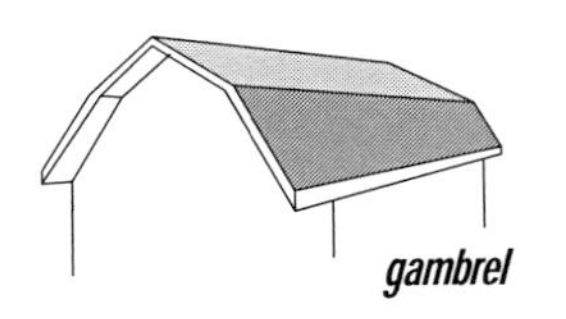

Hand-framed gambrel roof, per cubic foot of roof. Includes 7% waste. Roof area equals roof system width times height times length. Includes rafters, collar ties, ridge boards, valley rafters, rough fascia board, blocking, and wall fasteners. Does not include wall that must be built at the intersection of the top and bottom sloping roof sections.

	Craft@Hrs	Unit	Material	Labor	Total
Replace gambrel roof					
2" x 4" rafters	6C@.004	cf	.17	.24	.41
2" x 6" rafters	6C@.005	cf	.27	.30	.57
2" x 8" rafters	6C@.005	cf	.37	.30	.67
2" x 10" rafters	6C@.005	cf	.47	.30	.77
2" x 12" rafters	6C@.006	cf	.52	.35	.87
Remove, all sizes	1D@.002	cf	—	.09	.09
Add for each lf of valley	6C@1.14	lf	23.50	67.30	90.80

Hand-framed roof, per board foot. Includes 5% waste.

	Craft@Hrs	Unit	Material	Labor	Total
Replace hand-framed roof					
gable roof	6C@.023	bf	1.49	1.36	2.85
hip roof	6C@.023	bf	1.64	1.36	3.00
Dutch-hip roof	6C@.024	bf	1.66	1.42	3.08
mansard roof	6C@.026	bf	1.81	1.53	3.34
gambrel roof	6C@.025	bf	1.92	1.48	3.40
Remove, all sizes	1D@.007	bf	—	.33	.33

Minimum charge.

	Craft@Hrs	Unit	Material	Labor	Total
for roof framing	6C@3.55	ea	68.50	209.00	277.50

Hand-framed shed dormer. Includes 7% waste. Complete dormer including all side walls and roof framing. Length x width of dormer.

	Craft@Hrs	Unit	Material	Labor	Total
4/12 slope, replace	6C@.045	sf	1.47	2.66	4.13
remove	1D@.017	sf	—	.80	.80
6/12 slope, replace	6C@.047	sf	1.54	2.77	4.31
remove	1D@.018	sf	—	.85	.85
8/12 slope, replace	6C@.049	sf	1.61	2.89	4.50
remove	1D@.020	sf	—	.94	.94
10/12 slope, replace	6C@.054	sf	1.71	3.19	4.90
remove	1D@.025	sf	—	1.18	1.18
12/12 slope, replace	6C@.060	sf	1.88	3.54	5.42
remove	1D@.033	sf	—	1.56	1.56
14/12 slope, replace	6C@.062	sf	1.98	3.66	5.64
remove	1D@.040	sf	—	1.89	1.89
16/12 slope, replace	6C@.064	sf	2.14	3.78	5.92
remove	1D@.053	sf	—	2.50	2.50
18/12 slope, replace	6C@.067	sf	2.28	3.95	6.23
remove	1D@.081	sf	—	3.82	3.82

Hand-framed gable dormer. Includes framing lumber, framing nails, and installation. Installation includes all material and labor needed to place the dormer including ridge boards, rafters, jack rafters, rough fascia boards, side and front walls (when needed), roof sheathing, headers, and so on. Does not include modifications to the roof the dormer ties into, finish soffit and fascia, windows, or finish wall materials. Includes 7% waste. Measure length x width of dormer.

	Craft@Hrs	Unit	Material	Labor	Total
4/12 slope, replace	6C@.047	sf	1.62	2.77	4.39
remove	1D@.017	sf	—	.80	.80
6/12 slope, replace	6C@.049	sf	1.70	2.89	4.59
remove	1D@.018	sf	—	.85	.85
8/12 slope, replace	6C@.051	sf	1.81	3.01	4.82
remove	1D@.020	sf	—	.94	.94
10/12 slope, replace	6C@.056	sf	1.92	3.30	5.22
remove	1D@.025	sf	—	1.18	1.18
12/12 slope, replace	6C@.062	sf	2.11	3.66	5.77
remove	1D@.033	sf	—	1.56	1.56
14/12 slope, replace	6C@.065	sf	2.23	3.84	6.07
remove	1D@.040	sf	—	1.89	1.89
16/12 slope, replace	6C@.067	sf	2.38	3.95	6.33
remove	1D@.053	sf	—	2.50	2.50
18/12 slope, replace	6C@.070	sf	2.56	4.13	6.69
remove	1D@.081	sf	—	3.82	3.82

Hand-framed hip dormer. Includes framing lumber, framing nails, and installation. Installation includes all material and labor needed to place the dormer including ridge boards, rafters, hip rafters, jack rafters, rough fascia boards, side and front walls (when needed), roof sheathing, headers, and so on. Does not include modifications to the roof the dormer ties into, finish soffit and fascia, windows, or finish wall materials. Includes 7% waste. Measure length x width of dormer.

	Craft@Hrs	Unit	Material	Labor	Total
4/12 slope, replace	6C@.049	sf	1.68	2.89	4.57
remove	1D@.017	sf	—	.80	.80
6/12 slope, replace	6C@.051	sf	1.76	3.01	4.77
remove	1D@.018	sf	—	.85	.85
8/12 slope, replace	6C@.053	sf	1.88	3.13	5.01
remove	1D@.020	sf	—	.94	.94
10/12 slope, replace	6C@.058	sf	2.01	3.42	5.43
remove	1D@.025	sf	—	1.18	1.18
12/12 slope, replace	6C@.065	sf	2.15	3.84	5.99
remove	1D@.033	sf	—	1.56	1.56
14/12 slope, replace	6C@.067	sf	2.29	3.95	6.24
remove	1D@.040	sf	—	1.89	1.89
16/12 slope, replace	6C@.070	sf	2.69	4.13	6.82
remove	1D@.053	sf	—	2.50	2.50
18/12 slope, replace	6C@.073	sf	2.61	4.31	6.92
remove	1D@.081	sf	—	3.82	3.82

Hand-framed Dutch-hip dormer. Includes framing lumber, framing nails, and installation. Installation includes all material and labor needed to place the dormer including ridge boards, rafters, hip rafters, jack rafters, rough fascia boards, side and front walls (when needed), roof sheathing, headers, and so on. Does not include modifications to the roof the dormer ties into, finish soffit and fascia, windows, or finish wall materials. Includes 7% waste. Measure length x width of dormer.

	Craft@Hrs	Unit	Material	Labor	Total
4/12 slope, replace	6C@.051	sf	1.74	3.01	4.75
remove	1D@.017	sf	—	.80	.80
6/12 slope, replace	6C@.052	sf	1.81	3.07	4.88
remove	1D@.018	sf	—	.85	.85
8/12 slope, replace	6C@.054	sf	1.92	3.19	5.11
remove	1D@.020	sf	—	.94	.94
10/12 slope, replace	6C@.060	sf	2.06	3.54	5.60
remove	1D@.025	sf	—	1.18	1.18
12/12 slope, replace	6C@.067	sf	2.23	3.95	6.18
remove	1D@.033	sf	—	1.56	1.56
14/12 slope, replace	6C@.069	sf	2.35	4.07	6.42
remove	1D@.040	sf	—	1.89	1.89
16/12 slope, replace	6C@.072	sf	2.51	4.25	6.76
remove	1D@.053	sf	—	2.50	2.50
18/12 slope, replace	6C@.074	sf	2.67	4.37	7.04
remove	1D@.078	sf	—	3.68	3.68

Hand-framed gambrel dormer. Includes framing lumber, framing nails, and installation. Installation includes all material and labor needed to place the dormer including ridge boards, rafters, jack rafters, rough fascia boards, side and front walls (when needed), roof sheathing, headers, and so on. Does not include modifications to the roof the dormer ties into, finish soffit and fascia, windows, or finish wall materials. Includes 7% waste. Measure length x width of dormer.

	Craft@Hrs	Unit	Material	Labor	Total
replace	6C@.059	sf	1.79	3.48	5.27
remove	1D@.017	sf	—	.80	.80

Hand-framed roof dormer, per board foot. Includes framing lumber, framing nails, and installation. Installation includes all material and labor needed to place the dormer including ridge boards, rafters, hip rafters, jack rafters, rough fascia boards, studs in side and front walls (when needed), headers, and so on. Does not include roof sheathing, wall sheathing, modifications to the roof the dormer ties into, finish soffit and fascia, windows, or finish wall materials. Includes 7% waste.

	Craft@Hrs	Unit	Material	Labor	Total
Shed dormer, replace	6C@.034	bf	2.41	2.01	4.42
remove shed dormer	1D@.010	bf	—	.47	.47
Gable dormer, replace	6C@.035	bf	2.54	2.07	4.61
remove gable dormer	1D@.017	bf	—	.80	.80
Hip dormer, replace	6C@.036	bf	2.61	2.12	4.73
remove hip dormer	1D@.018	bf	—	.85	.85
Dutch hip dormer, replace	6C@.038	bf	2.66	2.24	4.90
remove Dutch hip dormer	1D@.019	bf	—	.90	.90
Gambrel dormer, replace	6C@.037	bf	2.73	2.18	4.91
remove gambrel dormer	1D@.023	bf	—	1.09	1.09

Roof truss, per linear foot. Add **10%** for treated trusses. Per lf of truss bottom chord. The lf price includes cat walks, diagonal braces, fasteners and rough fascia.

	Craft@Hrs	Unit	Material	Labor	Equip.	Total
Replace roof truss						
flat roof	6C@.023	lf	5.94	1.36	.28	7.58
4/12 slope	6C@.023	lf	6.24	1.36	.28	7.88
6/12 slope	6C@.023	lf	6.71	1.36	.28	8.35
8/12 slope	6C@.024	lf	9.07	1.42	.28	10.77
10/12 slope	6C@.024	lf	10.50	1.42	.28	12.20
12/12 slope	6C@.024	lf	12.20	1.42	.28	13.90
14/12 slope	6C@.025	lf	12.90	1.48	.28	14.66
16/12 slope	6C@.026	lf	13.90	1.53	.28	15.71
17/12 slope	6C@.028	lf	14.90	1.65	.28	16.83
Remove roof truss, all slopes	1D@.042	lf	—	1.98	—	1.98

Roof truss for heavy load, per linear foot. For slate or tile roofing. Add **10%** for treated trusses. Per lf of truss bottom chord. The lf price includes cat walks, diagonal braces, fasteners and rough fascia.

	Craft@Hrs	Unit	Material	Labor	Equip.	Total
Replace						
flat roof	6C@.023	lf	6.30	1.36	.28	7.94
4/12 slope	6C@.023	lf	6.71	1.36	.28	8.35
6/12 slope	6C@.023	lf	7.13	1.36	.28	8.77
8/12 slope	6C@.024	lf	9.52	1.42	.28	11.22
10/12 slope	6C@.024	lf	11.00	1.42	.28	12.70
12/12 slope	6C@.024	lf	12.60	1.42	.28	14.30
14/12 slope	6C@.025	lf	13.50	1.48	.28	15.26
16/12 slope	6C@.026	lf	14.40	1.53	.28	16.21
17/12 slope	6C@.028	lf	15.50	1.65	.28	17.43
Remove roof truss, all slopes	1D@.042	lf	—	1.98	—	1.98

Roof truss system, per square foot. Add **10%** for treated trusses. Per sf of area underneath the roof. For roofs with asphalt, metal, or wood roofing. Includes cat walks, diagonal braces, fasteners and rough fascia.

	Craft@Hrs	Unit	Material	Labor	Equip.	Total
4/12 slope, replace	6C@.012	sf	3.62	.71	.16	4.49
remove	1D@.017	sf	—	.80	—	.80
6/12 slope, replace	6C@.012	sf	3.91	.71	.16	4.78
remove	1D@.018	sf	—	.85	—	.85
8/12 slope, replace	6C@.012	sf	5.26	.71	.16	6.13
remove	1D@.021	sf	—	.99	—	.99
10/12 slope, replace	6C@.013	sf	6.06	.77	.16	6.99
remove	1D@.025	sf	—	1.18	—	1.18
12/12 slope, replace	6C@.013	sf	7.02	.77	.16	7.95
remove	1D@.034	sf	—	1.60	—	1.60
14/12 slope, replace	6C@.013	sf	7.48	.77	.16	8.41
remove	1D@.040	sf	—	1.89	—	1.89
16/12 slope, replace	6C@.013	sf	8.01	.77	.16	8.94
remove	1D@.053	sf	—	2.50	—	2.50
17/12 slope, replace	6C@.013	sf	8.69	.77	.16	9.62
remove	1D@.082	sf	—	3.87	—	3.87

Roof truss system for heavy load, per square foot. Add **10%** for treated trusses. Per sf of area underneath the roof. For roofs with slate, tile or other heavy roofing. Includes cat walks, diagonal braces, fasteners and rough fascia.

	Craft@Hrs	Unit	Material	Labor	Equip.	Total
4/12 slope, replace	6C@.012	sf	3.91	.71	.16	4.78
remove	1D@.017	sf	—	.80	—	.80
6/12 slope, replace	6C@.012	sf	4.14	.71	.16	5.01
remove	1D@.018	sf	—	.85	—	.85
8/12 slope, replace	6C@.012	sf	5.51	.71	.16	6.38
remove	1D@.021	sf	—	.99	—	.99
10/12 slope, replace	6C@.013	sf	6.33	.77	.16	7.26
remove	1D@.025	sf	—	1.18	—	1.18
12/12 slope, replace	6C@.013	sf	7.34	.77	.16	8.27
remove	1D@.034	sf	—	1.60	—	1.60
14/12 slope, replace	6C@.013	sf	7.78	.77	.16	8.71
remove	1D@.040	sf	—	1.89	—	1.89
16/12 slope, replace	6C@.013	sf	8.36	.77	.16	9.29
remove	1D@.053	sf	—	2.50	—	2.50
17/12 slope, replace	6C@.013	sf	9.01	.77	.16	9.94
remove	1D@.082	sf	—	3.87	—	3.87

Add for each lf of valley in truss roof, "California" fill. Includes materials and labor for "California" fill between truss roofs with ridge board and jack rafters. Does not include sheathing.

	Craft@Hrs	Unit	Material	Labor	Total
Replace only	6C@.812	lf	37.20	47.90	85.10

Add for each lf of hip or Dutch-hip in a truss roof system. Includes additional cost for trusses designed to create a hip. Dutch-hip is stacked and hand framed.

	Craft@Hrs	Unit	Material	Labor	Total
Replace only					
for hip	6C@.360	lf	33.70	21.20	54.90
for Dutch-hip	6C@.407	lf	36.80	24.00	60.80

	Craft@Hrs	Unit	Material	Labor	Total

Truss mansard roof system. Add **10%** for treated trusses. Roof area equals the roof system width times height times length. Includes cat walks, diagonal braces, fasteners and rough fascia.

	Craft@Hrs	Unit	Material	Labor	Total
Replace					
asphalt, metal, or wood roofing	6C@.003	cf	.53	.18	.71
slate or tile roofing	6C@.003	cf	.68	.18	.86
Remove, all types	1D@.006	sf	—	.28	.28
Add for each lf of valley in trussed mansard roof	6C@.952	lf	36.90	56.20	93.10

Truss gambrel roof system. Add **10%** for treated trusses. Roof area equals the roof system width times height times length. Includes cat walks, diagonal braces, fasteners and rough fascia.

	Craft@Hrs	Unit	Material	Labor	Total
Replace					
asphalt, metal, or wood roofing	6C@.003	cf	.72	.18	.90
slate or tile roofing	6C@.003	cf	.81	.18	.99
Remove, all types	1D@.006	sf	—	.28	.28

Add for each lf of valley in trussed gambrel roof. Includes materials and labor to fill between truss roofs with ridge boards and jack rafters. Does not include sheathing. Replace only.

	Craft@Hrs	Unit	Material	Labor	Total
replace	6C@1.16	lf	37.20	68.40	105.60

Remove truss for work, then reinstall.

	Craft@Hrs	Unit	Material	Labor	Total
replace, per lf	6C@.091	lf	—	5.37	5.37

Wall or roof sheathing. Includes sheathing, nails, and installation. Also includes 4% waste.

	Craft@Hrs	Unit	Material	Labor	Total
Fiberboard sheathing, replace					
1/2" thick, replace	6C@.015	sf	.63	.89	1.52
3/4" thick, replace	6C@.015	sf	.85	.89	1.74
Remove, all thicknesses	1D@.012	sf	—	.57	.57
Sound-deadening fiberboard sheathing					
1/2", replace	6C@.015	sf	.61	.89	1.50
remove	1D@.012	sf	—	.57	.57
Foil-faced foam sheathing, replace					
1/2" thick	6C@.015	sf	.74	.89	1.63
3/4" thick	6C@.015	sf	1.02	.89	1.91
1" thick	6C@.016	sf	1.25	.94	2.19
Remove, all thicknesses	1D@.012	sf	—	.57	.57
CDX plywood sheathing, replace					
3/8" thick	6C@.016	sf	.95	.94	1.89
1/2" thick	6C@.016	sf	.86	.94	1.80
5/8" thick	6C@.016	sf	1.28	.94	2.22
3/4" thick	6C@.016	sf	1.55	.94	2.49
Remove, all thicknesses	1D@.012	sf	—	.57	.57

	Craft@Hrs	Unit	Material	Labor	Total
Treated plywood sheathing, replace					
1/2" thick	6C@.016	sf	1.04	.94	1.98
5/8" thick	6C@.016	sf	1.48	.94	2.42
3/4" thick	6C@.016	sf	1.82	.94	2.76
Remove, all thicknesses	1D@.012	sf	—	.57	.57
Waferboard sheathing, replace					
1/2" thick, replace	6C@.016	sf	.82	.94	1.76
5/8" thick, replace	6C@.016	sf	1.41	.94	2.35
Remove, all thicknesses	1D@.012	sf	—	.57	.57
Oriented strand board (OSB) sheathing, replace					
1/2" thick, replace	6C@.016	sf	.89	.94	1.83
5/8" thick, replace	6C@.016	sf	1.41	.94	2.35
Remove, all thicknesses	1D@.012	sf	—	.57	.57
3/4" tongue-&-groove sheathing, replace					
plywood	6C@.016	sf	1.56	.94	2.50
waferboard	6C@.016	sf	1.64	.94	2.58
oriented strand board (OSB)	6C@.016	sf	1.66	.94	2.60
Remove, all types 3/4" T-&-G sheathing	1D@.012	sf	—	.57	.57
1" x 6" S4S board sheathing					
replace	6C@.017	sf	3.07	1.00	4.07
remove	1D@.014	sf	—	.66	.66
1" x 6" tongue-&-groove sheathing					
replace	6C@.017	sf	4.81	1.00	5.81
remove	1D@.014	sf	—	.66	.66
2" x 6" tongue-&-groove sheathing					
replace	6C@.017	sf	11.30	1.00	12.30
remove	1D@.016	sf	—	.76	.76
2" x 8" tongue-&-groove sheathing					
replace	6C@.017	sf	7.43	1.00	8.43
remove	1D@.015	sf	—	.71	.71

Add for shear-panel sheathing installation. Includes shear paneling installed with nailing patterns and splicing according to code specifications for a shear panel.

	Craft@Hrs	Unit	Material	Labor	Total
replace	6C@.003	sf	—	.18	.18

Frame interior soffit. Includes framing boards and nails. Interior soffits are often built around ducts, above cabinets, or for decorative purposes. Framework only. Includes fireblocking where required but does not include finish materials.

	Craft@Hrs	Unit	Material	Labor	Total
Frame interior soffit, two sides, replace					
typical	6C@.059	lf	7.08	3.48	10.56
with recess for indirect lighting	6C@.100	lf	10.50	5.90	16.40
Frame interior soffit, three sides, replace					
typical	6C@.080	lf	9.45	4.72	14.17
with recess for indirect lighting	6C@.129	lf	11.40	7.61	19.01
Remove interior soffit, all types	1D@.045	lf	—	2.12	2.12

	Craft@Hrs	Unit	Material	Labor	Equip.	Total

Factory designed and built structural panels. Includes manufactured structural panel and installation between site-built framing members. Structural panels are often called stress-skin panels. Wall panels are installed with dimensional lumber top and bottom plates. Roof and floor panels must be installed between structural members. Does not include the cost for these structural members.

	Craft@Hrs	Unit	Material	Labor	Equip.	Total
Factory-built structural wall panels, replace						
6" thick, replace	6C@.051	sf	7.69	3.01	.53	11.23
8" thick, replace	6C@.052	sf	8.69	3.07	.53	12.29
10" thick, replace	6C@.054	sf	9.45	3.19	.53	13.17
12" thick, replace	6C@.054	sf	10.50	3.19	.53	14.22
Remove, all thicknesses	1D@.007	sf	—	.33	—	.33
Factory-built structural floor panels, replace						
6" thick, replace	6C@.051	sf	8.28	3.01	.53	11.82
8" thick, replace	6C@.051	sf	9.24	3.01	.53	12.78
10" thick, replace	6C@.053	sf	10.20	3.13	.53	13.86
12" thick, replace	6C@.054	sf	11.20	3.19	.53	14.92
Remove, all thicknesses	1D@.008	sf	—	.38	—	.38
Factory-built structural roof panel, replace						
6" thick, replace	6C@.061	sf	9.50	3.60	.51	13.61
8" thick, replace	6C@.062	sf	10.80	3.66	.51	14.97
10" thick, replace	6C@.063	sf	11.10	3.72	.51	15.33
12" thick, replace	6C@.066	sf	12.30	3.89	.51	16.70
Remove, all thicknesses	1D@.015	sf	—	.71	—	.71

Wood-frame chimney cricket. Includes 5% waste.

	Craft@Hrs	Unit	Material	Labor	Total
replace	6C@.083	sf	2.08	4.90	6.98
remove	1D@.018	sf	—	.85	.85

chimney cricket

Mail box. Black, white, or grey.

	Craft@Hrs	Unit	Material	Labor	Total
replace	6C@.177	ea	15.90	10.40	26.30
remove	1D@.076	ea	—	3.59	3.59
replace, mail box post	1D@.817	ea	39.20	38.60	77.80
remove	1D@.429	ea	—	20.20	20.20
replace, mail box with post	1D@1.05	ea	55.00	49.60	104.60
remove	1D@.550	ea	—	26.00	26.00
remove, then reinstall	1D@.360	ea	.42	17.00	17.42

	Craft@Hrs	Unit	Material	Labor	Total
Time & Material Charts (selected items)					
Rough Carpentry Materials					
Lumber, random lengths per 1,000 bf					
#2 pine or better 2" wide framing lumber, ($1,420.00 per mbf, 1,000 bf), 7% waste	—	bf	1.53	—	1.53
#2 pine or better 1" wide framing lumber, ($1,740.00 per mbf, 1,000 bf), 7% waste	—	bf	1.85	—	1.85
#2 treated pine or better 2" wide framing lumber, ($1,780.00 per mbf, 1,000 bf), 7% waste	—	bf	1.91	—	1.91
redwood con common 2" wide framing lumber, ($2,890.00 per mbf, 1,000 bf), 7% waste	—	bf	3.12	—	3.12
redwood con heart 2" wide framing lumber, ($3,380.00 per mbf, 1,000 bf), 7% waste	—	bf	3.62	—	3.62
heavy oak timber, ($6,350.00 per mbf, 1,000 bf), 7% waste	—	bf	6.80	—	6.80
heavy pine timber, ($2,600.00 per mbf, 1,000 bf), 7% waste	—	bf	2.79	—	2.79
Fasteners and metal braces					
16d gun nails, ($102.00 per box, 2,500 ea), 5% waste	—	ea	.04	—	.04
12d gun nails, ($95.00 per box, 2,500 ea), 5% waste	—	ea	.03	—	.03
2" gun staples, ($157.00 per box, 10,000 ea), 5% waste	—	ea	.01	—	.01
construction adhesive, per 29 ounce tube, ($10.20 per tube, 75 lf), 5% waste	—	lf	.15	—	.15
metal let-in brace with V shape for saw kerf, ($10.50 per 11.5' stick)	—	ea	10.50	—	10.50
metal X bracing for between joists, ($2.84 per 24" oc)	—	ea	2.84	—	2.84
Furring strips					
1" x 2" x 8' #2 or better pine, ($2.02 ea, 8 lf), 5% waste	—	lf	.27	—	.27
2" x 2" x 8' #2 or better pine, ($4.68 ea, 8 lf), 5% waste	—	lf	.63	—	.63
2" x 2" x 8' con heart redwood, ($7.00 ea, 8 lf), 5% waste	—	lf	.92	—	.92
2" x 2" x 8' con common redwood, ($6.06 ea, 8 lf), 5% waste	—	lf	.79	—	.79
#2 or better pine framing lumber					
2" x 4", random lengths, ($7.52 ea, 8 lf), 5% waste	—	lf	.99	—	.99
2" x 6", random lengths, ($10.90 ea, 8 lf), 5% waste	—	lf	1.44	—	1.44
2" x 8", random lengths, 6 ($13.60 ea, 8 lf), 5% waste	—	lf	1.77	—	1.77

	Craft@Hrs	Unit	Material	Labor	Total
2" x 10", random lengths, ($21.50 ea, 8 lf), 5% waste	—	lf	2.83	—	2.83
2" x 12", random lengths, ($24.50 ea, 8 lf), 5% waste	—	lf	3.22	—	3.22
Redwood framing lumber					
2" x 4" con heart, random lengths, ($16.80 ea, 8 lf), 5% waste	—	lf	2.19	—	2.19
2" x 6" con heart, random lengths, ($20.00 ea, 8 lf), 5% waste	—	lf	2.62	—	2.62
2" x 8" con heart, random lengths, ($29.80 ea, 8 lf), 5% waste	—	lf	3.92	—	3.92
2" x 4" con common, random lengths, ($14.80 ea, 8 lf), 5% waste	—	lf	1.94	—	1.94
2" x 6" con common, random lengths, ($17.00 ea, 8 lf), 5% waste	—	lf	2.22	—	2.22
2" x 8" con common, random lengths, ($25.90 ea, 8 lf), 5% waste	—	lf	3.39	—	3.39
Treated framing lumber					
2" x 4", random lengths, ($12.30 ea, 8 lf), 5% waste	—	lf	1.61	—	1.61
2" x 6", random lengths, ($18.30 ea, 8 lf), 5% waste	—	lf	2.42	—	2.42
2" x 8", random lengths, ($30.70 ea, 8 lf), 5% waste	—	lf	4.04	—	4.04
Posts					
4" x 4"					
#2 or better pine, ($30.40 per 8' post), 4% waste	—	lf	3.95	—	3.95
construction common redwood, ($37.40 per 8' post), 4% waste	—	lf	4.85	—	4.85
treated pine, ($36.10 per 8' post), 4% waste	—	lf	5.15	—	5.15
4" x 6"					
#2 or better pine, ($35.50 per 8' post), 4% waste	—	lf	4.61	—	4.61
construction common redwood, ($39.70 per 8' post), 4% waste	—	lf	5.16	—	5.16
treated pine, ($46.10 per 8' post), 4% waste	—	lf	6.00	—	6.00
6" x 6"					
#2 or better pine, ($74.60 per 8' post), 4% waste	—	lf	9.69	—	9.69
construction common redwood, ($64.30 per 8' post), 4% waste	—	lf	8.36	—	8.36
treated pine, ($96.90 per 8' post), 4% waste	—	lf	12.60	—	12.60
8" x 8"					
#2 or better pine, ($132.00 per 8' post), 4% waste	—	lf	17.30	—	17.30
con common redwood, ($114.00 per 8' post), 4% waste	—	lf	14.90	—	14.90
treated pine, ($171.00 per 8' post), 4% waste	—	lf	22.00	—	22.00

	Craft@Hrs	Unit	Material	Labor	Total
Adjustable steel jack post					
51" to 90" to 13,000 pound load, ($76.40 ea)	—	ea	76.40	—	76.40
20" to 36" to 16,000 pound load, ($50.00 ea)	—	ea	50.00	—	50.00
37" to 60" to 17,500 pound load, ($63.10 ea)	—	ea	63.10	—	63.10
56" to 96" to 25,000 pound load, ($98.20 ea)	—	ea	98.20	—	98.20
Lally column					
3-1/2" diameter, 6' to 8', ($112.00 ea)	—	ea	112.00	—	112.00
3-1/2" diameter, 8' to 10', ($125.00 ea)	—	ea	125.00	—	125.00
3-1/2" diameter, 10' to 12', ($157.00 ea)	—	ea	157.00	—	157.00
Glue-laminated beam					
per bf, ($6.33 bf, 1 bf), 3% waste	—	bf	6.52	—	6.52
Micro-laminated beam					
1-3/4" x 9-1/2", ($9.41 lf, 1 lf), 4% waste	—	lf	9.79	—	9.79
1-3/4" x 11-7/8", ($11.80 lf, 1 lf), 4% waste	—	lf	12.30	—	12.30
1-3/4" x 14", ($13.90 lf, 1 lf), 4% waste	—	lf	14.50	—	14.50
1-3/4" x 18", ($17.80 lf, 1 lf), 4% waste	—	lf	18.40	—	18.40
Beams, select structural grade, per bf					
pine, ($2.60 bf, 1 bf), 4% waste	—	bf	2.69	—	2.69
oak, ($6.35 bf, 1 bf), 4% waste	—	bf	6.60	—	6.60
Laminated lumber I joist					
1-3/4" flange, 9-1/2" deep, ($3.26 lf, 1 lf), 3% waste	—	lf	3.35	—	3.35
1-3/4" flange, 11-7/8" deep, ($3.96 lf, 1 lf), 3% waste	—	lf	4.08	—	4.08
1-3/4" flange, 14" deep, ($4.86 lf, 1 lf), 3% waste	—	lf	5.03	—	5.03
2-5/16" flange, 11-7/8" deep, ($5.63 lf, 1 lf), 3% waste	—	lf	5.80	—	5.80
Floor truss					
typical loading, ($5.94 lf)	—	lf	5.94	—	5.94
heavy loading, ($6.40 lf)	—	lf	6.40	—	6.40
Roof truss					
flat, ($5.92 lf)	—	lf	5.92	—	5.92
4 in 12 slope, ($6.24 lf)	—	lf	6.24	—	6.24
8 in 12 slope, ($9.07 lf)	—	lf	9.07	—	9.07
12 in 12 slope, ($12.20 lf)	—	lf	12.20	—	12.20
17 in 12 slope, ($15.00 lf)	—	lf	15.00	—	15.00
Roof or wall sheathing					
1/2" fiberboard, ($19.30 per 8' sheet, 32 sf), 4% waste	—	sf	.63	—	.63
3/4" fiberboard ($26.40 per 8' sheet, 32 sf), 4% waste	—	sf	.86	—	.86
1/2" foil-faced foam, ($23.20 per 8' sheet, 32 sf), 4% waste	—	sf	.75	—	.75
3/4" foil-faced foam, ($29.70 per 8' sheet, 32 sf), 4% waste	—	sf	.95	—	.95
1" foil-faced foam, ($38.70 per 8' sheet, 32 sf), 4% waste	—	sf	1.25	—	1.25

	Craft@Hrs	Unit	Material	Labor	Total
1/2" CDX plywood, ($26.50 per 8' sheet, 32 sf), 4% waste	—	sf	.86	—	.86
1/2" waferboard, ($24.60 per 8' sheet, 32 sf), 4% waste	—	sf	.81	—	.81
1/2" oriented strand board (OSB), ($28.00 per 8' sheet, 32 sf), 4% waste	—	sf	.91	—	.91
5/8" CDX plywood, ($39.20 per 8' sheet, 32 sf), 4% waste	—	sf	1.28	—	1.28
5/8" waferboard, ($43.00 per 8' sheet, 32 sf), 4% waste	—	sf	1.40	—	1.40
5/8" oriented strand board (OSB), ($43.30 per 8' sheet, 32 sf), 4% waste	—	sf	1.40	—	1.40
3/4" CDX plywood, ($47.90 per 8' sheet, 32 sf), 4% waste	—	sf	1.56	—	1.56
3/4" tongue-&-groove plywood, ($48.20 per 8' sheet, 32 sf), 4% waste	—	sf	1.57	—	1.57
3/4" tongue-&-groove waferboard, ($51.00 per 8' sheet, 32 sf), 4% waste	—	sf	1.67	—	1.67
3/4" tongue-&-groove oriented strand board (OSB), ($51.20 per 8' sheet, 32 sf), 4% waste	—	sf	1.67	—	1.67
1" x 6" tongue-&-groove sheathing, ($2.32 lf, .50 sf), 4% waste	—	sf	4.82	—	4.82
2" x 6" tongue-&-groove sheathing, ($5.44 lf, .50 sf), 4% waste	—	sf	11.30	—	11.30
Factory designed and fabricated building panel					
6" thick wall panel, ($7.69 sf)	—	sf	7.69	—	7.69
8" thick wall panel, ($8.69 sf)	—	sf	8.69	—	8.69
10" thick floor panel, ($10.20 sf)	—	sf	10.20	—	10.20
12" thick floor panel, ($11.20 sf)	—	sf	11.20	—	11.20
6" thick roof panel, ($9.50 sf)	—	sf	9.50	—	9.50
8" thick roof panel, ($10.80 sf)	—	sf	10.80	—	10.80

Rough Carpentry Labor

Laborer	base wage	paid leave	true wage	taxes & ins.	total
Carpenter	$38.10	2.97	$41.07	26.53	$67.60
Carpenter's helper	$27.40	2.14	$29.54	20.86	$50.40
Post & beam carpenter	$49.10	3.83	$52.93	32.27	$85.20
P & B carpenter's helper	$33.80	2.64	$36.44	24.26	$60.70
Equipment operator	$52.10	4.06	$56.16	32.74	$88.90
Demolition laborer	$25.90	2.02	$27.92	19.28	$47.20

Paid leave is calculated based on two weeks paid vacation, one week sick leave, and seven paid holidays. Employer's matching portion of **FICA** is 7.65 percent. **FUTA** (Federal Unemployment) is .8 percent. **Worker's compensation** for the rough carpentry trade was calculated using a national average of 17.21 percent. **Unemployment insurance** was calculated using a national average of 8 percent. **Health insurance** was calculated based on a projected national average for 2020 of $1,280 per employee (and family when applicable) per month. Employer pays 80 percent for a per month cost of $1,024 per employee. **Retirement** is based on a 401(k) retirement program with employer matching of 50 percent. Employee contributions to the 401(k) plan are an average of 6 percent of the true wage. **Liability insurance** is based on a national average of 12.0 percent.

	Craft@Hrs	Unit	Material	Labor	Total
Rough Carpentry Labor Productivity					
Demolition of rough carpentry					
remove 2" x 4" wall, 8' tall	1D@.045	lf	—	2.12	2.12
remove 2" x 6" wall, 8' tall	1D@.048	lf	—	2.27	2.27
remove 2" x 8" wall, 8' tall	1D@.049	lf	—	2.31	2.31
remove wood post	1D@.059	lf	—	2.78	2.78
remove glue-laminated beam	1D@.222	lf	—	10.50	10.50
remove micro-laminated beam	1D@.106	lf	—	5.00	5.00
remove sheathing	1D@.012	sf	—	.57	.57
remove beam	1D@.222	lf	—	10.50	10.50
remove deck planking	1D@.026	sf	—	1.23	1.23
remove deck railing	1D@.034	lf	—	1.60	1.60
remove deck stairway per step	1D@.061	ea	—	2.88	2.88
remove furring strips applied to wood	1D@.011	sf	—	.52	.52
remove joist system	1D@.014	sf	—	.66	.66
remove floor truss system	1D@.014	lf	—	.66	.66
remove hand-framed roof, 4 in 12 slope	1D@.016	sf	—	.76	.76
remove hand-framed roof, 8 in 12 slope	1D@.019	sf	—	.90	.90
remove hand-framed roof, 12 in 12 slope	1D@.032	sf	—	1.51	1.51
remove hand-framed roof, 16 in 12 slope	1D@.050	sf	—	2.36	2.36
remove hand-framed roof, 18 in 12 slope	1D@.077	sf	—	3.63	3.63
remove flat roof truss	1D@.016	sf	—	.76	.76
remove truss roof, 4 in 12 slope	1D@.017	sf	—	.80	.80
remove truss roof, 8 in 12 slope	1D@.021	sf	—	.99	.99
remove truss roof, 12 in 12 slope	1D@.034	sf	—	1.60	1.60
remove truss roof, 16 in 12 slope	1D@.053	sf	—	2.50	2.50
remove truss roof, 17 in 12 slope	1D@.082	sf	—	3.87	3.87
remove hand-framed dormer, 4 in 12 slope	1D@.017	sf	—	.80	.80
remove hand-framed dormer, 10 in 12 slope	1D@.025	sf	—	1.18	1.18
remove hand-framed dormer, 18 in 12 slope	1D@.081	sf	—	3.82	3.82
remove prefabricated structural wall panel	1D@.007	sf	—	.33	.33
remove prefabricated structural floor panel	1D@.008	sf	—	.38	.38
remove prefabricated structural roof panel	1D@.015	sf	—	.71	.71

Framing crew

rough carpentry	carpenter	$67.60
rough carpentry	carpenter's helper	$50.40
rough carpentry	framing crew	$59.00

Post & beam framing crew

post & beam framing	carpenter	$85.20
post & beam framing	carpenter's helper	$60.70
post & beam framing	post & beam crew	$73.00

	Craft@Hrs	Unit	Material	Labor	Total
Build interior partition wall					
2" x 4" 8' tall 16" on center	6C@.241	lf	—	14.20	14.20
2" x 4" 10' tall 16" on center	6C@.244	lf	—	14.40	14.40
2" x 6" 8' tall 16" on center	6C@.244	lf	—	14.40	14.40
2" x 6" 10' tall 16" on center	6C@.247	lf	—	14.60	14.60
2" x 8" 8' tall 16" on center	6C@.247	lf	—	14.60	14.60
2" x 8" 10' tall 16" on center	6C@.250	lf	—	14.80	14.80

	Craft@Hrs	Unit	Material	Labor	Total
Build interior bearing wall					
2" x 4" 8' tall 16" on center	6C@.261	lf	—	15.40	15.40
2" x 4" 10' tall 16" on center	6C@.265	lf	—	15.60	15.60
2" x 6" 8' tall 16" on center	6C@.263	lf	—	15.50	15.50
2" x 6" 10' tall 16" on center	6C@.267	lf	—	15.80	15.80
2" x 8" 8' tall 16" on center	6C@.266	lf	—	15.70	15.70
2" x 8" 10' tall 16" on center	6C@.270	lf	—	15.90	15.90
Build exterior wall					
2" x 4" 8' tall 16" on center	6C@.267	lf	—	15.80	15.80
2" x 4" 10' tall 16" on center	6C@.270	lf	—	15.90	15.90
2" x 6" 8' tall 16" on center	6C@.270	lf	—	15.90	15.90
2" x 6" 10' tall 16" on center	6C@.272	lf	—	16.00	16.00
2" x 8" 8' tall 16" on center	6C@.272	lf	—	16.00	16.00
2" x 8" 10' tall 16" on center	6C@.275	lf	—	16.20	16.20
Build interior partition wall per sf					
2" x 4" 16" on center	6C@.030	sf	—	1.77	1.77
2" x 6" 16" on center	6C@.030	sf	—	1.77	1.77
2" x 8" 16" on center	6C@.031	sf	—	1.83	1.83
Build sloping interior partition wall per sf					
2" x 4" 16" on center	6C@.041	sf	—	2.42	2.42
2" x 6" 16" on center	6C@.042	sf	—	2.48	2.48
2" x 8" 16" on center	6C@.042	sf	—	2.48	2.48
Build sloping interior bearing wall per sf					
2" x 4" 16" on center	6C@.050	sf	—	2.95	2.95
2" x 6" 16" on center	6C@.050	sf	—	2.95	2.95
2" x 8" 16" on center	6C@.051	sf	—	3.01	3.01
Build interior bearing wall per sf					
2" x 4" 16" on center	6C@.033	sf	—	1.95	1.95
2" x 6" 16" on center	6C@.033	sf	—	1.95	1.95
2" x 8" 16" on center	6C@.033	sf	—	1.95	1.95
Build exterior wall per sf					
2" x 4" 16" on center	6C@.034	sf	—	2.01	2.01
2" x 6" 16" on center	6C@.034	sf	—	2.01	2.01
2" x 8" 16" on center	6C@.034	sf	—	2.01	2.01
Build sloping exterior wall per sf					
2" x 4" 16" on center	6C@.046	sf	—	2.71	2.71
2" x 6" 16" on center	6C@.046	sf	—	2.71	2.71
2" x 8" 16" on center	6C@.047	sf	—	2.77	2.77
Build walls per bf					
interior wall	6C@.034	bf	—	2.01	2.01
sloping interior wall	6C@.047	bf	—	2.77	2.77
exterior wall	6C@.037	bf	—	2.18	2.18
sloping exterior wall	6C@.051	bf	—	3.01	3.01
Build and install wall header					
(Made from 2" x dimension lumber and 1/2" cdx plywood)					
4" x 10" header	6C@.043	lf	—	2.54	2.54
4" x 12" header	6C@.045	lf	—	2.66	2.66
6" x 10" header	6C@.069	lf	—	4.07	4.07
6" x 12" header	6C@.074	lf	—	4.37	4.37

	Craft@Hrs	Unit	Material	Labor	Total
Install posts					
4" x 4" post	6C@.042	lf	—	2.48	2.48
6" x 6" post	6C@.044	lf	—	2.60	2.60
8" x 8" post	6C@.048	lf	—	2.83	2.83
Install adjustable steel jackpost					
20" to 36" adjustable	6C@.243	ea	—	14.30	14.30
37" to 60" adjustable	6C@.272	ea	—	16.00	16.00
48" to 100" adjustable	6C@.386	ea	—	22.80	22.80
56" to 96" adjustable	6C@.352	ea	—	20.80	20.80
Install lally column					
3-1/2" diameter, 6' to 8'	6C@.526	ea	—	31.00	31.00
3-1/2" diameter, 8' to 10'	6C@.556	ea	—	32.80	32.80
3-1/2" diameter, 10' to 12'	6C@.587	ea	—	34.60	34.60
Install glue-laminated beam per bf	6C@.060	bf	—	3.54	3.54
Install micro-laminated beam					
1-3/4" x 7-1/4"	6C@.048	lf	—	2.83	2.83
1-3/4" x 9-1/2"	6C@.050	lf	—	2.95	2.95
1-3/4" x 11-7/8"	6C@.053	lf	—	3.13	3.13
1-3/4" x 14"	6C@.056	lf	—	3.30	3.30
1-3/4" x 18"	6C@.063	lf	—	3.72	3.72
Install beam					
4" x 10"	6C@.114	lf	—	6.73	6.73
4" x 12"	6C@.149	lf	—	8.79	8.79
6" x 10"	6C@.132	lf	—	7.79	7.79
6" x 12"	6C@.154	lf	—	9.09	9.09
8" x 10"	6C@.156	lf	—	9.20	9.20
8" x 12"	6C@.182	lf	—	10.70	10.70
10" x 10"	6C@.176	lf	—	10.40	10.40
10" x 12"	6C@.196	lf	—	11.60	11.60
12" x 12"	6C@.204	lf	—	12.00	12.00
12" x 16"	6C@.222	lf	—	13.10	13.10

Post & beam framing

	Craft@Hrs	Unit	Material	Labor	Total
Cut and assemble bent with truss					
single-story with simple truss	4B@.527	sf	—	38.50	38.50
two-story with simple truss	4B@.571	sf	—	41.70	41.70
single-story with king post truss	4B@.569	sf	—	41.50	41.50
two-story with king post truss	4B@.620	sf	—	45.30	45.30
single-story with king post & struts truss	4B@.618	sf	—	45.10	45.10
two-story with king post & struts truss	4B@.679	sf	—	49.60	49.60
single-story with queen post truss	4B@.609	sf	—	44.50	44.50
two-story with queen post truss	4B@.671	sf	—	49.00	49.00
single-story with hammer beam truss	4B@1.05	sf	—	76.70	76.70
two-story with hammer beam truss	4B@1.25	sf	—	91.30	91.30
single-story with scissors truss	4B@.989	sf	—	72.20	72.20
two-story with scissors truss	4B@1.04	sf	—	75.90	75.90
add for overhang drop in two-story bent	4B@20.0	ea	—	1,460.00	1,460.00

	Craft@Hrs	Unit	Material	Labor	Total
Cut and assemble connecting members between bent (per sf of area beneath)					
between single-story bents	4B@1.11	sf	—	81.00	81.00
Cut post & beam framing joint and join timbers					
sill corner half-lap	2B@1.10	ea	—	93.70	93.70
sill corner half-lap with through tenon	2B@2.77	ea	—	236.00	236.00
sill corner dovetail	2B@1.99	ea	—	170.00	170.00
sill corner dovetail with through tenon	2B@3.86	ea	—	329.00	329.00
sill corner tongue & fork	2B@1.79	ea	—	153.00	153.00
sill corner tongue & fork with through tenon	2B@3.44	ea	—	293.00	293.00
lapped dovetail	2B@2.05	ea	—	175.00	175.00
housed lapped dovetail	2B@2.55	ea	—	217.00	217.00
tusk tenon	2B@2.77	ea	—	236.00	236.00
half-lap	2B@1.08	ea	—	92.00	92.00
through mortise & tenon with shoulders	2B@2.26	ea	—	193.00	193.00
housed mortise & tenon	2B@2.37	ea	—	202.00	202.00
shouldered mortise & tenon	2B@2.34	ea	—	199.00	199.00
wedged dovetail mortise & tenon	2B@2.37	ea	—	202.00	202.00
mortise & tenon (90 degree)	2B@1.25	ea	—	107.00	107.00
anchor-beam	2B@2.93	ea	—	250.00	250.00
framed overhang	2B@10.0	ea	—	852.00	852.00
through half-lap	2B@1.20	ea	—	102.00	102.00
mortise & tenon knee brace (45 degree)	2B@1.56	ea	—	133.00	133.00
collar tie lapped half dovetail	2B@1.96	ea	—	167.00	167.00
stopped splayed scarf	2B@3.22	ea	—	274.00	274.00
stopped splayed scarf with through tenon	2B@5.00	ea	—	426.00	426.00
bladed scarf	2B@2.45	ea	—	209.00	209.00
bladed scarf with through tenon	2B@4.35	ea	—	371.00	371.00
rafter foot housed bird's mouth	2B@1.39	ea	—	118.00	118.00
rafter foot beveled shoulder bird's mouth	2B@1.30	ea	—	111.00	111.00
rafter foot bird's mouth, purlin, & post joint	2B@3.13	ea	—	267.00	267.00
rafter foot bird's mouth joint	2B@1.00	ea	—	85.20	85.20
rafter foot bird's mouth with tenon	2B@2.86	ea	—	244.00	244.00
rafter peak tongue & fork	2B@1.79	ea	—	153.00	153.00
Build joist system for outdoor deck					
(with support beams, posts and concrete footing)					
2" x 8" joists	6C@.071	sf	—	4.19	4.19
2" x 10" joists	6C@.077	sf	—	4.54	4.54
2" x 12" joists	6C@.083	sf	—	4.90	4.90
Install outdoor deck planking per sf					
2" x 4"	6C@.040	sf	—	2.36	2.36
2" x 6"	6C@.037	sf	—	2.18	2.18
2" x 8"	6C@.033	sf	—	1.95	1.95
Install deck railing and stairs					
deck railing	6C@.275	lf	—	16.20	16.20
3' wide deck stairs with balustrades	6C@.587	ea	—	34.60	34.60
deck stairway landing per sf	6C@.074	sf	—	4.37	4.37

	Craft@Hrs	Unit	Material	Labor	Total
Install furring strips					
on wood 16" on center	6C@.013	sf	—	.77	.77
on wood 24" on center	6C@.011	sf	—	.65	.65
on masonry or concrete 16" on center	6C@.016	sf	—	.94	.94
on masonry or concrete 24" on center	6C@.014	sf	—	.83	.83
Build joist systems with blocking and cross bracing					
per bf	6C@.021	bf	—	1.24	1.24
Install laminated lumber I truss					
1-3/4" flange,11-7/8" deep	6C@.036	lf	—	2.12	2.12
1-3/4" flange,14" deep	6C@.037	lf	—	2.18	2.18
2-5/16" flange,11-7/8" deep	6C@.036	lf	—	2.12	2.12
2-5/16" flange,14" deep	6C@.038	lf	—	2.24	2.24
2-5/16" flange,16" deep	6C@.039	lf	—	2.30	2.30
Install floor truss					
typical	6C@.091	lf	—	5.37	5.37
Hand-frame roof, 24" on center					
Hand-frame roof with 2" x 4" rafters					
4 in 12 slope	6C@.034	sf	—	2.01	2.01
8 in 12 slope	6C@.040	sf	—	2.36	2.36
12 in 12 slope	6C@.049	sf	—	2.89	2.89
16 in 12 slope	6C@.060	sf	—	3.54	3.54
Hand-frame roof with 2" x 6" rafters					
4 in 12 slope	6C@.035	sf	—	2.07	2.07
8 in 12 slope	6C@.042	sf	—	2.48	2.48
12 in 12 slope	6C@.051	sf	—	3.01	3.01
16 in 12 slope	6C@.062	sf	—	3.66	3.66
Hand-frame roof with 2" x 8" rafters					
4 in 12 slope	6C@.036	sf	—	2.12	2.12
8 in 12 slope	6C@.043	sf	—	2.54	2.54
12 in 12 slope	6C@.052	sf	—	3.07	3.07
16 in 12 slope	6C@.064	sf	—	3.78	3.78
Hand-frame roof with 2" x 10" rafters					
4 in 12 slope	6C@.037	sf	—	2.18	2.18
8 in 12 slope	6C@.044	sf	—	2.60	2.60
12 in 12 slope	6C@.054	sf	—	3.19	3.19
16 in 12 slope	6C@.066	sf	—	3.89	3.89
Hand-frame roof with 2" x 12" rafters					
4 in 12 slope	6C@.038	sf	—	2.24	2.24
8 in 12 slope	6C@.046	sf	—	2.71	2.71
12 in 12 slope	6C@.056	sf	—	3.30	3.30
16 in 12 slope	6C@.068	sf	—	4.01	4.01
Additional hand-framed roof labor costs					
add for each lf of valley	6C@.786	lf	—	46.40	46.40
add for each lf of hip width	6C@.212	lf	—	12.50	12.50
add for each lf of Dutch hip width	6C@.229	lf	—	13.50	13.50
Hand-frame mansard roof, 24" on center (per cf of roof)					
2" x 4" rafters	6C@.005	cf	—	.30	.30
2" x 8" rafters	6C@.005	cf	—	.30	.30
2" x 12" rafters	6C@.006	cf	—	.35	.35
add for each lf of valley	6C@1.08	lf	—	63.70	63.70

	Craft@Hrs	Unit	Material	Labor	Total
Hand-frame gambrel roof, 24" on center (per cf of roof)					
2" x 4" rafters	6C@.004	cf	—	.24	.24
2" x 8" rafters	6C@.005	cf	—	.30	.30
2" x 12" rafters	6C@.006	cf	—	.35	.35
add for each lf of valley	6C@1.14	lf	—	67.30	67.30
Hand-frame roof per bf					
gable roof	6C@.023	bf	—	1.36	1.36
hip roof	6C@.023	bf	—	1.36	1.36
Dutch hip roof	6C@.024	bf	—	1.42	1.42
mansard roof	6C@.026	bf	—	1.53	1.53
gambrel roof	6C@.025	bf	—	1.48	1.48
Hand-frame dormer, 24" on center					
shed dormer					
4 in 12 slope	6C@.045	sf	—	2.66	2.66
8 in 12 slope	6C@.049	sf	—	2.89	2.89
12 in 12 slope	6C@.060	sf	—	3.54	3.54
16 in 12 slope	6C@.064	sf	—	3.78	3.78
gable dormer					
4 in 12 slope	6C@.047	sf	—	2.77	2.77
8 in 12 slope	6C@.051	sf	—	3.01	3.01
12 in 12 slope	6C@.062	sf	—	3.66	3.66
16 in 12 slope	6C@.067	sf	—	3.95	3.95
hip dormer					
4 in 12 slope	6C@.049	sf	—	2.89	2.89
8 in 12 slope	6C@.053	sf	—	3.13	3.13
12 in 12 slope	6C@.065	sf	—	3.84	3.84
16 in 12 slope	6C@.070	sf	—	4.13	4.13
Dutch hip dormer					
4 in 12 slope	6C@.051	sf	—	3.01	3.01
8 in 12 slope	6C@.054	sf	—	3.19	3.19
12 in 12 slope	6C@.067	sf	—	3.95	3.95
16 in 12 slope	6C@.072	sf	—	4.25	4.25
gambrel dormer					
per length of ridge x width of dormer	6C@.059	sf	—	3.48	3.48
Hand-frame dormer, per bf					
shed dormer	6C@.034	bf	—	2.01	2.01
gable dormer	6C@.035	bf	—	2.07	2.07
hip dormer	6C@.036	bf	—	2.12	2.12
Dutch-hip dormer	6C@.038	bf	—	2.24	2.24
gambrel dormer	6C@.037	bf	—	2.18	2.18
Install roof truss					
flat roof	6C@.023	lf	—	1.36	1.36
4 in 12 slope	6C@.023	lf	—	1.36	1.36
8 in 12 slope	6C@.024	lf	—	1.42	1.42
12 in 12 slope	6C@.024	lf	—	1.42	1.42
16 in 12 slope	6C@.026	lf	—	1.53	1.53

	Craft@Hrs	Unit	Material	Labor	Total
Install truss system, 24" on center, per sf					
4 in 12 slope	6C@.012	sf	—	.71	.71
8 in 12 slope	6C@.012	sf	—	.71	.71
12 in 12 slope	6C@.013	sf	—	.77	.77
16 in 12 slope	6C@.013	sf	—	.77	.77
Additional truss roof costs					
add labor for each lf of valley	6C@.812	lf	—	47.90	47.90
add labor for each lf of width in a hip	6C@.360	lf	—	21.20	21.20
add labor for each lf of width in a Dutch hip	6C@.407	lf	—	24.00	24.00
Install mansard truss system, 24" on center, per cf					
per width x height x length	6C@.003	cf	—	.18	.18
add labor for each lf of valley	6C@.952	lf	—	56.20	56.20
Install gambrel truss system, 24" on center, per cf					
per width x height x length	6C@.003	cf	—	.18	.18
add labor for each lf of valley	6C@1.16	lf	—	68.40	68.40
Install sheathing					
builder board	6C@.015	sf	—	.89	.89
plywood, waferboard, or OSB	6C@.016	sf	—	.94	.94
1" x 6" sheathing	6C@.017	sf	—	1.00	1.00
1" x 6" tongue-&-groove	6C@.017	sf	—	1.00	1.00
2" x 6" tongue-&-groove	6C@.017	sf	—	1.00	1.00
2" x 8" tongue-&-groove	6C@.017	sf	—	1.00	1.00
add labor for shear-panel installation	6C@.003	sf	—	.18	.18
Build interior soffit					
two sides	6C@.059	lf	—	3.48	3.48
two sides with recess for indirect lighting	6C@.100	lf	—	5.90	5.90
three sides	6C@.080	lf	—	4.72	4.72
three sides with recess for indirect lighting	6C@.129	lf	—	7.61	7.61
Install factory designed and fabricated structural panel					
6" thick wall panel	6C@.051	sf	—	3.01	3.01
8" thick wall panel	6C@.052	sf	—	3.07	3.07
10" thick wall panel	6C@.054	sf	—	3.19	3.19
12" thick wall panel	6C@.054	sf	—	3.19	3.19
6" thick floor panel	6C@.051	sf	—	3.01	3.01
8" thick floor panel	6C@.051	sf	—	3.01	3.01
10" thick floor panel	6C@.053	sf	—	3.13	3.13
12" thick floor panel	6C@.054	sf	—	3.19	3.19
6" thick roof panel	6C@.061	sf	—	3.60	3.60
8" thick roof panel	6C@.062	sf	—	3.66	3.66
10" thick roof panel	6C@.063	sf	—	3.72	3.72
12" thick roof panel	6C@.066	sf	—	3.89	3.89
Wood-frame chimney cricket					
length of ridge x width	6C@.083	sf	—	4.90	4.90

Security Systems

	Craft@Hrs	Unit	Material	Labor	Total
Minimum charge.					
for security system work	7S@3.00	ea	36.40	179.00	215.40
add for system installed in finished structure	—	%	—	37.0	—
Control panel. Includes up to 50 lf of wiring.					
Replace security control panel					
standard grade	7S@5.01	ea	492.00	299.00	791.00
high grade	7S@6.67	ea	619.00	398.00	1,017.00
Remove, all grades	1D@.250	ea	—	11.80	11.80
Key pad. Includes up to 50 lf of wiring.					
Replace security panel key pad					
typical	7S@1.27	ea	252.00	75.80	327.80
LED read-out	7S@1.27	ea	322.00	75.80	397.80
Remove security panel key pad all types	1D@.200	ea	—	9.44	9.44
Key control. Includes up to 50 lf of wiring.					
replace, outside	7S@1.14	ea	149.00	68.10	217.10
remove	1D@.143	ea	—	6.75	6.75
Contact. Includes up to 50 lf of wiring.					
Mechanical contact, per opening					
replace	7S@.952	ea	83.60	56.80	140.40
remove	1D@.143	ea	—	6.75	6.75
Magnetic contact, per opening					
replace	7S@1.37	ea	98.50	81.80	180.30
remove	1D@.167	ea	—	7.88	7.88

	Craft@Hrs	Unit	Material	Labor	Total
Sound detector. Includes up to 50 lf of wiring.					
replace	7S@1.56	ea	225.00	93.10	318.10
remove	1D@.167	ea	—	7.88	7.88
Motion detector. Includes up to 50 lf of wiring.					
replace	7S@1.49	ea	433.00	89.00	522.00
remove	1D@.167	ea	—	7.88	7.88
Pressure mat. Includes up to 50 lf of wiring.					
replace	7S@1.52	lf	77.20	90.70	167.90
remove	1D@.185	lf	—	8.73	8.73
Smoke detector. Includes up to 50 lf of wiring.					
replace	7S@1.49	ea	275.00	89.00	364.00
remove	1D@.167	ea	—	7.88	7.88
Horn or siren. Exterior or interior. Includes up to 50 lf of wiring.					
replace	7S@1.28	ea	91.50	76.40	167.90
remove	1D@.143	ea	—	6.75	6.75
Panic button. Includes up to 50 lf of wiring.					
replace	7S@.644	ea	77.20	38.40	115.60
remove	1D@.125	ea	—	5.90	5.90

Time & Material Charts (selected items)

Security Systems Materials

See Security Systems material prices with the line items above.

Security Systems Labor

Laborer	base wage	paid leave	true wage	taxes & ins.	total
Security system installer	$35.70	2.78	$38.48	21.22	$59.70
Demolition worker	$25.90	2.02	$27.92	19.28	$47.20

Paid leave is calculated based on two weeks paid vacation, one week sick leave, and seven paid holidays. Employer's matching portion of **FICA** is 7.65 percent. **FUTA** (Federal Unemployment) is .8 percent. **Worker's compensation** for the security systems trade was calculated using a national average of 6.89 percent. **Unemployment insurance** was calculated using a national average of 8 percent. **Health insurance** was calculated based on a projected national average for 2020 of $1,280 per employee (and family when applicable) per month. Employer pays 80 percent for a per month cost of $1,024 per employee. **Retirement** is based on a 401(k) retirement program with employer matching of 50 percent. Employee contributions to the 401(k) plan are an average of 6 percent of the true wage. **Liability insurance** is based on a national average of 12.0 percent.

Security Systems Labor Productivity

	Craft@Hrs	Unit	Material	Labor	Total
Demolition of security systems					
remove control panel	1D@.250	ea	—	11.80	11.80
remove exterior key pad	1D@.200	ea	—	9.44	9.44
remove contact	1D@.143	ea	—	6.75	6.75
remove detector (all types)	1D@.167	ea	—	7.88	7.88
remove pressure mat detector	1D@.185	ea	—	8.73	8.73
remove horn or siren	1D@.143	ea	—	6.75	6.75
remove panic button	1D@.125	ea	—	5.90	5.90
Install security system					
control panel, standard grade	7S@5.01	ea	—	299.00	299.00
control panel, high grade	7S@6.67	ea	—	398.00	398.00
key pad	7S@1.27	ea	—	75.80	75.80
outside key control	7S@1.14	ea	—	68.10	68.10
mechanical contact per opening	7S@.952	ea	—	56.80	56.80
magnetic contact per opening	7S@1.37	ea	—	81.80	81.80
sound detector	7S@1.56	ea	—	93.10	93.10
motion detector	7S@1.49	ea	—	89.00	89.00
pressure mat detector	7S@1.52	lf	—	90.70	90.70
smoke detector	7S@1.49	ea	—	89.00	89.00
horn or siren	7S@1.28	ea	—	76.40	76.40
panic button	7S@.644	ea	—	38.40	38.40

Siding

	Craft@Hrs	Unit	Material	Labor	Total
Minimum charge.					
minimum charge for siding work	6S@3.00	ea	79.80	174.00	253.80
Fiberglass corrugated siding. Includes fiberglass panels, rust-resistant nails, purlins with corrugated pattern, end caps when needed, boots, roofing felt, valley metal, drip edge, ridge cap, fasteners, and installation. Weights are per sf. Includes 4% waste. Greenhouse style.					
replace, 6 ounce	6S@.028	sf	3.20	1.63	4.83
replace, 8 ounce	6S@.028	sf	3.43	1.63	5.06
replace, 12 ounce	6S@.028	sf	4.77	1.63	6.40
remove	1D@.007	sf	—	.33	.33
Aluminum siding. Includes siding with trim pieces, fasteners, and installation. Also includes 4% waste.					
replace, standard grade	6S@.034	sf	2.48	1.98	4.46
replace, high grade	6S@.034	sf	2.91	1.98	4.89
replace, deluxe grade	6S@.034	sf	3.43	1.98	5.41
replace, custom grade	6S@.034	sf	4.97	1.98	6.95
remove	1D@.008	sf	—	.38	.38
add for insulated siding	—	sf	.70	—	.70
Cement fiber shingle siding. Includes cement fiber shingle siding, rust-resistant nails, and installation. Includes 4% waste. Also called asbestos cement. Some older styles contain dangerous asbestos which must be removed using hazardous material removal techniques.					
replace, standard grade	6S@.045	sf	2.08	2.61	4.69
replace, high grade	6S@.045	sf	2.61	2.61	5.22
remove	1D@.013	sf	—	.61	.61
replace single shingle	6S@.477	sf	4.35	27.70	32.05
minimum charge to repair cement fiber siding	6S@3.00	ea	62.80	174.00	236.80
Shake or wood shingle siding. Includes shake or wood shingles, rust-resistant nails, and installation. Does not include backing or furring strips. Also includes 4% waste.					
replace, standard grade	6S@.045	sf	1.82	2.61	4.43
replace, high grade	6S@.045	sf	2.01	2.61	4.62
replace, deluxe grade	6S@.045	sf	2.43	2.61	5.04
replace, custom grade (fancy cut)	6S@.045	sf	3.29	2.61	5.90
remove	1D@.014	sf	—	.66	.66
replace single shingle	6S@.408	ea	5.25	23.70	28.95
minimum charge to replace shake or wood shingle	6S@3.25	ea	71.80	189.00	260.80
Vinyl siding. Includes siding with trim pieces, fasteners, and installation. Includes 4% waste.					
replace, standard grade	6S@.034	sf	2.89	1.98	4.87
replace, high grade	6S@.034	sf	3.30	1.98	5.28
replace, deluxe grade	6S@.034	sf	4.96	1.98	6.94
remove	1D@.007	sf	—	.33	.33
add for insulated siding	—	sf	.78	—	.78

	Craft@Hrs	Unit	Material	Labor	Total

7/16" thick hardboard siding. Includes hardboard lap siding, rust-resistant nails, and installation. Includes 4% waste.

	Craft@Hrs	Unit	Material	Labor	Total
replace, standard grade	6S@.024	sf	1.16	1.39	2.55
replace, painted-board finish	6S@.024	sf	1.29	1.39	2.68
replace, stained-board finish	6S@.024	sf	1.82	1.39	3.21
replace, simulated stucco finish	6S@.024	sf	1.42	1.39	2.81
remove	1D@.009	sf	—	.42	.42

Hardboard lap siding. Includes hardboard lap siding, rust-resistant nails, and installation. Also includes 4% waste.

	Craft@Hrs	Unit	Material	Labor	Total
replace	6S@.026	sf	1.36	1.51	2.87
remove	1D@.010	sf	—	.47	.47

5/8" thick plywood siding, texture 1-11. Includes texture 1-11 siding (commonly called T1-11), rust-resistant nails, and installation. Also includes 4% waste.

	Craft@Hrs	Unit	Material	Labor	Total
replace, cedar face	6S@.025	sf	2.43	1.45	3.88
replace, rough-sawn cedar face	6S@.025	sf	2.98	1.45	4.43
replace, fir face	6S@.025	sf	2.13	1.45	3.58
replace, rough-sawn fir face	6S@.025	sf	2.44	1.45	3.89
replace, southern yellow pine face	6S@.025	sf	1.78	1.45	3.23
replace, redwood face	6S@.025	sf	4.05	1.45	5.50
remove	1D@.010	sf	—	.47	.47
add for factory stained siding.	—	sf	.32	—	.32

Plywood with 1" x 4" stained cedar boards. Includes plywood siding with 1" x 4" cedar boards over joints, rust-resistant nails, and installation. Does not include finishing. Includes 4% waste.

	Craft@Hrs	Unit	Material	Labor	Total
replace, 1/2"	6S@.057	sf	6.17	3.31	9.48
replace, 5/8"	6S@.057	sf	6.29	3.31	9.60
remove	1D@.018	sf	—	.85	.85

Board-and-batten siding. Includes board-and-batten siding, rust-resistant nails, and installation. Board-and-batten siding is installed vertically. Includes 4% waste. For the wood grades shown in the price list item, deduct **9%** for next lower grade, deduct **14%** for third grade.

	Craft@Hrs	Unit	Material	Labor	Total
replace, pine, select grade	6S@.037	sf	5.83	2.15	7.98
replace, cedar, A grade	6S@.037	sf	6.10	2.15	8.25
remove	1D@.011	sf	—	.52	.52

board and batten

Board-on-board siding. Includes board-on-board siding, rust-resistant nails, and installation. Board-on-board siding is installed vertically. Includes 4% waste. For the wood grades shown in the price list item, deduct **9%** for next lower grade, deduct **14%** for third grade.

	Craft@Hrs	Unit	Material	Labor	Total
replace, pine, select grade	6S@.034	sf	5.96	1.98	7.94
replace, cedar, A grade	6S@.034	sf	6.24	1.98	8.22
remove	1D@.011	sf	—	.52	.52

board on board

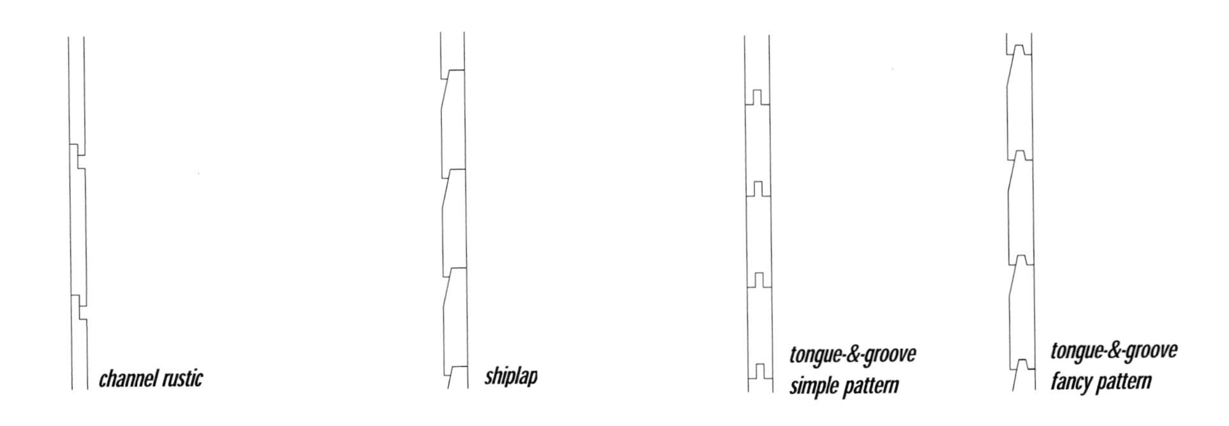

	Craft@Hrs	Unit	Material	Labor	Total

Channel rustic siding. Includes channel rustic siding, rust-resistant nails, and installation. Includes 4% waste. For the wood grades shown in the price list item, deduct **9%** for next lower grade, deduct **14%** for third grade.

	Craft@Hrs	Unit	Material	Labor	Total
replace, pine, select grade	6S@.039	sf	5.98	2.27	8.25
replace, cedar, A grade	6S@.039	sf	7.52	2.27	9.79
replace, redwood, clear all heart grade	6S@.039	sf	18.40	2.27	20.67
remove	1D@.011	sf	—	.52	.52

Shiplap siding. Includes shiplap siding, rust-resistant nails, and installation. Also includes 4% waste. For the wood grades shown in the price list item, deduct **9%** for next lower grade, deduct **14%** for third grade.

	Craft@Hrs	Unit	Material	Labor	Total
replace, pine, select grade	6S@.039	sf	6.06	2.27	8.33
replace, cedar, A grade	6S@.039	sf	7.61	2.27	9.88
replace, redwood, clear all heart grade	6S@.039	sf	18.50	2.27	20.77
remove	1D@.011	sf	—	.52	.52

Tongue-&-groove siding. Includes tongue-and-groove siding, rust-resistant nails, and installation. Includes 4% waste. For the wood grades shown in the price list item, deduct **9%** for next lower grade, deduct **14%** for third grade.

	Craft@Hrs	Unit	Material	Labor	Total
replace, pine, simple pattern, select grade	6S@.039	sf	6.20	2.27	8.47
replace, pine, fancy pattern, select grade	6S@.039	sf	6.62	2.27	8.89
replace, cedar, simple pattern, A grade	6S@.039	sf	7.76	2.27	10.03
replace, cedar, fancy pattern, A grade	6S@.039	sf	8.29	2.27	10.56
replace, redwood, simple pattern, clear all heart grade	6S@.039	sf	19.00	2.27	21.27
replace, redwood, fancy pattern, clear all heart grade	6S@.039	sf	19.90	2.27	22.17
remove	1D@.011	sf	—	.52	.52

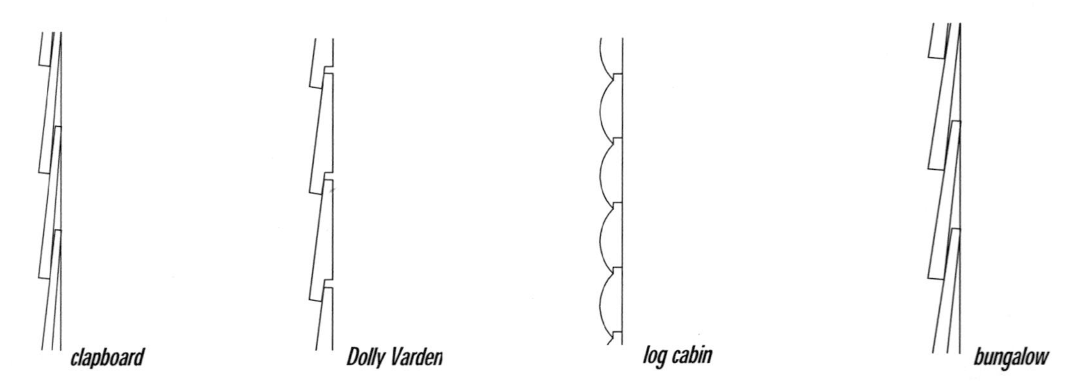

	Craft@Hrs	Unit	Material	Labor	Total
Radially sawn clapboard siding, eastern pine. Includes radially sawn clapboard, rust-resistant nails, and installation. Includes 4% waste. True clapboard. Based on grades marketed by Donnell's Clapboard Mill of Sedgwick, Maine.					
replace, #1 premium clear	6S@.055	sf	8.73	3.20	11.93
replace, #1 New England Cape	6S@.055	sf	7.60	3.20	10.80
remove	1D@.011	sf	—	.52	.52
Resawn beveled siding. Includes resawn, beveled clapboard, rust-resistant nails, and installation. Includes 4% waste. This siding looks like clapboard but is not radially sawn. The grain in a radially sawn board runs at a consistent 90-degree angle to the board face. In resawn boards grain orientation is inconsistent.					
replace, pine, select grade	6S@.041	sf	3.23	2.38	5.61
replace, cedar, select grade	6S@.041	sf	4.05	2.38	6.43
replace, redwood, clear all heart grade	6S@.041	sf	9.78	2.38	12.16
remove	1D@.011	sf	—	.52	.52
Bungalow siding. Includes bungalow siding, rust-resistant nails, and installation. For the wood grades shown in the price list item, deduct **9%** for next lower grade, deduct **14%** for third grade. Also includes 4% waste.					
replace, pine, select grade	6S@.039	sf	4.32	2.27	6.59
replace, cedar, select grade	6S@.039	sf	5.47	2.27	7.74
replace, redwood, clear all heart grade	6S@.039	sf	13.10	2.27	15.37
remove	1D@.011	sf	—	.52	.52
Dolly Varden siding. Includes Dolly Varden siding, rust-resistant nails, and installation. Includes 4% waste. For the wood grades shown in the price list item, deduct **9%** for next lower grade, deduct **14%** for third grade.					
replace, pine, select grade	6S@.039	sf	4.68	2.27	6.95
replace, cedar, select grade	6S@.039	sf	5.88	2.27	8.15
replace, redwood, clear all heart grade	6S@.039	sf	14.20	2.27	16.47
remove	1D@.011	sf	—	.52	.52

	Craft@Hrs	Unit	Material	Labor	Total

Log cabin siding. Includes log cabin siding, rust-resistant nails, and installation. Includes 4% waste. For the wood grades shown in the price list item, deduct **9%** for next lower grade, deduct **14%** for third grade.

	Craft@Hrs	Unit	Material	Labor	Total
replace, pine, select grade	6S@.039	sf	4.36	2.27	6.63
replace, cedar, select grade	6S@.039	sf	5.51	2.27	7.78
remove	1D@.011	sf	—	.52	.52

Add for factory stained siding. Replace only.

	Craft@Hrs	Unit	Material	Labor	Total
one coat	—	sf	.69	—	.69
two coat	—	sf	.99	—	.99

Wood siding repair. Replace only.

	Craft@Hrs	Unit	Material	Labor	Total
replace section	6S@.827	ea	35.20	48.00	83.20
minimum charge	6S@3.25	ea	96.20	189.00	285.20

Fascia. Includes fascia material, rust-resistant fasteners, and installation. Does not include soffits or rough fascia. Includes 4% waste.

	Craft@Hrs	Unit	Material	Labor	Total
replace, aluminum to 6" wide	6S@.034	lf	2.71	1.98	4.69
replace, aluminum 6" to 12" wide	6S@.034	lf	3.68	1.98	5.66
replace, vinyl to 6" wide	6S@.034	lf	3.70	1.98	5.68
replace, vinyl 6" to 12" wide	6S@.034	lf	5.13	1.98	7.11
replace, cedar to 6" wide	6S@.037	lf	3.37	2.15	5.52
replace, cedar 6" to 12" wide	6S@.037	lf	4.49	2.15	6.64
replace, redwood to 6" wide	6S@.037	lf	7.61	2.15	9.76
replace, redwood 6" to 12" wide	6S@.037	lf	10.10	2.15	12.25
remove, aluminum or vinyl	1D@.006	lf	—	.28	.28
remove, wood	1D@.007	lf	—	.33	.33
minimum charge	6S@3.00	ea	102.00	174.00	276.00

Soffit. Includes soffit material, rust-resistant fasteners, backing as needed, and installation. Does not include finish fascia. Includes 4% waste.

	Craft@Hrs	Unit	Material	Labor	Total
replace, aluminum	6S@.049	sf	4.62	2.85	7.47
replace, vinyl	6S@.049	sf	6.34	2.85	9.19
replace, cedar plywood	6S@.052	sf	3.22	3.02	6.24
replace, cedar rough-sawn plywood	6S@.052	sf	3.27	3.02	6.29
replace, fir plywood	6S@.052	sf	3.06	3.02	6.08
replace, fir rough-sawn plywood	6S@.052	sf	3.18	3.02	6.20
replace, redwood plywood	6S@.052	sf	8.45	3.02	11.47
replace, redwood rough-sawn plywood	6S@.052	sf	8.56	3.02	11.58
remove, aluminum or vinyl	1D@.009	sf	—	.42	.42
remove, wood	1D@.010	sf	—	.47	.47

Soffit, tongue-&-groove. Includes 4% waste.

	Craft@Hrs	Unit	Material	Labor	Total
replace, cedar, simple pattern	6S@.061	sf	6.10	3.54	9.64
replace, cedar, fancy pattern	6S@.061	sf	7.34	3.54	10.88
replace, fir, simple pattern	6S@.061	sf	5.69	3.54	9.23
replace, fir, fancy pattern	6S@.061	sf	5.98	3.54	9.52
replace, redwood, simple pattern	6S@.061	sf	11.70	3.54	15.24
replace, redwood, fancy pattern	6S@.061	sf	12.70	3.54	16.24
remove	1D@.012	sf	—	.57	.57

	Craft@Hrs	Unit	Material	Labor	Total

Rain gutter. Includes rain gutter with hidden rust-resistant fasteners and installation. Does not include downspouts. Includes 3% waste.

	Craft@Hrs	Unit	Material	Labor	Total
replace, aluminum	6S@.041	lf	6.56	2.38	8.94
replace, copper built-in box	6S@1.69	lf	60.50	98.20	158.70
replace, copper	6S@.041	lf	33.40	2.38	35.78
replace, galvanized steel	6S@.041	lf	16.60	2.38	18.98
replace, plastic	6S@.041	lf	5.28	2.38	7.66
replace, redwood	6S@.041	lf	74.90	2.38	77.28
replace, tin built-in box	6S@1.69	lf	43.10	98.20	141.30
replace, tin	6S@.041	lf	30.20	2.38	32.58
remove	1D@.011	lf	—	.52	.52
remove, built-in box	1D@.026	lf	—	1.23	1.23

built-in box

Rain gutter downspout. Includes manufactured downspout, fasteners, elbows, and installation. Does not include gutter. Also includes 3% waste.

	Craft@Hrs	Unit	Material	Labor	Total
replace, aluminum	6S@.038	lf	8.01	2.21	10.22
replace, copper built-in box	6S@.563	lf	38.10	32.70	70.80
replace, copper	6S@.038	lf	40.70	2.21	42.91
replace, galvanized steel	6S@.038	lf	20.60	2.21	22.81
replace, plastic	6S@.038	lf	6.45	2.21	8.66
replace, tin built-in box	6S@.038	lf	29.50	2.21	31.71
replace, tin	6S@.563	lf	36.70	32.70	69.40
remove	1D@.009	lf	—	.42	.42

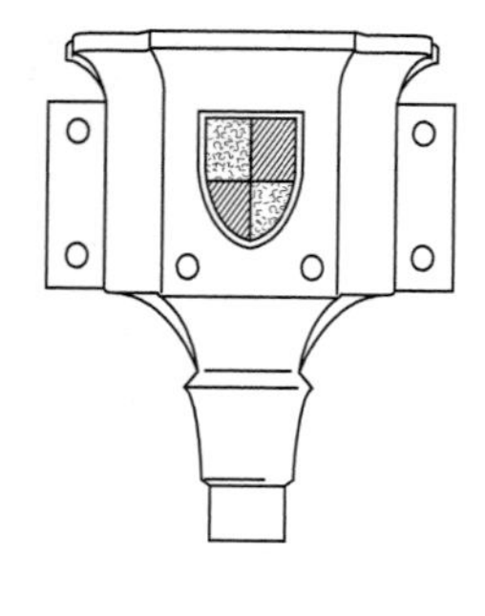

high grade

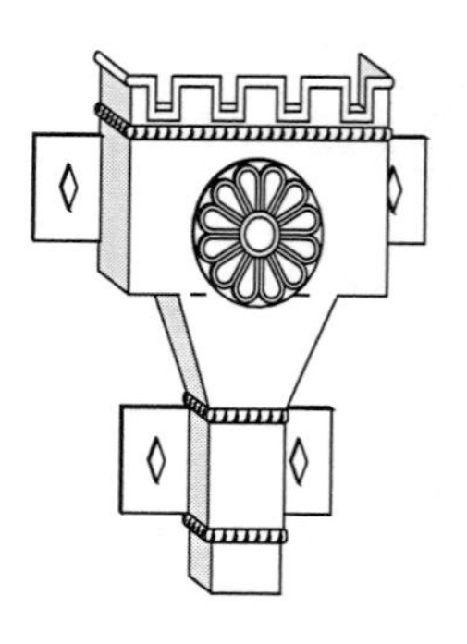

high grade

Rain gutter downspout conductor. Replace only. Includes manufactured downspout conductor, fasteners, and installation. Installation includes attachment to fascia and gutters but does not include the gutters or downspouts.

	Craft@Hrs	Unit	Material	Labor	Total
standard grade copper	6S@.432	ea	82.70	25.10	107.80
high grade copper	6S@.432	ea	154.00	25.10	179.10
standard grade galvanized steel	6S@.432	ea	67.70	25.10	92.80
high grade galvanized steel	6S@.432	ea	129.00	25.10	154.10

	Craft@Hrs	Unit	Material	Labor	Total

Polypropylene fixed shutter. Per 16" wide pair. For 12" wide deduct **42%**; for 18" wide add **8%**; for 20" wide add **14%**; for 24" wide add **26%**.

	Craft@Hrs	Unit	Material	Labor	Total
replace, 24" tall	6S@.323	ea	60.10	18.80	78.90
replace, 48" tall	6S@.323	ea	93.80	18.80	112.60
replace, 60" tall	6S@.323	ea	108.00	18.80	126.80
replace, 66" tall	6S@.323	ea	126.00	18.80	144.80
remove	1D@.156	ea	—	7.36	7.36

Wood fixed shutter. Includes wood shutter, mounting brackets and fasteners, and installation. Does not include painting. Per 16" wide pair. For 12" wide deduct **6%**; for 18" wide add **10%**; for 20" wide add **52%**; for 24" wide add **78%**.

	Craft@Hrs	Unit	Material	Labor	Total
replace, 36" tall	6S@.323	ea	196.00	18.80	214.80
replace, 48" tall	6S@.323	ea	223.00	18.80	241.80
replace, 60" tall	6S@.323	ea	250.00	18.80	268.80
replace, 72" tall	6S@.323	ea	272.00	18.80	290.80
remove	1D@.156	ea	—	7.36	7.36

Wood moveable shutter. Per 16" wide pair including hardware. For 12" wide deduct **15%**; for 18" wide add **7%**; for 20" wide add **14%**; for 24" wide add **27%**. For fixed half circle top add **$175** per pair. For moveable louvers with half-circle top add **$321** per pair. For raised panel shutter with half-circle top add **$199** per pair. For Gothic peaked top (Gothic arch) add **$334** per pair.

	Craft@Hrs	Unit	Material	Labor	Total
replace, 36" tall	6S@.333	ea	313.00	19.30	332.30
replace, 48" tall	6S@.333	ea	358.00	19.30	377.30
replace, 60" tall	6S@.333	ea	422.00	19.30	441.30
replace, 72" tall	6S@.333	ea	520.00	19.30	539.30
remove	1D@.158	ea	—	7.46	7.46
remove for work, then reinstall	6S@.451	ea	—	26.20	26.20

Time & Material Charts (selected items)
Siding Materials

See Siding material prices with the line items above.

Siding Labor

Laborer	base wage	paid leave	true wage	taxes & ins.	total
Siding installer	$37.60	2.93	$40.53	25.97	$66.50
Siding installer's helper	$27.10	2.11	$29.21	20.49	$49.70
Demolition laborer	$25.90	2.02	$27.92	19.28	$47.20

Paid leave is calculated based on two weeks paid vacation, one week sick leave, and seven paid holidays. Employer's matching portion of **FICA** is 7.65 percent. **FUTA** (Federal Unemployment) is .8 percent. **Worker's compensation** for the siding trade was calculated using a national average of 16.59 percent. **Unemployment insurance** was calculated using a national average of 8 percent. **Health insurance** was calculated based on a projected national average for 2020 of $1,280 per employee (and family when applicable) per month. Employer pays 80 percent for a per month cost of $1,024 per employee. **Retirement** is based on a 401(k) retirement program with employer matching of 50 percent. Employee contributions to the 401(k) plan are an average of 6 percent of the true wage. **Liability insurance** is based on a national average of 12.0 percent.

Siding Labor Productivity

	Craft@Hrs	Unit	Material	Labor	Total
Demolition of siding					
remove fiberglass corrugated siding	1D@.007	sf	—	.33	.33
remove aluminum siding	1D@.008	sf	—	.38	.38
remove cement fiber shingle siding	1D@.013	sf	—	.61	.61
remove shake or wood shingle siding	1D@.014	sf	—	.66	.66
remove vinyl siding	1D@.007	sf	—	.33	.33
remove hardboard siding	1D@.009	sf	—	.42	.42
remove hardboard beveled siding	1D@.010	sf	—	.47	.47
remove plywood siding	1D@.010	sf	—	.47	.47
remove wood siding	1D@.011	sf	—	.52	.52
remove aluminum or vinyl fascia	1D@.006	lf	—	.28	.28
remove wood fascia	1D@.007	lf	—	.33	.33
remove aluminum or vinyl soffit	1D@.009	sf	—	.42	.42
remove plywood soffit	1D@.010	sf	—	.47	.47
remove tongue-&-groove soffit	1D@.012	sf	—	.57	.57
remove rain gutter	1D@.011	lf	—	.52	.52
remove built-in box rain gutter	1D@.026	lf	—	1.23	1.23
remove downspout	1D@.009	lf	—	.42	.42
remove fixed shutter	1D@.156	ea	—	7.36	7.36
remove moveable shutter	1D@.158	ea	—	7.46	7.46

		Craft@Hrs	Unit	Material	Labor	Total
Siding installation crew						
install siding	siding installer	66.50				
install siding	installer's helper	49.70				
install siding	siding crew	58.10				

	Craft@Hrs	Unit	Material	Labor	Total
Install siding					
fiberglass corrugated	6S@.028	sf	—	1.63	1.63
aluminum	6S@.034	sf	—	1.98	1.98
cement shingle	6S@.045	sf	—	2.61	2.61
shake or wood shingle	6S@.045	sf	—	2.61	2.61
vinyl	6S@.034	sf	—	1.98	1.98
7/16" thick hardboard	6S@.024	sf	—	1.39	1.39
hardboard lap	6S@.026	sf	—	1.51	1.51
5/8" thick plywood	6S@.025	sf	—	1.45	1.45
plywood with 1" x 4" boards	6S@.057	sf	—	3.31	3.31
board-and-batten	6S@.037	sf	—	2.15	2.15
board-on-board	6S@.034	sf	—	1.98	1.98
drop siding	6S@.039	sf	—	2.27	2.27
clapboard	6S@.055	sf	—	3.20	3.20
bevel	6S@.041	sf	—	2.38	2.38
Install fascia					
aluminum or vinyl	6S@.034	lf	—	1.98	1.98
wood fascia	6S@.037	lf	—	2.15	2.15
Install soffit					
aluminum or vinyl	6S@.049	sf	—	2.85	2.85
plywood	6S@.052	sf	—	3.02	3.02
tongue-&-groove	6S@.061	sf	—	3.54	3.54
Install rain gutter					
typical	6S@.041	lf	—	2.38	2.38
built-in box	6S@1.69	lf	—	98.20	98.20
Install rain gutter downspout					
typical half-round or "K" style	6S@.038	lf	—	2.21	2.21
built-in box	6S@.563	lf	—	32.70	32.70
Install downspout conductor					
typical	6S@.432	ea	—	25.10	25.10
Install shutter					
fixed	6S@.323	ea	—	18.80	18.80
moveable	6S@.333	ea	—	19.30	19.30

Stairs

3' wide utility stairs. Per step. Includes three 2" x 12" stringers and 2" dimensional lumber treads. Does not include railing.

	Craft@Hrs	Unit	Material	Labor	Total
replace stairs	1C@.361	st	26.40	24.40	50.80
remove	1D@.101	st	—	4.77	4.77

Straight stairs. Per step. Includes four 2" x 12" stringers reinforced with plywood or waferboard. Does not include balustrade.

	Craft@Hrs	Unit	Material	Labor	Total
3' wide enclosed stairs					
replace covered treads and risers	1C@.474	st	47.30	32.00	79.30
replace with false oak tread and riser	1C@.563	st	98.80	38.10	136.90
replace with oak treads and paint-grade risers	1C@.474	st	99.70	32.00	131.70
replace with oak treads and oak risers	1C@.474	st	106.00	32.00	138.00
3' wide stairs, one side open					
replace covered treads and risers	1C@.485	st	47.30	32.80	80.10
replace with false oak tread and riser	1C@.575	st	120.00	38.90	158.90
replace with oak treads and paint-grade risers	1C@.485	st	99.70	32.80	132.50
replace with oak treads and oak risers	1C@.485	st	106.00	32.80	138.80
3' wide stairs, two sides open					
replace covered treads and risers	1C@.517	st	47.30	34.90	82.20
replace with false oak tread and riser	1C@.607	st	142.00	41.00	183.00
replace with oak treads and paint-grade risers	1C@.517	st	99.70	34.90	134.60
replace with oak treads and oak risers	1C@.517	st	106.00	34.90	140.90
remove	1D@.122	st	—	5.76	5.76

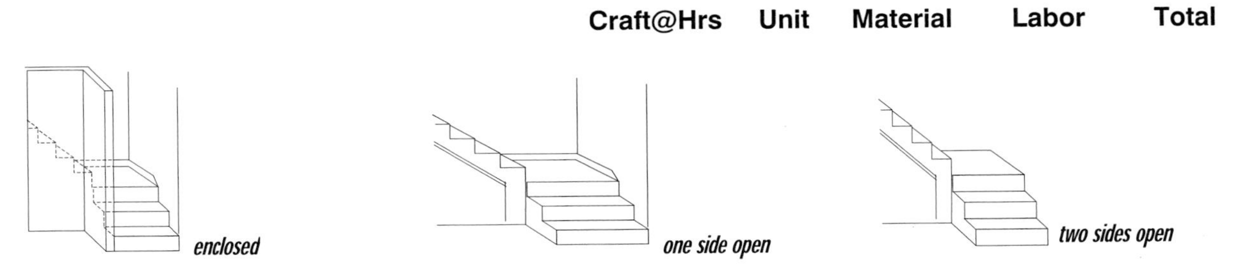

1/4 turn stairs. Per step. Ten step minimum. L-shaped stair system with 3' by 3' landing. Includes bearing walls, 2" x 8" joists, headers, and 3/4" tongue-&-groove sheathing. Includes four 2" x 12" stringers reinforced with plywood or waferboard. Does not include balustrade.

	Craft@Hrs	Unit	Material	Labor	Total
3' wide enclosed 1/4 turn stairs					
replace covered treads and risers	1C@.825	st	61.20	55.80	117.00
replace with false oak tread and riser	1C@1.16	st	117.00	78.40	195.40
replace with oak treads and paint-grade risers	1C@1.02	st	126.00	69.00	195.00
replace with oak treads and oak risers	1C@1.02	st	133.00	69.00	202.00
3' wide 1/4 turn stairs, one side open					
replace covered treads and risers	1C@.855	st	61.20	57.80	119.00
replace with false oak tread and riser	1C@1.20	st	141.00	81.10	222.10
replace with oak treads and paint-grade risers	1C@1.04	st	126.00	70.30	196.30
replace with oak treads and oak risers	1C@1.04	st	133.00	70.30	203.30
3' wide 1/4 turn stairs, two sides open					
replace covered treads and risers	1C@.885	st	61.20	59.80	121.00
replace with false oak tread and riser	1C@1.23	st	161.00	83.10	244.10
replace with oak treads and paint-grade risers	1C@1.05	st	126.00	71.00	197.00
replace with oak treads and oak risers	1C@1.05	st	133.00	71.00	204.00
remove	1D@.135	st	—	6.37	6.37

enclosed

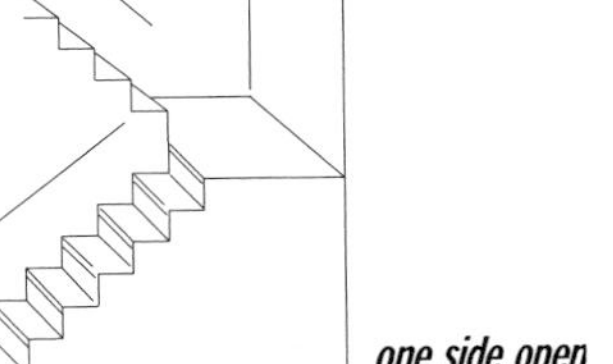

one side open

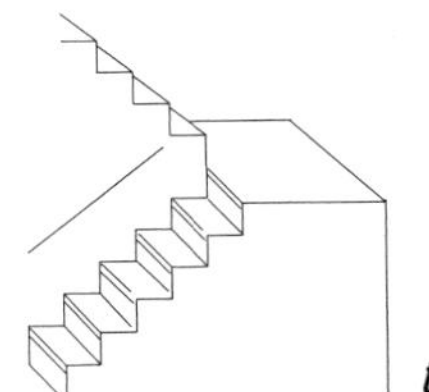

two sides open

1/2 turn stairs. Per step. Ten step minimum. U-shaped stair system with 6' by 3' landing. Landing includes bearing walls, 2" x 8" joists, headers, and 3/4" tongue-&-groove sheathing. Each stair system includes four 2" x 12" stringers reinforced with plywood or waferboard. Does not include balustrade.

	Craft@Hrs	Unit	Material	Labor	Total
3' wide enclosed 1/2 turn stairs					
replace covered treads and risers	1C@.908	st	74.80	61.40	136.20
replace with false oak tread and riser	1C@1.26	st	139.00	85.20	224.20
replace with oak treads and paint-grade risers	1C@1.12	st	147.00	75.70	222.70
replace with oak treads and oak risers	1C@1.12	st	151.00	75.70	226.70
3' wide 1/2 turn stairs, one side open					
replace covered treads and risers	1C@.943	st	74.80	63.70	138.50
replace with false oak tread and riser	1C@1.31	st	182.00	88.60	270.60
replace with oak treads and paint-grade risers	1C@1.15	st	147.00	77.70	224.70
replace with oak treads and oak risers	1C@1.15	st	151.00	77.70	228.70
3' wide 1/2 turn stairs, two sides open					
replace covered treads and risers	1C@.970	st	74.80	65.60	140.40
replace with false oak tread and riser	1C@1.34	st	182.00	90.60	272.60
replace with oak treads and paint-grade risers	1C@1.16	st	147.00	78.40	225.40
replace with oak treads and oak risers	1C@1.16	st	151.00	78.40	229.40
remove	1D@.150	st	—	7.08	7.08

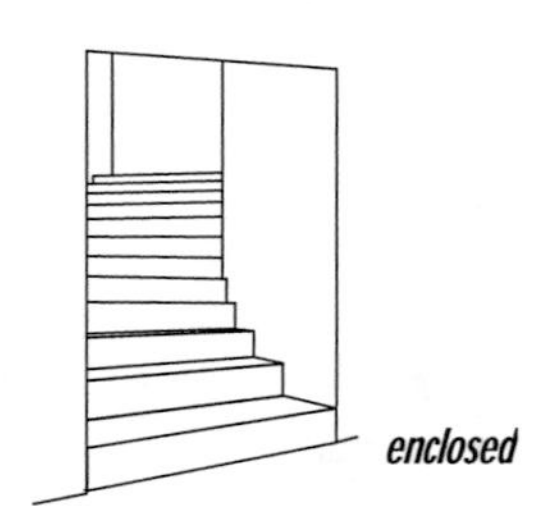
enclosed

one side open

two sides open

Circular stairs. Per step. Circular stairs usually form half of a circle or less. Based on 9' radius. Labor prices are for open or closed stringer stairs hand-framed on job site. But overall prices also apply for purchase and installation of prefabricated circular stairs. Does not include stair balustrade.

	Craft@Hrs	Unit	Material	Labor	Total
3' wide enclosed circular stairs					
replace covered treads and risers	1C@11.6	st	303.00	784.00	1,087.00
replace with false oak tread and riser	1C@11.8	st	355.00	798.00	1,153.00
replace with oak treads and paint-grade risers	1C@11.6	st	357.00	784.00	1,141.00
replace with oak treads and oak risers	1C@11.6	st	363.00	784.00	1,147.00
3' wide circular stairs, one side open					
replace covered treads and risers	1C@11.8	st	303.00	798.00	1,101.00
replace with false oak tread and riser	1C@12.0	st	378.00	811.00	1,189.00
replace with oak treads and paint-grade risers	1C@11.8	st	357.00	798.00	1,155.00
replace with oak treads and oak risers	1C@11.8	st	363.00	798.00	1,161.00
3' wide circular stairs, two sides open					
replace covered treads and risers	1C@11.9	st	303.00	804.00	1,107.00
replace with false oak tread and riser	1C@12.1	st	398.00	818.00	1,216.00
replace with oak treads and paint-grade risers	1C@11.9	st	357.00	804.00	1,161.00
replace with oak treads and oak risers	1C@11.9	st	363.00	804.00	1,167.00
remove	1D@.158	st	—	7.46	7.46

Additional costs for stairs.

	Craft@Hrs	Unit	Material	Labor	Total
add for 4' wide stairs	—	%	12.0	—	—
add for 4' wide circular stairs	—	%	16.0	—	—

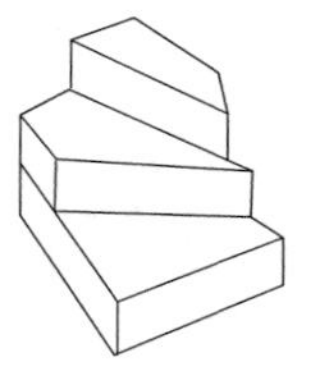

Add for winder on 1/4 and 1/2 turn platforms. Winders are pie-shaped stair treads, usually placed on top of the landing. Each winder is a box made from 2" x 8" boards ripped to the height of the riser. Additional ripped 2" x 8" boards are placed 12" on center inside the box for reinforcement. (Add **3%** for risers over 7-3/8" made from ripped 2" x 10" boards.) Winders are stacked on top of each other. Replace only.

	Craft@Hrs	Unit	Material	Labor	Total
particleboard tread, paint-grade riser	1C@.400	ea	69.70	27.00	96.70
oak tread, paint-grade riser	1C@.400	ea	121.00	27.00	148.00
oak tread, oak riser	1C@.400	ea	129.00	27.00	156.00

	Craft@Hrs	Unit	Material	Labor	Total

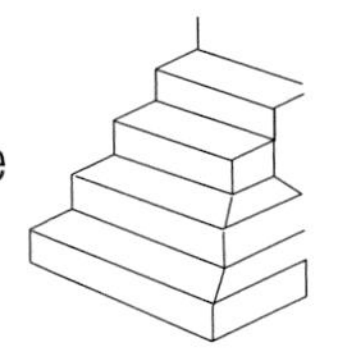

Add for mitered corners. For stairs where the bottom three steps turn at a 45-degree angle (or less) making three steps from the side and front. The three steps on the side are made from stacked boxes made from 2" x 8" boards ripped to the riser height, sheathed with 3/4" plywood then capped with the tread and riser material. Cost per side. (For example, for stairs with mitered corners on both sides, use this price twice.) Replace only.

	Craft@Hrs	Unit	Material	Labor	Total
bottom three treads	1C@2.32	ea	190.00	157.00	347.00

Add for bullnose on starting step. Red oak. Standard grade up to 4" wide. High grade up to 5" wide.

	Craft@Hrs	Unit	Material	Labor	Total
Starting step with bullnose one side					
standard grade	1C@.492	ea	349.00	33.30	382.30
high grade	1C@.492	ea	420.00	33.30	453.30
Starting step with bullnose two sides					
standard grade	1C@.492	ea	483.00	33.30	516.30
high grade	1C@.492	ea	583.00	33.30	616.30

Prefabricated spiral stairs. Per step. Prefabricated spiral stairs assembled and installed by carpenter on site. Treads are made from aluminum, steel, decorative cast iron (custom grade and above) or waferboard (for carpet). Includes balustrade. Add per lf for landing or balcony railing to match: Economy **$80**; standard **$107**; high **$120**; custom **$134**; deluxe **$148**; custom deluxe **$187**.

	Craft@Hrs	Unit	Material	Labor	Total
replace, economy grade	1C@.855	st	281.00	57.80	338.80
replace, standard grade	1C@.855	st	498.00	57.80	555.80
replace, high grade	1C@.855	st	548.00	57.80	605.80
replace, custom grade	1C@.855	st	598.00	57.80	655.80
replace, deluxe grade	1C@.855	st	722.00	57.80	779.80
replace, custom deluxe grade	1C@.855	st	789.00	57.80	846.80
remove	1D@.123	st	—	5.81	5.81

Disappearing attic stairs. Fold-up attic access stairs with side rails.

	Craft@Hrs	Unit	Material	Labor	Total
replace, standard grade	1C@2.39	ea	507.00	162.00	669.00
replace, high grade	1C@2.39	ea	851.00	162.00	1,013.00
replace, deluxe grade	1C@2.39	ea	1,440.00	162.00	1,602.00
remove	1D@.270	ea	—	12.70	12.70
remove for work, then reinstall	1C@3.33	ea	—	225.00	225.00

	Craft@Hrs	Unit	Material	Labor	Total

Pine stair balustrade. Balustrade built on stairs. Balusters connect directly to treads. Includes newel post. Add **9%** for red oak newel and hand rail. Add **85%** for installation on curved stairs. Deduct **10%** for balustrade installed on porch or other level application.

	Craft@Hrs	Unit	Material	Labor	Total
1-1/2" pine stair balusters					
replace, balusters	1C@1.32	lf	39.50	89.20	128.70
replace, fluted balusters	1C@1.32	lf	47.30	89.20	136.50
replace, spiral balusters	1C@1.32	lf	73.80	89.20	163.00
2-1/2" pine stair balusters					
replace, balusters	1C@1.33	lf	44.70	89.90	134.60
replace, fluted balusters	1C@1.33	lf	53.60	89.90	143.50
replace, spiral balusters	1C@1.33	lf	83.00	89.90	172.90
3-1/2" pine stair balusters					
replace, balusters	1C@1.35	lf	54.30	91.30	145.60
replace, fluted balusters	1C@1.35	lf	65.10	91.30	156.40
replace, spiral balusters	1C@1.35	lf	101.00	91.30	192.30
5-1/2" pine stair balusters					
replace, balusters	1C@1.41	lf	86.20	95.30	181.50
replace, fluted balusters	1C@1.41	lf	102.00	95.30	197.30
replace, spiral balusters	1C@1.41	lf	151.00	95.30	246.30
Remove	1D@.053	lf	—	2.50	2.50
Remove stair balustrade for work, then reinstall	1C@2.39	lf	—	162.00	162.00

Poplar stair balustrade. Add **9%** for red oak balustrades and newel post in poplar stair balustrade. Add **85%** for installation on curved stairs. Deduct **10%** for balustrade installed on porch or other level application.

	Craft@Hrs	Unit	Material	Labor	Total
1-1/2" poplar stair balusters					
replace, balusters	1C@1.32	lf	65.50	89.20	154.70
replace, fluted balusters	1C@1.32	lf	79.20	89.20	168.40
replace, spiral balusters	1C@1.32	lf	126.00	89.20	215.20
2-1/2" poplar stair balusters					
replace, balusters	1C@1.33	lf	74.00	89.90	163.90
replace, fluted balusters	1C@1.33	lf	89.40	89.90	179.30
replace, spiral balusters	1C@1.33	lf	142.00	89.90	231.90
3-1/2" poplar stair balusters					
replace, balusters	1C@1.35	lf	90.10	91.30	181.40
replace, fluted balusters	1C@1.35	lf	109.00	91.30	200.30
replace, spiral balusters	1C@1.35	lf	173.00	91.30	264.30
5-1/2" poplar stair balusters					
replace, balusters	1C@1.41	lf	141.00	95.30	236.30
replace, fluted balusters	1C@1.41	lf	167.00	95.30	262.30
replace, spiral balusters	1C@1.41	lf	258.00	95.30	353.30
Remove	1D@.053	lf	—	2.50	2.50
Remove stair balustrade for work, then reinstall	1C@2.39	lf	—	162.00	162.00

Redwood stair balustrade. Includes newel post. Add **85%** for installation on curved stairs. Deduct **10%** for balustrade installed on porch or other level application.

	Craft@Hrs	Unit	Material	Labor	Total
1-1/2" redwood stair balusters					
replace, balusters	1C@1.32	lf	73.10	89.20	162.30
replace, fluted balusters	1C@1.32	lf	88.20	89.20	177.40
replace, spiral balusters	1C@1.32	lf	140.00	89.20	229.20
2-1/2" redwood stair balusters					
replace, balusters	1C@1.33	lf	82.20	89.90	172.10
replace, fluted balusters	1C@1.33	lf	99.70	89.90	189.60
replace, spiral balusters	1C@1.33	lf	154.00	89.90	243.90
3-1/2" redwood stair balusters					
replace, balusters	1C@1.35	lf	100.00	91.30	191.30
replace, fluted balusters	1C@1.35	lf	121.00	91.30	212.30
replace, spiral balusters	1C@1.35	lf	190.00	91.30	281.30
5-1/2" redwood stair balusters					
replace, balusters	1C@1.41	lf	154.00	95.30	249.30
replace, fluted balusters	1C@1.41	lf	187.00	95.30	282.30
replace, spiral balusters	1C@1.41	lf	284.00	95.30	379.30
Stair balustrade					
remove	1D@.053	lf	—	2.50	2.50
remove stair balustrade for work, then reinstall	1C@2.39	lf	—	162.00	162.00

Red oak stair balustrade. Includes newel post. Add **85%** for installation on curved stairs. Deduct **10%** for balustrade installed on porch or other level application.

	Craft@Hrs	Unit	Material	Labor	Total
1-1/2" red oak stair balustrade					
replace, balusters	1C@1.32	lf	92.40	89.20	181.60
replace, fluted balusters	1C@1.32	lf	110.00	89.20	199.20
replace, spiral balusters	1C@1.32	lf	173.00	89.20	262.20
replace, hand-carved balusters	1C@1.32	lf	354.00	89.20	443.20
replace, heavy hand-carved balusters	1C@1.32	lf	380.00	89.20	469.20
2-1/2" red oak stair balustrade					
replace, balusters	1C@1.33	lf	104.00	89.90	193.90
replace, fluted balusters	1C@1.33	lf	126.00	89.90	215.90
replace, spiral balusters	1C@1.33	lf	196.00	89.90	285.90
replace, hand-carved balusters	1C@1.33	lf	359.00	89.90	448.90
replace, heavy hand-carved balusters	1C@1.33	lf	385.00	89.90	474.90
3-1/2" red oak stair balustrade					
replace, balusters	1C@1.35	lf	126.00	91.30	217.30
replace, fluted balusters	1C@1.35	lf	151.00	91.30	242.30
replace, spiral balusters	1C@1.35	lf	237.00	91.30	328.30
replace, hand-carved balusters	1C@1.35	lf	439.00	91.30	530.30
replace, heavy hand-carved balusters	1C@1.35	lf	476.00	91.30	567.30
5-1/2" red oak stair balustrade					
replace, balusters	1C@1.41	lf	201.00	95.30	296.30
replace, fluted balusters	1C@1.41	lf	237.00	95.30	332.30
replace, spiral balusters	1C@1.41	lf	359.00	95.30	454.30
replace, hand-carved balusters	1C@1.41	lf	426.00	95.30	521.30
replace, heavy hand-carved balusters	1C@1.41	lf	474.00	95.30	569.30
Stair balustrade					
remove	1D@.053	lf	—	2.50	2.50
remove for work, then reinstall	1C@2.39	lf	—	162.00	162.00

Sawn stair balustrade. Includes newel post. Deduct **10%** for balustrade installed on porch or other level application.

	Craft@Hrs	Unit	Material	Labor	Total
Pine sawn stair balustrade					
replace	1C@1.32	lf	66.10	89.20	155.30
replace, with panels	1C@1.32	lf	112.00	89.20	201.20
Poplar sawn stair balustrade					
replace	1C@1.32	lf	89.80	89.20	179.00
replace, with panels	1C@1.32	lf	138.00	89.20	227.20
Add for red oak hand rail and newel post in paint-grade sawn stair balustrade	—	lf	18.50	—	18.50
Redwood sawn stair balustrade					
replace	1C@1.32	lf	151.00	89.20	240.20
replace, with panels	1C@1.32	lf	241.00	89.20	330.20
Red oak sawn stair balustrade					
replace	1C@1.32	lf	122.00	89.20	211.20
replace, with panels	1C@1.32	lf	182.00	89.20	271.20
Stair balustrade					
remove	1D@.053	lf	—	2.50	2.50
remove for work, then reinstall	1C@2.39	lf	—	162.00	162.00

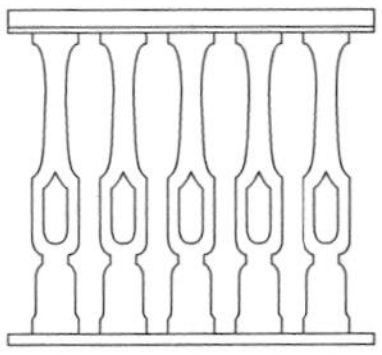

sawn balusters

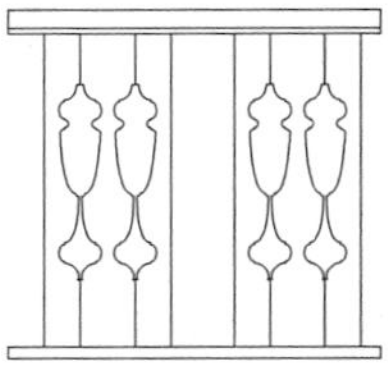

sawn baluster panels

volute end

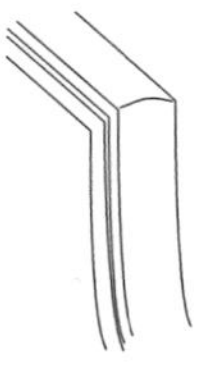

goose neck

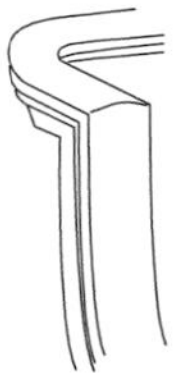

add for 1/4 turn in stair rail

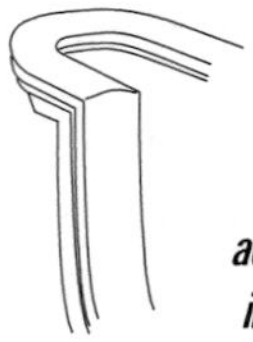

add for 1/2 turn in stair rail

Additional stair balustrade costs. Oak, poplar, or pine. Volute post includes additional balusters and labor to install volute end with circled balusters and newel. Usually not used on stairs with balusters larger than 2-1/2".

	Craft@Hrs	Unit	Material	Labor	Total
add for volute end	1C@3.03	ea	375.00	205.00	580.00
add for goose neck	1C@2.44	ea	425.00	165.00	590.00
add for 1/4 turn in stair rail	1C@2.27	ea	132.00	153.00	285.00
add for 1/2 turn in stair rail	1C@2.44	ea	303.00	165.00	468.00

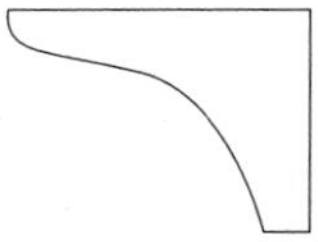

standard

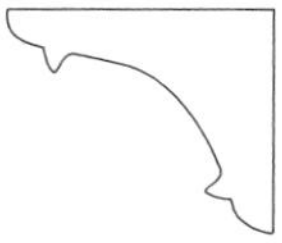

high grade

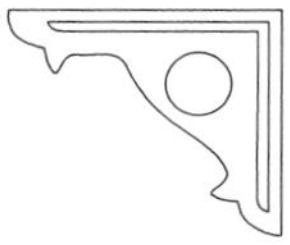

deluxe grade

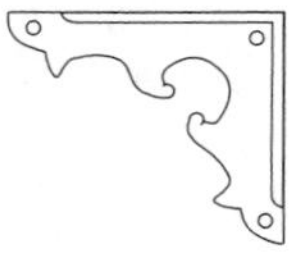

custom grade

Stair bracket. Brackets installed beneath open end of tread.

	Craft@Hrs	Unit	Material	Labor	Total
replace, standard grade	1C@.091	ea	14.80	6.15	20.95
replace, high grade	1C@.091	ea	17.20	6.15	23.35
replace, deluxe grade	1C@.091	ea	20.80	6.15	26.95
replace, custom grade	1C@.091	ea	23.20	6.15	29.35
remove	1D@.054	ea	—	2.55	2.55
remove for work, then reinstall	1C@.149	ea	—	10.10	10.10

	Craft@Hrs	Unit	Material	Labor	Total
Molding trim beneath tread. Up to 3/4" cove, quarter-round or similar installed beneath the open end of the tread.					
replace, pine	1C@.108	ea	.84	7.30	8.14
replace, red oak	1C@.108	ea	1.36	7.30	8.66
remove	1D@.019	ea	—	.90	.90
Pine newel.					
Pine newel for 1-1/2" balusters (newel is 3-1/2" to 4" wide)					
replace, newel post	1C@.961	ea	67.50	65.00	132.50
replace, fluted newel post	1C@.961	ea	89.20	65.00	154.20
replace, spiral newel post	1C@.961	ea	91.80	65.00	156.80
Pine newel for 2-1/2" balusters (newel is 4" to 4-1/2" wide)					
replace, newel post	1C@.961	ea	81.10	65.00	146.10
replace, fluted newel post	1C@.961	ea	107.00	65.00	172.00
replace, spiral newel post	1C@.961	ea	109.00	65.00	174.00
Pine newel for 3-1/2" balusters (newel is 4-1/2" to 5" wide)					
replace, newel post	1C@.961	ea	97.30	65.00	162.30
replace, fluted newel post	1C@.961	ea	129.00	65.00	194.00
replace, spiral newel post	1C@.961	ea	132.00	65.00	197.00
Pine newel for 5-1/2" balusters (newel is 6" to 7" wide)					
replace, newel post	1C@.961	ea	152.00	65.00	217.00
replace, fluted newel post	1C@.961	ea	206.00	65.00	271.00
replace, spiral newel post	1C@.961	ea	214.00	65.00	279.00
Remove	1D@.188	ea	—	8.87	8.87
Remove newel post for work, then reinstall	1C@1.41	ea	—	95.30	95.30
Poplar newel.					
Poplar newel for 1-1/2" balusters (newel is 3-1/2" to 4" wide)					
replace, newel post	1C@.961	ea	112.00	65.00	177.00
replace, fluted newel post	1C@.961	ea	149.00	65.00	214.00
replace, spiral newel post	1C@.961	ea	151.00	65.00	216.00
Poplar newel for 2-1/2" balusters (newel is 4" to 4-1/2" wide)					
replace, newel post	1C@.961	ea	134.00	65.00	199.00
replace, fluted newel post	1C@.961	ea	179.00	65.00	244.00
replace, spiral newel post	1C@.961	ea	183.00	65.00	248.00
Poplar newel for 3-1/2" balusters (newel is 4-1/2" to 5" wide)					
replace, newel post	1C@.961	ea	159.00	65.00	224.00
replace, fluted newel post	1C@.961	ea	218.00	65.00	283.00
replace, spiral newel post	1C@.961	ea	222.00	65.00	287.00

	Craft@Hrs	Unit	Material	Labor	Total
Poplar newel for 5-1/2" balusters					
(newel is 6" to 8" wide)					
replace, newel post	1C@.961	ea	257.00	65.00	322.00
replace, fluted newel post	1C@.961	ea	342.00	65.00	407.00
replace, spiral newel post	1C@.961	ea	352.00	65.00	417.00
Remove	1D@.188	ea	—	8.87	8.87
Remove newel post for work, then reinstall	1C@1.41	ea	—	95.30	95.30

Redwood newel.

	Craft@Hrs	Unit	Material	Labor	Total
Redwood newel for 1-1/2" balusters					
(newel is 3-1/2" to 4" wide)					
replace, newel post	1C@.961	ea	126.00	65.00	191.00
replace, fluted newel post	1C@.961	ea	166.00	65.00	231.00
replace, spiral newel post	1C@.961	ea	172.00	65.00	237.00
Redwood newel for 2-1/2" balusters					
(newel is 4" to 4-1/2" wide)					
replace, newel post	1C@.961	ea	151.00	65.00	216.00
replace, fluted newel post	1C@.961	ea	201.00	65.00	266.00
replace, spiral newel post	1C@.961	ea	208.00	65.00	273.00
Redwood newel for 3-1/2" balusters					
(newel is 4-1/2" to 5" wide)					
replace, newel post	1C@.961	ea	181.00	65.00	246.00
replace, fluted newel post	1C@.961	ea	241.00	65.00	306.00
replace, spiral newel post	1C@.961	ea	248.00	65.00	313.00
Redwood newel for 5-1/2" balusters					
(newel is 6" to 8" wide)					
replace, newel post	1C@.961	ea	285.00	65.00	350.00
replace, fluted newel post	1C@.961	ea	382.00	65.00	447.00
replace, spiral newel post	1C@.961	ea	393.00	65.00	458.00
Remove	1D@.188	ea	—	8.87	8.87
Remove newel post for work, then reinstall	1C@1.41	ea	—	95.30	95.30

Red oak newel.

	Craft@Hrs	Unit	Material	Labor	Total
Red oak newel for 1-1/2" balusters					
(newel is 3-1/2" to 4" wide)					
replace, newel post	1C@.961	ea	163.00	65.00	228.00
replace, fluted newel post	1C@.961	ea	220.00	65.00	285.00
replace, spiral newel post	1C@.961	ea	227.00	65.00	292.00
replace, hand-carved newel post	1C@.961	ea	303.00	65.00	368.00
replace, heavy hand-carved newel post	1C@.961	ea	328.00	65.00	393.00
Red oak newel for 2-1/2" balusters					
(newel is 4" to 4-1/2" wide)					
replace, newel post	1C@.961	ea	187.00	65.00	252.00
replace, fluted newel post	1C@.961	ea	249.00	65.00	314.00
replace, spiral newel post	1C@.961	ea	256.00	65.00	321.00
replace, hand-carved newel post	1C@.961	ea	344.00	65.00	409.00
replace, heavy hand-carved newel post	1C@.961	ea	370.00	65.00	435.00

	Craft@Hrs	Unit	Material	Labor	Total
Red oak newel for 3-1/2" balusters					
(newel is 4-1/2" to 5" wide)					
replace, newel post	1C@.961	ea	227.00	65.00	292.00
replace, fluted newel post	1C@.961	ea	297.00	65.00	362.00
replace, spiral newel post	1C@.961	ea	307.00	65.00	372.00
replace, hand-carved newel post	1C@.961	ea	375.00	65.00	440.00
replace, heavy hand-carved newel post	1C@.961	ea	399.00	65.00	464.00
Red oak newel for 5-1/2" balusters					
(newel is 6" to 8" wide)					
replace, newel post	1C@.961	ea	359.00	65.00	424.00
replace, fluted newel post	1C@.961	ea	475.00	65.00	540.00
replace, spiral newel post	1C@.961	ea	490.00	65.00	555.00
replace, hand-carved newel post	1C@.961	ea	666.00	65.00	731.00
replace, heavy hand-carved newel post	1C@.961	ea	719.00	65.00	784.00
Remove	1D@.188	ea	—	8.87	8.87
Remove newel post for work, then reinstall	1C@1.41	ea	—	95.30	95.30

Additional wood species. Add to the cost of poplar stair components.

	Craft@Hrs	Unit	Material	Labor	Total
add for mahogany	—	%	24.0	—	—
add for cherry	—	%	43.0	—	—
add for maple	—	%	11.0	—	—
add for birch	—	%	2.0	—	—

Time & Material Charts (selected items)

Stairs Materials

See Stairs material prices with the line items above.

Stairs Labor

Laborer	base wage	paid leave	true wage	taxes & ins.	total
Carpenter	$38.10	2.97	$41.07	26.53	$67.60
Demolition laborer	$25.90	2.02	$27.92	19.28	$47.20

Paid leave is calculated based on two weeks paid vacation, one week sick leave, and seven paid holidays. Employer's matching portion of **FICA** is 7.65 percent. **FUTA** (Federal Unemployment) is .8 percent. **Worker's compensation** for the stairs trade was calculated using a national average of 17.21 percent. **Unemployment insurance** was calculated using a national average of 8 percent. **Health insurance** was calculated based on a projected national average for 2020 of $1,280 per employee (and family when applicable) per month. Employer pays 80 percent for a per month cost of $1,024 per employee. **Retirement** is based on a 401(k) retirement program with employer matching of 50 percent. Employee contributions to the 401(k) plan are an average of 6 percent of the true wage. **Liability insurance** is based on a national average of 12.0 percent.

Stairs Labor Productivity

	Craft@Hrs	Unit	Material	Labor	Total
Demolition of stairs					
remove utility stairs	1D@.101	st	—	4.77	4.77
remove stairs	1D@.122	st	—	5.76	5.76
remove 1/4 turn stairs	1D@.135	st	—	6.37	6.37
remove 1/2 turn stairs	1D@.150	st	—	7.08	7.08
remove circular stairs	1D@.158	st	—	7.46	7.46

	Craft@Hrs	Unit	Material	Labor	Total
remove spiral stairs	1D@.123	st	—	5.81	5.81
remove disappearing attic stairs	1D@.270	ea	—	12.70	12.70
remove stair balustrade	1D@.053	lf	—	2.50	2.50
remove stair bracket	1D@.054	ea	—	2.55	2.55
remove newel post	1D@.188	ea	—	8.87	8.87
Build 3' wide stairs with three stringers					
utility stairs	1C@.361	st	—	24.40	24.40
Build 3' wide stairs with four stringers					
3' wide enclosed	1C@.474	st	—	32.00	32.00
3' wide, one side open	1C@.485	st	—	32.80	32.80
3' wide , two sides open	1C@.517	st	—	34.90	34.90
Build 3' wide 1/4 turn stairs with four stringers, and a 3' x 3' landing					
3' wide enclosed	1C@.825	st	—	55.80	55.80
3' wide enclosed & oak plank on landing	1C@1.02	st	—	69.00	69.00
3' wide, one side open	1C@.855	st	—	57.80	57.80
3' wide, one side open & oak plank on landing	1C@1.04	st	—	70.30	70.30
3' wide, two sides open	1C@.885	st	—	59.80	59.80
3' wide, two sides open & oak plank on landing	1C@1.05	st	—	71.00	71.00
Build 3' wide 1/2 turn stairs with four stringers, and a 6' x 3' landing					
3' wide enclosed	1C@.908	st	—	61.40	61.40
3' wide enclosed with oak plank on landing	1C@1.12	st	—	75.70	75.70
3' wide, one side open	1C@.943	st	—	63.70	63.70
3' wide, one side open & oak plank on landing	1C@1.15	st	—	77.70	77.70
3' wide, two sides open	1C@.970	st	—	65.60	65.60
3' wide, two sides open & oak plank on landing	1C@1.16	st	—	78.40	78.40
Additional stair costs					
add to build platform winders	1C@.400	ea	—	27.00	27.00
add to build mitered corner on bottom step	1C@2.32	ea	—	157.00	157.00
add to install bullnose starting step	1C@.492	ea	—	33.30	33.30
Install false treads and risers					
closed end	1C@.070	ea	—	4.73	4.73
open end	1C@.090	ea	—	6.08	6.08
Build 3' wide circular stairs					
enclosed	1C@11.6	st	—	784.00	784.00
one side open	1C@11.8	st	—	798.00	798.00
two sides open	1C@11.9	st	—	804.00	804.00
Install prefabricated spiral stairs					
all grades	1C@.855	st	—	57.80	57.80
Install disappearing attic stairs					
all grades	1C@2.39	ea	—	162.00	162.00
remove for work, then reinstall	1C@3.33	ea	—	225.00	225.00
Install wood stair balustrade					
with 1-1/2" balusters	1C@1.32	lf	—	89.20	89.20
with 2-1/2" balusters	1C@1.33	lf	—	89.90	89.90
with 3-1/2" balusters	1C@1.35	lf	—	91.30	91.30
with 5-1/2" balusters	1C@1.41	lf	—	95.30	95.30
remove for work, then reinstall	1C@2.39	lf	—	162.00	162.00
Install newel post					
install	1C@.961	ea	—	65.00	65.00
remove for work, then reinstall	1C@1.41	ea	—	95.30	95.30

Suspended Ceilings

12" x 12" concealed grid suspended ceiling tile. Includes 12" x 12" suspended ceiling tiles, and installation. Tiles are designed to be installed in a concealed grid. Does not include the concealed grid system.

	Craft@Hrs	Unit	Material	Labor	Total
smooth face	1S@.019	sf	1.69	1.03	2.72
fissured face	1S@.019	sf	2.03	1.03	3.06
textured face	1S@.019	sf	2.30	1.03	3.33
patterned face	1S@.019	sf	2.49	1.03	3.52
remove	1D@.006	sf	—	.28	.28

Concealed grid for ceiling tile. Includes main runners, cross tees, wires, and eyelet screws. Does not include tiles.

	Craft@Hrs	Unit	Material	Labor	Total
12" x 12" grid system	1S@.020	sf	1.89	1.08	2.97
remove	1D@.006	sf	—	.28	.28

Additional costs for 12" x 12" concealed grid system.

	Craft@Hrs	Unit	Material	Labor	Total
access panel	1S@.725	ea	90.30	39.20	129.50
remove	1D@.097	ea	—	4.58	4.58
relevel sagging grid	1S@.020	sf	—	1.08	1.08

2' x 4' suspended ceiling tile. Includes 2' x 4' suspended ceiling tiles, and installation. Tiles are designed to be installed in a suspended grid. Does not include the suspended grid system.

	Craft@Hrs	Unit	Material	Labor	Total
smooth face	1S@.011	sf	.74	.60	1.34
fissured face	1S@.011	sf	1.01	.60	1.61
textured face	1S@.011	sf	1.27	.60	1.87
patterned face	1S@.011	sf	1.66	.60	2.26
remove	1D@.005	sf	—	.24	.24
add for fire-rated tile installed with clips	1S@.008	sf	.55	.43	.98

2' x 4' luminous panels. Includes luminous panels and installation. Does not include grid.

	Craft@Hrs	Unit	Material	Labor	Total
polystyrene cracked-ice or mist white	1S@.010	sf	1.79	.54	2.33
acrylic cracked-ice or mist white	1S@.010	sf	3.49	.54	4.03
polystyrene egg crate	1S@.010	sf	1.73	.54	2.27
acrylic egg crate	1S@.010	sf	6.48	.54	7.02
acrylic egg crate with stainless steel finish	1S@.010	sf	7.08	.54	7.62
acrylic egg crate with brass finish	1S@.010	sf	7.18	.54	7.72
square crate polystyrene	1S@.010	sf	1.70	.54	2.24
square crate acrylic	1S@.010	sf	6.11	.54	6.65
square crate acrylic with stainless steel finish	1S@.010	sf	6.89	.54	7.43
square crate acrylic with brass finish	1S@.010	sf	6.98	.54	7.52
remove	1D@.005	ea	—	.24	.24

2' x 4' suspended grid for ceiling tile. Includes main runners 4' on center, cross tees, wire, and eyelet screws. Does not include tiles or luminous panels.

	Craft@Hrs	Unit	Material	Labor	Total
baked enamel	1S@.013	sf	.69	.70	1.39
colored baked enamel	1S@.013	sf	.79	.70	1.49
brass finish	1S@.013	sf	1.08	.70	1.78
stainless steel finish	1S@.013	sf	1.31	.70	2.01
remove	1D@.004	sf	—	.19	.19
add for fire-rated suspended grid	—	sf	.45	—	.45

	Craft@Hrs	Unit	Material	Labor	Total
2' x 4' narrow suspended grid. Includes main runners 4' on center, cross tees, wire, and eyelet screws. Does not include tiles or luminous panels.					
baked enamel	1S@.013	sf	.73	.70	1.43
colored baked enamel	1S@.013	sf	.73	.70	1.43
brass finish	1S@.013	sf	1.30	.70	2.00
stainless steel finish	1S@.013	sf	1.31	.70	2.01
remove	1D@.004	sf	—	.19	.19
add for fire-rated narrow suspended grid	1S@.013	sf	.49	.70	1.19
Additional 2' x 4' suspended ceiling system costs.					
relevel sagging suspended grid	1S@.014	sf	—	.76	.76
remove tiles for work, then reinstall	1S@.011	sf	—	.60	.60
remove tiles & blanket or insulation for work, then reinstall	1S@.013	sf	—	.70	.70
2' x 2' suspended ceiling tile. Includes 2' x 2' suspended ceiling tiles, and installation. Tiles are designed to be installed in a suspended grid. Does not include the suspended grid system.					
smooth face	1S@.009	sf	.92	.49	1.41
fissured face	1S@.009	sf	1.64	.49	2.13
textured face	1S@.009	sf	1.55	.49	2.04
patterned face	1S@.009	sf	2.02	.49	2.51
remove	1D@.004	sf	—	.19	.19
add for fire-rated tile installed with clips	1S@.009	sf	.67	.49	1.16
2' x 2' recessed edge suspended ceiling tile. Tile only, no grid.					
smooth face	1S@.009	sf	1.04	.49	1.53
fissured face	1S@.009	sf	1.34	.49	1.83
textured face	1S@.009	sf	1.77	.49	2.26
patterned face	1S@.009	sf	2.30	.49	2.79
remove	1D@.004	sf	—	.19	.19
add for fire-rated tile installed with clips	1S@.009	sf	.03	.49	.52
2' x 2' luminous panels. Includes 2' x 2' luminous panels and installation. Does not include grid.					
polystyrene cracked-ice or mist white	1S@.005	sf	2.16	.27	2.43
acrylic cracked-ice or mist white	1S@.005	sf	4.18	.27	4.45
polystyrene egg crate	1S@.005	sf	2.14	.27	2.41
acrylic egg crate	1S@.005	sf	7.77	.27	8.04
acrylic egg crate with stainless steel finish	1S@.005	sf	8.49	.27	8.76
acrylic egg crate with brass finish	1S@.005	sf	8.62	.27	8.89
square crate polystyrene	1S@.005	sf	2.07	.27	2.34
square crate acrylic	1S@.005	sf	7.36	.27	7.63
square crate acrylic with stainless steel finish	1S@.005	sf	8.29	.27	8.56
square crate acrylic with brass finish	1S@.005	sf	8.40	.27	8.67
remove	1D@.004	sf	—	.19	.19
2' x 2' suspended grid. Includes main runners 4' on center, cross tees, wire, and eyelet screws. Does not include tiles or luminous panels.					
baked enamel	1S@.013	sf	1.09	.70	1.79
colored baked enamel	1S@.013	sf	1.24	.70	1.94
brass finish	1S@.013	sf	1.60	.70	2.30
stainless steel finish	1S@.013	sf	1.64	.70	2.34
remove	1D@.004	sf	—	.19	.19
add for fire-rated grid	—	sf	.95	—	.95

	Craft@Hrs	Unit	Material	Labor	Total
2' x 2' narrow suspended grid. Includes main runners 4' on center, cross tees, wire, and eyelet screws. Does not include tiles or luminous panels.					
baked enamel	1S@.013	sf	1.14	.70	1.84
colored baked enamel	1S@.013	sf	1.17	.70	1.87
brass finish	1S@.013	sf	1.60	.70	2.30
stainless-steel finish	1S@.013	sf	2.08	.70	2.78
remove	1D@.004	sf	—	.19	.19
add for fire-rated narrow grid	—	sf	.95	—	.95
Additional 2' x 2' suspended ceiling system costs.					
relevel sagging suspended grid	1S@.016	sf	—	.87	.87
remove tiles for work, then reinstall	1S@.011	sf	—	.60	.60
remove tiles & blanket for work, then reinstall	1S@.014	sf	—	.76	.76
Minimum charge.					
minimum for suspended ceiling work	1S@3.00	ea	79.00	162.00	241.00
Drop-in HVAC panels for suspended ceiling systems. Includes manufactured HVAC panel and installation in suspended grid system. Does not include the suspended-grid system or the ductwork.					
cold-air return panel	1S@.609	ea	76.40	32.90	109.30
air diffuser panel	1S@.609	ea	76.40	32.90	109.30
remove	1D@.135	ea	—	6.37	6.37
Sound blanket. Includes sound-absorbing blanket installed above suspended ceiling grid. Does not include the grid system or panels.					
2" blanket above suspended grid	1S@.009	sf	.64	.49	1.13
remove	1D@.008	sf	—	.38	.38
3" blanket above suspended grid	1S@.009	sf	.97	.49	1.46
remove	1D@.009	sf	—	.42	.42

Time & Material Charts (selected items)
Suspended Ceilings Materials

	Craft@Hrs	Unit	Material	Labor	Total
Suspended ceiling with concealed grid					
wall angle, ($8.74 per 12' stick, 83 sf), 4% waste	—	sf	.09	—	.09
main tee, ($11.10 per 12' stick, 15.00 sf), 4% waste	—	sf	.78	—	.78
cross tee (12" x 12" installation), ($1.06 per 1' stick, 1.14 sf), 4% waste	—	sf	.96	—	.96
Suspended ceiling grid baked enamel finish					
wall angle, ($6.80 per 12' stick, 43.20 sf), 4% waste	—	sf	.17	—	.17
main tee, ($11.10 per 12' stick, 27.48 sf), 4% waste	—	sf	.42	—	.42
cross tee in 2' x 4' installation, ($3.73 per 4' stick, 41.60 sf), 4% waste	—	sf	.08	—	.08

	Craft@Hrs	Unit	Material	Labor	Total
cross tee in 2' x 2' installation, ($2.23 per 2' stick, 4.58 sf), 4% waste	—	sf	.51	—	.51
12" x 12" suspended tile for concealed grid system					
smooth face, ($1.61 per tile, 1 sf), 5% waste	—	sf	1.69	—	1.69
patterned face, ($2.38 per tile, 1 sf), 5% waste	—	sf	2.49	—	2.49
add for fire rating, ($.57 per tile, 1 sf), 5% waste	—	sf	.61	—	.61
2' x 4' suspended tile					
smooth face, ($5.65 per tile, 8 sf), 5% waste	—	sf	.74	—	.74
patterned face, ($12.50 per tile, 8 sf), 5% waste	—	sf	1.65	—	1.65
add for 2' x 4' tile with fire rating, ($3.67 per tile, 8 sf), 5% waste	—	sf	.48	—	.48
2' x 4' luminous panel					
polystyrene cracked-ice or mist white, ($13.80 per panel, 8 sf), 4% waste	—	sf	1.79	—	1.79
acrylic mist white, ($26.90 per panel, 8 sf), 4% waste	—	sf	3.50	—	3.50
polystyrene egg crate, ($13.50 per panel, 8 sf), 4% waste	—	sf	1.75	—	1.75
acrylic egg crate, ($50.00 per panel, 8 sf), 4% waste	—	sf	6.50	—	6.50
acrylic egg crate with stainless steel finish, ($54.40 per panel, 8 sf), 4% waste	—	sf	7.06	—	7.06
acrylic egg crate with brass finish, ($55.00 per panel, 8 sf), 4% waste	—	sf	7.15	—	7.15

Suspended Ceilings Labor

Laborer	base wage	paid leave	true wage	taxes & ins.	total
Installer	$35.70	2.78	$38.48	23.22	$61.70
Installer's helper	$25.80	2.01	$27.81	18.59	$46.40
Electrician	$43.80	3.42	$47.22	25.18	$72.40
Demolition laborer	$25.90	2.02	$27.92	19.28	$47.20

Paid leave is calculated based on two weeks paid vacation, one week sick leave, and seven paid holidays. Employer's matching portion of **FICA** is 7.65 percent. **FUTA** (Federal Unemployment) is .8 percent. **Worker's compensation** for the suspended trades was calculated using a national average of 11.96 percent . **Unemployment insurance** was calculated using a national average of 8 percent. **Health insurance** was calculated based on a projected national average for 2020 of $1,280 per employee (and family when applicable) per month. Employer pays 80 percent for a per month cost of $1,024 per employee. **Retirement** is based on a 401(k) retirement program with employer matching of 50 percent. Employee contributions to the 401(k) plan are an average of 6 percent of the true wage. **Liability insurance** is based on a national average of 12.0 percent.

	Craft@Hrs	Unit	Material	Labor	Total
Suspended Ceilings Labor Productivity					
Demolition of suspended ceiling systems					
remove 2' x 4' ceiling tiles	1D@.005	sf	—	.24	.24
remove 2' x 4' suspended grid only	1D@.004	sf	—	.19	.19
remove 2' x 4' suspended tiles & grid	1D@.009	sf	—	.42	.42
remove 2' x 2' suspended tiles only	1D@.004	sf	—	.19	.19
remove 2' x 2' grid only	1D@.004	sf	—	.19	.19
remove 2' x 2' tiles & grid	1D@.008	sf	—	.38	.38
remove 12" x 12" ceiling tiles only	1D@.006	sf	—	.28	.28
remove 12" x 12" grid only	1D@.006	sf	—	.28	.28
remove 12" x 12" tiles & grid	1D@.012	sf	—	.57	.57
Suspended ceiling installation crew					
install suspended ceiling — installer	$61.70				
install suspended ceiling — installer's helper	$46.40				
install suspended ceiling — installation crew	$54.10				
Install 2' x 4' suspended ceiling					
grid only	1S@.013	sf	—	.70	.70
grid and tiles	1S@.024	sf	—	1.30	1.30
ceiling tiles only	3S@.011	sf	—	.51	.51
luminous panels	3S@.010	sf	—	.46	.46
Install 12" x 12" concealed suspended ceiling					
grid only	1S@.020	sf	—	1.08	1.08
grid and tiles	1S@.039	sf	—	2.11	2.11
tiles only	1S@.019	sf	—	1.03	1.03

Swimming Pools

	Craft@Hrs	Unit	Material	Labor	Total
Minimum charge.					
for swimming pool work	8S@5.00	ea	62.80	381.00	443.80
Swimming pools. Per square feet of walls and bottom of pool. Includes all equipment (average to high quality) and 6" x 6" tile border. Does not include coping or deck.					
Gunite with plaster					
replace	8S@.203	sf	46.40	15.40	61.80
remove	1D@.245	sf	—	11.60	11.60
Concrete with vinyl liner					
replace	8S@.166	sf	32.40	12.60	45.00
remove	1D@.212	sf	—	10.00	10.00
Galvanized steel with vinyl liner					
replace	8S@.135	sf	27.70	10.30	38.00
remove	1D@.142	sf	—	6.70	6.70
Aluminum with vinyl liner					
replace	8S@.141	sf	27.90	10.70	38.60
remove	1D@.142	sf	—	6.70	6.70
Fiberglass					
replace	8S@.174	sf	43.30	13.20	56.50
remove	1D@.121	sf	—	5.71	5.71
Coping.					
replace, brick	8S@.228	lf	25.10	17.40	42.50
replace, precast concrete	8S@.198	lf	22.50	15.10	37.60
replace, flagstone	8S@.254	lf	27.10	19.30	46.40
remove	1D@.050	lf	—	2.36	2.36
remove, then reinstall	8S@.311	lf	1.25	23.70	24.95
Swimming pool decks.					
Concrete with epoxy aggregate surface					
replace	8S@.045	sf	4.02	3.42	7.44
remove	1D@.049	sf	—	2.31	2.31
Stamped and dyed concrete					
replace	8S@.040	sf	3.66	3.04	6.70
remove	1D@.049	sf	—	2.31	2.31
Bond beam repair.					
replace	8S@1.07	lf	15.00	81.40	96.40
Regrout tile swimming pool.					
replace	8S@.024	sf	.25	1.83	2.08
Leak detection.					
replace	8S@4.11	ea	4.73	313.00	317.73

	Craft@Hrs	Unit	Material	Labor	Total
Repair leak underwater.					
in vinyl lining	8S@3.62	ea	19.70	275.00	294.70
with hydraulic cement	8S@4.48	ea	32.40	341.00	373.40
rusted rebar & stain, patch with plaster	8S@7.01	ea	57.20	533.00	590.20
plaster popoff	8S@4.22	ea	25.10	321.00	346.10
Replaster swimming pool.					
replace	8S@.038	sf	2.49	2.89	5.38
Paint pool.					
replace, rubber base paint	8S@.027	sf	1.19	2.05	3.24
replace, epoxy paint	8S@.035	sf	1.34	2.66	4.00
Vinyl liner.					
replace	8S@.022	sf	4.04	1.67	5.71
Tile border.					
replace, 6" x 6"	8S@.232	lf	12.10	17.70	29.80
replace, 12" x 12"	8S@.316	lf	15.80	24.00	39.80
remove	1D@.062	lf	—	2.93	2.93
remove, then reinstall	8S@.631	lf	1.34	48.00	49.34
Acid wash.					
wash	8S@.004	sf	.23	.30	.53
minimum charge to acid wash	8S@3.99	ea	31.20	304.00	335.20
Drain pool.					
drain	8S@4.20	ea	—	320.00	320.00
drain and clean	8S@9.40	ea	26.60	715.00	741.60
Shock treatment.					
chemical shock	8S@1.25	ea	53.70	95.10	148.80
Open pool.					
open for summer	8S@3.26	ea	80.20	248.00	328.20
Close pool.					
close for winter	8S@5.01	ea	108.00	381.00	489.00
Caulk expansion joint.					
replace	8S@.029	lf	3.58	2.21	5.79

Temporary

Emergency board up. Prices are for work done between 7:00 a.m. and 6:00 p.m. Monday through Friday. Add **25%** for emergency work between 6:00 p.m. and 10:00 p.m. Monday through Friday. Add **65%** for emergency work between 10:00 p.m. and 7:00 a.m. Monday through Friday, on weekends, or on holidays.

	Craft@Hrs	Unit	Material	Labor	Total
Cover opening.					
with single sheet of plywood	1C@.715	ea	29.00	48.30	77.30
with two sheets of plywood	1C@.908	ea	58.10	61.40	119.50
install temporary framing and cover opening	1C@.022	sf	1.30	1.49	2.79
Cover roof or wall.					
with tarp	1C@.009	sf	.84	.61	1.45
with plastic	1C@.009	sf	.43	.61	1.04
Minimum charge.					
minimum for emergency board up	1C@4.00	ea	161.00	270.00	431.00
Temporary electric power.					
electrical hookup	1C@1.38	ea	148.00	93.30	241.30
power, per week	—	wk	127.00	—	127.00
Temporary heating.					
per week	—	wk	249.00	—	249.00

	Craft@Hrs	Unit	Material	Labor	Equip.	Total
Scaffolding. Per week. Per sf of wall that scaffolding covers. 60" wide scaffolding attached with scissor braces.						
per sf of wall covered by scaffold	—	sf	—	—	.32	.32
add for hook-end cat walk	—	ea	—	—	1.79	1.79
add for tented scaffolding	1C@.005	sf	—	.34	.42	.76
delivery, set-up, and take down	1C@.005	sf	—	.34	—	.34
minimum charge for scaffolding	1C@1.00	ea	—	67.60	83.90	151.50
Temporary chain-link fence. Per job, up to 12 months.						
5' tall	1C@.118	lf	—	7.98	4.55	12.53
6' tall	1C@.123	lf	—	8.31	4.71	13.02
minimum charge for chain-link fence	1C@3.00	ea	—	203.00	235.00	438.00
Office trailer.						
8' x 32', per week	—	ea	—	—	93.50	93.50
Storage trailer.						
per week	—	ea	—	—	55.30	55.30

	Craft@Hrs	Unit	Material	Labor	Equip.	Total
Portable toilet.						
chemical, per week	—	ea	—	—	47.10	47.10
Reflective barricades.						
folding, per week	—	ea	—	—	7.73	7.73
folding with flashing light, per week	—	ea	—	—	9.77	9.77
Traffic cone.						
28" high, per week	—	ea	—	—	.82	.82

	Craft@Hrs	Unit	Material	Labor	Total
Job site security. Per hour.					
security guard	1C@.346	hr	—	23.40	23.40
security guard with dog	1C@.477	hr	—	32.20	32.20
minimum charge for security guard	1C@7.00	ea	—	473.00	473.00
Orange plastic safety fence.					
4' tall	1C@.004	lf	.63	.27	.90
5' tall	1C@.004	lf	.96	.27	1.23
Safety tape.					
plastic, "do not cross" barricade tape	1C@.002	lf	.03	.14	.17

Time & Material Charts (selected items)

Temporary Items Materials

See Temporary material prices with the line items above.

Temporary Items Labor

Laborer	base wage	paid leave	true wage	taxes & ins.	total
Carpenter	$38.10	2.97	$41.07	26.53	$67.60

Paid leave is calculated based on two weeks paid vacation, one week sick leave, and seven paid holidays. Employer's matching portion of **FICA** is 7.65 percent. **FUTA** (Federal Unemployment) is .8 percent. **Worker's compensation** for carpentry work in the temporary trade was calculated using a national average of 17.21 percent. **Unemployment insurance** was calculated using a national average of 8 percent. **Health insurance** was calculated based on a projected national average for 2020 of $1,280 per employee (and family when applicable) per month. Employer pays 80 percent for a per month cost of $1,024 per employee. **Retirement** is based on a 401(k) retirement program with employer matching of 50 percent. Employee contributions to the 401(k) plan are an average of 6 percent of the true wage. **Liability insurance** is based on a national average of 12.0 percent.

Temporary Items Labor Productivity	Craft@Hrs	Unit	Material	Labor	Total
Temporary board up and fencing					
cover opening with single sheet of plywood	1C@.715	ea	—	48.30	48.30
cover opening with two sheets of plywood	1C@.908	ea	—	61.40	61.40
install temporary framing and cover opening	1C@.022	sf	—	1.49	1.49
cover roof or wall with tarp	1C@.009	sf	—	.61	.61
cover roof or wall with plastic	1C@.009	sf	—	.61	.61
install temporary 5' tall chain link fence	1C@.118	lf	—	7.98	7.98
install temporary 6' tall chain link fence	1C@.123	lf	—	8.31	8.31

Tile

with Cultured Marble

	Craft@Hrs	Unit	Material	Labor	Total
Minimum charge.					
minimum for tile work	1T@4.00	ea	73.10	251.00	324.10
Tile bathtub surround. Includes tile, mortar, grout, caulk, specialty tile pieces (such as corner pieces, soap dish, and so on), and installation. Surround includes coverage of the three walls surrounding bathtub: two widths and one length. Includes one soap dish. Does not include doors, or shower curtain rod.					
6' tall tile bathtub surround					
replace, adhesive set on moisture-resistant drywall	1T@5.26	ea	430.00	330.00	760.00
replace, adhesive set on tile backer board	1T@5.26	ea	475.00	330.00	805.00
remove	1D@2.86	ea	—	135.00	135.00
replace, mortar set	1T@7.69	ea	504.00	483.00	987.00
remove	1D@3.56	ea	—	168.00	168.00
7' tall tile bathtub surround					
replace, adhesive set on moisture-resistant drywall	1T@6.24	ea	498.00	392.00	890.00
replace, adhesive set on tile backer board	1T@6.24	ea	557.00	392.00	949.00
remove	1D@2.94	ea	—	139.00	139.00
replace, mortar set	1T@8.33	ea	592.00	523.00	1,115.00
remove	1D@3.70	ea	—	175.00	175.00
8' tall tile bathtub surround					
replace, adhesive set on moisture-resistant drywall	1T@7.15	ea	570.00	449.00	1,019.00
replace, adhesive set on tile backer board	1T@7.15	ea	635.00	449.00	1,084.00
remove	1D@3.03	ea	—	143.00	143.00
replace, mortar set	1T@9.09	ea	671.00	571.00	1,242.00
remove	1D@3.84	ea	—	181.00	181.00
Tile shower. Includes walls, floor, all trim pieces, and one tile soap dish. Does not include shower pan, fixtures, door or any plumbing.					
36" x 36" tile shower					
replace, adhesive set on tile backer board	1T@7.69	ea	708.00	483.00	1,191.00
remove	1D@3.33	ea	—	157.00	157.00
replace, mortar set	1T@9.09	ea	745.00	571.00	1,316.00
remove	1D@3.70	ea	—	175.00	175.00
add for tile ceiling	1T@.870	ea	110.00	54.60	164.60
36" x 48" tile shower					
replace, adhesive set on tile backer board	1T@9.09	ea	790.00	571.00	1,361.00
remove	1D@3.45	ea	—	163.00	163.00
replace, mortar set	1T@10.0	ea	825.00	628.00	1,453.00
remove	1D@3.45	ea	—	163.00	163.00
add for tile ceiling	1T@1.14	ea	131.00	71.60	202.60
Tile accessory. In shower or bathtub surround.					
add for soap holder or other accessory	1T@.418	ea	48.10	26.30	74.40

	Craft@Hrs	Unit	Material	Labor	Total

Cultured marble bathtub surround. Includes cultured marble bathtub surround, construction adhesive and/or attachment hardware, caulk, and installation. Surround includes coverage of the three walls surrounding bathtub: two widths and one length. Typically each wall is covered with one solid piece. Does not include wall framing, doors, or shower curtain rod.

	Craft@Hrs	Unit	Material	Labor	Total
replace, typical	1T@3.86	ea	525.00	242.00	767.00
remove	1D@1.28	ea	—	60.40	60.40

Tile window sill. Includes tile, mortar, grout, caulk, and installation.

	Craft@Hrs	Unit	Material	Labor	Total
replace, 4" wall	1T@.063	lf	8.66	3.96	12.62
replace, 6" wall	1T@.063	lf	9.26	3.96	13.22
replace, 8" wall	1T@.065	lf	10.10	4.08	14.18
remove	1D@.059	lf	—	2.78	2.78

Cultured marble window sill. Includes cultured marble sill, construction adhesive, caulk, and installation.

	Craft@Hrs	Unit	Material	Labor	Total
replace, 4" wall	1T@.053	lf	8.96	3.33	12.29
replace, 6" wall	1T@.053	lf	11.40	3.33	14.73
replace, 8" wall	1T@.054	lf	15.30	3.39	18.69
remove	1D@.018	lf	—	.85	.85

Tile repair.

	Craft@Hrs	Unit	Material	Labor	Total
regrout	1T@.040	sf	.42	2.51	2.93
replace, single tile	1T@.926	ea	—	58.20	58.20

Time & Material Charts (selected items)

Tile Materials

See Tile material prices with the line items above.

Tile Labor

Laborer	base wage	paid leave	true wage	taxes & ins.	total
Tile layer	$34.60	2.70	$37.30	25.50	$62.80
Demolition worker	$25.90	2.02	$27.92	19.28	$47.20

Paid leave is calculated based on two weeks paid vacation, one week sick leave, and seven paid holidays. Employer's matching portion of **FICA** is 7.65 percent. **FUTA** (Federal Unemployment) is .8 percent. **Worker's compensation** for the tile trade was calculated using a national average of 19.41 percent. **Unemployment insurance** was calculated using a national average of 8 percent. **Health insurance** was calculated based on a projected national average for 2020 of $1,280 per employee (and family when applicable) per month. Employer pays 80 percent for a per month cost of $1,024 per employee. **Retirement** is based on a 401(k) retirement program with employer matching of 50 percent. Employee contributions to the 401(k) plan are an average of 6 percent of the true wage. **Liability insurance** is based on a national average of 12.0 percent.

	Craft@Hrs	Unit	Material	Labor	Total

Wall Coverings

Strip wallpaper. Add **25%** for two layers, **36%** for three layers and **42%** for four layers. Remove only.

	Craft@Hrs	Unit	Material	Labor	Total
strippable	3H@.013	sf	—	.68	.68
nonstrippable	3H@.021	sf	—	1.10	1.10

Minimum charge.

	Craft@Hrs	Unit	Material	Labor	Total
for wall covering work	3H@3.00	ea	49.60	157.00	206.60

Wallpaper underliner. Includes sizing, blank stock underliner, wallpaper paste, and installation. Used to cover flaws in walls before wallpaper is installed. Sometimes used to cover paneling.

	Craft@Hrs	Unit	Material	Labor	Total
replace blank stock	3H@.013	sf	.81	.68	1.49
remove	1D@.017	sf	—	.80	.80

Grass cloth wallpaper. Includes wall sizing and 21% waste. Add **$.46** per sf to patch small holes, gouges, and cracks.

	Craft@Hrs	Unit	Material	Labor	Total
replace standard grade	3H@.034	sf	2.47	1.78	4.25
replace high grade	3H@.034	sf	3.97	1.78	5.75
remove	1D@.017	sf	—	.80	.80

Paper wallpaper. From flat to high gloss papers. Includes wall sizing and 21% waste. Add **$.46** per sf to patch small holes, gouges, and cracks.

	Craft@Hrs	Unit	Material	Labor	Total
replace standard grade	3H@.026	sf	1.31	1.36	2.67
replace high grade	3H@.026	sf	1.79	1.36	3.15
remove	1D@.017	sf	—	.80	.80

Vinyl-coated wallpaper. Vinyl with paper back. Includes wall sizing and 21% waste. Add **$.46** per sf to patch small holes, gouges, and cracks.

	Craft@Hrs	Unit	Material	Labor	Total
replace standard grade	3H@.028	sf	1.67	1.47	3.14
replace high grade	3H@.028	sf	2.23	1.47	3.70
remove	1D@.017	sf	—	.80	.80

Vinyl wallpaper. Seams overlapped, then cut. Includes wall sizing and 21% waste. Add **$.46** per sf to patch small holes, gouges, and cracks.

	Craft@Hrs	Unit	Material	Labor	Total
replace standard grade	3H@.028	sf	1.76	1.47	3.23
replace high grade	3H@.028	sf	2.33	1.47	3.80
remove	1D@.017	sf	—	.80	.80

Foil wallpaper. Includes wall sizing and 21% waste. Add **$.46** per sf to patch small holes, gouges, and cracks.

	Craft@Hrs	Unit	Material	Labor	Total
replace standard grade	3H@.027	sf	2.14	1.41	3.55
replace high grade	3H@.027	sf	2.71	1.41	4.12
remove	1D@.017	sf	—	.80	.80

Wallpaper border. Includes sizing, wallpaper border, wallpaper paste, and installation. All styles. Use remove price when removing only the border and no adjoining wallpaper on the same wall.

	Craft@Hrs	Unit	Material	Labor	Total
replace standard grade	3H@.023	lf	2.33	1.21	3.54
replace high grade	3H@.023	lf	3.27	1.21	4.48
remove	1D@.013	lf	—	.61	.61

Anaglypta embossed wall or ceiling covering. Includes Anaglypta style embossed wall or ceiling covering, adhesive, and installation. Does not include painting. Wall or ceiling coverings are heavy, highly embossed panels made from material similar to linoleum. Must be painted. Higher grades contain cotton fibers for extra strength and durability. Anaglypta is a registered trademark. Includes 21% waste.

	Craft@Hrs	Unit	Material	Labor	Total
replace standard grade	3H@.045	sf	1.91	2.36	4.27
replace high grade	3H@.045	sf	2.70	2.36	5.06
replace deluxe grade	3H@.045	sf	3.00	2.36	5.36
remove	1D@.023	sf	—	1.09	1.09

Lincrusta embossed wall or ceiling covering. Lincrusta style, deeply embossed wall or ceiling covering, adhesive, and installation. Lincrusta wall coverings are made from material similar to linoleum. Must be painted. Higher grades contain cotton fibers for extra strength and durability. Does not include painting. Lincrusta is a registered trademark. Lincrusta is typically much heavier than Anaglypta. Includes 21% waste.

	Craft@Hrs	Unit	Material	Labor	Total
replace standard grade	3H@.056	sf	5.48	2.93	8.41
replace high grade	3H@.056	sf	7.38	2.93	10.31
replace deluxe grade	3H@.056	sf	9.17	2.93	12.10
remove	1D@.033	sf	—	1.56	1.56

Anaglypta embossed frieze. Includes Anaglypta style, deeply embossed frieze, adhesive, and installation. Friezes are heavy, highly embossed trim pieces that are 12" to 21" wide. Friezes are made from material similar to linoleum. Must be painted. Higher grades contain cotton fibers for extra strength and durability. Does not include painting. Anaglypta is a registered trademark.

	Craft@Hrs	Unit	Material	Labor	Total
replace standard grade	3H@.043	lf	5.87	2.25	8.12
replace high grade	3H@.043	lf	6.65	2.25	8.90
remove	1D@.019	lf	—	.90	.90

Lincrusta embossed frieze. Includes Lincrusta style, deeply embossed frieze, adhesive, and installation. Friezes are heavy, highly embossed trim pieces that are approximately 21" wide. Friezes are made from material similar to linoleum. Must be painted. Higher grades contain cotton fibers for extra strength and durability. Does not include painting. Lincrusta is a registered trademark.

	Craft@Hrs	Unit	Material	Labor	Total
replace standard grade	3H@.053	lf	6.84	2.78	9.62
replace high grade	3H@.053	lf	8.51	2.78	11.29
remove	1D@.027	lf	—	1.27	1.27

Anaglypta embossed pelmet. Includes Anaglypta style embossed pelmet, adhesive, and installation. Pelmets are narrower and often less ornate than friezes and are from 4" to 7" wide. Pelmets are made from material similar to linoleum. Must be painted. Higher grades contain cotton fibers for extra strength and durability. Does not include painting. Anaglypta is a registered trademark.

	Craft@Hrs	Unit	Material	Labor	Total
replace standard grade	3H@.040	lf	3.68	2.10	5.78
replace high grade	3H@.040	lf	5.23	2.10	7.33
remove	1D@.019	lf	—	.90	.90

Anaglypta embossed dado. Includes Anaglypta style embossed dado wall covering, adhesive, and installation. Dados are heavy, highly embossed wall panels made from material similar to linoleum. Must be painted. Higher grades contain cotton fibers for extra strength and durability. Does not include painting. Anaglypta is a registered trademark.

	Craft@Hrs	Unit	Material	Labor	Total
replace standard grade	3H@.111	lf	28.30	5.82	34.12
replace high grade	3H@.111	lf	35.20	5.82	41.02
remove	1D@.128	lf	—	6.04	6.04

	Craft@Hrs	Unit	Material	Labor	Total

Time & Material Charts (selected items)
Wall Coverings Materials

Wallpaper materials were calculated based on double European size rolls with approximately 28 sf per single roll and 56 sf in a double roll. (Most manufacturers no longer make American size rolls which contained approximately 36 sf per single roll and 72 sf in a double roll.) Prices include sizing and glue application, even on pre-pasted rolls.

	Craft@Hrs	Unit	Material	Labor	Total
Wallpaper underliner					
blank stock,					
($35.20 double roll, 56 sf), 21% waste	—	sf	.75	—	.75
Grass cloth wallpaper					
standard grade grass cloth wallpaper,					
($109.00 double roll, 56 sf), 21% waste	—	sf	2.34	—	2.34
high grade grass cloth wallpaper,					
($175.00 double roll, 56 sf), 21% waste	—	sf	3.79	—	3.79
Paper wallpaper					
standard grade,					
($57.90 double roll, 56 sf), 21% waste	—	sf	1.24	—	1.24
high grade,					
($79.90 double roll, 56 sf), 21% waste	—	sf	1.72	—	1.72
Vinyl-coated wallpaper					
standard grade,					
($73.60 double roll, 56 sf), 21% waste	—	sf	1.59	—	1.59
high grade,					
($97.40 double roll, 56 sf), 21% waste	—	sf	2.12	—	2.12
Vinyl wallpaper					
standard grade,					
($78.10 double roll, 56 sf), 21% waste	—	sf	1.69	—	1.69
high grade,					
($103.00 double roll, 56 sf), 21% waste	—	sf	2.22	—	2.22
Foil wallpaper					
standard grade,					
($95.10 double roll, 56 sf), 21% waste	—	sf	2.05	—	2.05
high grade,					
($119.00 double roll, 56 sf), 21% waste	—	sf	2.57	—	2.57
Wallpaper border					
standard grade,					
($55.00 bolt, 30 lf), 21% waste	—	lf	2.22	—	2.22
high grade,					
($77.60 bolt, 30 lf), 21% waste	—	lf	3.12	—	3.12

	Craft@Hrs	Unit	Material	Labor	Total
Anaglypta embossed wall or ceiling covering					
standard grade, ($84.10 double roll, 56 sf), 21% waste	—	sf	1.81	—	1.81
high grade, ($119.00 double roll, 56 sf), 21% waste	—	sf	2.57	—	2.57
deluxe grade, ($134.00 double roll, 56 sf), 21% waste	—	sf	2.88	—	2.88
Lincrusta embossed wall or ceiling covering					
standard grade, ($244.00 double roll, 56 sf), 21% waste	—	sf	5.24	—	5.24
high grade, ($327.00 double roll, 56 sf), 21% waste	—	sf	7.05	—	7.05
deluxe grade, ($407.00 double roll, 56 sf), 21% waste	—	sf	8.78	—	8.78
Anaglypta embossed frieze					
standard grade, ($168.00 bolt, 33 lf), 13% waste	—	lf	5.73	—	5.73
high grade, ($191.00 bolt, 33 lf), 13% waste	—	lf	6.52	—	6.52
Lincrusta embossed frieze					
standard grade, ($197.00 bolt, 33 lf), 12% waste	—	lf	6.69	—	6.69
high grade, ($247.00 bolt, 33 lf), 12% waste	—	lf	8.34	—	8.34
Anaglypta embossed pelmet					
standard grade, ($105.00 bolt, 33 lf), 13% waste	—	lf	3.61	—	3.61
high grade, ($150.00 bolt, 33 lf), 13% waste	—	lf	5.13	—	5.13
Anaglypta embossed dado (five panels per box)					
standard grade, ($407.00 box, 16 lf), 10% waste	—	lf	28.00	—	28.00
high grade, ($505.00 box, 16 lf), 10% waste	—	lf	34.60	—	34.60

Wall Coverings Labor

Laborer	base wage	paid leave	true wage	taxes & ins.	total
Wallpaper hanger	$30.40	2.37	$32.77	19.63	$52.40

Paid leave is calculated based on two weeks paid vacation, one week sick leave, and seven paid holidays. Employer's matching portion of **FICA** is 7.65 percent. **FUTA** (Federal Unemployment) is .8 percent. **Worker's compensation** for the wall coverings trade was calculated using a national average of 8.58 percent. **Unemployment insurance** was calculated using a national average of 8 percent. **Health insurance** was calculated based on a projected national average for 2020 of $1,280 per employee (and family when applicable) per month. Employer pays 80 percent for a per month cost of $1,024 per employee. **Retirement** is based on a 401(k) retirement program with employer matching of 50 percent. Employee contributions to the 401(k) plan are an average of 6 percent of the true wage. **Liability insurance** is based on a national average of 12.0 percent.

	Craft@Hrs	Unit	Material	Labor	Total
Wall Coverings Labor Productivity					
Remove wallpaper					
strippable	3H@.013	sf	—	.68	.68
nonstrippable	3H@.021	sf	—	1.10	1.10
Install wallpaper					
underliner	3H@.013	sf	—	.68	.68
grass cloth wallpaper	3H@.034	sf	—	1.78	1.78
paper wallpaper	3H@.026	sf	—	1.36	1.36
vinyl coated	3H@.028	sf	—	1.47	1.47
vinyl	3H@.028	sf	—	1.47	1.47
foil	3H@.027	sf	—	1.41	1.41
Install wallpaper border					
all styles	3H@.023	lf	—	1.21	1.21
Install Anaglypta embossed wall or ceiling covering					
wall or ceiling	3H@.045	sf	—	2.36	2.36
Install Lincrusta embossed wall or ceiling covering					
wall or ceiling	3H@.056	sf	—	2.93	2.93
Install embossed frieze					
Anaglypta	3H@.043	lf	—	2.25	2.25
Lincrusta	3H@.053	lf	—	2.78	2.78
Install Anaglypta embossed pelmet					
all grades	3H@.040	lf	—	2.10	2.10
Install Anaglypta embossed dado					
all grades	3H@.111	lf	—	5.82	5.82

Water Extraction

Emergency Service. All prices are for work done between 7:00 a.m. and 6:00 p.m. Monday through Friday. Add **25%** for emergency work between 6:00 p.m. and 10:00 p.m. Monday through Friday. Add **65%** for emergency work between 10 p.m. and 7:00 a.m. Monday through Friday, on weekends, or on holidays.

	Craft@Hrs	Unit	Material	Labor	Equip.	Total
Minimum charge.						
for water extraction work	9S@3.50	ea	—	201.00	169.00	370.00
Equipment delivery, setup, and take-home charge. Per loss charge to deliver water extraction equipment, set the equipment up, and take the equipment home once the water is extracted and drying is complete.						
minimum	—	ea	—	—	52.30	52.30
Extract water from carpet. Lightly soaked: carpet and pad are wet but water does not rise around feet as carpet is stepped on. Typically wet: water in carpet and pad rises around feet as carpet is stepped on. Heavily soaked: water is visible over the surface of the carpet mat. Very heavily soaked: standing water over the surface of the carpet.						
lightly soaked	9S@.005	sf	—	.29	.09	.38
typically wet	9S@.007	sf	—	.40	.10	.50
heavily soaked	9S@.010	sf	—	.58	.14	.72
very heavily soaked	9S@.014	sf	—	.81	.20	1.01
Dehumidifiers. Does not include delivery, setup and take-home (see above). Does include daily monitoring and adjustment.						
10 gallon daily capacity	9S@.410	day	—	23.60	52.30	75.90
19 gallon daily capacity	9S@.410	day	—	23.60	87.40	111.00
24 gallon daily capacity	9S@.410	day	—	23.60	121.00	144.60
Drying fan. Does not include delivery, setup and take-home (see above). Does not include daily monitoring and adjustment.						
typical	9S@.104	day	—	5.98	43.60	49.58
Hang & dry carpet in plant. Square foot price to haul carpet to plant, hang, and dry. Includes transport to and from plant. Does not include carpet removal or reinstallation.						
carpet	9S@.005	sf	—	.29	.16	.45
oriental rug	9S@.008	sf	—	.46	.27	.73
Minimum charge.						
to hang and dry carpet in plant	9S@2.03	sf	—	117.00	83.90	200.90

	Craft@Hrs	Unit	Material	Labor	Total
Germicide and mildewcide treatment. Treatments to kill germs and mildew in wet carpet.					
germicide	9S@.003	sf	.03	.17	.20
mildewcide	9S@.003	sf	.03	.17	.20
Detach, lift & block carpet for drying.					
remove	9S@.004	sf	—	.23	.23

	Craft@Hrs	Unit	Material	Labor	Total
Remove wet carpet and pad. See Flooring chapter for prices to tear-out dry carpet and pad.					
remove wet carpet	9S@.080	sy	—	4.60	4.60
remove wet carpet pad	9S@.005	sf	—	.29	.29
Pad and block furniture.					
small room	9S@.399	ea	—	22.90	22.90
average room	9S@.558	ea	—	32.10	32.10
large room	9S@.720	ea	—	41.40	41.40
very large room	9S@.997	ea	—	57.30	57.30

Time & Material Charts (selected items)

Water Extraction Materials

See Water Extraction material prices above.

	Craft@Hrs	Unit	Material	Labor	Equip.	Total
Water Extraction Rental Equipment						
Dehumidifiers rental						
10 gallon daily capacity	—	day	—	—	52.30	52.30
19 gallon daily capacity	—	day	—	—	87.40	87.40
24 gallon daily capacity	—	day	—	—	121.00	121.00
Drying fan rental						
typical	—	day	—	—	43.60	43.60

Water Extraction Labor

Laborer	base wage	paid leave	true wage	taxes & ins.	total
Water extractor	$32.90	2.57	$35.47	22.03	$57.50

Paid Leave is calculated based on two weeks paid vacation, one week sick leave, and seven paid holidays. Employer's matching portion of **FICA** is 7.65 percent. **FUTA** (Federal Unemployment) is .8 percent. **Worker's compensation** for the water extraction trade was calculated using a national average of 12.35 percent. **Unemployment insurance** was calculated using a national average of 8 percent. **Health insurance** was calculated based on a projected national average for 2020 of $1,280 per employee (and family when applicable) per month. Employer pays 80 percent for a per month cost of $1,024 per employee. **Retirement** is based on a 401(k) retirement program with employer matching of 50 percent. Employee contributions to the 401(k) plan are an average of 6 percent of the true wage. **Liability insurance** is based on a national average of 12.0 percent.

	Craft@Hrs	Unit	Material	Labor	Total
Water Extraction Labor Productivity					
Extract water from carpet					
from lightly soaked carpet	9S@.005	sf	—	.29	.29
from very heavily soaked carpet	9S@.014	sf	—	.81	.81
Pad and block furniture for water extraction					
small room	9S@.399	ea	—	22.90	22.90
very large room	9S@.997	ea	—	57.30	57.30

	Craft@Hrs	Unit	Material	Labor	Total

Windows

Remove window. Tear-out and debris removal to a truck or dumpster on site. Does not include hauling, dumpster, or dump fees. No salvage value is assumed.

	Craft@Hrs	Unit	Material	Labor	Total
small (4 to 10 sf)	1D@.397	ea	2.18	18.70	20.88
average (11 to 16 sf)	1D@.407	ea	2.18	19.20	21.38
large (17 to 29 sf)	1D@.424	ea	2.18	20.00	22.18
very large (30 sf and larger)	1D@.433	ea	2.18	20.40	22.58

Remove window for work then reinstall.

	Craft@Hrs	Unit	Material	Labor	Total
small (4 to 10 sf)	1C@1.23	ea	2.18	83.10	85.28
average (11 to 16 sf)	1C@1.72	ea	2.18	116.00	118.18
large (17 to 29 sf)	1C@2.51	ea	2.18	170.00	172.18
very large (30 sf and larger)	1C@3.70	ea	2.18	250.00	252.18

Other window options. All windows in this chapter are double-glazed (insulated) units with clear glass. Use for all window types except skylights and roof windows. Higher quality windows may have Low-E or argon gas or both. Low-E: A coating that increases heat retention in winter and reduces the sun's ultraviolet rays that fade carpet and upholstery. Also called high-performance glass or other brand names. Argon gas: An odorless gas that improves the window's energy efficiency when injected between panes of glass. Argon is usually not available for windows with true divided lights. The windows in this chapter do not include Low-E or argon. To add, see table below; for skylights and roof windows, see page 442. Thermal Break: Aluminum is an excellent conductor, which means heat and cold easily flow through it, reducing efficiency. To cure this, a plastic or rubber spacer is introduced into the jamb to break the flow of hot or cold.

	Craft@Hrs	Unit	Material	Labor	Total
deduct for single-glazed window (all types)	—	sf	- 7.02	—	- 7.02
add for gray or bronze tinted glass (all window types)	—	sf	4.52	—	4.52
add 8% for Low-E between panes (all window types)					
add 5% for argon fill between panes (all window types)					
add 11% for thermal break (aluminum windows only)					

Add for alternative glazing (all window types)

	Craft@Hrs	Unit	Material	Labor	Total
tempered	—	sf	12.00	—	12.00
laminated	—	sf	7.87	—	7.87
obscure	—	sf	1.57	—	1.57
polished wire	—	sf	36.80	—	36.80

Aluminum. All aluminum windows are double-glazed, clear glass with white or bronze finish frames. For mill-finish windows deduct **13%**. All costs are for standard windows. Manufacturers' window sizes vary. For standard units round to the nearest size. For custom sizes add **80%**. **Grilles and Grid.** For aluminum grid between panes of glass or removable grilles outside of window add (per window): half-round **$29**, half-elliptical **$47**, round **$29**, elliptical **$39**, quarter-round **$29**, casement **$38**, awning **$38**, single-hung **$29**, sliding **$33**, fixed (picture) **$35**.

	Craft@Hrs	Unit	Material	Labor	Total
Half-round aluminum window top					
24" x 12"	1C@.991	ea	465.00	67.00	532.00
36" x 18"	1C@.991	ea	580.00	67.00	647.00
48" x 24"	1C@1.39	ea	687.00	94.00	781.00
72" x 36"	1C@2.00	ea	1,020.00	135.00	1,155.00
Half-elliptical aluminum window top					
36" x 15"	1C@.991	ea	889.00	67.00	956.00
48" x 16"	1C@.991	ea	986.00	67.00	1,053.00
60" x 19"	1C@1.39	ea	1,080.00	94.00	1,174.00
90" x 20"	1C@2.00	ea	1,590.00	135.00	1,725.00

half-round

half-elliptical

	Craft@Hrs	Unit	Material	Labor	Total
Round aluminum window					
20" x 20"	1C@.991	ea	942.00	67.00	1,009.00
30" x 30"	1C@.991	ea	1,160.00	67.00	1,227.00
40" x 40"	1C@1.39	ea	1,330.00	94.00	1,424.00
48" x 48"	1C@2.00	ea	1,460.00	135.00	1,595.00
Elliptical aluminum window					
15" x 24"	1C@.991	ea	1,750.00	67.00	1,817.00
24" x 36"	1C@.991	ea	1,950.00	67.00	2,017.00
25" x 40"	1C@1.39	ea	2,080.00	94.00	2,174.00
30" x 48"	1C@2.00	ea	2,260.00	135.00	2,395.00
Quarter-round aluminum window					
24" x 24"	1C@.991	ea	767.00	67.00	834.00
36" x 36"	1C@.991	ea	1,090.00	67.00	1,157.00
18" wide aluminum casement window					
36" tall	1C@.991	ea	557.00	67.00	624.00
48" tall	1C@.991	ea	603.00	67.00	670.00
60" tall	1C@.991	ea	690.00	67.00	757.00
24" wide aluminum casement window					
36" tall	1C@.991	ea	601.00	67.00	668.00
48" tall	1C@.991	ea	649.00	67.00	716.00
60" tall	1C@1.39	ea	735.00	94.00	829.00
30" wide aluminum casement window					
48" tall	1C@1.39	ea	705.00	94.00	799.00
60" tall	1C@1.39	ea	785.00	94.00	879.00
36" wide aluminum casement window					
48" tall	1C@1.39	ea	833.00	94.00	927.00
60" tall	1C@1.39	ea	956.00	94.00	1,050.00
24" wide aluminum awning windows					
20" tall	1C@.991	ea	514.00	67.00	581.00
24" tall	1C@.991	ea	579.00	67.00	646.00
30" tall	1C@.991	ea	631.00	67.00	698.00
30" wide aluminum awning window					
20" tall	1C@.991	ea	599.00	67.00	666.00
24" tall	1C@.991	ea	652.00	67.00	719.00
30" tall	1C@.991	ea	712.00	67.00	779.00
36" wide aluminum awning window					
24" tall	1C@.991	ea	708.00	67.00	775.00
30" tall	1C@.991	ea	793.00	67.00	860.00
40" wide aluminum awning window					
24" tall	1C@.991	ea	781.00	67.00	848.00
30" tall	1C@.991	ea	882.00	67.00	949.00

round

elliptical

quarter-round

casement

	Craft@Hrs	Unit	Material	Labor	Total
24" wide aluminum double-hung window					
36" tall	1C@.991	ea	301.00	67.00	368.00
48" tall	1C@.991	ea	336.00	67.00	403.00
60" tall	1C@.991	ea	379.00	67.00	446.00
72" tall	1C@.991	ea	424.00	67.00	491.00
36" wide aluminum double-hung window					
36" tall	1C@.991	ea	370.00	67.00	437.00
48" tall	1C@.991	ea	411.00	67.00	478.00
60" tall	1C@.991	ea	443.00	67.00	510.00
72" tall	1C@1.39	ea	487.00	94.00	581.00
42" wide aluminum double-hung window					
36" tall	1C@.991	ea	406.00	67.00	473.00
48" tall	1C@.991	ea	444.00	67.00	511.00
60" tall	1C@1.39	ea	487.00	94.00	581.00
72" tall	1C@1.39	ea	529.00	94.00	623.00
24" wide aluminum single-hung window					
36" tall	1C@.991	ea	249.00	67.00	316.00
48" tall	1C@.991	ea	285.00	67.00	352.00
60" tall	1C@.991	ea	328.00	67.00	395.00
72" tall	1C@.991	ea	374.00	67.00	441.00
36" wide aluminum single-hung window					
36" tall	1C@.991	ea	318.00	67.00	385.00
48" tall	1C@.991	ea	363.00	67.00	430.00
60" tall	1C@.991	ea	394.00	67.00	461.00
72" tall	1C@1.39	ea	436.00	94.00	530.00
42" wide aluminum single-hung window					
36" tall	1C@.991	ea	359.00	67.00	426.00
48" tall	1C@.991	ea	395.00	67.00	462.00
60" tall	1C@1.39	ea	436.00	94.00	530.00
72" tall	1C@1.39	ea	477.00	94.00	571.00
36" wide aluminum sliding window					
24" tall	1C@.991	ea	108.00	67.00	175.00
36" tall	1C@.991	ea	120.00	67.00	187.00
48" tall	1C@.991	ea	168.00	67.00	235.00
60" tall	1C@.991	ea	221.00	67.00	288.00
48" wide aluminum sliding window					
24" tall	1C@.991	ea	121.00	67.00	188.00
36" tall	1C@.991	ea	134.00	67.00	201.00
48" tall	1C@.991	ea	186.00	67.00	253.00
60" tall	1C@1.39	ea	247.00	94.00	341.00
60" wide aluminum sliding window					
24" tall	1C@.991	ea	133.00	67.00	200.00
36" tall	1C@.991	ea	147.00	67.00	214.00
48" tall	1C@1.39	ea	204.00	94.00	298.00
60" tall	1C@1.39	ea	271.00	94.00	365.00
72" wide aluminum sliding window					
24" tall	1C@.991	ea	158.00	67.00	225.00
36" tall	1C@1.39	ea	173.00	94.00	267.00
48" tall	1C@1.39	ea	242.00	94.00	336.00
60" tall	1C@2.00	ea	318.00	135.00	453.00

double-hung

	Craft@Hrs	Unit	Material	Labor	Total
24" wide aluminum fixed (picture) window					
36" tall	1C@.991	ea	155.00	67.00	222.00
48" tall	1C@.991	ea	223.00	67.00	290.00
60" tall	1C@.991	ea	263.00	67.00	330.00
72" tall	1C@.991	ea	285.00	67.00	352.00
30" wide aluminum fixed (picture) window					
36" tall	1C@.991	ea	184.00	67.00	251.00
48" tall	1C@.991	ea	235.00	67.00	302.00
60" tall	1C@.991	ea	254.00	67.00	321.00
72" tall	1C@.991	ea	328.00	67.00	395.00
36" wide aluminum fixed (picture) window					
36" tall	1C@.991	ea	201.00	67.00	268.00
48" tall	1C@.991	ea	266.00	67.00	333.00
60" tall	1C@.991	ea	271.00	67.00	338.00
72" tall	1C@1.39	ea	381.00	94.00	475.00
42" wide aluminum fixed (picture) window					
36" tall	1C@.991	ea	223.00	67.00	290.00
48" tall	1C@.991	ea	317.00	67.00	384.00
60" tall	1C@1.39	ea	384.00	94.00	478.00
72" tall	1C@1.39	ea	441.00	94.00	535.00
48" wide aluminum fixed (picture) window					
36" tall	1C@.991	ea	266.00	67.00	333.00
48" tall	1C@.991	ea	328.00	67.00	395.00
60" tall	1C@1.39	ea	419.00	94.00	513.00
72" tall	1C@1.39	ea	479.00	94.00	573.00
60" wide aluminum fixed (picture) window					
36" tall	1C@.991	ea	271.00	67.00	338.00
48" tall	1C@1.39	ea	339.00	94.00	433.00
60" tall	1C@1.39	ea	395.00	94.00	489.00
72" tall	1C@2.00	ea	526.00	135.00	661.00
72" wide aluminum fixed (picture) window					
36" tall	1C@2.00	ea	381.00	135.00	516.00
48" tall	1C@2.00	ea	482.00	135.00	617.00
60" tall	1C@2.94	ea	526.00	199.00	725.00
72" tall	1C@2.94	ea	626.00	199.00	825.00

Vinyl. All vinyl windows are double-glazed with clear glass. All costs are for standard windows. Manufacturers' window sizes vary. For standard units round measurements to the nearest size. Add approximately **80%** for custom sizes. For double-hung windows add approximately **10%** to the cost of single-hung. Some manufacturers provide a less rigid, lower grade of windows for **15%** less. For aluminum grid between panes of glass or removable grilles outside of window add: half-round **$29**, half-elliptical **$47**, round **$29**, elliptical **$39**, quarter-round **$29**, casement **$38**, awning **$38**, single-hung **$29**, sliding **$33**, fixed (picture) **$35**.

	Craft@Hrs	Unit	Material	Labor	Total
Half-round vinyl window top					
24" x 12"	1C@.991	ea	330.00	67.00	397.00
36" x 18"	1C@.991	ea	401.00	67.00	468.00
48" x 24"	1C@1.39	ea	582.00	94.00	676.00
72" x 36"	1C@2.00	ea	862.00	135.00	997.00
Half-elliptical vinyl window top					
36" x 15"	1C@.991	ea	748.00	67.00	815.00
48" x 16"	1C@.991	ea	680.00	67.00	747.00
60" x 19"	1C@1.39	ea	622.00	94.00	716.00
90" x 20"	1C@2.00	ea	1,330.00	135.00	1,465.00

half-round

half-elliptical

	Craft@Hrs	Unit	Material	Labor	Total
Round vinyl window					
20" x 20"	1C@.991	ea	794.00	67.00	861.00
30" x 30"	1C@.991	ea	789.00	67.00	856.00
40" x 40"	1C@1.39	ea	1,110.00	94.00	1,204.00
48" x 48"	1C@2.00	ea	1,220.00	135.00	1,355.00
Elliptical vinyl window					
15" x 24"	1C@.991	ea	1,490.00	67.00	1,557.00
24" x 36"	1C@.991	ea	1,340.00	67.00	1,407.00
25" x 40"	1C@1.39	ea	1,740.00	94.00	1,834.00
30" x 48"	1C@2.00	ea	1,890.00	135.00	2,025.00
Quarter-round vinyl window					
24" x 24"	1C@.991	ea	646.00	67.00	713.00
36" x 36"	1C@.991	ea	758.00	67.00	825.00
18" wide vinyl casement window					
36" tall	1C@.991	ea	469.00	67.00	536.00
48" tall	1C@.991	ea	506.00	67.00	573.00
60" tall	1C@.991	ea	583.00	67.00	650.00
24" wide vinyl casement window					
36" tall	1C@.991	ea	504.00	67.00	571.00
48" tall	1C@.991	ea	546.00	67.00	613.00
60" tall	1C@1.39	ea	619.00	94.00	713.00
30" wide vinyl casement window					
48" tall	1C@1.39	ea	595.00	94.00	689.00
60" tall	1C@1.39	ea	661.00	94.00	755.00
36" wide vinyl casement window (double)					
48" tall	1C@1.39	ea	706.00	94.00	800.00
60" tall	1C@1.39	ea	811.00	94.00	905.00
24" wide vinyl awning window					
20" tall	1C@.991	ea	435.00	67.00	502.00
24" tall	1C@.991	ea	487.00	67.00	554.00
30" tall	1C@.991	ea	534.00	67.00	601.00
30" wide vinyl awning window					
20" tall	1C@.991	ea	502.00	67.00	569.00
24" tall	1C@.991	ea	547.00	67.00	614.00
30" tall	1C@.991	ea	603.00	67.00	670.00
36" wide vinyl awning window					
24" tall	1C@.991	ea	600.00	67.00	667.00
30" tall	1C@.991	ea	669.00	67.00	736.00
40" wide vinyl awning window					
24" tall	1C@.991	ea	516.00	67.00	583.00
30" tall	1C@.991	ea	746.00	67.00	813.00

round

elliptical

quarter-round

casement

awning

	Craft@Hrs	Unit	Material	Labor	Total
24" wide vinyl single-hung window					
36" tall	1C@.991	ea	210.00	67.00	277.00
48" tall	1C@.991	ea	242.00	67.00	309.00
60" tall	1C@.991	ea	276.00	67.00	343.00
72" tall	1C@.991	ea	317.00	67.00	384.00
36" wide vinyl single-hung window					
36" tall	1C@.991	ea	271.00	67.00	338.00
48" tall	1C@.991	ea	308.00	67.00	375.00
60" tall	1C@.991	ea	333.00	67.00	400.00
72" tall	1C@1.39	ea	367.00	94.00	461.00
42" wide vinyl single-hung window					
36" tall	1C@.991	ea	301.00	67.00	368.00
48" tall	1C@.991	ea	334.00	67.00	401.00
60" tall	1C@1.39	ea	396.00	94.00	490.00
72" tall	1C@1.39	ea	403.00	94.00	497.00
48" wide vinyl single-hung window					
36" tall	1C@.991	ea	313.00	67.00	380.00
48" tall	1C@1.39	ea	335.00	94.00	429.00
60" tall	1C@1.39	ea	381.00	94.00	475.00
72" tall	1C@1.39	ea	429.00	94.00	523.00
48" wide vinyl sliding window					
24" tall	1C@.991	ea	223.00	67.00	290.00
36" tall	1C@.991	ea	271.00	67.00	338.00
48" tall	1C@.991	ea	310.00	67.00	377.00
60" tall	1C@.991	ea	367.00	67.00	434.00
60" wide vinyl sliding window					
36" tall	1C@.991	ea	308.00	67.00	375.00
48" tall	1C@.991	ea	359.00	67.00	426.00
60" tall	1C@1.39	ea	438.00	94.00	532.00
72" wide vinyl sliding window					
48" tall	1C@1.39	ea	405.00	94.00	499.00
60" tall	1C@1.39	ea	474.00	94.00	568.00
24" wide vinyl fixed (picture) window					
36" tall	1C@.991	ea	132.00	67.00	199.00
48" tall	1C@.991	ea	184.00	67.00	251.00
60" tall	1C@.991	ea	223.00	67.00	290.00
72" tall	1C@.991	ea	242.00	67.00	309.00
30" wide vinyl fixed (picture) window					
36" tall	1C@.991	ea	159.00	67.00	226.00
48" tall	1C@.991	ea	195.00	67.00	262.00
60" tall	1C@.991	ea	216.00	67.00	283.00
72" tall	1C@.991	ea	276.00	67.00	343.00

single-hung

sliding

fixed (picture)

	Craft@Hrs	Unit	Material	Labor	Total
36" wide vinyl fixed (picture) window					
36" tall	1C@.991	ea	215.00	67.00	282.00
48" tall	1C@.991	ea	226.00	67.00	293.00
60" tall	1C@.991	ea	230.00	67.00	297.00
72" tall	1C@1.39	ea	321.00	94.00	415.00
42" wide vinyl fixed (picture) window					
36" tall	1C@.991	ea	214.00	67.00	281.00
48" tall	1C@.991	ea	308.00	67.00	375.00
60" tall	1C@1.39	ea	374.00	94.00	468.00
72" tall	1C@1.39	ea	430.00	94.00	524.00
48" wide vinyl fixed (picture) window					
36" tall	1C@.991	ea	226.00	67.00	293.00
48" tall	1C@.991	ea	276.00	67.00	343.00
60" tall	1C@1.39	ea	359.00	94.00	453.00
72" tall	1C@1.39	ea	405.00	94.00	499.00
60" wide vinyl fixed (picture) window					
36" tall	1C@.991	ea	230.00	67.00	297.00
48" tall	1C@1.39	ea	287.00	94.00	381.00
60" tall	1C@1.39	ea	334.00	94.00	428.00
72" tall	1C@2.00	ea	442.00	135.00	577.00
72" wide vinyl fixed (picture) window					
36" tall	1C@2.00	ea	321.00	135.00	456.00
48" tall	1C@2.00	ea	405.00	135.00	540.00
60" tall	1C@2.94	ea	442.00	199.00	641.00
72" tall	1C@2.94	ea	529.00	199.00	728.00

Wood. All wood windows are double-glazed with clear glass. Interiors are natural wood and exteriors are primed wood or aluminum clad or vinyl clad. Hardware on wood windows is bright brass, brushed brass, antique brass or white. All costs are for standard windows. Manufacturers' window sizes vary. For standard units, round measurements to the nearest size. Add approximately **80%** for custom sizes.

	Craft@Hrs	Unit	Material	Labor	Total
Half-round wood window top					
24" x 12"	1C@.991	ea	966.00	67.00	1,033.00
36" x 18"	1C@.991	ea	1,240.00	67.00	1,307.00
48" x 24"	1C@1.39	ea	1,220.00	94.00	1,314.00
60" x 30"	1C@2.00	ea	1,700.00	135.00	1,835.00
Half-elliptical wood window top					
36" x 15"	1C@.991	ea	1,520.00	67.00	1,587.00
48" x16"	1C@.991	ea	1,680.00	67.00	1,747.00
60" x 17"	1C@1.39	ea	1,760.00	94.00	1,854.00
72" x 19"	1C@2.00	ea	2,140.00	135.00	2,275.00
Additional costs for half-round and half-elliptical wood windows					
add 114% for true divided light, spoke pattern (half-round or half-elliptical)					
add for 3 spoke hub grille (half-round window) all sizes	—	ea	29.30	—	29.30

half-round

half-elliptical

	Craft@Hrs	Unit	Material	Labor	Total
Round wood window					
30" x 30"	1C@.991	ea	2,100.00	67.00	2,167.00
48" x 48"	1C@.991	ea	2,450.00	67.00	2,517.00
Elliptical (oval) wood window					
25" x 30"	1C@.991	ea	2,460.00	67.00	2,527.00
30" x 36"	1C@.991	ea	2,750.00	67.00	2,817.00
Additional round and elliptical wood window costs					
add for 4-lite grille in round or elliptical window					
all sizes	—	ea	26.10	—	26.10
add for 9-lite grille in round or elliptical window					
all sizes	—	ea	33.50	—	33.50
add for sunburst grille in round or elliptical window					
all sizes	—	ea	40.80	—	40.80
Quarter-round wood window					
24" x 24"	1C@.991	ea	1,200.00	67.00	1,267.00
36" x 36"	1C@.991	ea	1,400.00	67.00	1,467.00

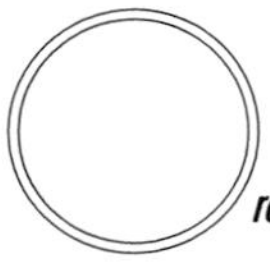

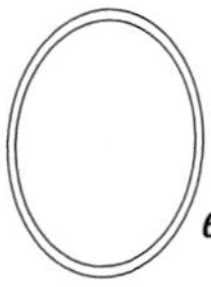

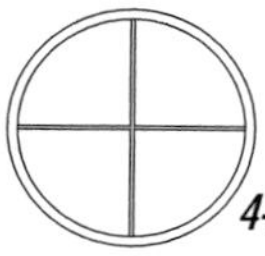

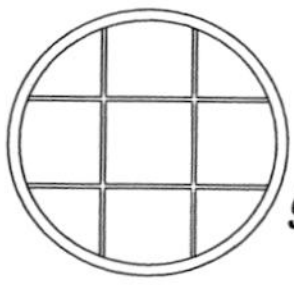

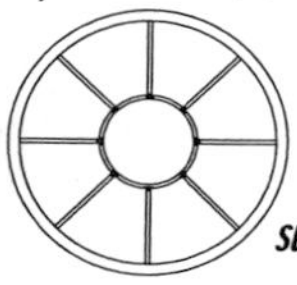

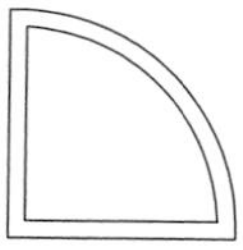

round

elliptical

4-lite grille

9-lite grille

sunburst grille

quarter-round

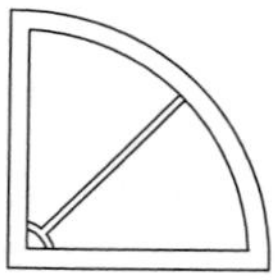

single spoke grille

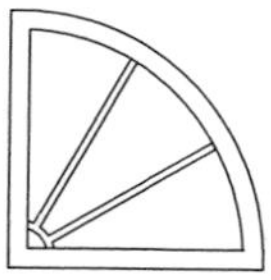

double spoke grille

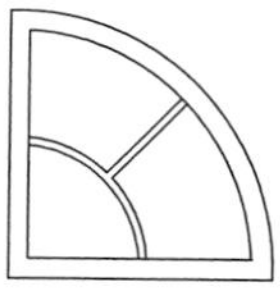

radial bar grille

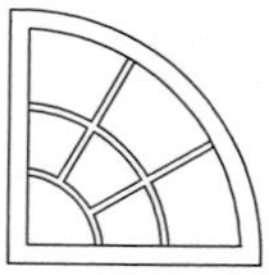

double radial grille

	Craft@Hrs	Unit	Material	Labor	Total
Additional quarter-round wood window costs					
add for single spoke grille					
all sizes	—	ea	36.60	—	36.60
add for double spoke grille					
all sizes	—	ea	42.30	—	42.30
add for single radial bar grille					
all sizes	—	ea	49.70	—	49.70
add for double radial bar grille					
all sizes	—	ea	55.50	—	55.50

	Craft@Hrs	Unit	Material	Labor	Total
18" wide wood casement window					
36" tall	1C@.991	ea	318.00	67.00	385.00
48" tall	1C@.991	ea	429.00	67.00	496.00
60" tall	1C@.991	ea	510.00	67.00	577.00
72" tall	1C@.991	ea	601.00	67.00	668.00
24" wide wood casement window					
36" tall	1C@.991	ea	405.00	67.00	472.00
48" tall	1C@.991	ea	461.00	67.00	528.00
60" tall	1C@1.39	ea	539.00	94.00	633.00
72" tall	1C@.991	ea	626.00	67.00	693.00
30" wide wood casement window					
48" tall	1C@1.39	ea	518.00	94.00	612.00
60" tall	1C@1.39	ea	610.00	94.00	704.00
72" tall	1C@.991	ea	661.00	67.00	728.00
36" wide wood casement window					
48" tall	1C@1.39	ea	584.00	94.00	678.00
60" tall	1C@1.39	ea	687.00	94.00	781.00
72" tall	1C@1.39	ea	750.00	94.00	844.00
Additional casement window costs					
add 87% for authentic divided lites					
add for removable wood grille all sizes	—	ea	17.70	—	17.70
add for aluminum grille between panes all sizes	—	ea	16.50	—	16.50
24" wide wood awning window					
20" tall	1C@.991	ea	339.00	67.00	406.00
24" tall	1C@.991	ea	369.00	67.00	436.00
30" tall	1C@.991	ea	396.00	67.00	463.00
30" wide wood awning window					
20" tall	1C@.991	ea	361.00	67.00	428.00
24" tall	1C@.991	ea	384.00	67.00	451.00
30" tall	1C@1.39	ea	409.00	94.00	503.00
36" wide wood awning window					
20" tall	1C@.991	ea	381.00	67.00	448.00
24" tall	1C@1.39	ea	409.00	94.00	503.00
30" tall	1C@1.39	ea	435.00	94.00	529.00
48" wide wood awning window					
20" tall	1C@.991	ea	458.00	67.00	525.00
24" tall	1C@1.39	ea	470.00	94.00	564.00
30" tall	1C@1.39	ea	523.00	94.00	617.00
Additional awning window costs					
add 80% for authentic divided lites					
add for removable wood grille all sizes	—	ea	20.40	—	20.40
add for aluminum grille between panes all sizes	—	ea	18.90	—	18.90

awning

	Craft@Hrs	Unit	Material	Labor	Total
24" wide wood double-hung window					
36" tall	1C@.991	ea	345.00	67.00	412.00
48" tall	1C@.991	ea	402.00	67.00	469.00
60" tall	1C@.991	ea	452.00	67.00	519.00
72" tall	1C@.991	ea	509.00	67.00	576.00
36" wide wood double-hung window					
36" tall	1C@.991	ea	419.00	67.00	486.00
48" tall	1C@.991	ea	485.00	67.00	552.00
60" tall	1C@.991	ea	547.00	67.00	614.00
72" tall	1C@1.39	ea	620.00	94.00	714.00
48" wide wood double-hung window					
36" tall	1C@.991	ea	536.00	67.00	603.00
48" tall	1C@.991	ea	536.00	67.00	603.00
60" tall	1C@1.39	ea	617.00	94.00	711.00
72" tall	1C@1.39	ea	714.00	94.00	808.00
Additional double-hung window costs					
add 105% for authentic divided lites					
add for removable wood grille					
all sizes	—	ea	20.40	—	20.40
add for aluminum grille between panes					
all sizes	—	ea	18.90	—	18.90
24" wide wood single-hung window					
36" tall	1C@.991	ea	256.00	67.00	323.00
48" tall	1C@.991	ea	313.00	67.00	380.00
60" tall	1C@.991	ea	365.00	67.00	432.00
72" tall	1C@.991	ea	420.00	67.00	487.00
36" wide wood single-hung window					
36" tall	1C@.991	ea	330.00	67.00	397.00
48" tall	1C@.991	ea	395.00	67.00	462.00
60" tall	1C@.991	ea	459.00	67.00	526.00
72" tall	1C@1.39	ea	533.00	94.00	627.00
40" wide wood single-hung window					
36" tall	1C@.991	ea	445.00	67.00	512.00
48" tall	1C@.991	ea	445.00	67.00	512.00
60" tall	1C@1.39	ea	530.00	94.00	624.00
72" tall	1C@1.39	ea	624.00	94.00	718.00
Additional single-hung window costs					
add 105% for authentic divided lites					
add for removable wood grille					
all sizes	—	ea	20.40	—	20.40
add for aluminum grille between panes					
all sizes	—	ea	18.90	—	18.90
add 97% for segmented top					

double-hung window

single-hung window

	Craft@Hrs	Unit	Material	Labor	Total
48" wide wood sliding window					
24" tall	1C@.991	ea	503.00	67.00	570.00
36" tall	1C@.991	ea	542.00	67.00	609.00
48" tall	1C@.991	ea	641.00	67.00	708.00
60" tall	1C@.991	ea	730.00	67.00	797.00
60" wide wood sliding window					
36" tall	1C@.991	ea	648.00	67.00	715.00
48" tall	1C@.991	ea	733.00	67.00	800.00
60" tall	1C@1.39	ea	839.00	94.00	933.00
72" wide wood sliding window					
48" tall	1C@1.39	ea	825.00	94.00	919.00
60" tall	1C@1.39	ea	932.00	94.00	1,026.00
Additional sliding window costs					
add 90% for authentic divided lites					
add for removable wood grille					
all sizes	—	ea	20.40	—	20.40
add for aluminum grille between panes					
all sizes	—	ea	18.90	—	18.90
24" wide wood fixed (picture) window					
36" tall	1C@.991	ea	293.00	67.00	360.00
48" tall	1C@.991	ea	347.00	67.00	414.00
60" tall	1C@.991	ea	400.00	67.00	467.00
72" tall	1C@.991	ea	458.00	67.00	525.00
36" wide wood fixed (picture) window					
36" tall	1C@.991	ea	367.00	67.00	434.00
48" tall	1C@.991	ea	430.00	67.00	497.00
60" tall	1C@.991	ea	496.00	67.00	563.00
72" tall	1C@1.39	ea	568.00	94.00	662.00
48" wide wood fixed (picture) window					
36" tall	1C@.991	ea	482.00	67.00	549.00
48" tall	1C@.991	ea	482.00	67.00	549.00
60" tall	1C@1.39	ea	566.00	94.00	660.00
72" tall	1C@1.39	ea	661.00	94.00	755.00
60" wide wood fixed (picture) window					
48" tall	1C@1.39	ea	571.00	94.00	665.00
60" tall	1C@1.39	ea	658.00	94.00	752.00
72" tall	1C@2.00	ea	767.00	135.00	902.00
72" wide wood fixed (picture) window					
60" tall	1C@2.00	ea	750.00	135.00	885.00
72" tall	1C@2.00	ea	856.00	135.00	991.00
Additional fixed window costs.					
add 80% for authentic divided lites					
add for removable wood grille					
all sizes	—	ea	21.80	—	21.80
add for aluminum grille between panes					
all sizes	—	ea	20.40	—	20.40
add 97% for segmented top					

sliding

fixed (picture)

	Craft@Hrs	Unit	Material	Labor	Total

Skylight. Includes skylight, frame, nails, flashing, peel-and-stick moisture barrier, framed curb (when required by manufacturer), and installation. Flashing is for a roof with low-profile roofing such as asphalt shingles. For tile shingles or ribbed metal roofing, use the price list item that adds an additional cost for a high-profile roof. Curbs are usually required on low-slope roofs. A special sloped curb is usually required on flat roofs. Does not include roof framing, shingles, or interior finishes.

	Craft@Hrs	Unit	Material	Labor	Total
Additional skylight costs					
add for laminated glass over tempered glass in skylight or roof window, all types	—	ea	52.80	—	52.80
add for Low-E in skylight or roof window all types	—	ea	13.20	—	13.20
add for argon gas filled skylight or roof window all types	—	ea	11.70	—	11.70
add for bronze or gray tint (all sizes) all types	—	ea	21.80	—	21.80
add for skylight or roof window installed on high-profile roof, all types	—	ea	33.50	—	33.50
add for skylight or roof window installed on low-slope roof, all types	—	ea	441.00	—	441.00
add for skylight or roof window installed on flat roof, all types	—	ea	458.00	—	458.00
Single dome skylight					
24" x 24"	1C@2.62	ea	95.10	177.00	272.10
24" x 48"	1C@2.62	ea	188.00	177.00	365.00
30" x 30"	1C@2.62	ea	149.00	177.00	326.00
48" x 48"	1C@2.62	ea	477.00	177.00	654.00
Double dome skylight					
24" x 24"	1C@2.62	ea	128.00	177.00	305.00
24" x 48"	1C@2.62	ea	261.00	177.00	438.00
30" x 30"	1C@2.62	ea	245.00	177.00	422.00
48" x 48"	1C@2.62	ea	563.00	177.00	740.00
Triple dome skylight					
24" x 24"	1C@2.62	ea	165.00	177.00	342.00
24" x 48"	1C@2.62	ea	378.00	177.00	555.00
30" x 30"	1C@2.62	ea	300.00	177.00	477.00
48" x 48"	1C@2.62	ea	720.00	177.00	897.00
Single dome ventilating skylight					
24" x 24"	1C@2.78	ea	201.00	188.00	389.00
24" x 48"	1C@2.78	ea	394.00	188.00	582.00
30" x 30"	1C@2.78	ea	314.00	188.00	502.00
48" x 48"	1C@2.78	ea	993.00	188.00	1,181.00
Double dome ventilating skylight					
24" x 24"	1C@2.78	ea	264.00	188.00	452.00
24" x 48"	1C@2.78	ea	546.00	188.00	734.00
30" x 30"	1C@2.78	ea	507.00	188.00	695.00
48" x 48"	1C@2.78	ea	1,170.00	188.00	1,358.00
Triple dome ventilating skylight					
24" x 24"	1C@2.78	ea	338.00	188.00	526.00
24" x 48"	1C@2.78	ea	785.00	188.00	973.00
30" x 30"	1C@2.78	ea	622.00	188.00	810.00
48" x 48"	1C@2.78	ea	1,490.00	188.00	1,678.00

	Craft@Hrs	Unit	Material	Labor	Total
22" wide fixed skylight					
28" long	1C@2.69	ea	345.00	182.00	527.00
38" long	1C@2.69	ea	403.00	182.00	585.00
48" long	1C@2.69	ea	450.00	182.00	632.00
55" long	1C@2.69	ea	492.00	182.00	674.00
30" wide fixed skylight					
38" long	1C@2.69	ea	465.00	182.00	647.00
55" long	1C@2.69	ea	569.00	182.00	751.00
44" wide fixed skylight					
28" long	1C@2.69	ea	506.00	182.00	688.00
48" long	1C@2.69	ea	646.00	182.00	828.00
22" wide fixed skylight with ventilation flap					
28" long	1C@2.69	ea	391.00	182.00	573.00
38" long	1C@2.69	ea	450.00	182.00	632.00
48" long	1C@2.69	ea	500.00	182.00	682.00
55" long	1C@2.69	ea	529.00	182.00	711.00
30" wide fixed skylight with ventilation flap					
38" long	1C@2.69	ea	526.00	182.00	708.00
55" long	1C@2.69	ea	614.00	182.00	796.00
44" wide fixed skylight with ventilation flap					
28" long	1C@2.69	ea	588.00	182.00	770.00
48" long	1C@2.69	ea	708.00	182.00	890.00
22" wide ventilating skylight					
28" long	1C@2.94	ea	650.00	199.00	849.00
38" long	1C@2.94	ea	708.00	199.00	907.00
48" long	1C@2.94	ea	773.00	199.00	972.00
55" long	1C@2.94	ea	805.00	199.00	1,004.00
30" wide ventilating skylight					
38" long	1C@2.94	ea	785.00	199.00	984.00
55" long	1C@2.94	ea	906.00	199.00	1,105.00
44" wide ventilating skylight					
28" long	1C@2.94	ea	826.00	199.00	1,025.00
48" long	1C@2.94	ea	1,010.00	199.00	1,209.00
22" wide roof window					
28" long	1C@2.94	ea	888.00	199.00	1,087.00
48" long	1C@2.94	ea	980.00	199.00	1,179.00
30" wide roof window					
38" long	1C@2.94	ea	986.00	199.00	1,185.00
55" long	1C@2.94	ea	1,120.00	199.00	1,319.00
44" wide roof window					
55" long	1C@2.94	ea	1,220.00	199.00	1,419.00
Add for cord-operated roller shades					
28" long	1C@.832	ea	93.50	56.20	149.70
38" long	1C@.832	ea	93.50	56.20	149.70
48" long	1C@.832	ea	93.50	56.20	149.70
55" long	1C@.832	ea	93.50	56.20	149.70

	Craft@Hrs	Unit	Material	Labor	Total
Add for motorized rod cord control for roller shades					
all sizes	—	ea	292.00	—	292.00
Add for manually controlled Venetian blinds					
28" long	—	ea	132.00	—	132.00
38" long	—	ea	157.00	—	157.00
48" long	—	ea	165.00	—	165.00
55" long	—	ea	208.00	—	208.00
Add for electrically controlled skylight system					
28" long	1C@2.00	ea	214.00	135.00	349.00
38" long	1C@2.00	ea	214.00	135.00	349.00
48" long	1C@2.00	ea	214.00	135.00	349.00
55" long	1C@2.00	ea	214.00	135.00	349.00
Add for electrically controlled window or skylight opener					
28" long	1C@1.39	ea	214.00	94.00	308.00
38" long	1C@1.39	ea	214.00	94.00	308.00
48" long	1C@1.39	ea	214.00	94.00	308.00
55" long	1C@1.39	ea	214.00	94.00	308.00
Add for infrared remote control system for opener					
28" long	1C@1.11	ea	474.00	75.00	549.00
38" long	1C@1.11	ea	474.00	75.00	549.00
48" long	1C@1.11	ea	474.00	75.00	549.00
55" long	1C@1.11	ea	474.00	75.00	549.00

Storm window.

	Craft@Hrs	Unit	Material	Labor	Total
24" wide aluminum storm window					
24" tall	1C@1.00	ea	102.00	67.60	169.60
36" tall	1C@1.00	ea	148.00	67.60	215.60
48" tall	1C@1.00	ea	184.00	67.60	251.60
60" tall	1C@1.00	ea	208.00	67.60	275.60
36" wide aluminum storm window					
24" tall	1C@1.00	ea	148.00	67.60	215.60
36" tall	1C@1.00	ea	195.00	67.60	262.60
48" tall	1C@1.00	ea	230.00	67.60	297.60
60" tall	1C@1.08	ea	261.00	73.00	334.00
48" wide aluminum storm window					
24" tall	1C@1.00	ea	184.00	67.60	251.60
36" tall	1C@1.00	ea	230.00	67.60	297.60
48" tall	1C@1.08	ea	273.00	73.00	346.00
60" tall	1C@1.08	ea	317.00	73.00	390.00
60" wide aluminum storm window					
24" tall	1C@1.00	ea	208.00	67.60	275.60
36" tall	1C@1.08	ea	261.00	73.00	334.00
48" tall	1C@1.08	ea	317.00	73.00	390.00
60" tall	1C@1.08	ea	370.00	73.00	443.00
72" wide aluminum storm window					
24" tall	1C@1.00	ea	230.00	67.60	297.60
36" tall	1C@1.08	ea	296.00	73.00	369.00
48" tall	1C@1.08	ea	362.00	73.00	435.00
60" tall	1C@1.22	ea	426.00	82.50	508.50

	Craft@Hrs	Unit	Material	Labor	Total
24" wide wood storm window					
24" tall	1C@1.00	ea	122.00	67.60	189.60
36" tall	1C@1.00	ea	173.00	67.60	240.60
48" tall	1C@1.00	ea	201.00	67.60	268.60
60" tall	1C@1.00	ea	228.00	67.60	295.60
36" wide wood storm window					
24" tall	1C@1.00	ea	173.00	67.60	240.60
36" tall	1C@1.00	ea	210.00	67.60	277.60
48" tall	1C@1.00	ea	247.00	67.60	314.60
60" tall	1C@1.08	ea	284.00	73.00	357.00
48" wide wood storm window					
24" tall	1C@1.00	ea	201.00	67.60	268.60
36" tall	1C@1.00	ea	247.00	67.60	314.60
48" tall	1C@1.08	ea	296.00	73.00	369.00
60" tall	1C@1.08	ea	341.00	73.00	414.00
60" wide wood storm window					
24" tall	1C@1.00	ea	228.00	67.60	295.60
36" tall	1C@1.08	ea	284.00	73.00	357.00
48" tall	1C@1.08	ea	341.00	73.00	414.00
60" tall	1C@1.08	ea	402.00	73.00	475.00
72" wide wood storm window					
24" tall	1C@1.00	ea	247.00	67.60	314.60
36" tall	1C@1.08	ea	318.00	73.00	391.00
48" tall	1C@1.08	ea	391.00	73.00	464.00
60" tall	1C@1.22	ea	461.00	82.50	543.50

Window screen. Includes prefabricated window screen with aluminum frame and installation. Screens are made to match existing windows and include a tinted or mill-finish frame with fiberglass mesh.

	Craft@Hrs	Unit	Material	Labor	Total
24" wide window screen					
24" tall	1C@.125	ea	16.80	8.45	25.25
36" tall	1C@.125	ea	20.30	8.45	28.75
48" tall	1C@.125	ea	23.10	8.45	31.55
60" tall	1C@.125	ea	27.60	8.45	36.05
36" wide window screen					
24" tall	1C@.125	ea	19.60	8.45	28.05
36" tall	1C@.125	ea	21.10	8.45	29.55
48" tall	1C@.125	ea	29.30	8.45	37.75
60" tall	1C@.125	ea	32.10	8.45	40.55

Reglaze window. Includes removal of damaged glass, replacement of gaskets or glazing compound, and glass piece to fit. Does not include replacement parts in window or door frame.

	Craft@Hrs	Unit	Material	Labor	Total
Reglaze windows with clear glass					
1/8" thick	1C@.063	sf	6.12	4.26	10.38
3/16" thick	1C@.063	sf	8.18	4.26	12.44
1/4" thick	1C@.063	sf	6.69	4.26	10.95
3/8" thick	1C@.063	sf	21.50	4.26	25.76
Reglaze window with tempered glass					
1/8" thick	1C@.063	sf	10.20	4.26	14.46
3/16" thick	1C@.063	sf	12.00	4.26	16.26
1/4" thick	1C@.063	sf	12.80	4.26	17.06

	Craft@Hrs	Unit	Material	Labor	Total
Reglaze window with laminated glass					
1/4" thick	1C@.063	sf	15.20	4.26	19.46
3/8" thick	1C@.063	sf	20.90	4.26	25.16
Reglaze window with polished wire glass					
1/4" thick	1C@.063	sf	40.20	4.26	44.46
Minimum charge for reglazing					
all types of glass	1C@2.50	ea	50.90	169.00	219.90
Repair antique double-hung window.					
rehang sash weight	1C@2.27	ea	26.10	153.00	179.10
refurbish	1C@3.13	ea	46.50	212.00	258.50
recondition	1C@2.08	ea	23.60	141.00	164.60

Time & Material Charts (selected items)

Windows Materials

See Windows material prices with the line items above.

Windows Labor

Laborer	base wage	paid leave	true wage	taxes & ins.	total
Carpenter	$38.10	2.97	$41.07	26.53	$67.60
Demolition laborer	$25.90	2.02	$27.92	19.28	$47.20

Paid leave is calculated based on two weeks paid vacation, one week sick leave, and seven paid holidays. Employer's matching portion of **FICA** is 7.65 percent. **FUTA** (Federal Unemployment) is .8 percent. **Worker's compensation** for the windows trade was calculated using a national average of 17.21 percent. **Unemployment insurance** was calculated using a national average of 8 percent. **Health insurance** was calculated based on a projected national average for 2020 of $1,280 per employee (and family when applicable) per month. Employer pays 80 percent for a per month cost of $1,024 per employee. **Retirement** is based on a 401(k) retirement program with employer matching of 50 percent. Employee contributions to the 401(k) plan are an average of 6 percent of the true wage. **Liability insurance** is based on a national average of 12.0 percent.

Windows Labor Productivity	Craft@Hrs	Unit	Material	Labor	Total
Remove window					
small (4 to 10 sf)	1D@.397	ea	—	18.70	18.70
average size (11 to 16 sf)	1D@.407	ea	—	19.20	19.20
large (17 to 29 sf)	1D@.424	ea	—	20.00	20.00
very large (30 sf and larger)	1D@.433	ea	—	20.40	20.40
Remove window for work & reinstall					
small (4 to 10 sf)	1C@1.23	ea	—	83.10	83.10
average size (11 to 16 sf)	1C@1.72	ea	—	116.00	116.00
large (17 to 28 sf)	1C@2.51	ea	—	170.00	170.00
very large (30 sf and larger)	1C@3.70	ea	—	250.00	250.00

	Craft@Hrs	Unit	Material	Labor	Total
Install window					
small (4 to 10 sf)	1C@.991	ea	—	67.00	67.00
average size (11 to 16 sf)	1C@1.39	ea	—	94.00	94.00
large (17 to 28 sf)	1C@2.00	ea	—	135.00	135.00
very large (30 sf and larger)	1C@2.94	ea	—	199.00	199.00
Install skylight or roof window					
dome ventilating	1C@2.78	ea	—	188.00	188.00
fixed	1C@2.69	ea	—	182.00	182.00
ventilating	1C@2.94	ea	—	199.00	199.00
cord-operated roller shades	1C@.832	ea	—	56.20	56.20
electrically controlled system	1C@2.00	ea	—	135.00	135.00
electrically controlled opener	1C@1.39	ea	—	94.00	94.00
infrared remote control system for opener	1C@1.11	ea	—	75.00	75.00
Install storm window					
small (4 to 14 sf)	1C@1.00	ea	—	67.60	67.60
average size (15 to 25 sf)	1C@1.08	ea	—	73.00	73.00
large storm (26 sf and larger)	1C@1.22	ea	—	82.50	82.50
Reglaze window					
remove and replace glass	1C@.063	sf	—	4.26	4.26
Repair antique double-hung window					
rehang sash weight	1C@2.27	ea	—	153.00	153.00
refurbish	1C@3.13	ea	—	212.00	212.00
recondition	1C@2.08	ea	—	141.00	141.00

QuickCalculators

QuickCalculators

QuickCalculators are designed to help you quickly calculate quantities and the surface areas of some basic room shapes. Pages 450-463 contain *QuickCalculators* that are designed to calculate quantities, and pages 464-469 contain *QuickCalculators* that are designed to calculate the surface area of rooms.

Instruction for each *QuickCalculator* sheet are contained above the page.

QuickCalculator mathematics

CIRCLE

Area

❶ = Pi x $radius^2$

❷ = .7854 x $diameter^2$

❸ = .0796 x $perimeter^2$

Perimeter

❶ = Pi x diameter

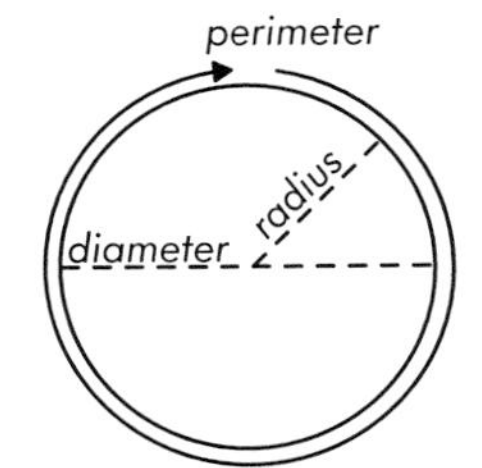

PARALLELOGRAM

Area

❶ = base x height

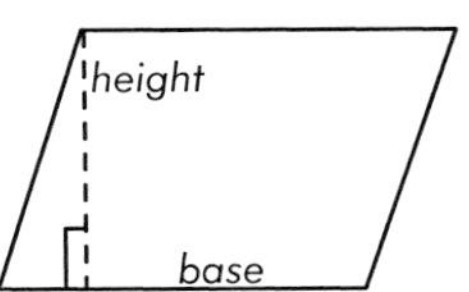

PRISM

Area

❶ = sum of sides ÷ 2 x radius

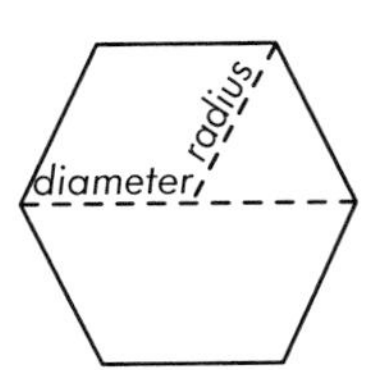

RECTANGLE

Area

❶ = length x width

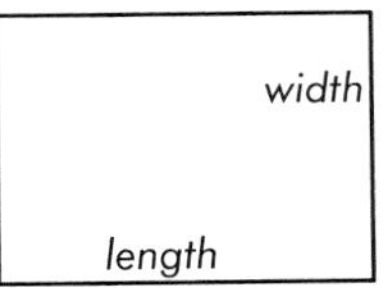

TRAPEZOID

Area

❶ = base + top ÷ 2 x height

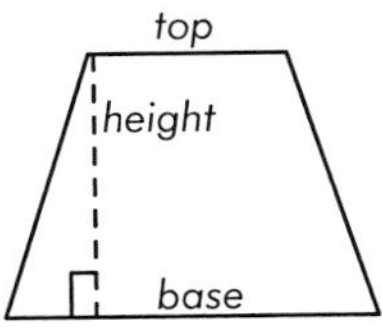

TRIANGLE

Area

❶ = base x height ÷ 2

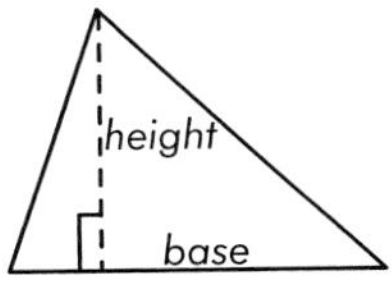

PYTHAGOREAN THEOREM

Rafter length

❶ rafter $length^2$ = $rise^2$ + run^2

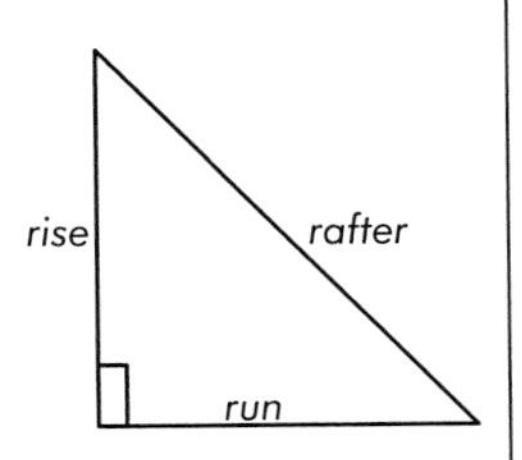

CONVERSION FACTORS

DIVIDE	BY	TO CONVERT TO
square inches	144	square feet
square inches	144	board feet
cubic inches	1,728	cubic feet
cubic inches	46,656	cubic yards
square feet	9	square yards
square feet	100	squares
cubic feet	27	cubic yards
feet	16.5	rods
rods	40	furlongs
feet	5,280	miles
meters	1,609	miles
yards	1,760	miles

MULTIPLY	BY	TO CONVERT TO
inches	2.54	centimeters
square inches	6.4516	square centimeters
square inches	.0069	square feet
square feet	9.29	square centimeters
square feet	.0929	square meters
square yards	.8361	square meters
square meters	1.1959	square yards
centimeters	.3937	inches
feet	.3048	meters
meters	3.281	feet
yards	91.44	centimeters
yards	.9144	meters
cubic feet	.0283	cubic meters
cubic meters	35.3145	cubic feet
cubic meters	1.3079	cubic yards
cubic yards	.7646	cubic meters
cubic inches	16.3872	cubic centimeters
cubic centimeters	.0610	cubic inches
cubic meters	1,000	cubic liters

USING THE BEARING WALL
QuickCalculator

A Calculates the lineal feet of studs in a wall *before* adding for openings and corners. The length of the wall (Answer 1) is converted into inches by multiplying it by 12. The number of studs is calculated by dividing the wall length in inches by the stud centers (Answer 8), then adding 1 for the first stud. The calculation should now be rounded *up*, then multiplied by the wall height (Answer 4).

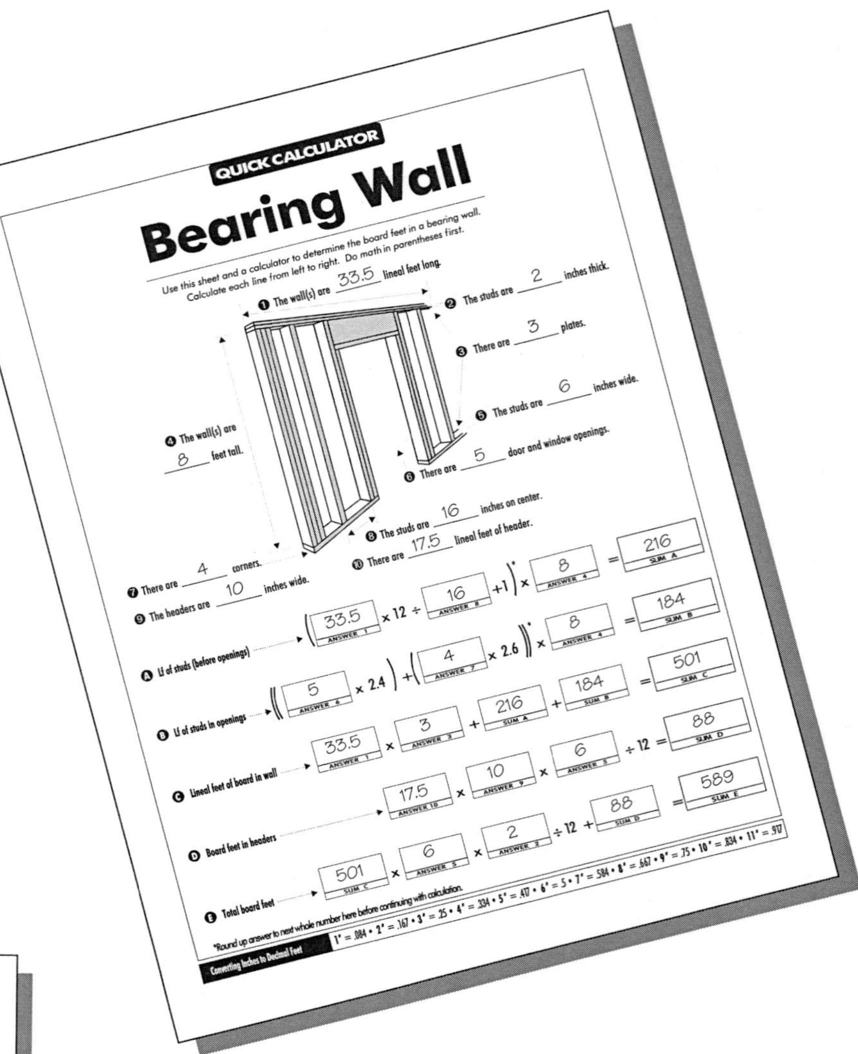

B Calculates the lineal feet of studs typically added for openings and corners. The number of openings (Answer 6) is multiplied by 2.4 (the *average* additional studs typically needed for an opening). The number of corners (Answer 7) is multiplied by 2.6 (the *average* additional studs typically needed at corners). The studs for openings and corners are added together, then should be rounded *up*. The total additional studs is then multiplied by the wall height (Answer 4).

C Calculates the total lineal feet of lumber in the wall. The length of the wall (Answer 1) is multiplied by the number of plates (Answer 3). The lineal feet of plates is added to the lineal feet of studs (Sum A) and additional studs (Sum B).

D Calculates the board feet in the headers. The lineal feet of headers (Answer 10) is multiplied by the header width (Answer 9), then multiplied by the wall thickness (Answer 5). This total is divided by 12 to convert to board feet.

E Calculates the total board feet in the wall. The lineal feet of board in the wall (Sum C) is multiplied by the wall width (Answer 5) then by the stud thickness (Answer 2). The total is divided by 12 to convert to board feet then added to the board feet in the headers (Sum D).

QUICK FACTS

☞ See page 449 for more information about the geometric formulas used in this *QuickCalculator*. See page 357 for wall framing priced per board foot.

QUICK CALCULATOR

Bearing Wall

Use this sheet and a calculator to determine the board feet in a bearing wall. Calculate each line from left to right. Do math in parentheses first.

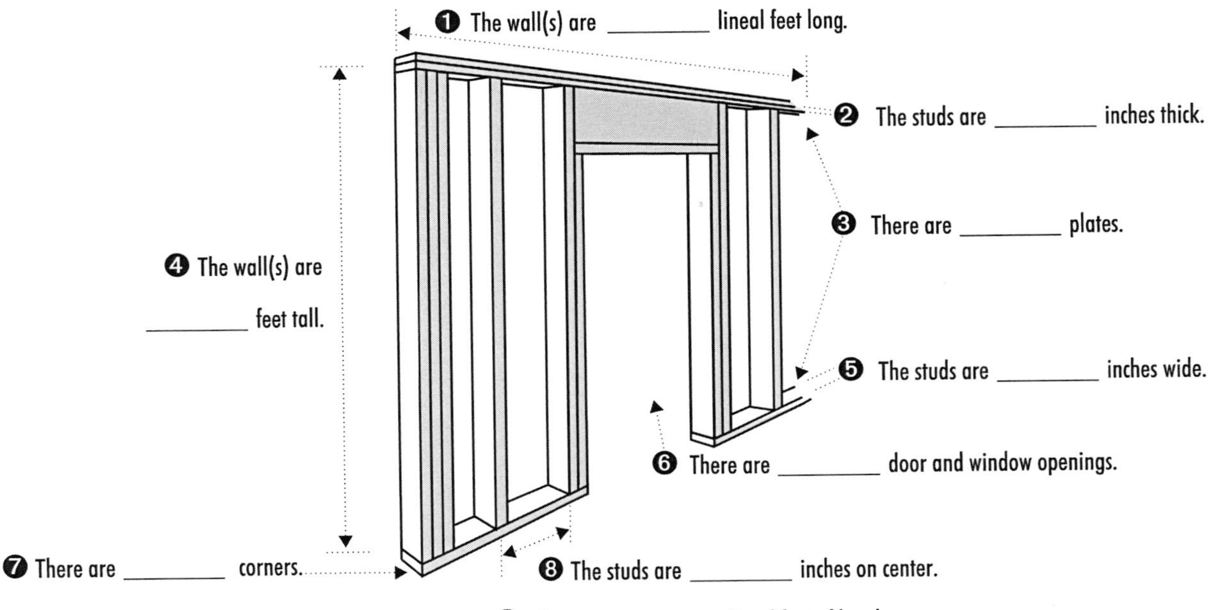

❾ The headers are ________ inches wide.

❿ There are ________ lineal feet of header.

A Lf of studs (before openings) → ([ANSWER 1] × 12 ÷ [ANSWER 8] +1)* × [ANSWER 4] = [SUM A]

B Lf of studs in openings → (([ANSWER 6] × 2.4) + ([ANSWER 7] × 2.6))* × [ANSWER 4] = [SUM B]

C Lineal feet of board in wall → [ANSWER 1] × [ANSWER 3] + [SUM A] + [SUM B] = [SUM C]

D Board feet in headers → [ANSWER 10] × [ANSWER 9] × [ANSWER 5] ÷ 12 = [SUM D]

E Total board feet → [SUM C] × [ANSWER 5] × [ANSWER 2] ÷ 12 + [SUM D] = [SUM E]

*Round up answer to next whole number here before continuing with calculation.

Converting Inches to Decimal Feet 1″ = .084 • 2″ = .167 • 3″ = .25 • 4″ = .334 • 5″ = .417 • 6″ = .5 • 7″ = .584 • 8″ = .667 • 9″ = .75 • 10″ = .834 • 11″ = .917

USING THE CONCRETE WALL & FOOTING
QUICKCALCULATOR

A Calculates the cubic yards of footing by multiplying Answer 2 by 12 to convert the length of the footing to inches. The length in inches is then multiplied by the thickness (Answer 4) and by the width (Answer 5) to determine the cubic inches in the footing. The cubic inches are converted to cubic feet by dividing by 1,728. The cubic feet are then converted to cubic yards by dividing by 27.

B Calculates the cubic yards of foundation wall by multiplying Answer 2 by 12 to convert the length of the wall to inches. The length in inches is then multiplied by the thickness (Answer 1) and by the height (Answer 3) to determine the cubic inches in the wall. The cubic inches are converted to cubic feet by dividing by 1,728. The cubic feet are then converted to cubic yards by dividing by 27.

C Calculates the total cubic yards of concrete in the wall & footing by adding the cubic yards of footing (Sum A) to the cubic yards of foundation wall (Sum B).

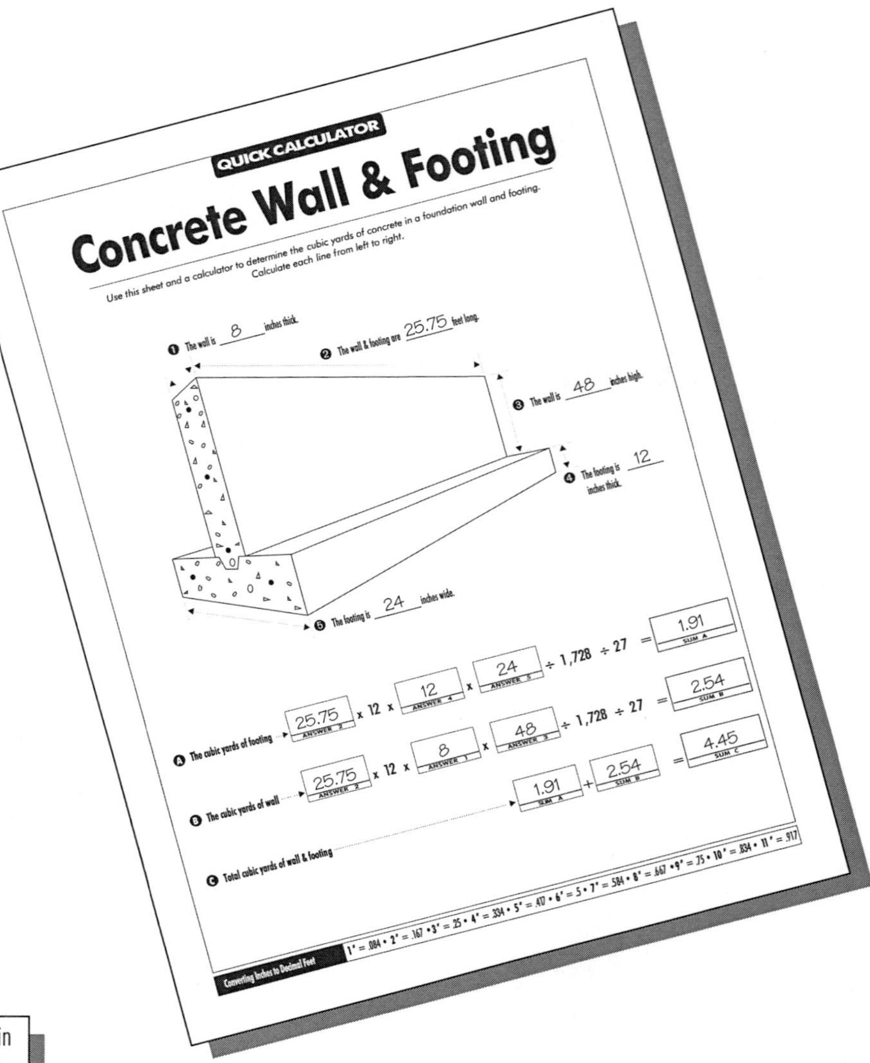

QUICK FACTS

☞ The sample *QuickCalculator* sheet shown above (in the printed book) contains quantities that have been rounded. We suggest rounding to two decimal places. We also suggest rounding the final total (Sum C) up to the next 1/4 yards since this is how concrete must usually be ordered. (In the example the Sum C total of 4.45 should be rounded up to 4.5 cubic yards.) See pages 56-59 for concrete walls and footings per cubic yard.

☞ See page 449 for more information about the geometric formulas used in this *QuickCalculator*.

QUICK CALCULATOR

Concrete Wall & Footing

Use this sheet and a calculator to determine the cubic yards of concrete in a foundation wall and footing. Calculate each line from left to right.

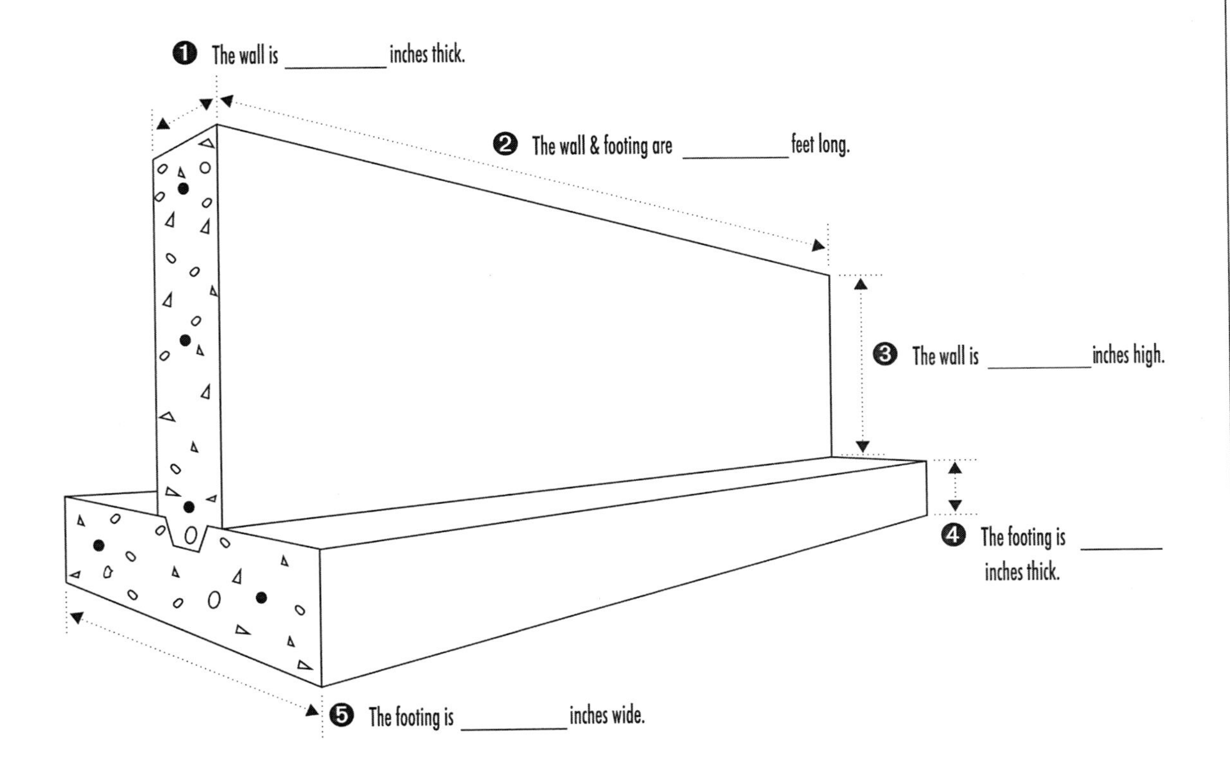

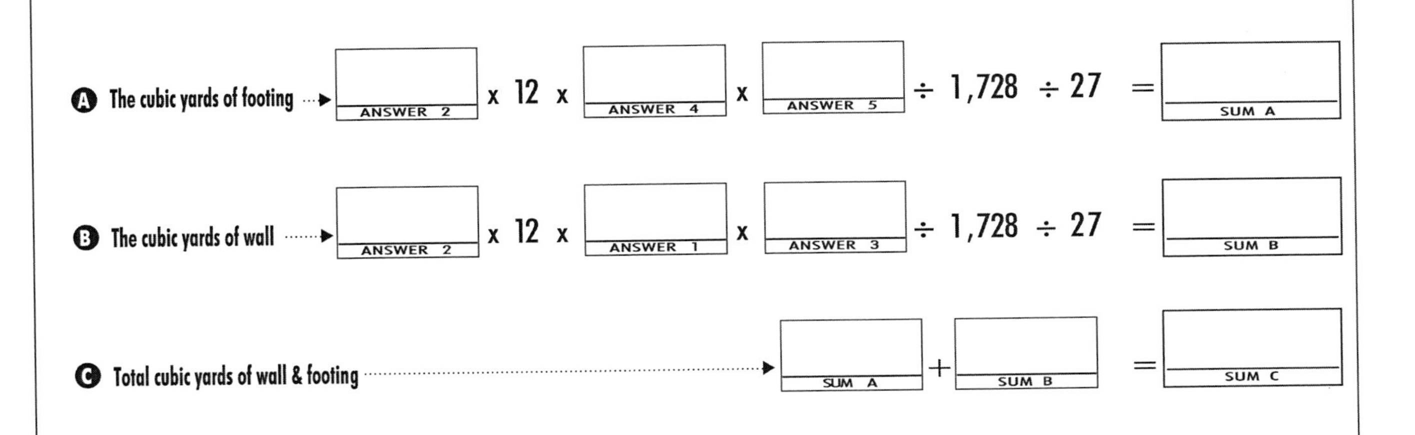

Converting Inches to Decimal Feet 1″ = .084 • 2″ = .167 • 3″ = .25 • 4″ = .334 • 5″ = .417 • 6″ = .5 • 7″ = .584 • 8″ = .667 • 9″ = .75 • 10″ = .834 • 11″ = .917

USING GABLE ROOF RAFTERS

QUICKCALCULATOR

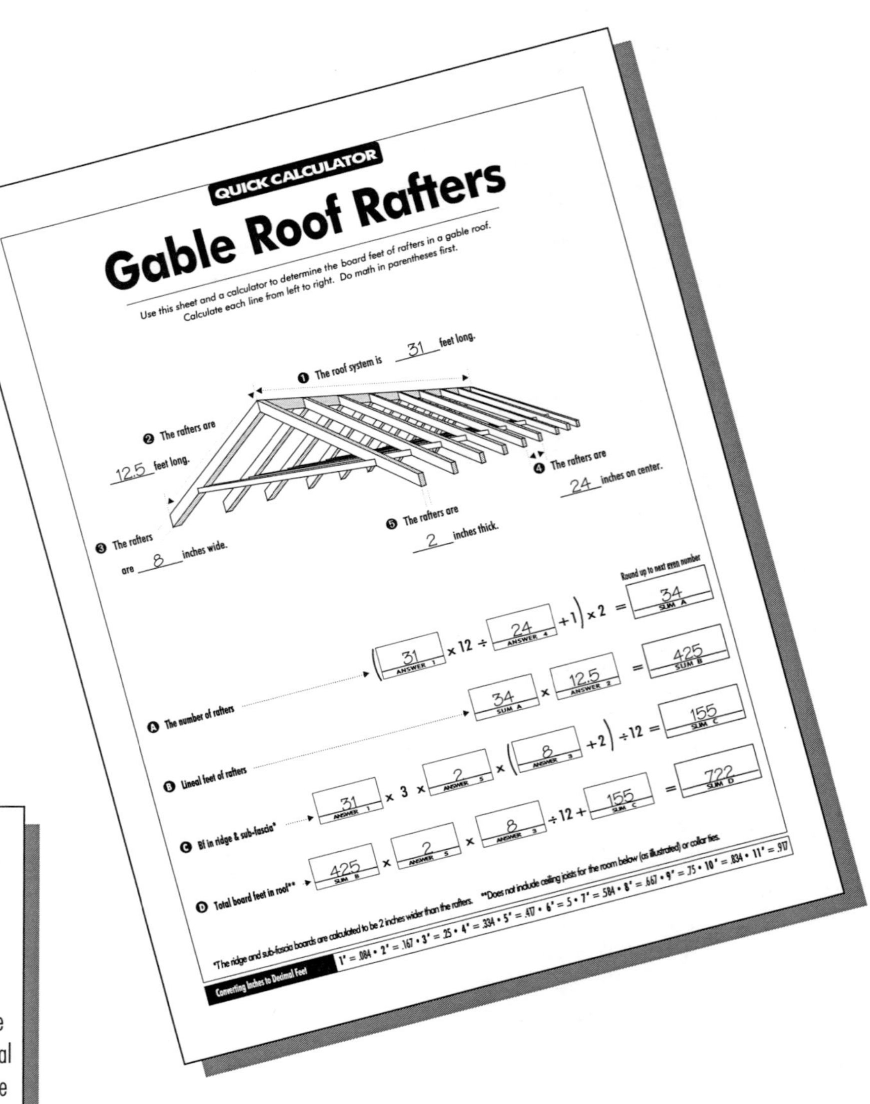

A Calculates the total number of rafters in the roof. The length of the roof (Answer 1) is converted into inches by multiplying it by 12. The number of rafters in the roof is calculated by taking the length in inches and dividing by the rafter centers (Answer 4) then adding 1 for the first rafter. Because you can never have a fraction of a rafter, the final answer should always be rounded *up* to the next *even* number. This calculates the total for one side of the roof only so the total is multiplied by 2.

B Calculates the lineal feet of rafters in the roof by multiplying the number of rafters (Sum A) by the rafter length (Answer 2).

C Calculates the board feet in the ridge and sub-fascia. The roof length (Answer 1) is multiplied by 3 (2 sub-fascia boards and 1 ridge) to get the total lineal feet of sub-fascia and ridge. This total is multiplied by the rafter thickness (Answer 5), then by the rafter width (Answer 3) plus 2" (e.g. a roof with 2" x 8" rafters will have 10" wide ridge and sub-fascia boards). This total is divided by 12 to get the total board feet in the ridge and sub-fascia boards.

D Calculates the total board feet in the roof system. The total lineal feet of rafters (Sum B) is *multiplied* by the rafter thickness (Answer 5) then by the rafter width (Answer 3). This total is *divided* by 12 to get the total board feet in the rafters. The total board feet of rafters is then added to the total board feet in the ridge and sub-fascia boards (Sum C) for the total board feet in the entire roof system.

QUICK FACTS

☞ The sample *QuickCalculator* sheet shown above contains quantities that have been rounded. The total number of rafters should always be rounded up to the next even number. We also suggest rounding lineal feet and board feet to the nearest whole number.

☞ See page 449 for more information about the geometric formulas used in this *QuickCalculator*. See page 372 for rafter systems priced per board foot.

QUICK CALCULATOR

Gable Roof Rafters

Use this sheet and a calculator to determine the board feet of rafters in a gable roof. Calculate each line from left to right. Do math in parentheses first.

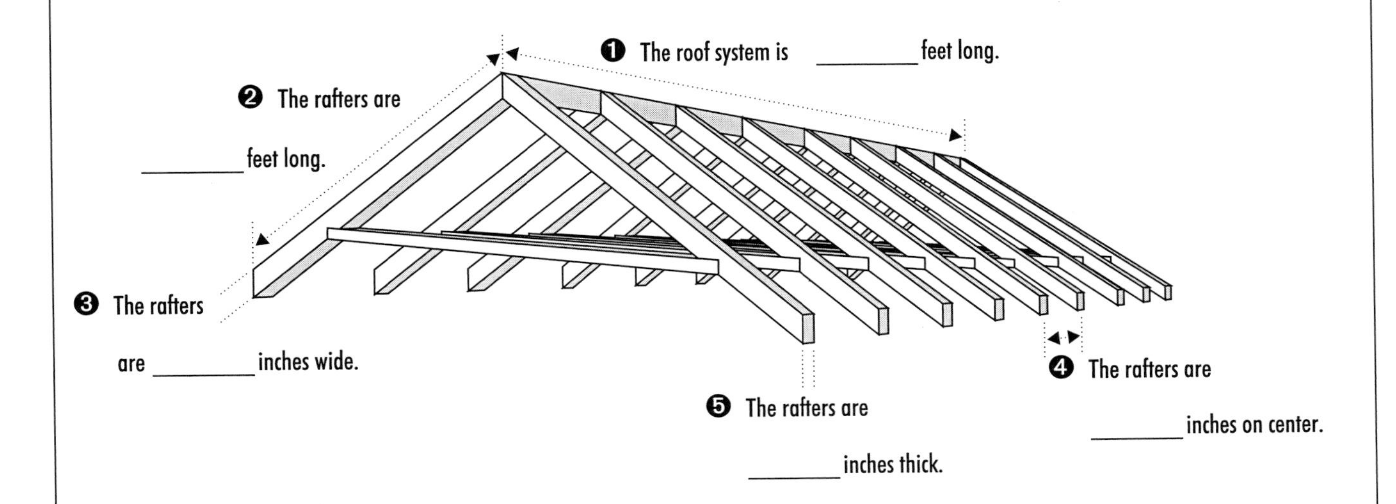

Round up to next even number

Ⓐ **The number of rafters** .... ([ANSWER 1] × 12 ÷ [ANSWER 4] + 1) × 2 = [SUM A]

Ⓑ **Lineal feet of rafters** .... [SUM A] × [ANSWER 2] = [SUM B]

Ⓒ **Bf in ridge & sub-fascia*** .... [ANSWER 1] × 3 × [ANSWER 5] × ([ANSWER 3] + 2) ÷ 12 = [SUM C]

Ⓓ **Total board feet in roof**** .... [SUM B] × [ANSWER 5] × [ANSWER 3] ÷ 12 + [SUM C] = [SUM D]

*The ridge and sub-fascia boards are calculated to be 2 inches wider than the rafters. **Does not include ceiling joists for the room below (as illustrated) or collar ties.

Converting Inches to Decimal Feet	**1″** = .084 • **2″** = .167 • **3″** = .25 • **4″** = .334 • **5″** = .417 • **6″** = .5 • **7″** = .584 • **8″** = .667 • **9″** = .75 • **10″** = .834 • **11″** = .917

USING THE GABLE ROOF TRUSSES
QuickCalculator

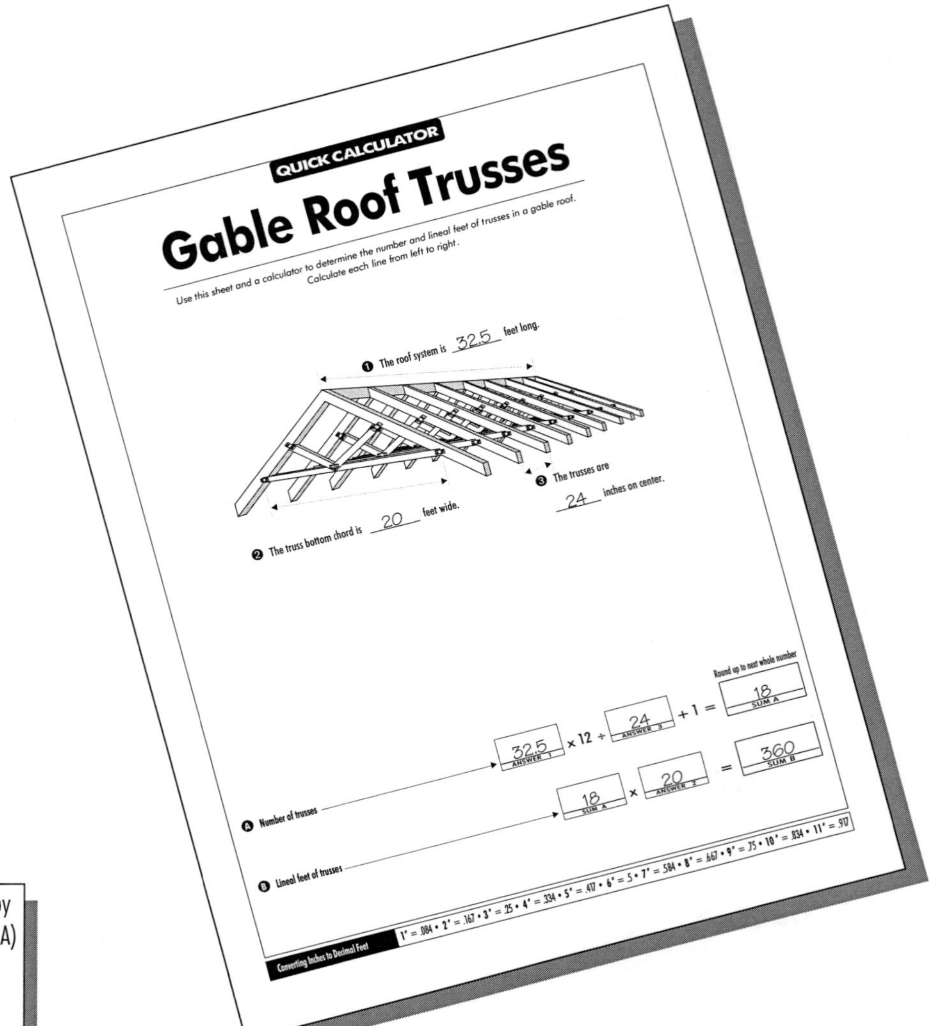

A Calculates the number of trusses in the roof by converting the length of the roof (Answer 1) into inches by multiplying it by 12. The number of trusses is then calculated by taking this length in inches, dividing it by the truss centers (Answer 3), then adding 1 for the first truss.

B Calculates the lineal feet of trusses by multiplying the number of trusses (Sum A) by the length of the truss bottom chord (Answer 2).

QUICK FACTS

☞ The sample *QuickCalculator* sheet shown above contains quantities that have been rounded. The total number of trusses should always be rounded up to the next whole number. We also suggest rounding lineal feet lineal feet to the nearest whole number.

☞ See page 449 for more information about the geometric formulas used in this *QuickCalculator*. See page 374 for truss systems priced per lineal foot.

QUICK CALCULATOR

Gable Roof Trusses

Use this sheet and a calculator to determine the number and lineal feet of trusses in a gable roof. Calculate each line from left to right.

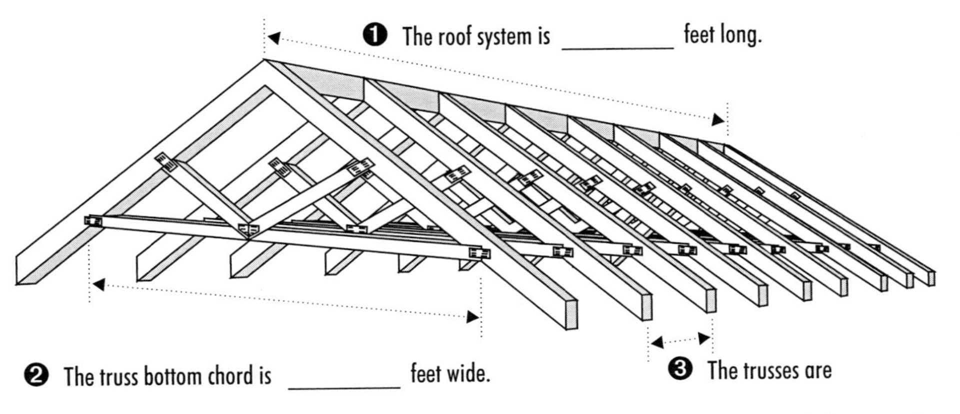

Round up to next whole number

Ⓐ **Number of trusses** .... [ANSWER 1] × 12 ÷ [ANSWER 3] + 1 = [SUM A]

Ⓑ **Lineal feet of trusses** .... [SUM A] × [ANSWER 2] = [SUM B]

Converting Inches to Decimal Feet	**1″** = .084 • **2″** = .167 • **3″** = .25 • **4″** = .334 • **5″** = .417 • **6″** = .5 • **7″** = .584 • **8″** = .667 • **9″** = .75 • **10″** = .834 • **11″** = .917

USING THE GABLE ROOFING QUICKCALCULATOR

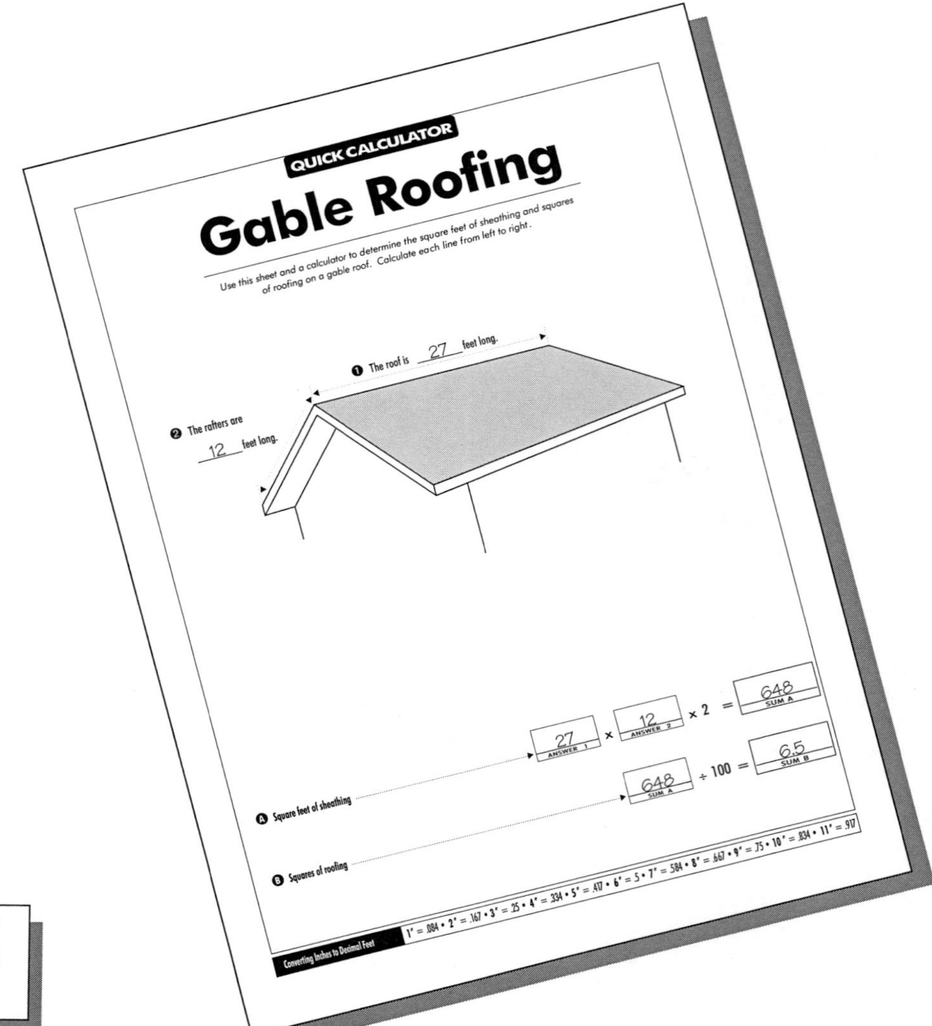

A Calculates the square feet of sheathing on the roof by multiplying the roof length (Answer 1) by the rafter length (Answer 2), then multiplying this total by 2 to calculate both sides of the roof.

B Calculates the squares of roofing by dividing the square feet of roof (Sum A) by 100.

QUICK FACTS

☞ The sample *QuickCalculator* sheet shown above contains quantities that have been rounded. We suggest rounding sheathing to the nearest square foot and rounding squares *up* to the next 1/5 square for shakes, the next 1/4 square for wood shingles and clay tile, and the next 1/3 square for asphalt shingles.

☞ See page 449 for more information about the geometric formulas used in this *QuickCalculator*. Also see Roofing beginning on page 339.

QUICK CALCULATOR

Gable Roofing

Use this sheet and a calculator to determine the square feet of sheathing and squares of roofing on a gable roof. Calculate each line from left to right.

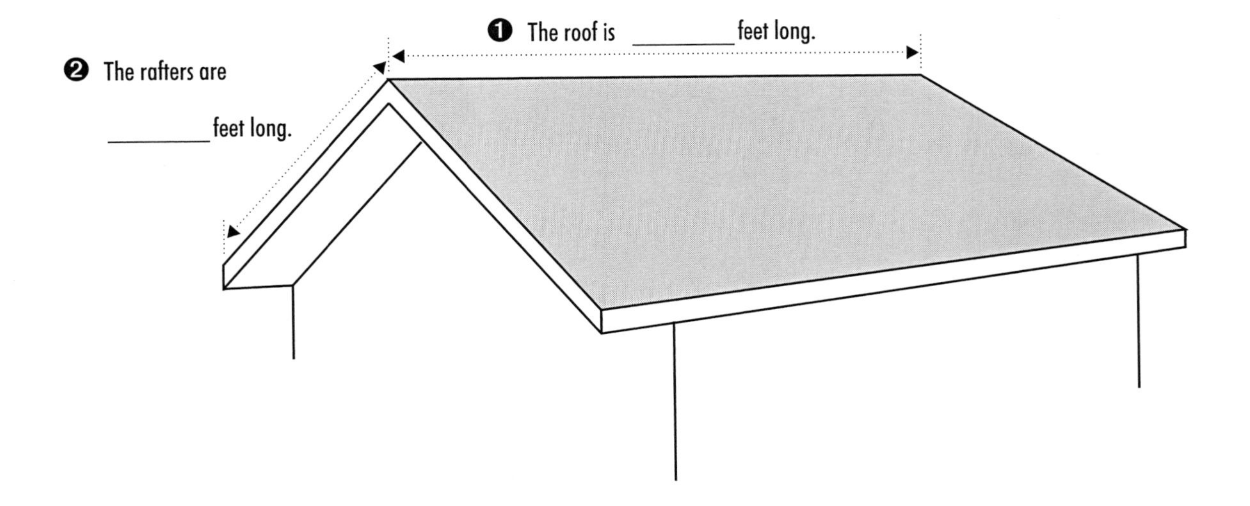

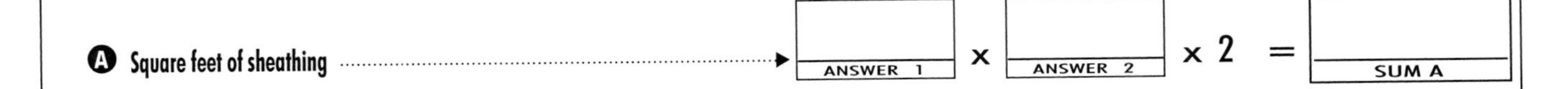

Ⓑ Squares of roofing ········ [SUM A] ÷ 100 = [SUM B]

Converting Inches to Decimal Feet 1″ = .084 • 2″ = .167 • 3″ = .25 • 4″ = .334 • 5″ = .417 • 6″ = .5 • 7″ = .584 • 8″ = .667 • 9″ = .75 • 10″ = .834 • 11″ = .917

USING THE JOIST SYSTEM
QUICKCALCULATOR

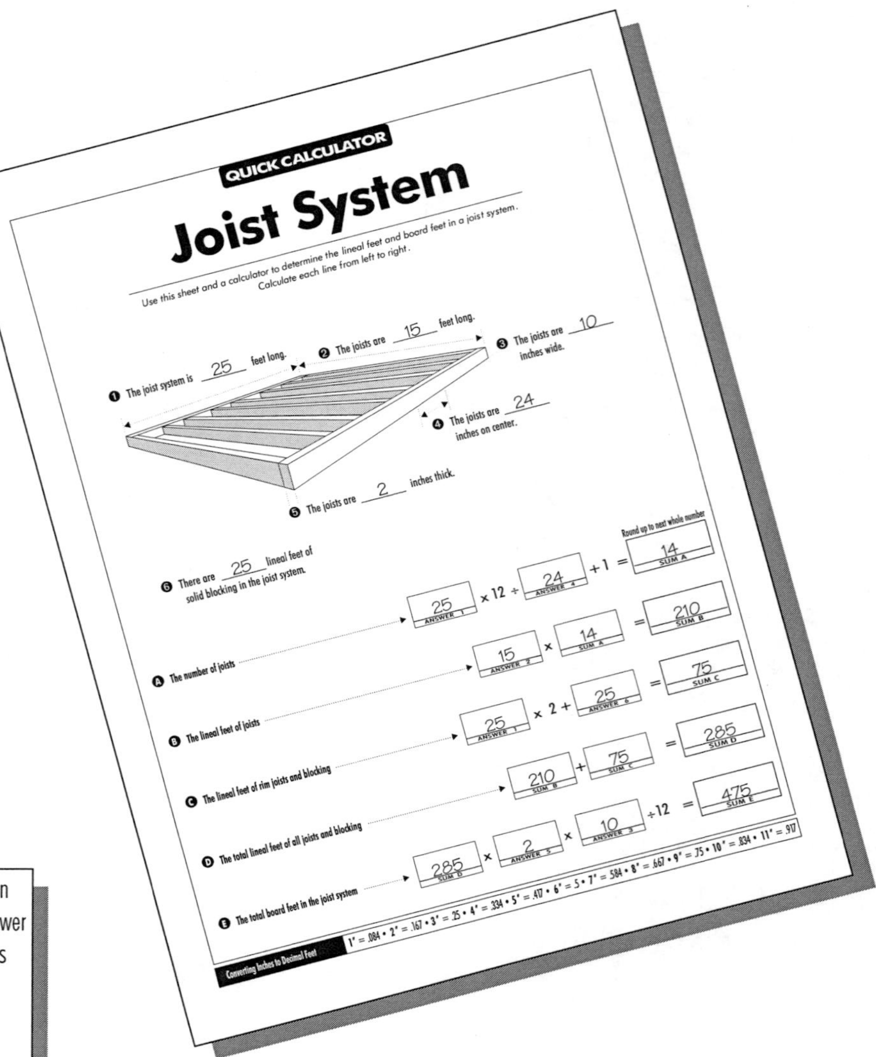

A Calculates the total number of joists. The length of the joist system (Answer 1) is converted into inches by multiplying it by 12. The number of joists is calculated by taking the joist system length in inches and dividing it by the joist centers (Answer 4) then adding 1 for the first joist.

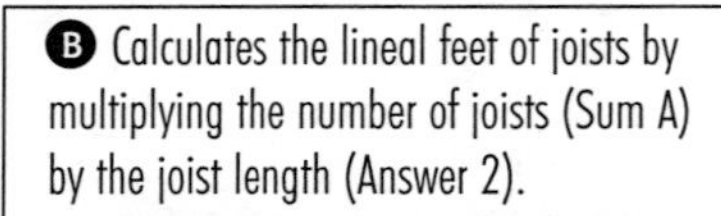

B Calculates the lineal feet of joists by multiplying the number of joists (Sum A) by the joist length (Answer 2).

C Calculates the lineal feet in rim joists and in solid blocking. The length of the rim joist (Answer 1) is multiplied by 2 for rim joists on both sides of the joist system. The total lineal feet of rim joists are then added to the lineal feet of solid blocking (Answer 6).

D Calculates total lineal feet of lumber in the joist system by adding the lineal feet of joists (Sum B) to the lineal feet of rim joists and solid blocking (Sum C).

E Calculates the board feet in the joist system by multiplying the total lineal feet of joists (Sum D) by the joist thickness (Answer 5), then by the joist width (Answer 3). This total is then divided by 12.

QUICK FACTS

☞ The sample *QuickCalculator* sheet shown above contains quantities that have been rounded. The total number of joists should always be rounded up to the next whole number. We also suggest rounding lineal feet and board feet to the nearest whole number.

☞ page 449 for more information about the geometric formulas used in this *QuickCalculator*. See page 369 for joist systems priced per board foot.

QUICK CALCULATOR

Joist System

Use this sheet and a calculator to determine the lineal feet and board feet in a joist system. Calculate each line from left to right.

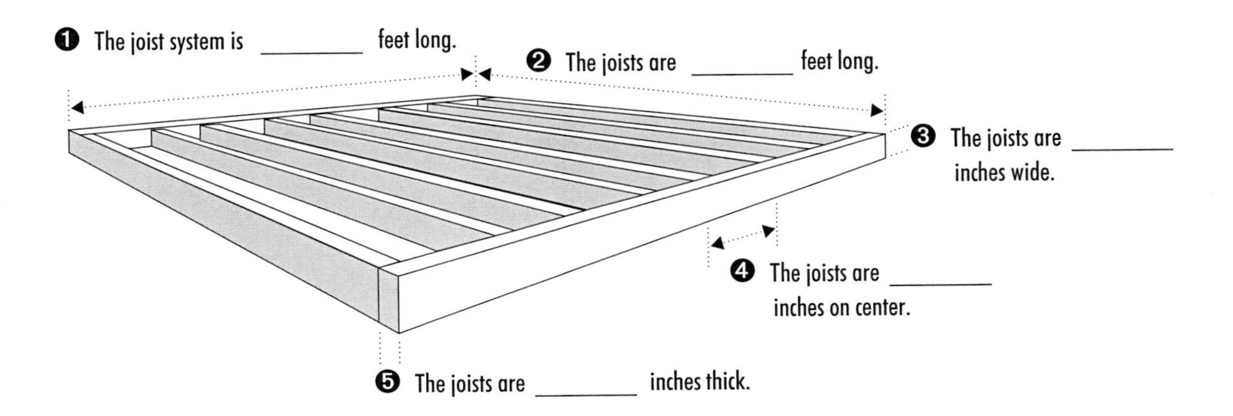

❻ There are ________ lineal feet of solid blocking in the joist system.

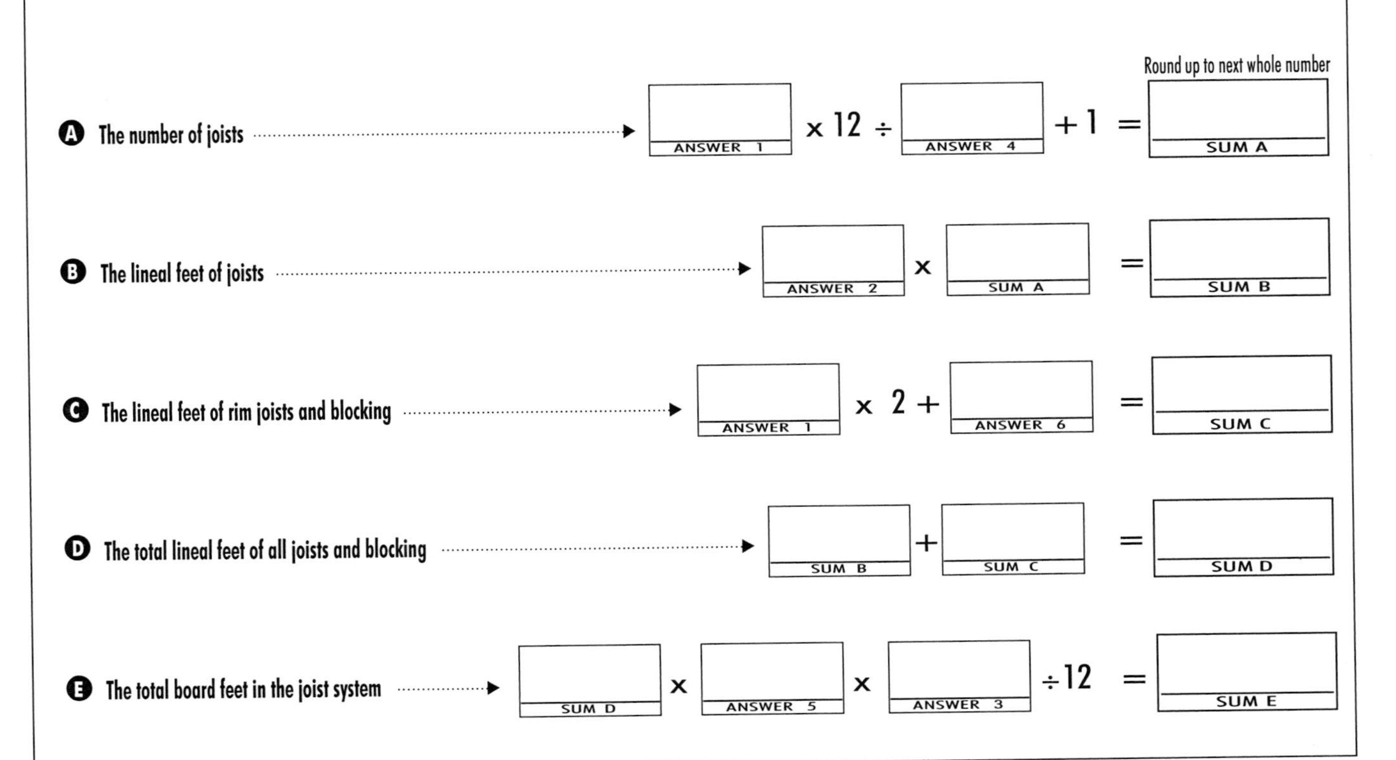

Converting Inches to Decimal Feet 1″ = .084 • 2″ = .167 • 3″ = .25 • 4″ = .334 • 5″ = .417 • 6″ = .5 • 7″ = .584 • 8″ = .667 • 9″ = .75 • 10″ = .834 • 11″ = .917

USING THE NON-BEARING WALL
QUICKCALCULATOR

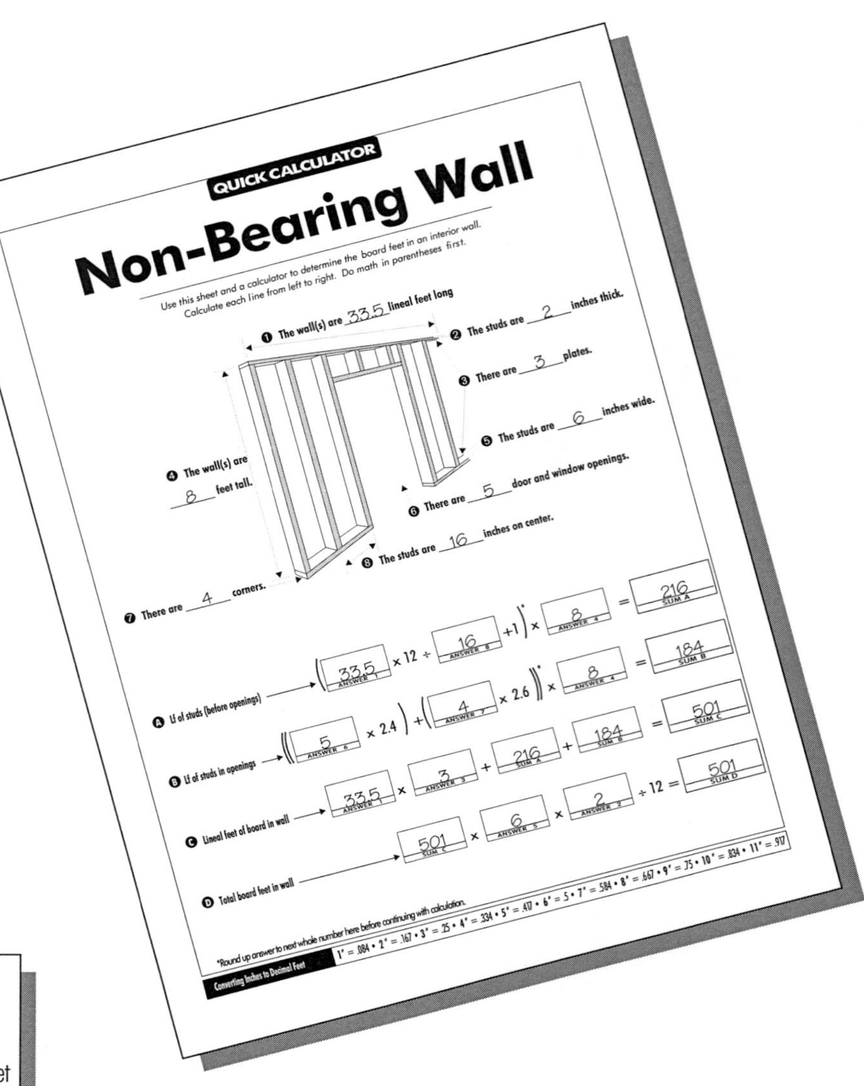

❶ Calculates the lineal feet of studs in a wall *before* adding for openings and corners. The length of the wall (Answer 1) is converted into inches by multiplying it by 12. The number of studs is calculated by dividing the wall length in inches by the stud centers (Answer 8), then adding 1 for the first stud. The calculation should now be rounded *up*, then multiplied by the wall height (Answer 4).

❷ Calculates the lineal feet of studs typically added for openings and corners. The number of openings (Answer 6) is multiplied by 2.4 (the *average* additional studs typically needed for an opening). The number of corners (Answer 7) is multiplied by 2.6 (the *average* additional studs typically needed at corners). The studs for openings and corners are added together, then should be rounded *up*. The total additional studs is then multiplied by the wall height (Answer 4).

❸ Calculates the total lineal feet of lumber in the wall. The length of the wall (Answer 1) is multiplied by the number of plates (Answer 3). The lineal feet of plates is added to the lineal feet of studs (Sum A) and additional studs (Sum B).

❹ Calculates the total board feet in the wall. The lineal feet of board in the wall (Sum C) is multiplied by the wall width (Answer 5) then by the stud thickness (Answer 2). The total is divided by 12 to convert to board feet.

QUICK FACTS

☞ The sample *QuickCalculator* sheet shown above contains quantities that have been rounded. When calculating Sum A and Sum B we suggest rounding the calculation up to the next whole number at the asterisk. Board feet and lineal feet should be rounded.

☞ See page 449 for more information about the geometric formulas used in this *QuickCalculator*. See page 357 for walls priced per board foot.

QUICK CALCULATOR

Non-Bearing Wall

Use this sheet and a calculator to determine the board feet in an interior wall. Calculate each line from left to right. Do math in parentheses first.

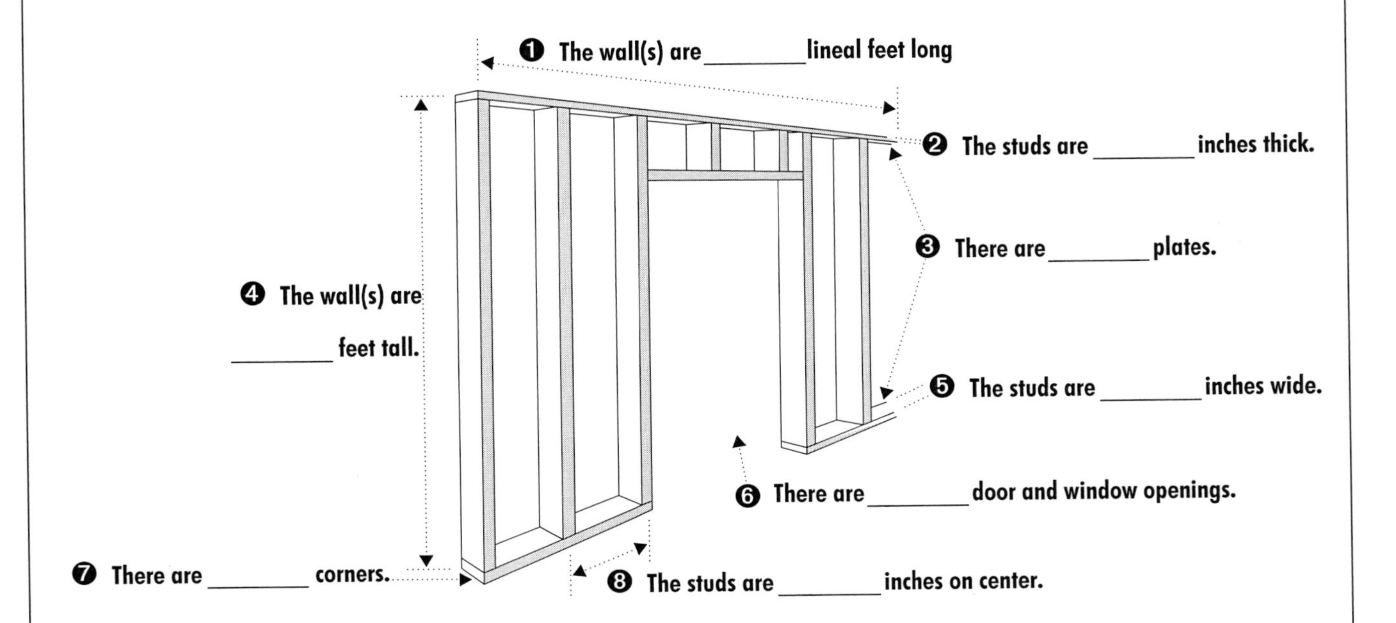

Ⓐ Lf of studs (before openings) ⟶ ([ANSWER 1] × 12 ÷ [ANSWER 8] +1)* × [ANSWER 4] = [SUM A]

Ⓑ Lf of studs in openings ⟶ (([ANSWER 6] × 2.4) + ([ANSWER 7] × 2.6))* × [ANSWER 4] = [SUM B]

Ⓒ Lineal feet of board in wall ⟶ [ANSWER 1] × [ANSWER 3] + [SUM A] + [SUM B] = [SUM C]

Ⓓ Total board feet in wall ⟶ [SUM C] × [ANSWER 5] × [ANSWER 2] ÷ 12 = [SUM D]

*Round up answer to next whole number here before continuing with calculation.

Converting Inches to Decimal Feet	1″ = .084 • 2″ = .167 • 3″ = .25 • 4″ = .334 • 5″ = .417 • 6″ = .5 • 7″ = .584 • 8″ = .667 • 9″ = .75 • 10″ = .834 • 11″ = .917

USING THE PRISMATIC ROOM
QUICKCALCULATOR

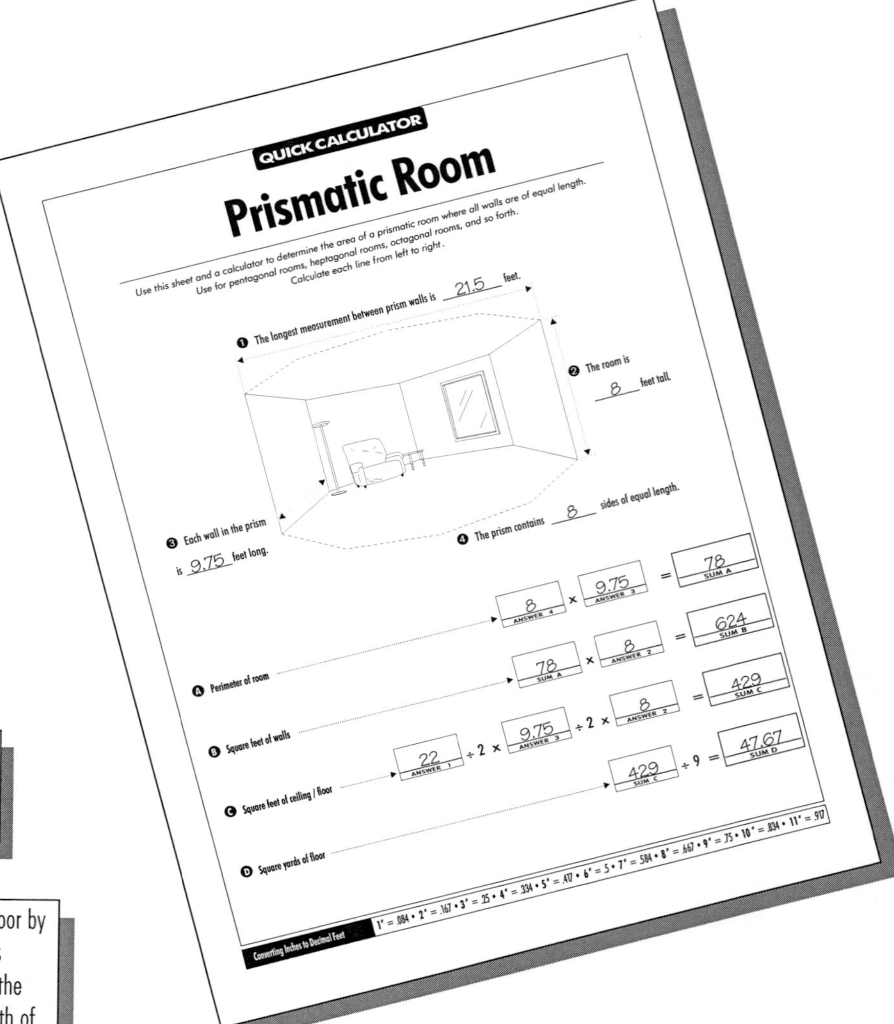

A Calculates the perimeter of the prism by multiplying the number of prism walls (Answer 4) by the length of the prism walls (Answer 3).

B Calculates the square feet of walls by multiplying the perimeter of the room (Sum A) by the room height (Answer 2).

C Calculates the square feet of ceiling / floor by taking the longest dimension between walls (Answer 1) and dividing by 2 to determine the radius. The radius is multiplied by the length of each prism wall (Answer 3), divided by 2 and multiplied by the height of the walls (Answer 2).

D Calculates the square yards of flooring by dividing the square feet of floor (Sum C) by 9.

QUICK FACTS

☞ The sample *QuickCalculator* sheet shown above contains quantities that have been rounded. We suggest rounding lineal feet to two decimal places then to the nearest inch when converting to inches. Square feet should probably be rounded to the nearest square foot and square yards should be rounded up to the nearest 1/3 or 1/4 yard.

☞ See page 449 for more information about the geometric formulas used in this *QuickCalculator*.

QUICK CALCULATOR

Prismatic Room

Use this sheet and a calculator to determine the area of a prismatic room where all walls are of equal length.
Use for pentagonal rooms, heptagonal rooms, octagonal rooms, and so forth.
Calculate each line from left to right.

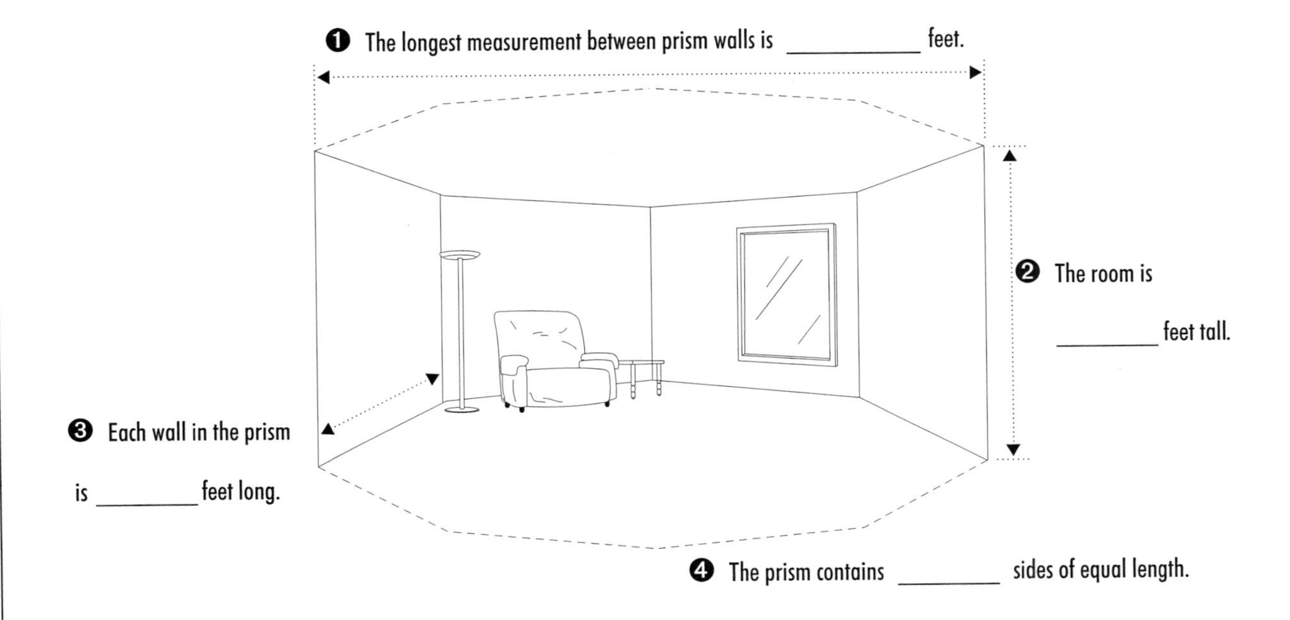

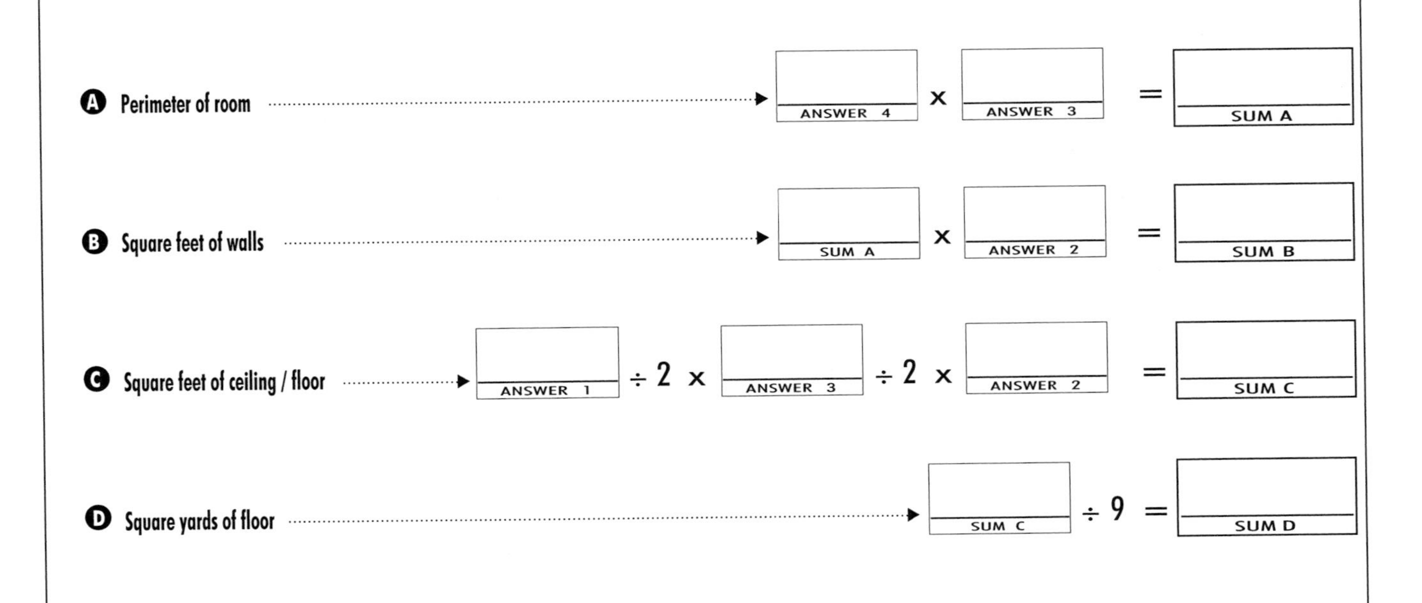

Converting Inches to Decimal Feet	1″ = .084 • 2″ = .167 • 3″ = .25 • 4″ = .334 • 5″ = .417 • 6″ = .5 • 7″ = .584 • 8″ = .667 • 9″ = .75 • 10″ = .834 • 11″ = .917

USING THE RECTANGULAR ROOM
QuickCalculator

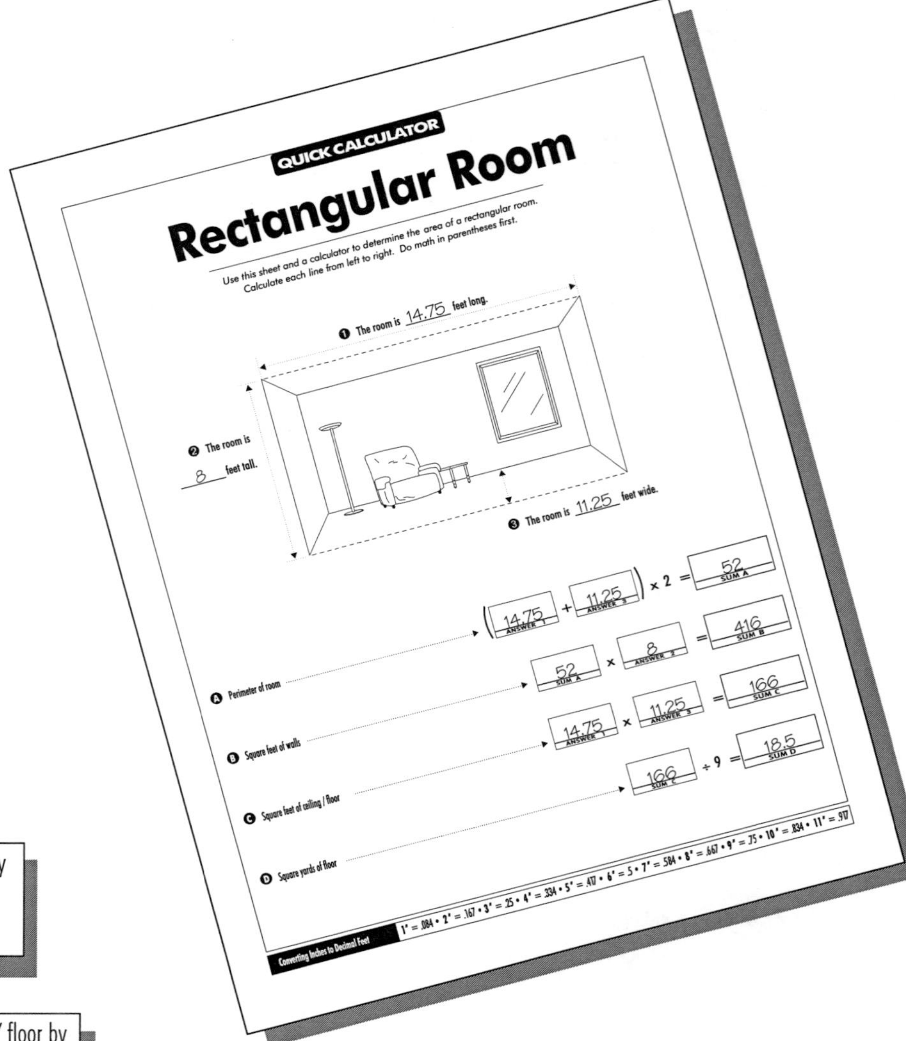

A Calculates the perimeter of the room by adding the length (Answer 1) to the width (Answer 3) and multiplying by 2.

B Calculates the square feet of walls by multiplying the perimeter of the room (Sum A) by the room height (Answer 2).

C Calculates the square feet of ceiling / floor by multiplying the length of the room (Answer 1) by the width (Answer 3).

D Calculates the square yards of flooring by dividing the square feet of floor (Sum C) by 9.

QUICK FACTS

☞ The sample *QuickCalculator* sheet shown above contains quantities that have been rounded. We suggest rounding lineal feet to two decimal places then to the nearest inch when converting to inches. Square feet should probably be rounded to the nearest square foot and square yards should be rounded to the nearest 1/3 or 1/4 yard.

☞ See page 449 for more information about the geometric formulas used in this *QuickCalculator*.

QUICK CALCULATOR

Rectangular Room

Use this sheet and a calculator to determine the area of a rectangular room. Calculate each line from left to right. Do math in parentheses first.

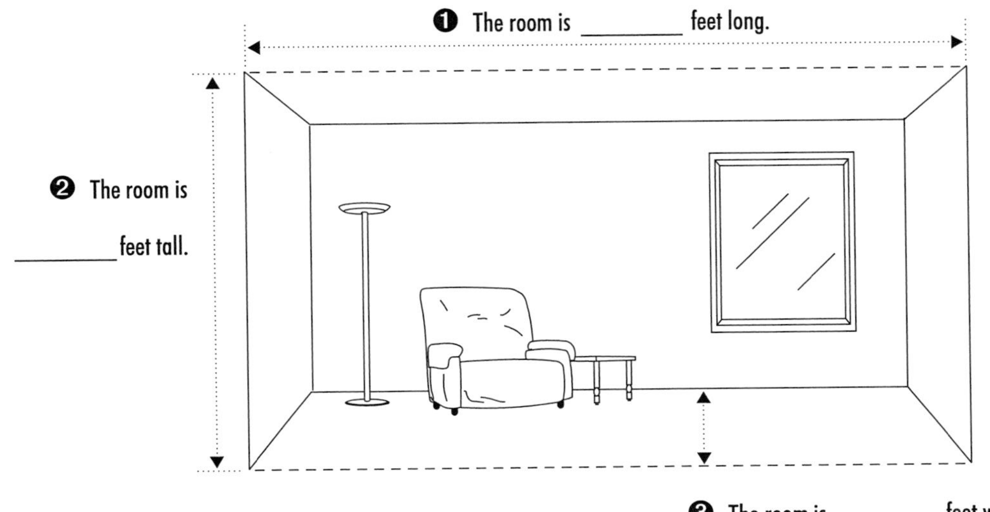

Ⓐ Perimeter of room ([ANSWER 1] + [ANSWER 3]) × 2 = [SUM A]

Ⓑ Square feet of walls [SUM A] × [ANSWER 2] = [SUM B]

Ⓒ Square feet of ceiling / floor [ANSWER 1] × [ANSWER 3] = [SUM C]

Ⓓ Square yards of floor [SUM C] ÷ 9 = [SUM D]

Converting Inches to Decimal Feet	1″ = .084 • 2″ = .167 • 3″ = .25 • 4″ = .334 • 5″ = .417 • 6″ = .5 • 7″ = .584 • 8″ = .667 • 9″ = .75 • 10″ = .834 • 11″ = .917

USING THE ROUND ROOM
QuickCalculator

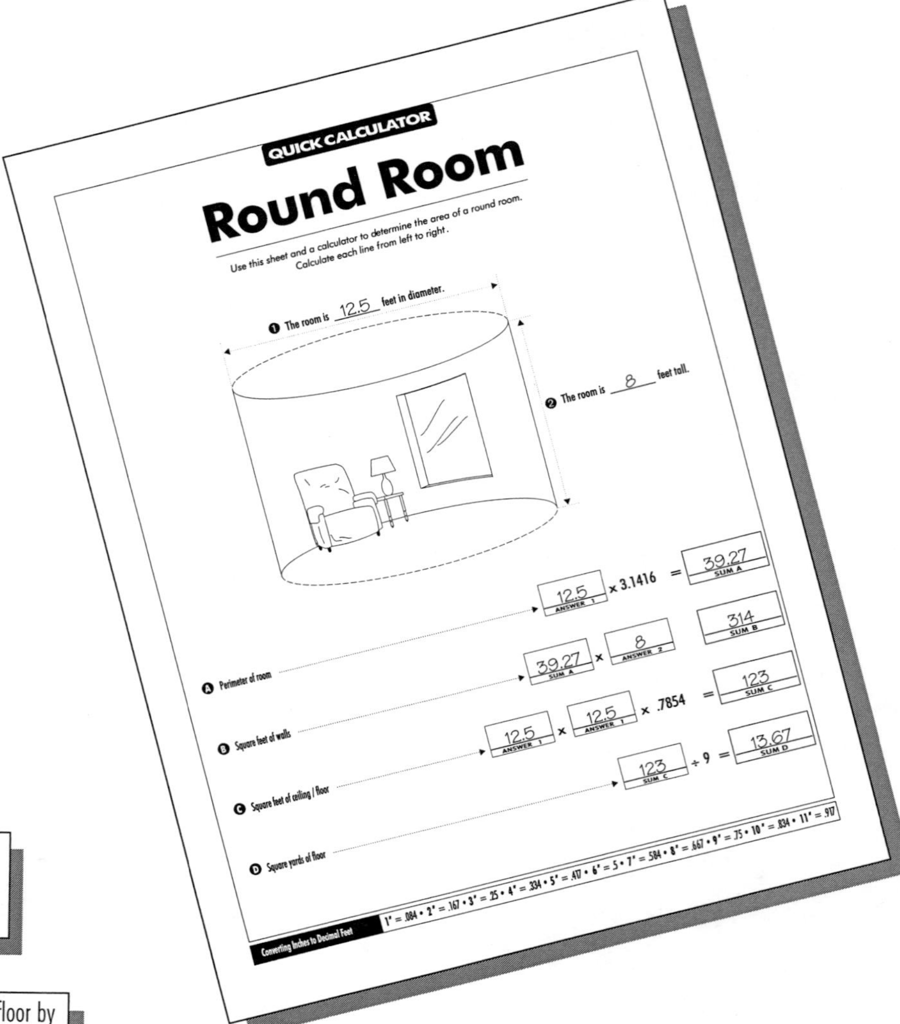

A Calculates the perimeter of the circle by multiplying the diameter (Answer 1) by Pi (3.1416).

B Calculates the square feet of walls by multiplying the perimeter of the room (Sum A) by the room height (Answer 2).

C Calculates the square feet of ceiling / floor by squaring the diameter (Answer 1 times Answer 1) and multiplying by .7854.

D Calculates the square yards of flooring by dividing the square feet of floor (Sum C) by 9.

QUICK FACTS

☞ The sample *QuickCalculator* sheet shown above contains quantities that have been rounded. We suggest rounding lineal feet to two decimal places then to the nearest inch when converting to inches. Square feet should probably be rounded to the nearest square foot and square yards should be rounded to the nearest 1/3 or 1/4 yard.

☞ See page 449 for more information about the geometric formulas used in this *QuickCalculator*.

QUICK CALCULATOR

Round Room

Use this sheet and a calculator to determine the area of a round room. Calculate each line from left to right.

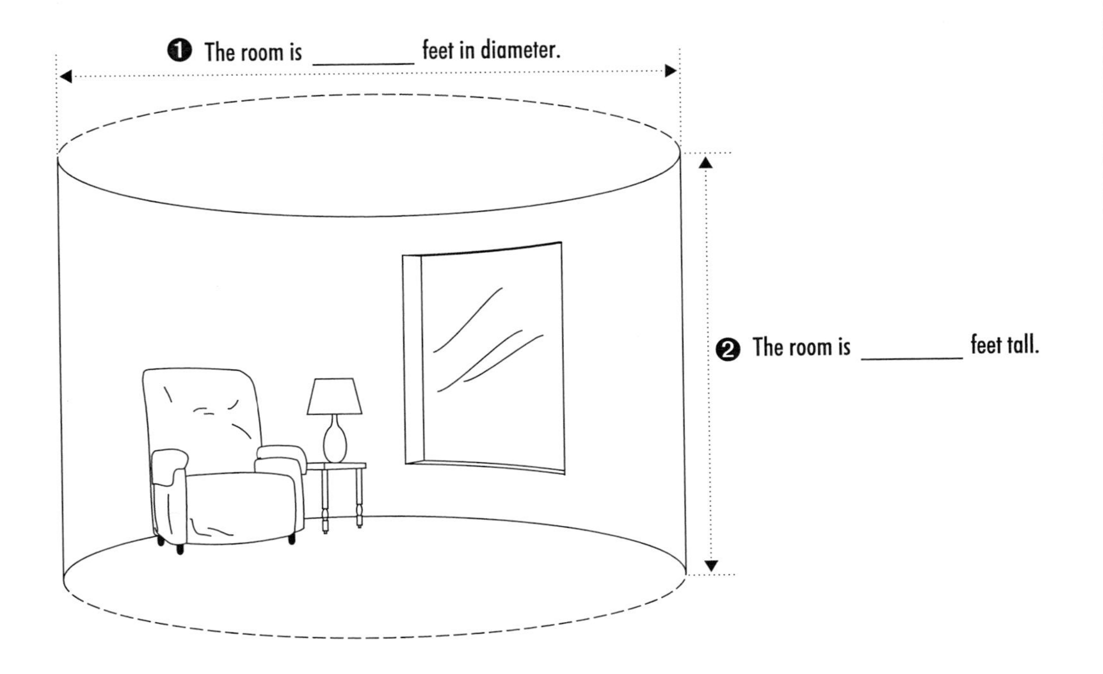

Ⓐ Perimeter of room .... [ANSWER 1] x 3.1416 = [SUM A]

Ⓑ Square feet of walls .... [SUM A] x [ANSWER 2] [SUM B]

Ⓒ Square feet of ceiling / floor .... [ANSWER 1] x [ANSWER 1] x .7854 = [SUM C]

Ⓓ Square yards of floor .... [SUM C] ÷ 9 = [SUM D]

Converting Inches to Decimal Feet 1″ = .084 • 2″ = .167 • 3″ = .25 • 4″ = .334 • 5″ = .417 • 6″ = .5 • 7″ = .584 • 8″ = .667 • 9″ = .75 • 10″ = .834 • 11″ = .917

Index

A

B

C

D

E

F

G

H

I

J

K

L

M

N

O

P

Q

R

S

XYZ

Practical References for Builders

Construction Forms for Contractors

This practical guide contains 78 practical forms, letters and checklists, guaranteed to help you streamline your office, organize your jobsites, gather and organize records and documents, keep a handle on your subs, reduce estimating errors, administer change orders and lien issues, monitor crew productivity, track your equipment use, and more. Includes accounting forms, change order forms, forms for customers, estimating forms, field work forms, HR forms, lien forms, office forms, bids and proposals, subcontracts, and more. All are also on the CD-ROM included, in Excel spreadsheets, as formatted Rich Text that you can fill out on your computer, and as PDFs. **360 pages, 8½ x 11, $48.50**

***eBook (PDF) also available;* $24.25 at** www.craftsman-book.com

CD Estimator

CD Estimator puts at your fingertips over 150,000 construction costs for new construction, remodeling, renovation & insurance repair, home improvement, framing & finish carpentry, electrical, concrete & masonry, painting, earthwork & heavy equipment and plumbing & HVAC. Quarterly cost updates are available at no charge on the Internet. You'll also have the National Estimator program — a stand-alone estimating program for *Windows*™ that *Remodeling* magazine called a "computer wiz," and *Job Cost Wizard*, a program that lets you export your estimates to QuickBooks Pro for actual job costing. A 60-minute interactive video teaches you how to use this CD-ROM to estimate construction costs. And to top it off, to help you create professional-looking estimates, the disk includes over 40 construction estimating and bidding forms in a format that's perfect for nearly any *Windows*™ word processing or spreadsheetprogram.
CD Estimator is $149.50

Construction Contract Writer

Relying on a "one-size-fits-all" boilerplate construction contract to fit your jobs can be dangerous — almost as dangerous as a handshake agreement. *Construction Contract Writer* lets you draft a contract in minutes that precisely fits your needs and the particular job, and meets both state and federal requirements. You just answer a series of questions — like an interview — to construct a legal contract for each project you take on. Anticipate where disputes could arise and settle them in thecontract before they happen. Include the warranty protection you intend,the payment schedule, and create subcontracts from the prime contractby just clicking a box. Includes a feedback button to an attorney on the Craftsman staff to help should you get stumped — *No extra charge.* **$149.95.** Download the *Construction Contract Writer* at:
http:/www.constructioncontractwriter.com

National Home Improvement Estimator

Current labor and material prices for home improvement projects. Provides manhours for each job, recommended crew size, and the labor cost for removal and installation work. Material prices are current, with location adjustment factors and free monthly updates on the Web. Gives step-by-step instructions for the work, with helpful diagrams, and home improvement shortcuts and tips from experts. Includes a free download of an electronic version of the book, and *National Estimator*, a stand-alone *Windows*™ estimating program. Additional information and *National Estimator* ShowMe tutorial video is available on our website under the "Support" dropdown tab.
568 pages, 8½ x 11, $98.75. Revised annually

***eBook (PDF) also available;* $49.38 at** www.craftsman-book.com

Plumber's Handbook Revised

This new edition explains simply and clearly, in non-technical, everyday language, how to install all components of a plumbing system to comply not only with recent changes in the *International Plumbing Code* and the *Uniform Plumbing Code*, but with the requirements of the Americans with Disabilities Act. Originally written for working plumbers to assure safe, reliable, code-compliant plumbing installations that pass inspection the first time, Plumber's Handbook, because of its readability, accuracy and clear, simple diagrams, has become the textbook of choice for numerous schools preparing plumbing students for the plumber's exams. Now, with a set of questions for each chapter, full explanations for the answers, and with a 200-question sample exam in the back, this handbook is one of the best tools available for preparing for almost any plumbing journeyman, master or state-required plumbing contracting exam.
384 pages, 8½ x 11, $44.50

***eBook (PDF) also available;* $22.25 at** www.craftsman-book.com

National Repair & Remodeling Estimator

The complete pricing guide for dwelling reconstruction costs. Reliable, specific data you can apply on every repair and remodeling job. Up-to-date material costs and labor figures based on thousands of jobs across the country. Provides recommended crew sizes; average production rates; exact material, equipment, and labor costs; a total unit cost and a total price including overhead and profit. Separate listings for high- and low-volume builders, so prices shown are specific for any size business. Estimating tips specific to repair and remodeling work to make your bids complete, realistic, and profitable. Includes a free download of an electronic version of the book with *National Estimator.*Additional information and *National Estimator* ShowMe tutorial video is available on our website under the "Support" dropdown tab.
512 pages, 8½ x 11, $98.50. Revised annually

***eBook (PDF) also available;* $49.25 at** www.craftsman-book.com

National Building Cost Manual

Square-foot costs for residential, commercial, industrial, military, schools, greenhouses, churches and farm buildings. Includes important variables that can make any building unique from a cost standpoint. Quickly work up a reliable budget estimate based on actual materials and design features, area, shape, wall height, number of floors, and support requirements. Now includes free download of Craftsman's easy-to-use software that calculates total in-place cost estimates or appraisals. Use the regional cost adjustment factors provided to tailor the estimate to any jobsite in the U.S. Then view, print, email or save the detailed PDF report as needed.
280 pages, 8½ x 11, $88.00. Revised annually

***eBook (PDF) also available;* $44.00 at** www.craftsman-book.com

National Painting Cost Estimator

A complete guide to estimating painting costs for just about any type of residential, commercial, or industrial painting, whether by brush, spray, or roller. Shows typical costs and bid prices for fast, medium, and slow work, including material costs per gallon, square feet covered per gallon, square feet covered per manhour, labor, material, overhead, and taxes per 100 square feet, and how much to add for profit. Includes a free download of an electronic version of the book, with *National Estimator*, a stand-alone *Windows*™ estimating program. Additional information and *National Estimator* ShowMe tutorial video is available on our website under the "Support" dropdown tab.
448 pages, 8½ x 11, $98.00. Revised annually

***eBook (PDF) also available;* $49.00 at** www.craftsman-book.com

Insurance Restoration Contracting: Startup to Success

Insurance restoration — the repair of buildings damaged by water, fire, smoke, storms, vandalism and other disasters — is an exciting field of construction that provides lucrative work that's immune to economic downturns. And, with insurance companies funding the repairs, your payment is virtually guaranteed. But this type of work requires special knowledge and equipment, and that's what you'll learn about in this book. It covers fire repairs and smoke damage, water losses and specialized drying methods, mold remediation, content restoration, even damage to mobile and manufactured homes. You'll also find information on equipment needs, training classes, estimating books and software, and how restoration leads to lucrative remodeling jobs. It covers all you need to know to start and succeed as the restoration contractor that both homeowners and insurance companies call on first for the best jobs.
640 pages, 8½ x 11, $69.00

***eBook (PDF) also available;* $34.50 at** www.craftsman-book.com

National Electrical Estimator

This year's prices for installation of all common electrical work: conduit, wire, boxes, fixtures, switches, outlets, loadcenters, panelboards, raceway, duct, signal systems, and more. Provides material costs, manhours per unit, and total installed cost. Explains what you should know to estimate each part of an electrical system. Includes a free download of an electronic version of the book with *National Estimator*, a stand-alone *Windows*™ estimating program. Additional information and *National Estimator* ShowMe tutorial video is available on our website under the "Support" dropdown tab. **552 pages, 8½ x 11, $97.75. Revised annually**

***eBook (PDF) also available;* $48.88 at** www.craftsman-book.com